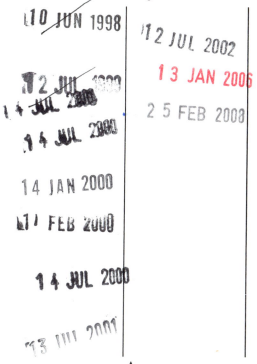

HOST PLANT RESISTANCE TO INSECTS

Dedicated to M.S. and Mina Swaminathan

HOST PLANT RESISTANCE
TO INSECTS

Niranjan Panda

former Professor and Head
Department of Entomology
Orissa University of Agriculture and Technology
Bhubaneswar
India

and

Gurdev S. Khush

Principal Plant Breeder and Head
Plant Breeding, Genetics and Biochemistry Division
International Rice Research Institute
Philippines

CAB INTERNATIONAL
in association with the
International Rice Research Institute

CAB INTERNATIONAL   Tel: +44 (0)1491 832111
Wallingford   Telex: 847964 (COMAGG G)
Oxon OX10 8DE   E-mail: cabi@cabi.org
UK   Fax: +44 (0)1491 833508

Published in association with:
International Rice Research Institute
PO Box 933
1099 Manila
Philippines

A catalogue entry for this book is available from the British Library.

ISBN 0 85198 963 2

Typeset by Solidus (Bristol) Limited
Printed and bound in the UK by Biddles Ltd, Guildford

CONTENTS

Contents

FOREWORD

Plants and insects have coexisted for eons. Harmful insects were suppressed either by other insects or through the natural plant defense mechanisms. Crop intensification for higher yields led to disturbance of the balance between the host and insect communities, leading to insect outbreaks and serious crop losses. To overcome the pest problems in modern crop production systems, synthetic pesticides were introduced. The age of pesticides began with the commercial introduction of DDT in 1946. After the Second World War, dependence upon chemical control of insects increased with the availability of inexpensive broad-spectrum insecticides. Adverse effects of the excessive dependence upon pesticides for insect control soon became evident, however. Insects developed resistance to pesticides. Useful insects were killed resulting in the resurgence of harmful insects. Environmental pollution and insecticide poisoning of agricultural workers, especially in developing countries, became a serious problem.

To minimize the dependence upon chemicals as the sole insect control strategy and to overcome such limitations in the use of insecticides, scientists developed an ecological approach for pest control, popularly known as Integrated Pest Management or IPM. This approach uses a combination of host plant resistance, cultural, biological and chemical methods. IPM is designed to suppress pest numbers to below crop-damaging levels. The salient feature of IPM is the manipulation of the environment to make it unfavorable to insects and less harmful to their natural enemies. A resistant crop variety provides the basic foundation for IPM.

International agricultural research centers (IARCs) supported by the Consultative Group on International Agricultural Research (CGIAR), charged with developing environmentally friendly technologies for sustainable food production, have been in the forefront of developing high-yielding crop varieties with resistance to insects and diseases. High-yielding varieties of rice with multiple resistance to insects and diseases developed at the International Rice Research Institute (IRRI) in the Philippines, and of wheat at the International Wheat and Maize Improvement Center (CIMMYT) in Mexico, are now planted to millions of hectares worldwide, and gave rise to the so-called Green Revolution.

The current concepts of host plant resistance to insects are reviewed in this book. The broad coverage includes crop plant and insect diversity, mechanisms of insect–plant interaction, host plant selection, components of resistance, the biochemical basis of resistance, mechanisms of resistance, factors affecting expression of resistance, screening techniques for resistance, the role of insect resistance in IPM programs, the genetics of resistance, and methods of developing insect-resistant crop varieties, including modern techniques of biotechnology. The book should prove useful to students of entomology and plant breeding, and researchers and practitioners of the art and science of plant breeding. It should give further impetus to the development of host plant resistance programs.

The authors are eminently qualified to write on such a complex subject. Dr N. Panda, Entomologist and former Head of the Department of Entomology at Orissa University of Agriculture and Technology, has taught courses on host plant resistance for more than 25 years. He was a visiting scientist at IRRI for four years. Dr G. S. Khush, Principal Plant Breeder and Head of the Division of Plant Breeding, Genetics and Biochemistry at IRRI, has spear-headed the Institute's program on developing insect- and disease-resistant varieties of rice for the last 28 years. The authors are to be congratulated for their labor of love in writing this comprehensive textbook for entomologists, plant breeders and all those concerned with plant protection.

George Rothschild
Director General
International Rice Research Institute
PO Box 933, Manila, Philippines

PREFACE

Insect damage to food and fiber crops costs farmers and consumers millions of dollars each year. Many insect control strategies have been employed during this century. It is increasingly being realized that a reliance upon a single control strategy is not advisable. For example, heavy reliance upon insecticides during the 1950s and 1960s created a host of problems: resurgence of pest populations; emergence of secondary pest outbreaks as a consequence of elimination of beneficial insect species; increased cost of food production; and serious environmental pollution. Because most insect pests have high reproductive rates, they can eventually adapt to natural and synthetic pesticides, making complete eradication highly unlikely. Control measures therefore must aim at containment of insect populations rather than their eradication. This thinking has led to the development of integrated pest management or IPM. IPM aims at keeping pest populations below economic threshold levels through an ecologically sound, economically practical, socially acceptable and environmentally sustainable technology. Host plant resistance, along with natural, biological and cultural control measures, is the basic component of an IPM system.

Host plant resistance is the cheapest technology. It is easiest to introduce, least inimical to the environment, and is compatible with other control tactics such as biological, cultural and chemical control. It is recognized as the most effective component of IPM. As a result, there has been a resurgence of interest and major investment in research on host plant resistance during the last 30–40 years. Techniques for screening germplasm for insect resistance have been developed, donors for resistance have been identified, genetics of resistance have been investigated and numerous insect-resistant varieties of food and fiber crops have been developed. These are grown in millions of hectares of cropland worldwide and are the foundation stones of our food security.

Literature on host plant resistance is scattered in numerous journal papers, review articles and symposium volumes. In this book we review current concepts of host plant resistance to insects. Chapters are arranged to meet the requirements of a course on host plant resistance. We hope the volume will also serve as a useful reference for researchers on host plant resistance.

Chapter 1 contains an introduction to the subject and provides a brief description of the development of insect-resistant varieties. Crop and insect diversity are covered in Chapter 2. Plants and insects that coevolved during different geological eras are highlighted in this chapter. Information on the biosynthetic pathways for the production of secondary plant metabolites is presented in Chapter 3, including details on the role of these compounds in imparting resistance to insects. Insect–plant interactions follow in Chapter 4, outlining how plants defend themselves against insects, and how insects affect plants. One of the striking features of insect–plant interaction is the high degree of food specialization of phytophagous insects. The theories and mechanisms governing host plant selection via olfactory function and chemosensory coding system are covered in Chapter 5. Mechanisms of plant resistance to insects through various defense strategies including endogenous chemicals or morphological or phenological attributes are elaborated in Chapter 6. The expression of resistance, however, is affected by various biotic and abiotic factors (Chapter 7). Techniques for mass rearing of insects and for artificial insect infestation of plants are prerequisites for screening the germplasm for insect resistance (Chapter 8). The place of host plant resistance in IPM is reviewed in Chapter 9. Information on the inheritance of resistance is useful for selecting breeding strategies for insect resistance. Techniques of genetic analysis are included in Chapter 10. The final chapter covers methods for developing insect-resistant varieties, including the role of biotechnology.

We are indebted to several colleagues who were kind enough to read the earlier drafts of some of the sections: Drs George G. Kennedy, Fred Gould, E. A. Heinrichs, Dale Norris, Dale Bottrell, Duncan Vaughan, Heidi Hernandez, D. S. Brar and Kuldeep Singh made useful comments for the improvement of various chapters. Abner Maruzzo helped with much of the library work and typing of the first draft. Elma Nicolas typed subsequent versions of the manuscript. We are grateful to many individuals who provided original prints of photographs, and numerous publishers for permission to reproduce figures from copyrighted material. Special thanks are due to our beloved spouses, Mrs Swarna Panda and Mrs Harwant Khush for their unstinted support during the preparation of the manuscript.

Niranjan Panda
Bhubaneswar, India

Gurdev S. Khush
Los Baños, Laguna, Philippines

May 1, 1995

1 Introduction

In natural ecosystems, phytophagous insects have coexisted in a complex relationship with plant communities that vary in phenology, abundance, and association with other living organisms. In plant life, insects are intimately associated with a number of beneficial activities including pollination. Some insects may, on the other hand, be harmful to the plant and even cause its death. In the biological warfare between plants and insects, harmful insects are suppressed either by other insects or through natural plant defense mechanisms. These factors prevent the excessive population growth of harmful insect species.

To increase crop production, natural ecosystems are disturbed when land is cleared to plant crops. Crop intensification for higher yields leads to a disturbance of the balance between the host and insect communities. The introduction of crop cultivars for a higher production of food, fiber, spices, and cosmetics in agricultural regions provides an opportunity for specialist herbivores to interact with hosts. Those herbivores that produce fertile offspring become dominant in the cropping systems.

The very technologies that help to increase crop yield per unit area of land are often the cause of an increased threat to crops by the triple alliance of insects, pathogens, and weeds. Instances of pest outbreak are not uncommon in crop production intensification systems. In the rice zones of Southeast and South Asia, the brown planthopper and whitebacked planthopper became a serious threat to the rice economy after crop intensification in the 1970s. The recent epidemics of southern corn leaf blight in the corn belt of the United States provide additional evidence of crop devastation in modern agricultural systems.

To overcome pest problems in modern crop production systems, agro-chemicals – generally synthetic pesticides – were and still are used. The age of pesticides began with the commercial introduction of DDT (dichloro-diphenyltrichloroethane) in agriculture in 1946, and the ensuing three decades can be divided into the Era of Optimism (1946–1962), the Era of Doubt (1962–1976), and the Era of Integrated Pest Management (1976 to present) (Metcalf 1980). These successive epochs have witnessed the enthusiastic overuse and misuse of insecticides leading towards 'ecocatastrophe'. In the United States, three commercial crops, e.g., cotton, maize, and apple, have been protected by insecticides against pests since 1900. Yet there is no evidence that overall crop damage from insect infestation has decreased.

The use of conventional insecticides is highly efficient in terms of the number of insects destroyed when the insect population is high, but highly ineffective when the population is low (Knipling 1966). Despite the use of about 450 million kg of pesticides annually in the United States (45% of world pesticide production), pests destroy more than 40% of the food and about 37% of food crops are lost before harvest (Pimentel 1981). The dollar value of crop losses is calculated to be US$50 billion annually, which does not include external costs to the environment.

Broad-spectrum pesticides can kill most of the target organisms. Indeed the farmers demand such products. Experimental results have shown that given adequate insecticidal pressure, nearly every species of insect is capable of developing resistance, at least to a particular insecticide (Georghiou 1972). In the tropics, resurgence of insect pests is another hidden cost in the use of insecticides. Such resurgence has been documented for the brown plant-hopper and whitebacked planthopper as a consequence of continued use of certain insecticides in rice production. Many brown planthopper and white-backed planthopper outbreaks in farmers' fields throughout Asia have been induced by insecticides (Heinrichs *et al* 1982, Salim and Heinrichs 1987). Insecticides also pollute the environment; thus, the end result is increasing insecticidal poisoning of agricultural workers, especially in developing countries, destruction of natural enemies, upsurge of secondary pests, and major disruption to the ecosystem. Insecticides have failed, not because of any inherent weakness in the concept of reducing insect populations by chemicals, but because of their misuse (improper insecticidal coverage) and overuse (frequent use of insecticides).

In the face of such limitations in the use of insecticides against crop pests, plant protection specialists have developed an ecological approach to pest control, popularly known as integrated pest management (IPM) (Geier and Clark 1961). The IPM approach utilizes a combination of host plant resistance, and cultural, biological, and chemical control methods. The integrated system is designed to suppress pest numbers to below crop-damaging levels. The salient feature of IPM is manipulation of the environment to make it unfavorable to pests, and less harmful to natural enemies and the environment itself (National Academy of Science 1969). In this context, a resistant crop variety provides the basic foundation on which structures of IPM for different pests can be built.

In the last four decades, crop improvement programs have emphasized the breeding of crop varieties with multiple resistance to pests and diseases. Resistant crop varieties developed in recent years represent some of the greatest achievements of modern agriculture in increasing and stabilizing the supplies of food and fibers.

A brief historical overview of the development of crop varieties resistant to insect pests from the earliest farming period up to the present stage of advanced agricultural production systems is presented in the following section.

Historical Overview of Host Plant Resistance

Natural plant resistance

The natural plant mechanisms that reduce insect attack are: (1) escape in time and space, (2) association with other species, and (3) accommodation and/or confrontation of the arthropod. All these mechanisms operate collectively in nature; each plant species exhibits a particular mix of these factors which, in turn, varies with time and space. Thus, over long periods of time, plants both wild and cultivated have developed a great diversity of mechanisms for warding off or tolerating attack by plant-feeding insects. An extensive pool of genetic factors for such resistance exists in our domestic cultivars as well as in wild progenitors. Before domestication, most of the susceptible individuals in natural populations were probably destroyed. This resulted in a gradual increase in the frequency of genes for resistance in wild populations.

Earlier observations

Genetic resistance is one of the oldest recognized bases of plant pest control. Theophrastus recognized differences in disease susceptibility among crop cultivars in the third century BC (Allard 1960). The earliest documentation on host plant resistance (HPR) dates back to the earlier days of applied entomology and plant pathology. A wheat variety resistant to the Hessian fly, *Mayetiola destructor* (Say), was reported in 1782 when Havens published a paper regarding Hessian fly resistance in Underhill, a wheat cultivar in the United States. The apple variety Winter Majetin in the UK was reported to be resistant to the woolly aphid *Eriosma lanigerum* (Hausmann) (Lindley 1831), and is still known to retain its resistance. Another outstanding early success in utilizing HPR in pest management was the control of the grape phylloxera *Phylloxera vitifoliae* (Fitch) in France. Balachowsky (1951) described the 'brutal appearance' of this North American pest species in France in 1861 and its spread to the vineyards of European and Mediterranean countries. The entire French wine industry was on the verge of annihilation by 1880, but the pest was successfully controlled by 1890 after French vineyards were grafted with North American rootstocks of the grapevine *Vitis* spp. The entire operation cost France 10 billion francs, but the French wine industry was saved. Application of such HPR persists today in the wine-producing districts of Western Europe.

Another notable example is the resistance of sorghum to the grasshopper *Melanoplus* spp. C. V. Riley first reported in 1817 that *Sorghum vulgare* was more resistant to grasshoppers than corn. This difference has repeatedly been shown during each grasshopper outbreak in some areas of North America. Despite the fact that the crop area under the resistant sorghums has increased progressively from a few thousand to millions of acres, the grasshoppers have not yet 'learned' to eat sorghum.

Other examples of resistance as a major method of insect control are also well documented. Cotton is grown in over thousands of acres in South Africa because of the selection of a single plant that was resistant to leafhoppers of the genus *Empoasca* (Parnell 1935). The progenies of this plant later became the variety U4, which not only made possible the growing of cotton in South Africa but also led to the development of other better adapted resistant varieties. In India, Husain and Lal (1940) evolved hairy cotton varieties resistant to *Empoasca devastans* Distant, and by 1943 resistant varieties such as Punjab 4F, L.S.S., and 289F covered extensive areas where jassids had posed a serious threat to cotton production.

Scientific approach

Although plant resistance to insects and diseases was recognized in the 19th century, the breeding of pest-resistant cultivars was undertaken only after the rediscovery of Mendel's law of heredity in 1900. An experimental report by Biffen in 1907 that resistance to yellow rust in wheat is controlled by a single recessive gene stimulated plant breeders and geneticists to search genes for disease resistance in wheat and other crops. This was the start of an era of modern plant breeding for resistance to pests. Breeding plants resistant to insects is a 20th century phenomenon that stems from the knowledge of basic genetics and the methodology of selecting, crossing, and hybridizing plants.

Despite this spectacular beginning, plant resistance to insects attracted little research interest until the pioneering and systematic research in the field of HPR was inaugurated in the 1920s by the late Professor Reginald H. Painter of Kansas State University. Development of Hessian fly-resistant wheat varieties is considered a landmark achievement in pest control technology. The Hessian fly has been a serious pest of wheat *Triticum* spp. in the United States since 1779 when it was first observed on Long Island. The fly gradually moved westward and became a potential hazard in most of the wheat-growing areas. It reached Kansas in 1871 and has caused economic losses of US$5–25 million in that state alone (Painter 1966). Until the development of resistant varieties, the other control methods were delayed planting and destruction of infested stubble and volunteer wheat. The development of wheat varieties resistant to Hessian fly in California, Kansas, and Indiana through the cooperation of state and federal entomologists and plant breeders gave a promise of removing the fly pest from the list of important pests of wheat. Since 1942, 60 Hessian fly-resistant wheat varieties have been released in the United States; in 1964, they were grown on 8.5 million acres in 34 states. The area under resistant wheat varieties Big Club 43 and POS042 increased in California; consequently, the populations decreased so much so that the insect was difficult to find and infestations were reduced to less than 1% (Luginbill 1969). A similar success was reported in central Kansas after the release of Pawnee. Breeding for resistance to nine biotypes of the Hessian fly (Everson and Gallun 1980) was successfully accomplished and more than 30 resistant cultivars were released and planted on approximately 20 million acres across the United States in 1974 (Reitz and Hamlin 1978). In the same year solid-stemmed wheat varieties resistant to the wheat stem sawfly were grown on 1.5 million acres, providing the sole protection against this insect pest. To date, 12 resistant varieties have been released by the Canadian and US breeding programs (Weiss and Morrill 1992).

Post-green revolution approach

The potentialities of HPR as an insect control method, however, were not fully appreciated until the mid-1960s, partially because of overdependence on persistent insecticides against pests of high-yielding crop varieties. Crop failures due to pest epidemics encouraged entomologists to explore alternative strategies for pest control. Pest management programs were formulated with HPR as a means of control or in concert with other pest suppression measures such as cultural management, biological control, and need-based use of pesticides. Breeding for pest resistance was intensified and donors for resistance from germplasm collections were identified.

Considerable progress in breeding plants for resistance to insects has been made worldwide, and the use of insect-resistant cultivars has been instrumental in controlling insect pests and reducing the use of insecticides. It has been estimated that almost 63.3 million pounds (28.7 thousand tonnes) of insecticides are saved annually in the United States through the planting of corn, barley, grain sorghum, and alfalfa cultivars resistant to chinch bug, corn leaf aphid, European corn borer, greenbug, and spotted alfalfa aphid. Work on insect-resistant vegetables and fruit crops has also progressed well. According to Wiseman (1990), more than 200 cultivars with resistance to 50 insect species have been released and are in commercial production in the United States.

Commercially adapted materials

The international agricultural research centers (IARCs) sponsored by the Consultative Group on International Agricultural Research (CGIAR) in the tropics as well as several national crop improvement programs have taken the lead in the development of pest-resistant crop varieties and IPM programs. The International Rice Research Institute (IRRI) established at Los Baños in the Philippines in 1960 was the first of the eight centers devoted to improving crop production through the development of high-yielding varieties (HYV) and management practices. Spectacular success was achieved with HYV of rice and wheat developed at the IRRI and at the International Wheat and Maize Improvement Center (CIMMYT), respectively. Changes in rice and wheat production technology accompanied by the introduction of HYVs resulted in an increased incidence of diseases and insects. Both the IRRI and CIMMYT therefore embarked on the incorporation of genetic resistance to major diseases and insects into the improved germplasm.

Vast germplasm collections maintained at these centers were screened to identify sources of resistance and the donors for resistance were identified. Genes for resistance were incorporated and germplasm for multiple resistance was developed rapidly. Earlier HYVs of rice such as IR8, IR22, and IR24 were susceptible to most of the diseases and insects. However, succeeding varieties such as IR26, IR28, IR30, IR36, etc., have multiple resistance to as many as five diseases and five insects. IR36 is resistant to the brown planthopper, green leafhopper, yellow stem borer, striped stem borer, and gall midge. It was rapidly accepted by farmers in tropical Asia and became the most widely grown variety of rice or any other crop the world has ever known. It was planted on 11 million ha of riceland in the early 1980s. Subsequent rice varieties developed at IRRI and by national programs are equally resistant. After the widescale adoption of multiple-resistance varieties, it was easy to introduce IPM concepts. Most of the rice-growing countries in Asia have made IPM part of their national policy and the use of insecticides has been drastically reduced.

Biochemical basis of resistance

The most celebrated classic example of HPR in corn was the discovery of 2,4-dihydroxy-7-methoxy-1,4-benzoxazin-3-one (DIMBOA), which imparts chemical defense against the first generation of the European corn borer *Ostrinia nubilalis*. A direct correlation between plant age and concentration of 6-methoxy-2-benzoxazolinone (MBOA) – a more stable degradation product of DIMBOA – and resistance to feeding by the corn borer has been established (Klun and Brindley 1966). Chemical antibiosis was demonstrated in lyophilized corn silk and plant extracts from corn varieties resistant to the corn earworm *Helicoverpa zea*. The evidence indicates that these toxic compounds are polyphenols.

Another classic work on the characterization of cucurbitacins (Cucs) has been done elaborately by Metcalf *et al* (1982). At least 20 chemically different Cucs from roots, stems, cotyledons, leaves, and fruits of Cucurbitaceae have been characterized. Oxygenated tetracyclic triterpenes, commonly called cucurbitacins A to S, function as defense against herbivores of Cucurbitaceae and other non-adapted herbivores. However, this bitter compound is a feeding stimulant for diabroticite beetles.

Biotechnology in resistance breeding

Recent advances in biotechnology, especially tissue culture and molecular biology, have opened new opportunities for HPR. Through the use of embryo rescue and protoplast fusion techniques, it is now possible to produce distant hybrids and transfer genes for pest resistance from wild relatives to crop plants, as in the case of rice. Through genetic engineering techniques, novel genes for pest resistance can be introduced into the gene pools of crop species. For example, the *Bt* gene coding for a toxin, which is produced by the bacterium *Bacillus thuringiensis*, has been transferred into tomato, tobacco, potato, and cotton where it imparts a high level of resistance to insects in both laboratory and field tests (Hilder *et al* 1987).

Promoters primarily responsible for gene regulation have been isolated, characterized, and modified to achieve a strong expression in specific tissues of different plants. Hence, it may be possible to produce transgenic plants that express multiple forms of endotoxins and thus enhance their resistance to beetles, flies, and Lepidoptera.

Molecular markers such as restriction fragment length polymorphism (RFLP) markers, hold good promise in resistance breeding. Molecular genetic maps have become available in maize, rice, potato, tomato, barley, lettuce and many other crops. Naturally occurring and novel genes when tagged with RFLP markers would facilitate marker-based selection.

Although these tools of biotechnology enable genetically engineered plants to produce their own toxins, molecular geneticists will need empirical information to give themselves assurance that the efficacy of a specific type of toxin-gene expression in a plant will not be circumvented by extant insect behavior and physiology. Understanding such phenomena will be most fruitful in IPM programs.

2

CROP PLANT AND INSECT DIVERSITY

All the major food and fiber crops of the world are of ancient origin. They were domesticated by early man some thousands of years ago and had become staple food for humans long before recorded history. In the process of evolution, an apparently unpromising group of weedy-looking wild plants became the successful crops of the present era. Darwin (1882) was the first to point out the variability of plants and animals under domestication. He was struck by the enormous divergence of some domesticated forms from their wild relatives and presumptive progenitors. The domesticated plants have evolved rapidly, and indeed are still in the process of evolving, which is not only a subject of theoretical interest, but has great practical implications for crop improvement.

A knowledge of the origins of agriculture, the process by which cultivated plants have evolved from their wild progenitors, the genetic variability of crops, insect variability in the crop environment, and the evolution of insect–plant systems is of immense value to plant breeders, entomologists, and other specialists engaged in resistance breeding. Such knowledge should provide a better understanding of the appropriate plant materials to use in plant breeding, and insect populations to be used in screening the breeding materials for development of cultivars with durable resistance to insects.

Crop Plant Evolution

Crop plants have not always existed in their present forms and their development has been one of the greatest achievements of humankind. Almost all crop plants have been derived from wild progenitors. 'The origin of a cultivated plant is a process not an event' (Anderson 1960). Cultivated plants must, therefore, be traced back to their original wild forms by finding the relationships among various wild species and by judging similarities that exist between them. Using that method, Alphonse de Candolle in his classic work, *Origine des Plantes Cultivees* (1886), was the first to answer the question of the origin of a number of important cultivated plants. De Candolle emphasized the importance of archaeological evidence for establishing a plant's origin. Although there were no truly fossilized cultivated plants, many well-preserved and dried seed, fruit, and pollen remains that were found enabled him to follow the course of evolution in such crops as maize and wheat.

Before looking further into the differences between wild and cultivated plants, we must know how man began to intentionally plant certain wild species and create conditions for the gradual transformation of wild forms into actual cultivated plants. Man seems to have existed as a species for a period of one or two million years. During most of that period, he gathered wild plants in the form of seeds, roots, juicy stems, and fruits, together with animals that he could catch by hunting and fishing. That was the so-called hunter–gatherer stage of our cultural prehistory, which still continues today among some aboriginals. At some time in the late Mesolithic or early Neolithic age or about 10,000 years ago, man seems to have begun to domesticate plants. His life and activities were revolutionized by agriculture, which enabled him to abandon his nomadic lifestyle and settle down in permanent habitations. The first few plants that man cultivated came from 'gathered plants,' plants in their wild forms, that man selected for his vital food.

The landraces that evolved as plants were domesticated came into genetic equilibrium with all sorts of herbivorous pests. Genes for virulence in pests were matched by genes for resistance in the host so that neither hosts nor pests were threatened. Insect outbreaks were rare because of the protection provided by diverse genes for resistance present in the mixed heterogeneous plant populations. The wild ancestors of crop plants still exist today, because both have been evolving in parallel ever since their initial divergence. The wild progenitors hybridize with the crop plants and introduce new genetic variability by introgression. Most of our cultivated plants have companion weed races, and without the genetic support of the latter, the former would never have succeeded as domesticates (Harlan 1975a).

In dynamic interacting genetic systems, the landraces have evolved in a stepwise gradual evolutionary process: wild race→weed race→landrace. The landraces are the basic raw materials from which the present-day uniform crops have evolved. The crops and the wild and weedy races are sometimes in the same agroecological zones.

An example showing the origin of a cultivated plant from a wild form is wheat. Emmer wheat *Triticum dicoccum* is one of the oldest cultivated plants of the Old World. Its original wild form was unknown for a long time. However, in the middle of the 19th century, first in Syria, then in Palestine and Iran, a wild species of grass that was discovered proved to be the wild ancestor of emmer wheat. The wild form was generally smaller and more delicate than the more robust cultivated form. The spikelets of the wild grass were very hairy and the spikes shattered spontaneously into individual spikelets when ripe (Fig. 2.1). Considering the close similarity of the two forms, it was concluded that the wild grass *T. dicoccoides* was the progenitor of cultivated emmer or *T. durum*.

Several Old World grain and oil crops and vegetables seem to have evolved from weed species. The cultivated oat, *Avena sativa*, is derived from the wild oat *Avena fatua* ssp. *fatua*, native to Eastern Europe and the Middle East, where it became a weed among emmer wheat and barley and arrived in Central Europe together with the cultivated plants.

Millet *Panicum miliaceum* and Italian millet *Setaria italica* are cultivated plants. The original form of millet is perhaps *Panicum spontaneum* which grows as a weed in Central Asia. The barnyard grass *Setaria viridis*, a weed of worldwide distribution, is probably the progenitor of Italian millet. In tropical and subtropical agriculture, three other cereal species are important: sorghum *Sorghum bicolor* and African millet *Pennisetum spicatum*, both of which

Fig. 2.1. Ears of wild wheat *Triticum dicoccoides* (left) and cultivated *T. dicoccum* (right). The ear of the wild form is smaller in virtually all parts and shatters into individual spikelets when ripe. From Schwanitz (1966).

are derived polyphyletically from the hybridization of a large number of wild African species; and raggee *Eleusine coracana*, which probably originated from the weed species *E. indica*. Presumably all three are of African origin, but today they are distributed as cultivated plants in the warm zones of the entire world.

Vavilov, a Russian geneticist and agronomist, was interested in the study of plant origins and diversity for the purpose of breeding new varieties for the varied ecological conditions of the former USSR. He launched a vigorous worldwide plant exploration program for the first time in the 1920s. In 1926 he wrote the essay 'Studies on the origin of cultivated plants' (Vavilov 1926) dedicated to Alphonse de Candolle, in which his pioneering work on the genetic diversity of crop plants was demonstrated. He emphasized the importance of the genetic diversity of our crop plants as well as related wild species for crop improvement. He stressed the importance of weed races of crops not from the point of theoretical interest but because weeds serve as reservoirs of variability and are able to exchange genes with the crop through occasional hybridization.

Crop domestication

The age of wild plants as a main source of human nutrition ended when man, instead of collecting edible wild plants, started intentionally planting and maintaining those plants. The process of crop cultivation began with the gradual transformation of wild forms into cultivated plants. Crop evolution through the millennia has been shaped by complex interactions involving mutations, natural and artificial selection pressures, and the alternate isolation of stock followed by migration and seed exchanges. Such phenomena have brought stock into new environments and that permitted new hybridizations and recombination of characteristics. This period of crop evolution ultimately produced field varieties or landraces, which gave relatively low but stable yields and were well adapted to the climatic conditions of their region as well as being tolerant of epiphytotic diseases and insect damage.

Landraces are adapted to rather primitive agronomic practices as well as to low fertility. Most important, they are genetically diverse, at equilibrium with disease and insect pressures, and genetically dynamic. They are the products of millennia of natural and artificial selection. Landrace populations were the basis for subsistence agriculture that spread slowly over the earth. Their distribution was by no means uniform and certain regions accumulated a much greater diversity of landraces than others. They tend to accumulate in regions where effective agriculture has been practiced for very long periods of time (Harlan 1969).

The evolution of cultivated plants did not reach a second stage until the 19th century. Better crop varieties were developed by assembling the landrace populations, and pure lines were then selected out of them. Generally these lines were higher yielding than the landraces. The third stage of plant breeding started in the 20th century with the creation of new variability by intraspecific and interspecific hybridization. Many highly productive cultivars were developed through this process. It also became possible to introduce genes for disease and insect resistance into these cultivars through hybridization with landraces or pure line donors.

Wild relatives are extraordinary source materials for disease and insect resistance. In several examples genes from wild relatives stand between man and starvation or economic ruin, even though wild races are poorly represented in our germplasm collection. Of late, wild relatives have been used more for resistance breeding than for any other purpose. For example, tomato is a major world food crop, but it could not have been a commercial crop without the genetic support of its wild relatives. Resistance to *Verticillium*, *Fusarium* wilt, bacterial canker, curly top virus, mosaic virus, three kinds of spotted wilt virus, and root knot nematode were incorporated into tomato varieties from wild relatives.

Instances of the transfer of insect resistance from wild species to cultivated forms can be found in the work of Schooler and Anderson (1979), who transferred pubescence characteristics for aphid resistance by cytogenetic manipulations from *Hordeum bogdanii* to cultivated barley. Steidi *et al* (1979) have incorporated cereal leaf beetle antibiosis from wild oats into cultivated oats. Baksh (1979) has transferred resistance to borers from wild *Solanum incanum* to cultivated *S. melongena*. High levels of resistance to the brown planthopper and whitebacked planthopper have been transferred from *Oryza officinalis* to cultivated rice *O. sativa* L. (Jena and Khush 1990).

Evolution of the Insect–Plant System

Evolution is a continuing process involving eons of gradual changes in the morphology, anatomy, and the interactive process among diversified taxa of animals, plants, and all other organisms in the surrounding ecosystem. The interlocking chain of events in the evolutionary process of the insect–plant system has come to the present status of the dominance of insects and angiosperms in the modern agroecosystem. This is not to say that the role of other organisms at various trophic levels has been less important in this interactive process. On the contrary, scientists have identified other equally important interactive biotic and abiotic agents, which are associated with the evolutionary system.

The following discussion presents an insight into the insect–plant system comprising phytophagous insects, their entomophagous enemies, and angiosperm hosts. One striking feature of these insects and plant groups is their extraordinary speciation rates. Another feature is the tremendous food-web complexity in the insect–plant system. A third feature is the diversity of chemical and morphological structures of plants that provide physiologic and behavioral adaptations in the insect life patterns. There is now a fast growing body of literature dealing with the structure and evolution of insect–plant systems (Zwolfer 1978, Strong 1979). This interactive system can be elucidative in tracing the phylogenetic history of insect and plant taxa.

Insect evolution and diversity

The phyllum Arthropoda evolved during the Carboniferous period with the class Insecta evolving particularly rapidly and became well adapted to terrestrial habitats. Current estimates suggest that insects make up more than one-half of all living species; phytophagous insects comprise approximately one-fourth and their hosts, the green plants, make up a second quarter (Fig. 2.2). For every species of phytophagous insects there is approximately one predatory, parasitic, or saprophagous insect species, and these comprise a third quarter of macroscopic organisms (Price 1977, Southwood 1978a).

The existing insect fauna is the result of rapid evolution. When they first arose as a group, insects were undoubtedly few in numbers of both individuals and species and had only a limited range of habitats. Subsequently, as the entire organic world developed, the insects, having diversified adapted themselves to newly emerging ecological niches.

Several long eras passed in the geological sense before insects appeared. The first forms of life were unicellular organisms, which probably arose over 2.5 billion years ago. From these evolved many kinds of multicellular plants and animals. Through pre-Cambrian times, from which no good fossils are known, life was presumably marine. The animals were multicellular through

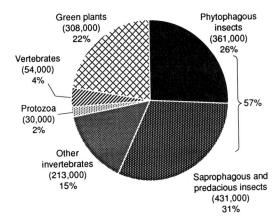

Fig. 2.2. The number and proportion of species in the major taxa, excluding fungi, algae, and microbes. Redrawn from Strong *et al* (1984).

Table 2.1. Condensed geological timetable represented by fossil records (modified from Benson 1976).

Mya	Geological period	Notable events that occurred for plants and animals	
		Plant life	Animal life
0	Cenozoic era—modern life		
1	Quaternary	Dominance of flowering plants	Modern insect life
63	Tertiary	Dominance of angiosperms	Lepidoptera and mammals diversified
	Mesozoic era—medieval life		
135	Cretaceous	Dominance of flowering plants	Isoptera, Lepidoptera, Cecidomyiidae, social insects
181	Jurassic	Climax of gymnosperms, including cycads	Era of dinosaurs; origin of mammals; Diptera, Odonata
230	Triassic		Hymenoptera
	Paleozoic era—ancient life		
280	Permian	Gymnosperms	Thysanoptera, Hemiptera, Coleoptera, rise of reptiles
350	Carboniferous	Mosses, first great tropical forests—warm, moist	Orthoptera, dominance of amphibians
405	Devonian	Algae, ferns	Beginning of land animals
425	Silurian	Rise of land plants—early pteridophytes	Predominance of marine invertebrates
500	Ordovician	Algae	First chordates
600	Cambrian	Algae	Invertebrates

Mya, millions of years ago.

the Cambrian and Ordovician periods of the Paleozoic, when the seas were shallow, extensive, and warm. At this time, the Arthropods were well developed, represented by the Trilobita, complex Crustacea, and Eurypterida. All the animal phyla as we know them today are represented in the fossils of those periods, including the first vertebrate fossils and primitive fish. At this time the land was presumably barren, at best with encrustations or local pockets of lower plants, and virtually had no signs of living matter. In the Silurian came the first feeble beginnings of a land biota, which appeared in the fossil records of land plants. The first known air-breathing animals were scorpions and myriapods. It seems almost certain that primeval wingless insects also evolved in that era. All these animals were probably tropical and shade-loving, occurring along the beaches and in the swamps of the warm moist Silurian climate. Table 2.1 represents the period and events of the evolution of plant and animal life for the last 600 million years.

The Devonian period

The earliest known fossil insect is *Rhyniella praecursor*, a collembolan from the Devonian period of Scotland (Whalley and Jarzembowski 1981). Although most living Collembola are not herbivorous, Kevan *et al* (1975) suggested that *Rhyniella* might have fed on 'plant juice obtained through punctured wounds'. The characteristic spiny systems of Devonian plants were considered as defense against insect attack. The success of early land plants was due to the evolution of the aromatic acid acylation of the polysaccharide components of the cell wall. The phenolic acid acylation of cell-wall polysaccharides is of even greater importance in evolutionary terms than protection against pathogens (Friend 1976).

The Carboniferous period

The Carboniferous saw the rise of all modern plant divisions except the angiosperms (Chaloner 1970). The main herbivores of the Carboniferous were probably the insects which, by then, had acquired flight ability. All the modern groups of insects (Ephemeroptera, Orthoptera, ancestral Hemiptera, and Mecoptera) except the paraneopterous and mecopteroid stocks were in existence by that period (Smart and Hughes 1973). It is clear that the insects had achieved great diversity by the end of that period (Carpenter 1977). In spite of this evolutionary success, it is obvious from the extensive coal-swamp remains of the Carboniferous, that plants were not totally consumed by herbivores. There are probably three reasons:

1 The coal-swamp flora were out of reach for most potential spore eaters and, by producing tannins, plants protected themselves against fungal attack.
2 Protection of the haploid thallus against attack by fungi was due to

increased production of cinnamic acid derivatives.

3 Production of ecdysone analogs would greatly interfere with insect metamorphosis (Swain and Cooper-Driver 1974).

Most of the insect orders (Dictyoneurida, Michopterida, and Diaphanopterida) that flourished in the Carboniferous period are now extinct. All three of those orders had beak-like, apparently piercing mouth parts, assumed to be reminiscent of the mouth parts of modern Hemiptera. These insects were believed to be phytophagous, feeding on the reproductive organs of fruits of Carboniferous and Permian plants (Smart and Hughes 1973). But late Carboniferous insects never successfully overcame all these chemical defenses and the world had to wait for the more versatile herbivores.

The Permian period

In the Permian period, which had a considerably longer duration than the Upper Carboniferous, 19 orders of insects were in existence. Ten of these orders appeared for the first time, including the Hemiptera–Homoptera in the Lower Permian and Hemiptera–Heteroptera in the Upper Permian. It is not clear as to what the Permian Heteroptera fed on, because phytophagous Heteroptera were not known until the early Jurassic. However, all evidence suggests that Permian Homoptera were then, as now, predominantly phytophagous, with fossils attributed to two modern groups, i.e., Auchenorrhyncha and Sternorrhyncha (Wootton 1981).

The characteristic homopterous beak was well developed, indicating that, like the Palaeodictyoptera and their relatives, the Homoptera could find abundant sources of liquid food from the then existing plants. These plants had a thin cortex and phloem close to the outer surface of the stem. In that period, there was a shift from Pteridophyte dominance to Gymnosperm dominance in the vegetation. Consequently, insects like Odonata with chewing mouth parts decreased considerably (Smart and Hughes 1973). Thus the most striking feature of the Permian insect fauna was the presence of four existing orders of Endopterygota: the Neuroptera, Mecoptera, Trichoptera, and Coleoptera. The first Coleoptera of the Permian was mainly a primitive type with an elytrous venation. They apparently remained dominant until the Jurassic, when an assortment of more modern families appeared, some of which still exist. In terms of diversity of form and the association of generalized and specialized species, the fauna of the Permian was probably the most diverse in the evolutionary history of insects (Carpenter 1977).

The Mesozoic era

The most notable feature of the Mesozoic fauna was the apparent absence of the orders found in the Paleozoic. Although some primitive Hymenoptera and

Diptera have been found in that period, only a few insects were recorded from the Triassic period. By the early Jurassic period, with the addition of the Dermaptera, Phasmatidea, and Thysanoptera, and with the loss of the extinct orders, the fauna was essentially a modern one, with the presence of existing families of several orders, such as the Odonata, Orthoptera, Diptera, and Hymenoptera (Carpenter 1977). It is very likely that at the beginning of the Mesozoic, all the orders of insects usually thought of as being associated with flowering plants were present, except Lepidoptera with their exceptional degree of morphological adaptation to feeding on the flowers of advanced angiosperms (Smart and Hughes 1973).

In the middle Jurassic, the climate of the world favored the abundance of insects. It is probable that the plant-feeding habit became well established on the varied flora which developed during that period. Some flowering plants produced conspicuous flowers, which were visited by many insects belonging to many families. The most important groups that emerged during the Mesozoic were the Hymenoptera, Cecidomyiidae, and Lepidoptera.

Primitive Hymenoptera were present in the Triassic with normal mandibulate mouth parts. In the next geological period, the Jurassic, fossils of other sawfly superfamilies were found, whose larvae might well have fed on plant tissues, like their present day relatives (Henning 1981). The other main suborder of the Hymenoptera includes phytophages, particularly the Cynipidae, which are specialist gall-makers on oaks (Quercus) and Rosaceae. In addition, some members of the superfamily Chalcidoidea are seed feeders; this habit arose from feeding on the embryos of plants. A good example is the fig wasp belonging to Agaonidae.

The possible evolutionary pathway of the gall midges (Cecidomyiidae) has been reconstructed in detail by Mamaev (1968). Fossils of Diptera were known from the Lower Triassic, but they did not include any phytophages. Mamaev believed that the ancestors of gall midges that diverged in the Upper Jurassic or Lower Cretaceous had their larvae feeding on the litter of wet forests. The path to phytophagy might have begun with species exploiting fungal hyphae among dead leaves or fungal necroses on living plants.

The Lepidoptera also diversified in the Cretaceous and the earliest fossils were primitive Micropterygidae from Lower Cretaceous amber (Whalley 1977). Caterpillars of extant micropterygids feed on mosses and liverworts, whereas their adults feed on pollen. The two major extant suborders of Lepidoptera, Monotrysia and Ditrysia, evolved after the radiation of angiosperms in the Cretaceous. In contrast to the situation with sawflies, there is no evidence for an initial evolution of early Lepidoptera on primitive plants, the pteridophytes. One of the most interesting contributions of the Cretaceous and Tertiary fossils is to our understanding of the antiquity of social behavior in insects, such as Hodotermitidae, Vespidae, *Bombus*, true *Apis*, and ants (Carpenter 1977).

The Rise of Modern Insect Life on Angiosperms

Much of the flowering plant and modern insect life was established during the Cretaceous period. The origins of the flowering angiosperms have now been pinpointed near the Jurassic–Cretaceous boundary (Knoll and Rothwell 1981). Angiosperms dominated the terrestrial ecosystems by the mid-Cretaceous. The angiosperm flowering plants dominated over the spore-bearing floras and the archaic gymnosperms (the cycads and their allies), so that by the end of the Cretaceous period, 90% of the plant genera were of the woody kinds found today, including elm, oak, fig, magnolia, beech, birch, and maple. The angiosperms emerged from the gymnosperms with the evolution of enclosed seed characteristics. The protected seed ensures the success of sexual reproduction and consequently the opportunity for more evolutionary advancements. The primitive angiosperms were woody and the evolution of herbaceous angiosperms (such as sedges and grasses) was a later phenomenon.

Although most of the angiosperms are wind pollinated, many genera known from the Cretaceous period were undoubtedly insect pollinated as they are today. The gymnosperms are largely wind pollinated. For efficient and economical wind pollination, individuals of a kind must be growing close together; otherwise, a huge amount of metabolically precious pollen is likely to be blown away by wind without finding its mark. Pollination by insects requires a much smaller amount of pollen and does not require such large populations for cross-pollination. For attracting insects for pollination, angiosperms have developed a new evolutionary strategy of producing nectar and colorful scented petals.

The interrelationship of plants and animals on land may be regarded as lying somewhere between development from the rudimentary in the early Devonian time to the complex circumstances of today (Smart and Hughes 1973). The progressive evolution of the plant–insect relationship occurred in the early Cretaceous (Fig. 2.3). The rise of many favorable host plants in the Cretaceous may have encouraged great evolutionary development among phytophagous insects. The appearance of numerous parasitic Hymenoptera during the Triassic indicates that populations of their insect host species were at least fairly large and that the present day interrelationships of the groups were already well established in the ecosystem. The reptiles, which were the dominant animals up to the end of the Triassic, were rapidly being supplanted by the early land mammals. The extinction of the dinosaurs is inversely related to some of the toxic phytochemicals found in flowering plants during the Cretaceous (Swain 1976). But the tremendous success of varying groups of insects in surviving and adapting to the rise of the angiosperms is striking.

As unadapted species of plants disappeared, many insects must have become extinct along with their hosts. Among the important extant phytophages, only the Hemiptera, sawflies, and Orthoptera certainly pre-date the

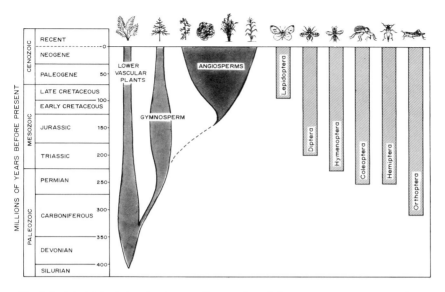

Fig. 2.3. Model depicting the degree of integration of plant—insect progressive evolution in different geological periods.

rise of the angiosperms. Moreover, some of the most successful phytophagous superfamilies within these three early groups did not appear until after the evolution of angiosperms (Strong *et al* 1984). These insects, including leafhoppers, planthoppers, and shield bugs, are presently exploiting our food crops and are also serious vectors of plant virus diseases.

Functionary Adaptations in Insect—Plant Systems

Insects have lived on earth for about 350 million years compared with the less than 2 million years for man. During this period, insects have evolved in many directions to become adapted to terrestrial life in almost every type of habitat within the limits of geography and physical environment (Borer *et al* 1976).

The evolutionary success of insects can be attributed to four anatomical characteristics (Hinton 1977a). The first was their development of a tracheal system, which involved the invagination of a relatively enormous surface area permeable to both oxygen and water. In this system, water loss, so critical to small terrestrial animals, can be greatly reduced because contact with the external environment is only via the spiracles, which in many cases can be closed in the face of adverse conditions. The development of the tracheal system paved the way for the second feature, which was the acquisition of a more or less impermeable exoskeleton that enabled the insects to invade dry terrestrial environments. The cuticle is less permeable to water than to oxygen and gives protection from desiccation and mechanical injury. The third feature occurred when the extension of the mesothorax and the metathorax evolved into flapping wings so that sustained flight became possible for the first time. This feature occurred during the Carboniferous. The development of wings brought a new dimension to the location of food supplies, mates, and to both escape and predation. The fourth characteristic appeared in the Upper Carboniferous or Lower Permian when, in some forms, the developing wings were invaginated into the body of the larvae, and the last-instar larvae evolved into pupae. The resulting differentiation in form and function of the larvae as a feeding stage and the adult as a reproductive and dispersal stage is a feature that permitted the same individual to divide resources effectively between the two stages. This resulted in considerable selection towards different directions within the life cycle. Indeed the success of this development is reflected in the fact that 88% of the known species of insects are Endopterygota.

Most insects are relatively small, and are able to live in places that would not be available to larger animals. Although the insect exoskeleton tends to limit body size, it is more efficient for smaller animals. Surface–volume ratios suggest that smaller insects are less exposed to the risk of desiccation. Body size is related to generation time; smaller organisms, in general, have shorter generation times than larger organisms. Body size is also related to the quantity of food required. Small quantities of food could be exploited by insects from all possible ecological niches. The efficiency of the insect tracheal system has virtually eliminated the transportation of respiratory gases, supposedly the function of blood. Thus the insect circulatory system, known as an 'open circulatory system' is simple and may be an additional factor limiting body size. The sophisticated chemosensory mechanisms associated with a diversified mode of feeding and mating and other forms of communication in the ecosystem have also been major features of the evolution of insects.

The rise of a given plant taxon has been more or less accompanied by a concurrent rise of an exploiting insect taxon (Zwolfer 1978). For many phytophagous and some entomophagous insect taxa, plants provide microhabitats as well as food resources. In addition to their trophic function, the plants provide token stimuli, shelter, and rendezvous places for male and female insects.

Only nine insect orders out of 29 living orders exploit the living tissues of higher plants for food. Life on higher plants presents a formidable evolutionary hurdle for insects. Southwood (1973) suggests that the step from the original primitive scavenging/microorganism feeding to plant feeding is nutritionally more difficult than that to predation. As such, phytophagous insects had at least three major hurdles to overcome before they could exploit plants successfully: the problems of desiccation, attachment, and nutrition.

These hurdles are overcome by insects with their built-in structural, physiologic, and behavioral adaptations. The cuticle–tracheal system reduces water loss or desiccation and provides tolerance and control of the changes in hemolymph osmolality that inevitably accompany water loss (Willmer 1980). The problems of attachment on the host have been solved: (1) by structural modifications or secretions of the insect's adhesion, and (2) living inside the plant as in the case of tissue borers, gall formers, and leaf miners. The pretarsal structures of many insects aid in gripping plant surfaces. Sawfly and lepidopteran caterpillars have independently evolved similar sucker-like abdominal prolegs. Many caterpillars also spin silk threads to aid attachment.

Problems of obtaining adequate nourishment arise because plants and animals differ markedly in protein content. The plant tissue critical for phytophages is protein nitrogen. Given the importance of protein nitrogen levels in the diets of many species, insect paleontologists have suggested two main routes for the evolution of phytophagous insects. One route is via pollen feeding (Rohdendorf and Raznitsin 1980). The proportion of protein in fresh pollen is high and is similar to that in insects. The other route enlists the aid of external microorganisms in digesting dead and dying plant tissue. A phytophage that relies upon microorganisms in plant wounds is the onion fly *Delia antiqua*, which eats cellulytic bacteria in feeding lesions in the host plant. An evolutionary shift to more intimate mutualism between microbes and phytophages is the culture of yeasts or bacteria within the insect's gut (Jones 1983). The bacteria not only synthesize the requirement of amino acids and proteins but also synthesize vitamins and sterols. Such symbiotic micro-organisms are associated with a wide range of Coleoptera and many groups of Homoptera. The evolution of such specialized and adapted means for overcoming all sorts of hurdles, including nutritional ones, is proof of insect diversification in green vegetation.

The preceding discussion of the insect–plant evolutionary relationship shows that insect diversity in the modern era is not a paleontological artifact, but is due to a rise in plant diversity. The variety of resources made available by an 'average plant' has tended to increase during evolution. The diversity of plant species and the architectural complexity of plants has increased during evolution. Likewise, insects have exploited plants in a number of ways and have diversified themselves in the ecosystem (Schoener 1974). The interplay of diversified defense mechanisms and diversified exploitation strategies seems to provide systems of positive feedback loops (Zwolfer 1978). Reciprocal adaptation and counteradaptation between insect phytophages and plants have been important mechanisms driving a steady increase in plant and insect diversity over the broad sweep of the fossil record (Strong *et al* 1984). The interactions of these two phyla are elaborated in Chapter 4.

3

Secondary Plant Metabolites for Insect Resistance

Green plants utilize principal metabolic pathways and biochemical cofactors for converting carbon dioxide and water to sugars, nitrogen to amino acids, and to synthesize nucleotides, lipids, and simple organic acids. These ubiquitous substances are formed in plants and animals and are referred to as *primary metabolites*. All organisms, from bacteria to plants, synthesize the same range of primary metabolites. The primary plant metabolites are the starting materials for the biosynthesis of specific, genetically controlled, and enzymatically catalyzed complex compounds known as *secondary metabolites* (Geissman and Crout 1969). It is believed that secondary plant metabolites play an important role in the defense against insect herbivores as they act as insect repellents, feeding inhibitors, and/or toxins, thus protecting plants at different phases of growth. Antiherbivory functions of secondary plant metabolites are elaborated in Chapters 4, 5, and 6.

Secondary compounds are in no way restricted to plant species; many such compounds have also been isolated from microorganisms and animals. Nevertheless, over four-fifths of all presently known secondary compounds are of plant origin (Robinson 1980). Both the terms secondary metabolite and secondary compound are currently in general use. The distinction between primary and secondary metabolites is not always clear. The original emphasis that secondary metabolites are not ubiquitous is no longer applicable. Abundant evidence supports the hypothesis that many secondary metabolites have played an indispensable role in maintaining the plant species during the course of evolution (Swain 1977). These metabolites are synthesized in a dynamic state, as evidenced by their rapid rate of turnover and degradation. Some of the toxic secondary compounds may act as storage for carbon and nitrogen; in their absence metabolic activity may be locally limited. For example, nonprotein amino acids in legume seeds, dhurrin (a cyanogenic compound) in sorghum species, and cyanolipids in the seeds of the Mexican buckeye *Ungnadia speciosa* are sources of nutrients for germinating seedlings (Janzen 1969). The quantities of alkaloids that are translocated to the leaves from the roots to effect an increase in leaf alkaloid titer for insect defensive purposes also serve as the nitrogen transport system. In this context, nicotine alkaloid provides an additional support for the enrichment of the nitrogen content of leaves (Baldwin 1991). This evidence suggests that secondary metabolites are intimately involved in the primary metabolic functions of the plant.

Nature of Secondary Metabolites

It has been demonstrated that the formation of secondary metabolites and their storage sites are restricted to certain developmental phases of the plant, specific organs, tissues, or specialized cells.

Compartmentation

Plant cells have developed two methods of storing secondary metabolites. They may be classified as *intracytoplasmic* where vacuoles and plastids are the storage sites, and *extracytoplasmic* where the cell walls, pollen walls, sub-cuticular spaces, and cuticular surfaces are involved in storing the secondary compounds. Thus, the accumulation of secondary metabolites may correlate with the differentiation of specific accumulation sites.

Vacuoles and chloroplasts are the most important storage sites for many hydrophilic secondary metabolites. The vacuoles in the cells of sorghum seedlings are the major sites of accumulation of the cyanogenic glucoside dhurrin. Chloroplasts of various species of plants contain simple phenyl-propanes and flavonoids. Lipophilic secondary metabolites (lignin) may accumulate in the cell membranes of plants (Schulze *et al* 1967). Plants produce specialized containment structures for these compounds. For example, terpenes are generally restricted to resin ducts and glandular hairs, and accumulation of furanocoumarins is restricted to oil tubes (Camm *et al* 1976).

Once the secondary metabolites are secreted by plant cells, they are held back in their respective storage compartments, but it does not mean that further metabolism or degradation of these products is not possible. Many secondary compounds including chlorogenic acids, cyanogenic glycosides, and alkaloids undergo rapid turnovers as evidenced from radioactive-labeled precursors in pulse-chase experiments (Strack and Reznik 1976).

Distribution

Secondary metabolites are not distributed evenly throughout the plant, either qualitatively or quantitatively, in space and time. The enormous variation in the distribution of terpenoid substances in plants is apparent at several stages of plant growth. Each plant population often has its own distinctive terpenoid profile and within individual plants, terpenoid levels usually vary among organs, tissues, and cells. Finally, age, season, or environmental conditions frequently influence terpenoid composition (Gershenzon and Croteau 1991). Secondary metabolites are not synthesized or accumulated in any specific group of cells at all times during the life of the plant. For example, the cyanogenic glucoside dhurrin in *Sorghum bicolor* is located in the epidermal layer of leaves, whereas catabolic dhurrin resides in the mesophyll tissue (Kojima *et al* 1979).

The concentration of secondary compounds in most plants vary diurnally. Furthermore, the quantities of these compounds are likely to be altered by climatic and edaphic factors, and exposure to microorganisms, grazing herbivores, or even air-borne pollutants (McKey 1974). Even within a single plant, a single toxin can vary both in space and time, for example production of cyanide in many plants is extremely variable.

Enzyme regulation

The expression of secondary metabolites is regulated by metabolic enzymes. The enzymatic activities of secondary metabolism appear before or during the *de novo* synthesis of RNA and proteins. In many instances, the beginning of the formation of secondary metabolites is directly coupled with the synthesis of the corresponding enzymes (Wiermann 1981). For example, the biosynthesis of flavonoids in the cell culture of *Petroselinum hortense* is controlled by phenylalanine ammonia lyase (PAL), the key enzyme in the biosynthetic pathway of flavonoid compounds (Hahlbrock *et al* 1976). Subsequently, it has been shown that PAL biosynthesis depends on the RNA and protein biosynthesis. This demonstrates that the expression of flavonoid metabolism is regulated by transcription.

Functions of Secondary Metabolites

Secondary metabolites perform many useful functions for the plants. Some are plant growth hormones such as auxins, cytokinins, abscisic acids, and gibberellins. The wound hormone traumatic acid is very effective in healing damaged plant tissues. There are reports that not only the wound hormones can help heal plant damage but the toxic secondary compounds also play a similar role. It was recently reported that although the tetracyclic triterpene

Table 3.1. Major classes of secondary plant metabolites involved in host plant resistance to insects.

Class	Subclass	Biosynthetic origin	Approximate number of structures	Distribution
Plant surface	Alkanes, aldehydes, ketones, waxes (long chain)	Acetate–malonate	na	Mostly on plant leaf surface as cuticular waxes
Carbohydrates/polymers	Lignins and tannins (phenolic polymers)	Shikimic acid	na	Integral cell-wall constituent of all vascular plants
Terpenoids	Monoterpenoids	Acetate–mevalonate		Widely distributed in essential oils
	Iridoids		600	Restricted to 57 families in Dicotyledoneae
	Others		700	Well known in conifers
	Sesquiterpenoids			Widespread
	Phytojuvenile hormones			
	Sesamin and sesamolin			From sesamin oil
	Juvocimenes			From plants of sweet basil *Ocimum basilicum*
	Juvenile hormone III			From Malaysian sedge *Cyperus iria* (Cyperaceae)
	Antijuvenile hormones			
	Precocenes 1 and 2			From *Ageratum houstonianum* (Compositae)
	Sesquiterpene lactones		3500	Mostly in Compositae, localized in glandular hairs
	Others			Widespread
	Diterpenoids		3000	Widely distributed, especially in latex and plant resins of conifers

Table 3.1. Continued.

Class	Subclass	Biosynthetic origin	Approximate number of structures	Distribution
	Clerodanes		400	Found in Lamiaceae, Asteraceae, and others
	Others		2500	Widespread
	Triterpenoids			
	Cucurbitacins		20	Confined mainly to Cucurbitaceae family
	Cardenolides		150	Asclepiadaceae and others
	Limonoids		300	Mainly in Rutaceae, Mellaceae
	Quassinoids		200	Simaroubaceae
	Steroidal and triterpene saponins		1200	Widespread
	Others		1500	Widespread
Antihormones (steroids)	Phytoecdysones	Acetate–mevalonate and other pathways	70	Found in more than 100 plant families
Phenolics	Simple phenols	Shikimic acid	200	Universal in leaf, often in other tissues as well
	Coumarins	Shikimate–chorismate	300	From 70 families of dicot
Flavonoids	Anthocyanins Flavonols Flavones Isoflavonoids	Shikimate–malonate	4000	Universal in angiosperms, gymnosperms, and ferns

Quinones	Benzoquinone Naphthoquinone Anthraquinone Extended quinone	Shikimate–mevalonate	800	Widely distributed in all plant families except ferns and mosses
Alkaloids	Benzylisoquinoline Monoterpene indole Simple indole Pyrrolizidine Quinolizidine Polyhydroxy	Heterogeneous group	6500	Widely distributed in angiosperms, in roots, leaves, and fruits
Nonprotein amino acids	L-Canavanine L-Arginine	Amino acid derivatives	400	Especially seeds of legumes but relatively widespread
Cyanogenic glycosides	Dhurrin Amygdalin	Amino acid derivatives	60	Recorded in 2500 plants of more than 130 families
Glucosinolates	Isothiocyanate Allylisothiocyanate Sinigrin	Amino acid derivatives	100	Cruciferous and 10 other Acrid families

na, data not available.

cucurbitacin of *Cucurbita* plants is highly toxic against many generalist insects, it is equally responsible for healing the leaf-feeding damage caused by the *Epilachna* squash beetle (Tallamy and McCloud 1991).

The interactions between plants and insects are of particular ecological significance. Some insect pollinators are attracted to plants by the colored flavonoids and carotenoids, while other secondary metabolites, such as the volatile terpenoids, repel potential invaders. However, the major functions of secondary metabolites in plants are as chemical signals in the ecosystem and as antibiosis agents against insects and pathogens. To perform these functions, a vast array of these compounds are synthesized by plants. Secondary compounds can be grouped into different classes based on their biosynthetic pathways (Table 3.1). The following sections describe secondary compounds that are specific to insect–plant interactions.

Biosynthetic Pathways of Secondary Metabolites

Excluding the primary process of sugar and protein biosynthesis, there are three major pathways for the production of secondary metabolites in plants. They are: (1) the acetate–malonate pathway, (2) the acetate–mevalonate pathway, and (3) the shikimic acid pathway. Besides these, three more

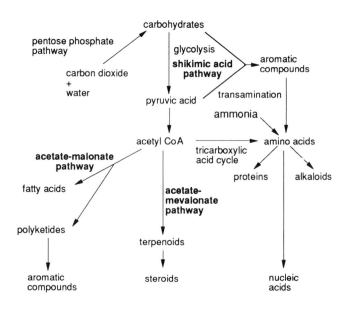

Fig. 3.1. Major biochemical pathways involved in the production of secondary plant metabolites.

diversified pathways lead to the production of many other groups of secondary compounds: (1) mixed biogenesis, (2) heterogeneous groups, and (3) amino acids. These biosynthetic pathways are complex and interrelated (Fig. 3.1).

Acetate–malonate pathway

Acetyl coenzyme A (CoA) (activated acetic acid) is the precursor of the acetate–malonate pathway. Acetyl CoA is also the precursor of a host of primary and secondary metabolites, including fatty acids, terpenoids, steroids, polyketides, aromatic compounds, acetyl esters, and amides (Fig. 3.2). The conversion of acetyl CoA to citrate and other tricarboxylic acids leads to the formation of amino acids, nucleic acids, and alkaloids. The important derivatives of this pathway are the fatty acids and their metabolites, as well as plant aliphatic and aromatic compounds, which are biosynthesized through the formation of polyketides. Little is known about the function of higher plant polyketide derivatives, but they seem to protect plants from bacterial and fungal infections. Some of the naturally occurring phototoxins are polyketides (polyenes, thiophenes, quinones, and chromenes). The first described plant photosensitizers are the red anthraquinone derivatives, hypericin and pseudohypericin from the glands of leaves, flowers, and stems of the plant St John's wort (*Hypericum perforatum*) and many other species of the same genus (Arnason *et al* 1992). Some plants of the family Hypericaceae

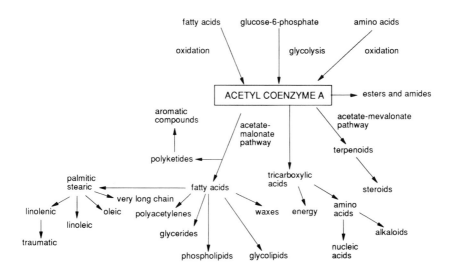

Fig. 3.2. Acetyl coenzyme A as the precursor of primary and secondary plant metabolites.

are phytotoxic to insect herbivores due to the presence of furanocoumarins and furanochromones (Champagne *et al* 1986).

Fatty acids and lipids

Fatty acid synthesis in a variety of biological systems is catalyzed by two enzyme systems, which function sequentially: acetyl CoA carboxylase and fatty acid synthetase. All the carbon atoms of the fatty acids produced are derived from acetyl CoA.

Plant fatty acids

There are about 20 common fatty acids in plants. All are ubiquitous and are categorized according to their concentrations. C_{16} and C_{18} fatty acids are the most common fatty acids. The saturated acids are palmitic (16:0) and stearic (18:0) while the major unsaturated acids are oleic (18:9), linoleic (18:9, 12), and linolenic (18:9, 12, 15). It has been recognized that unsaturated fatty acids are the sources of volatile compounds with characteristic odors in various plants, e.g., the leaf alcohols and aldehydes (Eriksson 1975). The initial products of linolenic acid are the volatile aldehydes *cis*-3-noneal and -hexenal. Analagous volatile products from linolenic acid are *cis*-3, *cis*-6-nonadienal and *cis*-3-hexenal (Phillips *et al* 1979). The characteristic smell of green leaves and grass is due to *cis*-2-hexenal (leaf aldehyde) and the corresponding alcohol (leaf alcohol).

One of the major constituents of plant waxes are long-chain alkanes ranging from 25 to 35 carbon atoms. Alkanes are derived from fatty acids by chain elongation and subsequent decarboxylation. This is the reason why alkanes in plants often have an odd number of carbons.

Plant lipids

Fatty acids are antimetabolites. They react with protein and destroy enzymes necessary for cell processes. Hence, in plants most of the fatty acids occur mainly in a bound form, as lipids such as triglycerides, phospholipids, or glycolipids.

Other types of lipids present in plants include the steroidal esters, sphingolipids (cerebrosides and phytoglycolipids), wax esters, and cutins. The steroidal glucoside lipids are unique to plants. Cuticular wax is generally a mixture of relatively simple hydrocarbons, wax esters, fatty alcohols, ketones, and acids.

Cutins are polymeric substances in which fatty acids exist as an interesterified rigid meshwork. They cover the epidermal cells (cuticle) of plants. They are mainly formed from polyhydroxy acids (palmitic and oleic acid) but possess also phenolic compounds. Cutins are embedded in a matrix of wax and are degraded by exoenzymes (cutinase) with esterase activity.

Biological functions of fatty acids and lipids

The unsaturated fatty acids, linoleic and linolenic acid, are utilized in the healing of damaged plant tissue through their degradation to the wound hormone, traumatic acid. This compound helps the cells to enlarge around the wound, thereby healing the wound. Some fatty acids and lipids act as insect attractants. The unsaturated acid (18:3) present in the pollen of clover *Trifolium* spp. (Leguminosae) acts as a food marker for bees (Thompson 1980).

One of the established functions of plant lipids in seed oils is for storage of energy. Lipids are important constituents of plant membranes and are characterized by semipermeability. Membranes are also the site of many enzymatic reactions involved in the photosynthetic electron transport system and respiratory system of mitochondria (Galliard and Mercer 1975).

The plant cuticle contains fatty acids, wax esters, and hydrocarbons. Sometimes it may contain terpenoids and flavonoids, as in Leguminosae. The hydrocarbons present in the plant cuticle are formed by decarboxylation of very long-chain fatty acids. The alkanes present in the wax, as well as the cutin of the leaf surface tissues of the plant, act as a protective layer preventing insect, fungal, and bacterial attack. Details about the anti-herbivory functions of the plant cuticle are given in Chapter 6.

Acetate–mevalonate pathway

Several important groups of natural compounds classified as terpenoids and steroids are biosynthesized by the acetate–mevalonate pathway. Angiosperms produce all types of terpenoids and steroids as a major defense strategy against herbivores.

Terpenoids

The largest and most structurally diversified class of naturally occurring organic products of secondary plant metabolites are the terpenes, the allied terpenoid compounds, and the steroids. Although terpenoids are complex in nature, they are all composed of the isoprene repeating unit. It is difficult to make any generalizations about the chemical properties of terpenoids, except that they are lipophilic substances. However, certain types of terpenoids occur as conjugates with sugar (e.g., iridoids) or with other molecules.

Biosynthesis

The terpenes are constructed from isoprene (2-methylbutadiene) units (Ruzicka 1953). The terms terpenoid and isoprenoid are interchangeable; isoprenoid refers to the five-carbon isoprene unit ($CH_2=C-CH_3-CH=CH_2$) from which all terpenoids are derived. These compounds are classified by the number of isoprene units in their structure.

There is a common isoprene unit in the biosynthetic pathway of all terpenoids. The isoprene units themselves are derived from the branched condensation of three molecules of acetyl CoA to form the six-carbon compound, 3-hydroxy-3-methylglutaryl CoA, which is then reduced to mevalonic acid (MVA). Thus MVA is the primary precursor of all the plant terpenoids and steroids. MVA is pyrophosphorylated and then undergoes decarboxylation to yield isopentenyl pyrophosphate (IPP) (Fig. 3.3).

The next step in the biosynthesis of terpenoids is the condensation of IPP and dimethylallyl pyrophosphate to form geranyl pyrophosphate (C_{10}). The pyrophosphorylated precursors can further undergo an enormous range of cyclizations and rearrangements to produce the parent carbon skeletons, which by oxidations and other transformations provide a large number of terpenoids present in plants. The original five-carbon units do not always exist in different subclasses of terpenoid. For example, several groups of triterpenoids, including the steroids, limonoids (C_{26}), and cardenolides (C_{23}), have fewer than 30 carbon atoms.

In the case of sterol biosynthesis, the first C_{30} hydrocarbon precursor squalene is formed from two molecules of farnesyl pyrophosphate (C_{15}). The first C_{30} compound produced by squalene synthetase is presqualene pyrophosphate, which is then converted into squalene in the presence of NADPH (reduced form of nicotinamide adenine dinucleotide phosphate). Steroids

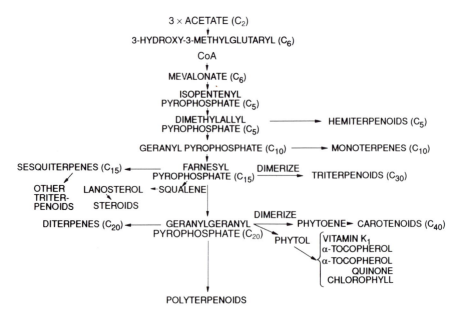

Fig. 3.3. Biosynthetic pathway of terpenoids in green tissues.

originate from lanosterol, which itself is formed from cyclization of squalene. The first C_{40} hydrocarbon precursor of the carotenoids is phytoene. It is produced from the C_{40} prephytoene pyrophosphate, which is itself formed by the condensation of two geranylgeranyl pyrophosphate (C_{20}) molecules (Fig. 3.3). Phytol is ubiquitous in higher plants as a side chain of chlorophyll. It is also a precursor of the tocopherols (vitamin E) and phylloquinone (vitamin K_1). For a comprehensive account of terpenoid biosynthesis in plants, see Porter and Spurgeon (1981).

Biological functions
There are several studies about the effects of plant terpenoids on herbivorous insects. Terpenoids may act as toxicants, feeding deterrents, and/or oviposition deterrents. Some selected examples of the antiherbivore nature of terpenoids are examined in the following sections.

Monoterpenoids. Monoterpenes are typically found and widely distributed in higher plants and possess a basic skeleton of 10 carbon atoms. Monoterpenes are volatile, lipophilic, and give characteristic odors to many plants. These compounds accumulate in resin ducts, secretory cavities, and epidermal glands. They are important constituents of essential oils present in conifers, mints, composites, and citrus. Some commercial monoterpenes are menthol, camphor, and α-pinene (Fig. 3.4).

One of the best examples of monoterpene derivatives which are toxic to insects are the pyrethroids found in the leaves and flowers of some *Chrysanthemum* species (Casida 1973). Pyrethroids are neurotoxins, causing hyperexcitation, uncoordinated movement, and paralysis of insects. The naturally occurring pyrethroids and their synthetic analogs are important commercial insecticides. Monoterpenes in conifer resins such as α-pinene, β-pinene, limonene, and myrcene (Fig. 3.4) are potential defense chemicals against bark beetles (Coleoptera: Scolytidae) (Paine and Stephen 1988). Monoterpenes such as limonene and geraniol are constituents of flower scents, which attract pollinators. However, a high concentration of monoterpenes in plants act as a repellent to potential predators. The monoterpenoid compound citronellol acts as an oviposition deterrent to the leafhopper *Amrasca devastans* (Homoptera: Cicadellidae) (Saxena and Basit 1982).

Iridoids. Iridoids are synthesized in plants via the MVA pathway (Inouye 1978). They are cyclopentanoid monoterpene-derived compounds. Their nomenclature is derived from the structural similarity and biosynthetic relationship to the compounds iridodial and iridomyrmecin, which are the defensive secretions of ants belonging to the genus *Iridomyrmex*. The iridoid glycosides are restricted to 57 plant families of the Dicotyledoneae, where they usually occur in a glycosidic form derived from the O-heterocyclic form of iridodial (Fig. 3.4). Some may contain a heterocyclic nitrogen atom (iridoid

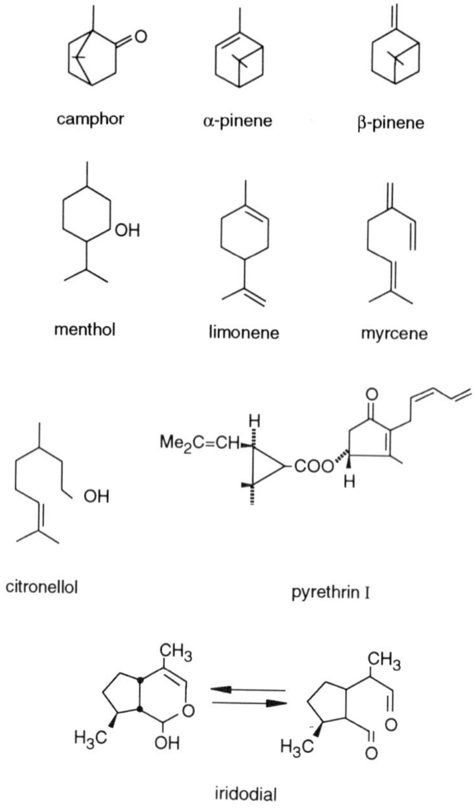
camphor
α-pinene
β-pinene
OH
menthol
limonene
myrcene
OH
Me2C=CH
H
COO
H
O
citronellol
pyrethrin I
CH3
O
H3C
OH
CH3
O
H3C
O
iridodial

Fig. 3.4. Structure of some monoterpenoids.

alkaloids) and are precursors of several classes of alkaloid. Most plant iridoids are bitter substances. In the past decade, the number of known iridoids has doubled and the importance of these compounds in insect–plant interactions has been investigated (Bowers 1991a). Iridoid glycosides potentially act as deterrents or are toxic to a variety of generalist insect herbivores and have been shown to act as antifeedants to grasshoppers and lepidopteran larvae (Bowers and Puttick 1988). The deterrency or physiologic toxicity is dosage-dependent. Higher doses of iridoids inhibit insect growth (Bowers and Puttick 1989). On the other hand, iridoid glycosides act as feeding stimulants to many specialist insects including the larvae of *Euphydryas* sp. (Bowers 1983).

Sesquiterpenoids. Sesquiterpenes are aliphatic or cyclic, isoprenoid C_{15}-compounds comprising an almost bewildering array of structural types.

These compounds display many skeletal types that have been implicated in herbivore defense.

Drimane skeletal type. These types of sesquiterpenes are the most potent insect-feeding deterrents. They include polygodial from *Polygonum hydropiper* (Polygonaceae) and warburganal, isolated from *Warburgia* sp. (Canellaceae) (Fig. 3.5). The drimane dialdehydes inhibit the feeding of the armyworms *Spodoptera exempta, S. littoralis,* and *Leptinotarsa decemlineta* as well as the aphid *Myzus persicae* (Asakawa *et al* 1988). The feeding inhibition activity of sesquiterpenes against certain lepidopteran larvae seems to be due to the blocking of the stimulatory effects of glucose, sucrose, and inositol on chemosensory receptor cells located in the insect mouth parts (Frazier 1986).

Sesquiterpene lactones. These are the largest single group of sesqui-terpenes in plants. Plant species of the composite family (Asteraceae) are rich in this compound, which is localized in the glandular hairs or in the latex ducts (Rees and Harborne 1985). Sesquiterpene lactones are poisonous to several lepidopterans, flour beetles, and grasshoppers (Isman and Rodriguez 1983). Some of the toxic compounds of this group are alantolactone (Fig. 3.5) from *Inula helenium* and a cumambranolide derivative from *Helianthus maximiiliani* (Gershenzon *et al* 1985). The exact basis of the toxicity of sesquiterpene lactone is not known.

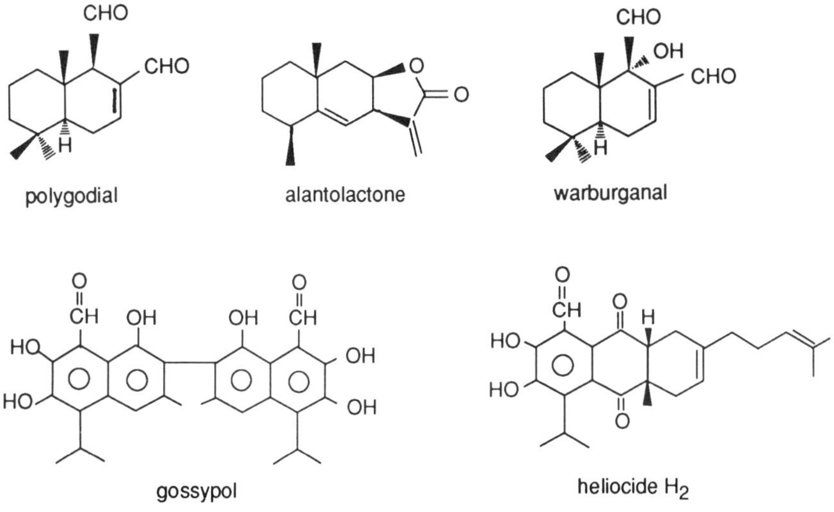

polygodial   alantolactone   warburganal

gossypol   heliocide H_2

Fig. 3.5. Structure of some sesquiterpenoids.

Gossypol.   Gossypol is a phenolic, cadinene-type sesquiterpene dimer with two aldehyde residues (Fig. 3.5) that is predominant in the pigment glands of cotton leaves, flowers, and other parts of the plant. Other terpenoids present in the pigment glands include a group of C_{25} aldehydes, known as heliocides. Gossypol is toxic to a number of herbivorous insects, causing an antibiosis effect on many insect pests of cotton including the tobacco budworm *Heliothis virescens* and cotton leafworm *Spodoptera littoralis*. The heliocides are toxic to the tobacco budworm and to other insects (Stipanovic *et al* 1986). The toxicity of gossypol to insects is believed to result from its complexation with proteins

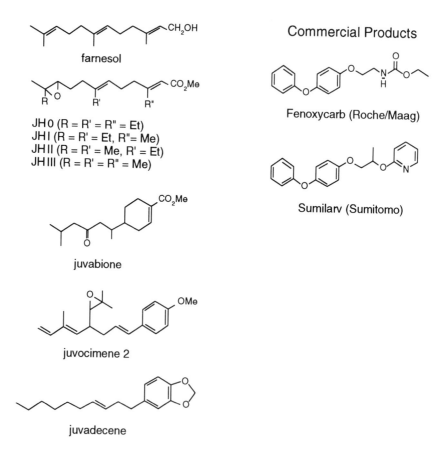

farnesol

JH0 (R = R' = R" = Et)
JHI (R = R' = Et, R"= Me)
JHII (R = R' = Me, R' = Et)
JHIII (R = R' = R" = Me)

juvabione

juvocimene 2

juvadecene

Commercial Products

Fenoxycarb (Roche/Maag)

Sumilarv (Sumitomo)

Fig. 3.6. Phytochemical farnesol provides the defined chemistry of insect juvenile hormones (JH) O, I, II, and III; JH III is common in all insects. Phytojuvenile hormone, juvabione, has structural kinship with insect JH III. Active phytojuvenoids, juvocimene 2 and juvadecene, were discovered in plants. Some commercial products of juvenile hormones are Fenoxycarb and Sumilarv.

in the gastrointestinal tract, causing a reduction in their digestibility, or with the digestive enzymes themselves, resulting in a loss of enzymatic activity (Meisner *et al* 1977a).

Phytojuvenile hormones. The possibility of ecologically important interactions between insects and plant terpenoids is offered by the juvenile hormone (JH) of insects and plant products that mimic its action (Williams 1970). The structure of JH suggests its terpenoid origin via a farnesol backbone (Fig. 3.6). It has a sesquiterpenoid structure related to farnesol (Schmialek 1961).

The discovery of plants with insect JH activity came from the investigations of Carroll Williams. In his research work, the cotton bug *Pyrrhocoris apteru* persistently molted into nymphal–adultoid forms as a result of using paper toweling derived from the paper pulp of the balsam fir tree *Abies balsamea* (Slama and Williams 1965). Investigations revealed that the paper toweling contained an active JH substance, originally referred to as 'paper factor'. The principal active compound was isolated and identified as a monocyclic sesquiterpenoid comprising a ketone and α,β-unsaturated methyl ester. The ketone component of the hormonal compound juvabione the natural insect hormone, was found to be a structural analog of *trans, trans*-10,11-epoxy farnesenic acid methyl ester (Bowers *et al* 1966). In biological tests, this compound was identified as the principal JH of most insects. This compound is called JH III (Judy *et al* 1973). From the literature, it appears that there are four compounds, known as JH O, JH I, JH II, and JH III, which represent the JH activity of the majority of insects. Later on, the *bis*-epoxide of JH III termed JH B$_3$ was revealed as a JH of *Drosophila melanogaster* (Richard *et al* 1989). It has now been reported that large amounts of natural JH III and its unepoxidized precursor methyl farnesenate are extracted from the Malaysian sedge *Cyperus iria*. This JH compound caused abnormal molting when fed to grasshoppers (Toong *et al* 1988).

Numerous phytojuvenoids (i.e., sesamin, sesamolin, juvocimene, juvadecene, and others) of modest to very high JH activity have recently been isolated from several plant families (Bowers 1991b). The discovery of juvocimenes in sweet basil *Ocimum basilicum* (Bowers and Nishida 1980) that disrupted the insect hormonal process has paved the way for the development of a second generation of JH-active commercial products such as Fenoxycarb and Sumilarv, which have no structural resemblance to the original product (Fig. 3.6). The effectiveness of these hormone-based insect growth regulators seems to be stable against several crop pests (El-Ibrashy 1987).

Antijuvenile hormones from plants. Even more remarkable is the discovery that anti-JH substances in plants provide protection against insects (Bowers *et al* 1976). Two chromenes, called precocene I and II (Fig. 3.7), have been isolated from the composite plant *Ageratum houstoniatum*. These two

precocene I   precocene II

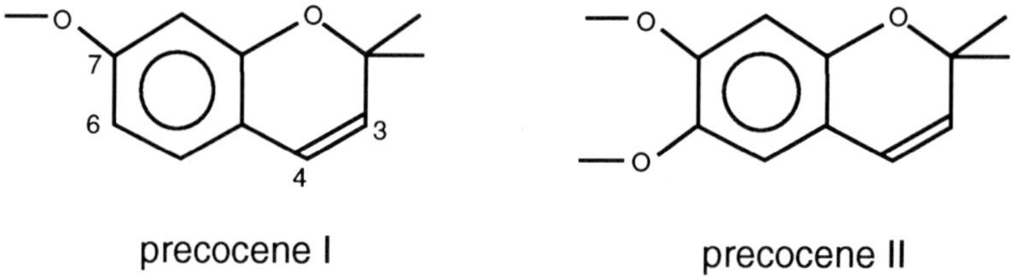

Fig. 3.7. Structures of antijuvenile hormones, precocene I and precocene II.

compounds and their analogs interfere with the JH activity that causes precocious metamorphosis and sterilization in the Hemiptera, Homoptera, and Orthoptera groups of insects (Bowers 1991b).

By virtue of the ability of precocene to induce a chemical allatectomy, the use of the precocenes can inhibit as well as terminate JH secretion in the corpus allatum situated in the insect's brain. Diminished neurosecretory activity in precocene-treated insects may account for delayed molting in several insects (Unnithan *et al* 1978). Ovarian development in precocene sterilized females of *Dysdercus bimaculatus* can be completely reversed by treatment with JH III. These results indicate that the antihormone affects the corpus allatum directly. In contrast to these natural materials, a number of synthetic compounds including fluoromevalonate are similar in action to the precocenes in inhibiting JH biosynthesis in insects (Staal 1986).

Diterpenoids. Diterpenes comprise four C_5 isoprenoid units leading to a C_{20} compound. Geranylgeranyl pyrophosphate can be regarded as the primary precursor of the large class of natural C_{20} terpenoid compounds. In general, diterpenes are nonvolatile and are found in the resins of higher plants. For instance, resin diterpenes of pine are the key factors in resistance to the southern pine bark beetle *Dendroctonus frontalis* (Hodges *et al* 1979). Higher levels of diterpene resin acids such as abietic and levopimaric acids were found to increase mortality, reduce growth, and extend development time of several species of sawfly larvae (Wagner *et al* 1983). Clerodane diterpenes have been demonstrated to inhibit the feeding of a number of species of lepidopteran larvae. Such diterpenes have been isolated mostly from plants of the families Lamiaceae, Verbenaceae, and Asteraceae (Cole *et al* 1990).

Triterpenoids. Triterpenes are composed of six C_5 isoprene units. They can be divided into two main classes: tetracyclic compounds and pentacyclic compounds. Many triterpenoids do not obey the classic isoprene rule and they can be seen to be composed of discrete isoprene units. Some of the major types of triterpenes related to plant defense are cucurbitacins, limonoids, and saponins.

Cucurbitacins. The cucurbitacins (Cucs) are a group of about 20 tetracyclic triterpenoids isolated from the squash family Cucurbitaceae (Guha and Sen 1975). Cucs are extremely bitter and are typically found in the roots, stems, cotyledons, leaves, and fruits of the plants of the Cucurbitaceae. They are characterized by extensive oxidation, including oxidation at C-2 and C-3 of the A-ring. Cuc B is the predominant compound (Fig. 3.8) and is found in about 91% of all the species characterized, followed by Cucs D, G, H, E, I, and J, and Cucs A, C, F, and L. Cucs B and E appear to be the primary Cucs, and other Cucs are formed by enzymatic processes during plant development and maturation (Metcalf 1986a).

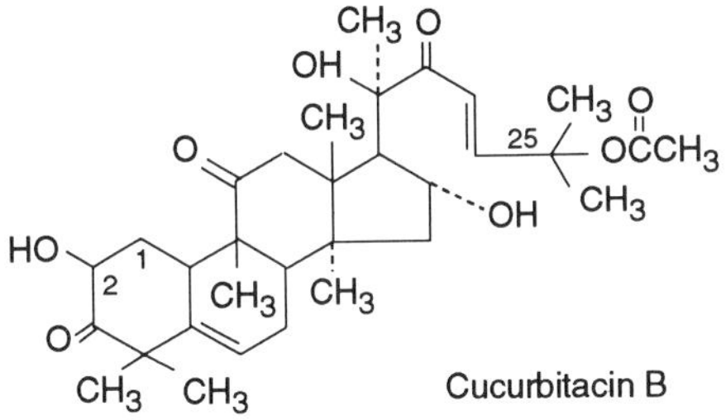

Fig. 3.8. Structure of cucurbitacin B.

Cucurbitacins are feeding deterrents for a number of arthropods including cucumber leaf beetles *Phyllotreta* spp., *Phaedon* spp., and *Cerotoma trifurcata*; stem borer *Margonia hyalinata*; and red spider mites (Da Costa and Jones 1971). In contrast, beetles of the tribe Luperini, such as the spotted cucumber beetle *Diabrotica undecimpunctata howardi* are immune to the toxic effects of dietary Cucs (Tallamy and Krischik 1989). Cuc B acts as a powerful host-recognition cue for luperine beetles and stimulates feeding even at very low concentrations (Metcalf *et al* 1982).

Limonoids. The limonoids are another major type of triterpenes. They are highly oxygenated substances with a basic skeleton of 26 instead of 30 carbon atoms, and thus they are tetranotriterpenoids. Azadirachtin from the neem tree *Azadirachta indica* (Meliaceae) is possibly the most promising botanical

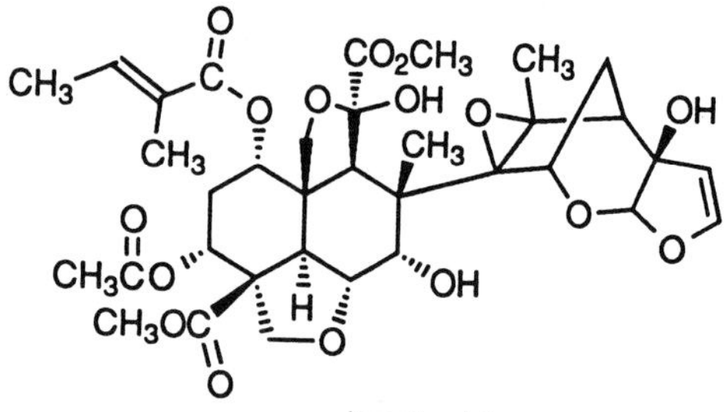

Azadirachtin

Fig. 3.9. Structure of azadirachtin.

pesticide against several insects, nematodes, and pathogens. This compound, first isolated by Butterworth and Morgan (1971), is effective at doses as low as 0.1 ppm and has been shown to act as a feeding deterrent to more than 100 species of herbivorous insects (Saxena 1989). Besides the deterrent properties to insects, azadirachtin increases larval mortality and interferes with normal growth and development by disrupting the molting process and inducing pronounced morphological deformities (Schmutterer 1990). The skeletal structure and stereochemistry of azadirachtin (Fig. 3.9) has warranted its evaluation as a model for a new commercial insect control agent, because its potent toxicity to insects is coupled with low toxicity to mammals (Saxena 1989).

Saponins.   The saponins are a large group of triterpenoid-derived glycosides. The presence of both triterpene and sugar elements in one molecule gives

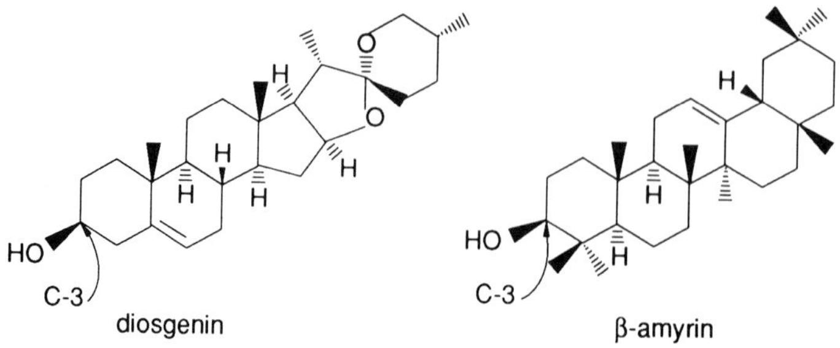

diosgenin   β-amyrin

Fig. 3.10. Structure of some saponins.

saponins the properties of a detergent. There may be as many as 10 monosaccharide units in the sugar moieties, which are often branched. Saponins are well distributed in the plant kingdom. They are classified into two major classes: (1) steroidal saponins with a C_{27} skeleton such as diosgenin, and (2) triterpene saponins with a C_{30} skeleton such as β-amyrin. Both diosgenin and β-amyrin are glycosylated at C-3 (Fig. 3.10).

Saponins have been shown to act as toxins and feeding deterrents to species of mites, lepidopterans, beetles, and many other insects (Ishaaya *et al* 1969). Saponin toxicity to insects seems to be due to the binding of saponins to free sterols in the gut, thereby reducing the rate of sterol uptake into the hemolymph. In reducing the sterol supply, saponins could interfere with the insect molting process (Ishaaya *et al* 1969).

Steroids

Steroids are also triterpenoid derived, having lost one or more carbon atoms of the original 30 carbons. Sterols have the same ring structure as the triterpenoids, but are distinguished from triterpenoids by lack of methyl groups at C-4 and C-14. However, there are exceptions to this rule, and there seems to be no overall pathway of steroid biosynthesis applicable to all higher plants.

In higher plants, sitosterol (C_{29}) is the most common of the group belonging to 4-demethylsterols although stigmasterol and campesterol are also common. Cholesterol, which was once considered an animal sterol, has now been detected in many higher plant tissues, including *Solanum* and *Nicotiana* spp. (Keller *et al* 1969).

Biosynthesis
The formation of squalene is shown in Fig. 3.3. Squalene is first converted into squalene epoxide by a monooxygenase; cyclization of squalene epoxide then leads to the formation of cycloartenol (2-A). Cycloartenol is then converted to 24-methylenecycloartenol or cholesterol. It is generally accepted that cycloartenol, not lanosterol, is the first cyclic product in plants. Thus cycloartenol is the precursor of a host of compounds collectively known as phytosteroids, which contain varying numbers of carbon atoms (Fig. 3.11). Sitosterol, the common plant steroid, is biosynthesized from cycloartenol through the formation of 24-methylenecycloartenol. Stigmasterol is subsequently produced through dehydrogenation of sitosterol. Stigmasterol is also a widespread constituent of higher plants. Cholesterol has been isolated from a number of higher plants and is possibly a precursor of other steroids (Heftmann 1973). Many steroids and steroidal alkaloids (e.g., solanidine, α-tomatine) are directly or indirectly derived from cholesterol (Fig. 3.11).

Chapter 3

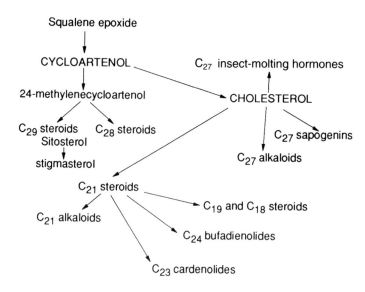

Fig. 3.11. Formation of phytosteroids through squalene epoxide.

Phytoecdysones

The ecdysteroids are a group of recently discovered terpenoids. Most steroids are nonpolar but the ecdysteroids are polar (Heftmann 1973). The term 'ecdysone' or 'molting hormone' is used for this group of steroids because α- and β-ecdysone isolated from insects by Butenandt and Karlson (1954) could cause ecdysis or molting in insects. These steroids have now been discovered in plants, and the term 'ecdysone,' which is related to insect hormone, has been used for these plant steroids which are named 'ecdysteroid' or 'phytoecdysone'. The insect-molting hormones (ecdysone) clearly resemble plant cholesterol in structure (Fig. 3.12).

Insect-molting hormone analogs have been found in various plants. During a search for antitumor compounds in plants, Nakanishi *et al* (1966) reported the presence of massive amounts of β-ecdysone in the leaves and roots of the common yew *Taxus baccata*. They further reported the isolation of a plant steroid hormone, ponasterone A, from *Podocarpus nakaii* plants. A fundamental relationship between the zooecdysones and phytoecdysones was established when Hüppi and Siddall (1968) showed that α-ecdysone (insect hormone) and ponasterone A (plant hormone) are identical in absolute configuration at all 10 asymmetric centers, except for the presence of a hydroxyl group at C-20 in ponasterone A rather than at C-25 (Fig. 3.12). Seventy ecdysteroids have been discovered in about 100 plant families, and they possess much greater hormonal activities than α- and β-ecdysone (Bergamasco and Horn 1983).

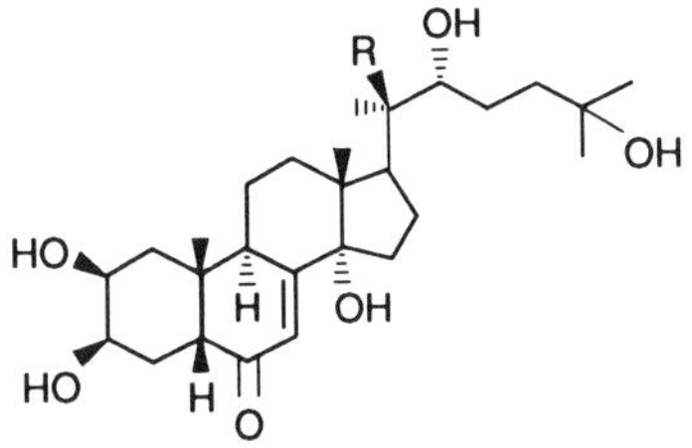

α-ecdysone, R = H
β-ecdysone, R = OH

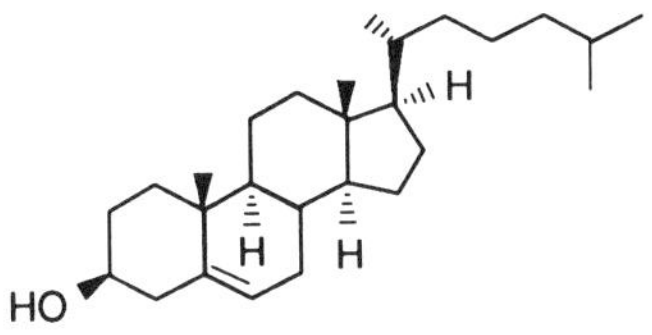

plant cholesterol

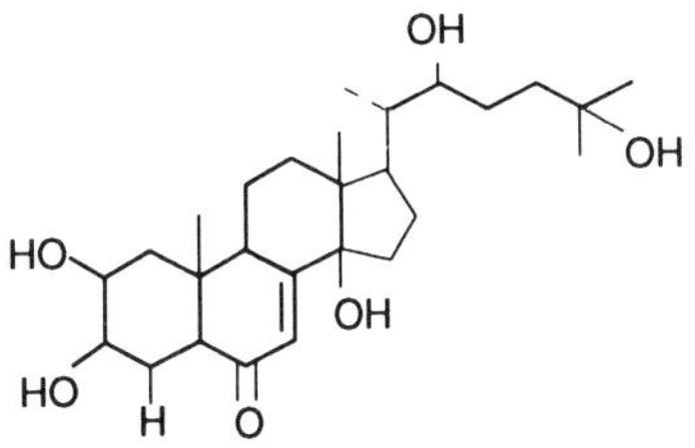

plant ecdysone

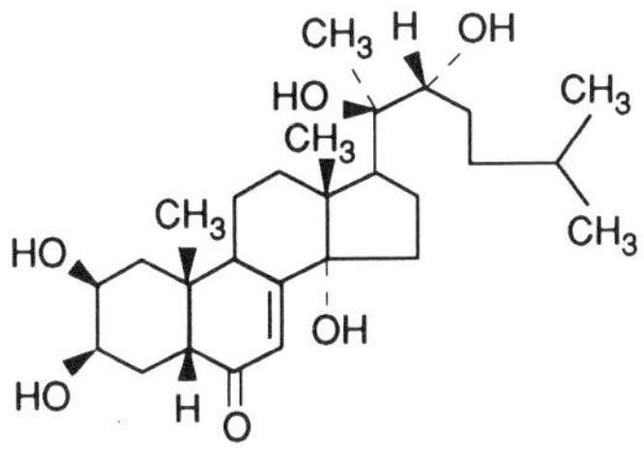

ponasterone A

Fig. 3.12. The chemical structure of insect α-ecdysone is comparable to plant ecdysone, plant cholesterol, and ponasterone A.

The high concentrations of ecdysteroids in plants at strategic locations suggest a defensive function against nonadapted insect herbivores. It has been shown that phytoecdysones and several synthetic ecdysone analogs severely inhibit growth, development, and reproduction when fed to several species of insects (Singh *et al* 1982). Phytoecdysones need not be lethal to insects, but minor effects on metamorphosis or reproduction would probably be sufficient to reduce the fitness of the pest and suppress its predation.

Shikimic acid pathway

Shikimic acid, first isolated from the plant *Illicium anisatum* (Illiciaceae), is now recognized as a discrete plant component. The shikimate pathway is restricted to the plant kingdom. It provides the main biosynthetic route by which aromatic compounds are produced from carbohydrates. The pathway is important not only because it provides the essential aromatic acids for protein synthesis, but also because these same acids are further employed for the production of almost all the secondary aromatic compounds. The pathway is long and complex and includes rather unusual biochemical reactions. Its intermediates play the part of precursors for a large range of secondary metabolites in plants. For extensive coverage of the subject, Haslam (1974) should be consulted.

Biosynthesis of shikimic acid

The pathway owes its name to shikimic acid, one of the prearomatic intermediates. Some ten steps are involved in the production of phenylalanine and tyrosine by combining phosphoenolpyruvate (PEP) with erythrose-4-phosphate, a product of glycolysis (Fig. 3.13). The biosynthesis of aromatic amino acids takes place in plants via the shikimate pathway. The overall sequence consists of a common branch, the prechorismate pathway, followed by three terminal pathways in the biosynthesis of the three aromatic amino acids.

Chorismic acid is an important branch point leading to either tryptophan

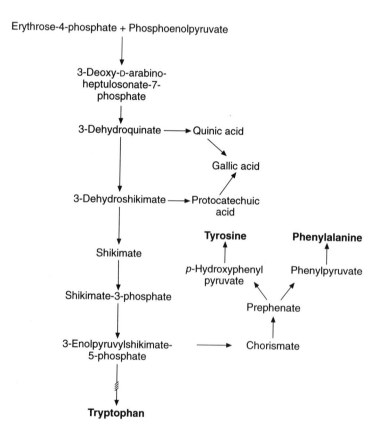

Fig. 3.13. Biosynthetic pathway showing chorismate as a branch point of aromatic compounds, tyrosine and phenylalanine.

or through phenylpyruvic acid to phenylalanine and tyrosine (Fig. 3.13). Chorismic acid is the key compound, which by virtue of its many reaction possibilities, opens up routes to the formation of a wide spectrum of aromatic compounds. Tryptophan is the precursor of the indole alkaloids and the plant hormone indole-3-acetic acid. Tyrosine and phenylalanine are both biosynthesized from prephenic acid, but via independent pathways. These two amino acids are specific precursors for a variety of phenolic compounds.

Phenolic compounds

Higher plants possess a wide variety of phenolic compounds (Harborne 1980). These compounds are aromatic in nature and possess one or more phenolic hydroxy groups. Of this group, the phenylpropanoids are the most widely distributed and are found in essentially all higher plants, whereas other members of the group, i.e., the hydroxyl-containing indole, quinoline, and isoquinoline alkaloids and the hydroxylated quinones, are of limited distribution. The phenylpropanoids form a biosynthetically homogeneous group formed at least in part via the shikimate pathway. The phenylpropanes are generally denoted as C_6–C_3 compounds.

Cinnamic acid

The phenylpropanes are derivatives of cinnamic acid from phenylalanine by the action of PAL. The common cinnamic acids are stored in plants as esters with glucose, as quinic acid, or as glycosides. Cinnamic acid forms the starting point from which an enormous number of secondary metabolic processes begin, as shown below in the phenylalanine–cinnamate pathway:

Phenylalanine $\xrightarrow{\text{PAL}}$ Cinnamic acid $\rightarrow$ *p*-coumaric acid $\rightarrow$
Caffeic acid $\rightarrow$ Ferulic acid $\rightarrow$ 5-hydroxyferulic acid $\rightarrow$ Sinapic acid

Practically all higher plant polyphenols (polypeptides, flavonoids, lignins, etc.) are formed from shikimate via the shikimic acid pathway and are produced through the intermediacy of phenylalanine, involving multienzyme complexes including PAL, cinnamate-4-hydroxylase, and a series of subsequent hydroxylation and methylation reactions (Stafford 1974).

Chlorogenic acid

The most common hydroxycinnamic acid of plants is caffeic acid, which is widely distributed in higher plants in the form of its esters with quinic acid, e.g., the 3-caffeylquinic acid known as chlorogenic acid. The formation of chlorogenic acid in higher plants from phenylalanine (cinnamic acid) has been established by experiments using a ^{14}C-labeled precursor. Chlorogenic acid is the end product of several biosynthetic pathways in plants (Fig. 3.14) (Harborne 1980).

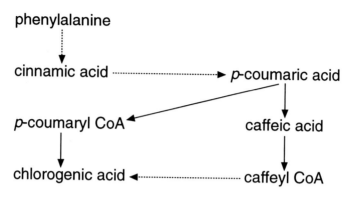

Fig. 3.14. Biosynthetic route for chlorogenic acid.

The enzymes catalyzing the formation of chlorogenic acid in tomato *Lycopersicon esculentum* fruits have been isolated and identified. High concentrations of rutin (see flavonoids p. 50) and chlorogenic acids have been found in the trichomes of tomato leaves. These compounds are toxic to corn earworm (Isman and Duffey 1982).

Lignins

The two main types of phenolic polymers in plants are the lignins and the tannins. Although both lignins and tannins are synthesized at least partially from the metabolic products of the shikimic acid pathway, they greatly differ in their chemical and biological properties. Lignin is an integral cell-wall component of all vascular plants including herbaceous species. It is a product of the phenylpropanoid pathway. This pathway involves a series of reactions in which phenylalanine is converted to cinnamic acid and its hydroxylated and methoxylated derivatives: *p*-coumaric acid, ferulic acid, and sinapic acid. The three acids are then hydrogenated to their corresponding alcohols. The alcohols serve as substrates for a peroxidase-catalyzed oxidation. The free radicals generated by the peroxidase activity condense to form lignin (Sarkanen and Ludwig 1971). But the final stage in the formation of lignin is the polymerization of the cinnamyl alcohols. Thus, no single unique structure can be established for lignin, because elongation of the polymer is a random process.

The possible role of lignification in the evolution of plants has been reviewed by Raven (1977). Lignification provides mechanical strength for aerial shoots. Both the mechanical barrier and unpalatability due to lignification of plant cells help to resist the attack of microorganisms and herbivores. Lignification is a common response to infection or wounding in plants. By reducing the digestibility of cell-wall carbohydrates, lignin may protect the plants from herbivores.

Tannins

Tannins are compounds with an astringent taste, which is due to lower oligomers reacting with proteins. By definition, tannins are protein-binding agents. They are usually of two categories: (1) the condensed tannins, and (2) the hydrolyzable tannins. Condensed tannins are oligomers and polymers of flavanoid units, linked by carbon–carbon bonds that are not susceptible to hydrolysis. In condensed tannins there is a single bond between C-2 and C-3. These structures are called flavanoids rather than flavonoids. In contrast, hydrolyzable tannins are joined by carboxylic ester linkages and are hydrolyzed readily by acidic or basic conditions. The two groups also differ in that hydrolyzable tannins usually contain glucose as an integral part of their structures. The most common of the hydrolyzable tannins are esters of gallic acid and ellagic acid with sugars. Gallic acid can be biosynthesized in plants by two distinct pathways involving shikimic acid metabolites; the second pathway to gallic acid is from cinnamic acid derivatives.

Regardless of their *raison d'être*, tannins appear to play a role in plant defense against herbivory. Insects tend to avoid astringent food, which affects their digestion because of the coagulation of mucoproteins in their oral cavity (Harborne 1988a). Inhibition of dietary protein digestion may not be the major reason for the reduced growth rate caused by tannins. Recent evidence suggests that postabsorptive inhibition, rather than inhibition of digestion, is the primary factor responsible for poor insect growth caused by dietary tannins (Butler 1989).

Coumarins

Coumarins are generally localized in roots and seed coats. More than 300 simple coumarins have been reported from nine families of monocots and more than 70 families of dicots. Although 7-hydroxycoumarins are widely distributed, furanocoumarins reportedly occur in only 15 families (Murray *et al* 1982). Linear furanocoumarins are more widely distributed than angular furanocoumarins; the latter are found predominantly in the Umbelliferae and in the genus *Psoralea* of the Leguminosae (Games and James 1972).

Biosynthesis. Coumarins are derived at least in part from the shikimate–chorismate pathway. Catalyzed by PAL, phenylalanine is transformed to *trans*-cinnamic acid (Murray *et al* 1982). The pathway up to this stage is shared with lignins and other phenylpropanoid products. In most cases there is one further shared step: the 4-hydroxylation of *trans*-cinnamic acid to yield *p*-coumaric acid (Fig. 3.15). At the point of divergence the first committed step in the elaboration of coumarins is the 2-hydroxylation of *p*-coumaric acid, a unique feature of the pathway. In the biosynthesis of coumarin itself, 2-hydroxylation occurs on cinnamic acid to form *o*-coumaric acid, so that no hydroxyl function appears in position 7 of the final coumarin product. Once

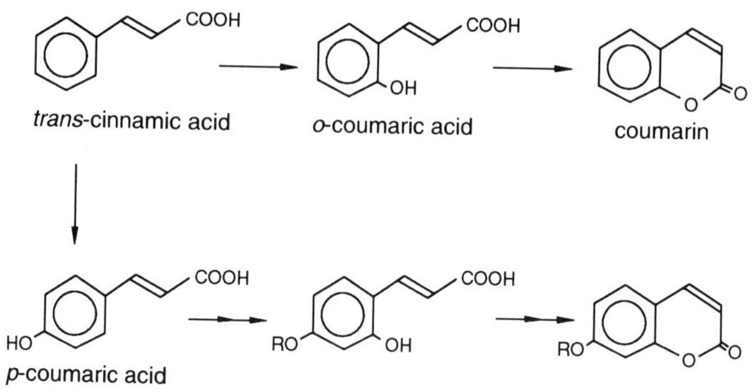

Fig. 3.15. Biosynthetic route for coumarins.

the 2-hydroxyl is in place, the stage is set for the final step in the synthesis. The 2-hydroxyl is isomerized to the *cis* form in the presence of ultraviolet (UV) light (Towers 1984).

Coumarin itself occurs in fresh plant tissues in a bound form as the corresponding *trans-o*-glucosyloxycinnamic acid. When tissue is damaged, this bound form undergoes enzymic loss of sugar, *trans→cis* isomerization, and ring closure. Volatile coumarin is thus released from the leaf surface. Some methoxycoumarins occur naturally, bound in the same way. In contrast, the common hydroxycoumarins occur in plants as such, with an O-glycosidic attachment. The three most common hydroxycoumarins of plants are umbelliferone, aesculetin, and scopoletin; they correspond in structure to *p*-coumaric, caffeic, and ferulic acids.

Furanocoumarins. There are many varieties of complex coumarins, other than the hydroxycoumarins, known as furanocoumarins. Furanocoumarins have a furan ring fused to the benzene ring in one of two orientations, linear or angular (Fig. 3.16). Though furanocoumarins are closely related to the coumarins, their biosynthetic pathway involves the acetate–mevalonate metabolite, dimethylallyl pyrophosphate (DMPP) (Murray *et al* 1982). Furanocoumarins are derived from umbelliferone by condensation of umbelliferone and DMPP, mediated by a transferase, to yield 7-demethylsuberosin and, further, to yield linear furanocoumarins. Analogous reactions via osthenol and coelumbianetin give the angular furanocoumarin series (Fig. 3.17).

Biochemical properties of coumarins. Coumarins are phototoxic in the presence of UV light. They bind with DNA as evidenced from the *in vitro* work with purified DNA and coumarins. The phototoxicity of furanocoumarins is associated with the binding of pyrimidine bases of DNA (Berenbaum 1983). Simple coumarins are capable of adduct formation via 3,4-bond photocycliza-

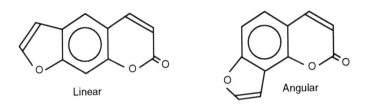

Fig. 3.16. Structures of two types of furanocoumarin, linear and angular.

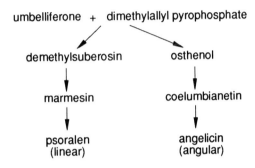

Fig. 3.17. Biosynthetic route of furanocoumarins.

tion (Musajo *et al* 1965). Photosensitizing of furanocoumarins is due to the formation of an excited triplet state on absorption of a photon. The excited-triplet-state furanocoumarin can react directly with the molecules of the pyrimidine base which is associated with furanocoumarins, i.e., 8-methoxypsoralen (8-MOP). Another type of photosensitization can react with oxygen which occurs in polyacetylenes, i.e., α-terthienyl (Arnason *et al* 1983).

Antiherbivory functions of coumarins. Coumarins are toxic to a broad range of organisms including bacteria, viruses, fungi, vertebrates, and invertebrates. There is a growing body of reports on the effect of coumarin on herbivorous insects. Simple coumarin is ovicidal to the Colorado potato beetles, and toxic to mustard beetles and a variety of insects. Furanocoumarins are capable of photosensitizing insects. Xanthotoxin is a furanocoumarin. This compound, when incorporated in an artificial diet (0.1%) and UV irradiated, could cause 100% mortality to the southern armyworm (Berenbaum 1991). Alpha-terthienyl and related furanocoumarin compounds are among the most promising chemicals for insect control in field conditions (Arnason *et al* 1992).

Secondary compounds with mixed biogenesis

Most of the aromatic compounds in plants are biosynthesized by one of the three main pathways: the acetate–malonate, acetate–mevalonate, or shikimic acid pathways. Besides metabolites biosynthesized via a specific pathway, many aromatic compounds have a mixed biogenesis, derived from products of two or more of the main pathways. The flavonoids are biosynthesized via the shikimic acid as well as the acetate–malonate pathways, whereas many plant quinones are products of the shikimic acid and acetate–mevalonate pathways.

Flavonoids

Flavonoids are aromatic compounds. They all absorb UV light and some visible light as well, and hence are brightly colored, e.g. anthocyanidins. Other flavonoids are colorless, e.g., isoflavonoids, and some have yellow color. According to a recent estimate, there are about 4000 known structures of naturally occurring flavonoids (Harborne 1988b). Flavonoids occur universally throughout higher plants, both angiosperms and gymnosperms. Flower colors rich in flavonoids promote flower pollination and fruit and seed dispersal. Isoflavonoids are biologically active compounds and often show antifungal properties.

Biosynthesis

All flavonoids are based on the same C_{15} skeleton. This skeleton is a derivative of the mixed pathway comprising three acetate units condensing with a phenylalanine-derived C_6–C_3 precursor. Hence, the flavonoid and isoflavonoid skeleton consists of a ring A derived from three acetate units (malonate), while ring B and three carbons of the central ring are derived from cinnamic acid (Fig. 3.18). The central three-carbon fragment with the attached B-ring can occur in various forms. Under different oxidation levels, they can form different flavonoid classes. When the B-ring is attached at C-3 instead of C-2, isoflavonoid classes are formed. The acetate units are first converted to malonyl CoA, and addition of malonyl CoA units to cinnamic acid derivatives (mostly *p*-coumaric acid) is catalyzed by the enzyme flavanone synthase. The first product of the addition reaction is a C_{15} intermediate. The central C_{15}

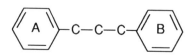

Fig. 3.18. Flavonoid skeleton.

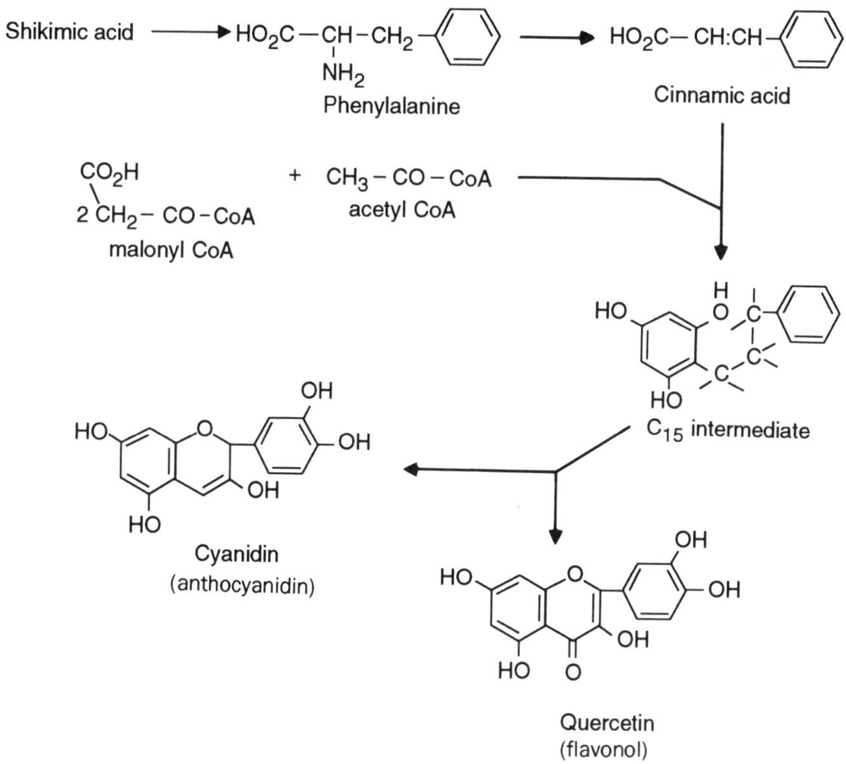

Fig. 3.19. Biosynthetic pathway of flavonoids.

intermediate is chalcone (Wong 1968). Once the chalcone is formed, it is modified in a variety of ways to yield different flavonoid classes, e.g., cyanidin, quercetin (Fig. 3.19).

Different flavonoid compounds and antiherbivory functions

Flavonoids can be divided into 14 classes according to the oxidation level of the central pyran ring (Harborne 1988b). Of these compounds, the four most widespread and important are anthocyanins, flavonols, flavones and iso-flavonoids.

Anthocyanins. Anthocyanins are glycosides of anthocyanidins. Three anthocyanidins – scarlet-colored pelargonidin, crimson cyanidin, and mauve delphinidin – or their simple methyl ethers occur widely in colored flowers and fruits. The major function of these colored pigments is to attract insects and animals for pollination and seed dispersal. Colorless flavonoids also contribute to pollination by exhibiting extensive UV absorption. Such flavonoids are visible to bees and are thus sometimes used as honey or nectar guides (Harborne 1988b). Pigmented leaves containing cyanidin can act as a defense against insect herbivores.

Flavonols.   Flavonols are common constituents of leaves and lignin of wood of higher plants. Important flavonols are kaempferol, quercetin, and myricetin. The first two flavonols occur throughout the angiosperms. Myricetin is restricted mainly to the leaves of woody plants, which often contain tannins. Although myricetin and tannins occur in the same plant, they are not chemically linked. All these compounds are antiherbivorous in nature.

Flavones.   Flavones lack the 3-hydroxy group present in anthocyanidins and flavonols. Two structures, apigenin and luteolin, are common in many herbaceous plants. Flavones, unlike the other common flavonoids, are found frequently in C-glycosidic combinations.

   Both flavonols and flavones usually occur in living cells as glycosides. For example, the flavonol quercetin usually occurs as a 3-glycoside; thus, rutin is quercetin 3-rhamnosyl-glucoside. When fed at low concentration, quercetin and three of its glycosides repelled the feeding of cotton borers *Helicoverpa zea*, *H. virescens*, and *Pectinophora gossypol*. When quercetin was mixed in the diet at concentrations of 0.2% or more, these common flavonoids killed the larvae of cotton borers (Shaver and Lukefahr 1969).

Isoflavonoids.   In contrast to most other flavonoids, isoflavonoids have a rather limited taxonomic distribution, and are mainly confined to the Leguminosae. Isoflavones possess 'rearranged' flavonoid skeletons, the B-ring being attached to the 3-position of the central ring. Like the flavonoids, the isoflavonoids are biosynthesized from chalcone precursors and ring migration has been postulated as an oxidation of a flavanone intermediate leading to an isoflavone (Fig. 3.20). Their chemical structures exhibit an unusually wide range of modifications. The isoflavones are the largest class of isoflavonoids; about 630 known structures are reported (Harborne 1988b). Fewer glycosides are characterized than for most classes of flavonoid.

   Complex isoflavones with isoprenoid substitution have been found in several tropical leguminous trees. Coumestrol has been detected in mungbean and soybean seedlings. These isoflavonoids have demonstrated antifeedant and antibiotic activities in the cabbage looper *Trichoplusia ni* feeding on soybean (Neupane and Norris 1990). Isoflavones are possible precursors of many other classes of isoflavonoids including rotenoids, isoflavones, coumestans, and pterocarpans. The pterocarpans occur in leguminous plants as phytoalexins. Pisatin and phaseollin are phytoalexins produced after infection in the pea *Pisum sativum* and French bean *Phaseolus vulgaris*, respectively (Perrin and Cruickshank 1965). Similarly, glyceollin is another major phytoalexin of soybean, which has strong antifeeding qualities against the

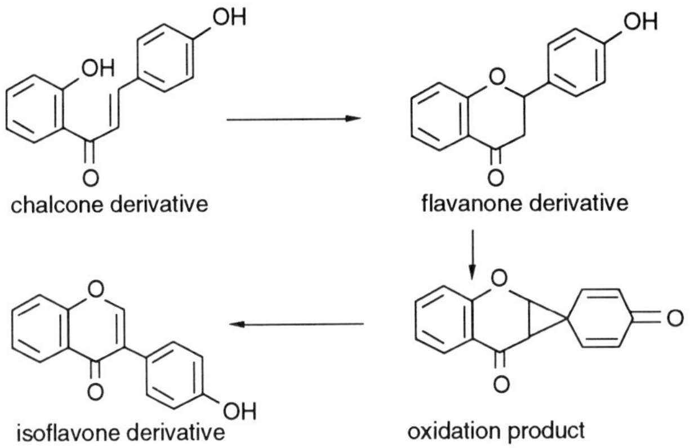

Fig. 3.20. Biosynthetic route of isoflavones.

Mexican bean beetle *Epilachna verivestis* (Burden and Norris 1992).

The only group of flavonoids that are known to be highly toxic to many insects are the isoflavonoid-based rotenoids, making rotenoids important insecticides. Rotenone is the active principle of derris obtained from the roots of *Derris elliptica* (Leguminosae). Rotenoids have a low mammalian toxicity, but are highly toxic to insects and fish. The toxic activity is almost certainly due to the resemblance in structure between rotenoids (rotenone) and steroidal saponins (diosgenin) and they act through the inhibition of mitochondrial oxidation. Some of the representative structures of iso-flavonoids are given in Fig. 3.21.

Quinones

Chemically, quinones are compounds with either a 1,4-diketocyclohexa-2,5-dienoid or a 1,2-diketocyclohexa-3,5-dienoid moiety. The former com-pounds are named *p*-quinones and the latter, *q*-quinones. Most naturally occurring quinones are *p*-quinones. Their structure is based on benzoqui-none, naphthoquinone, anthraquinone, and extended quinones. Most qui-nones are hydroquinones and have phenolic properties. They may occur *in vivo* combined with sugar, as in case of anthraquinones.

Quinones are widely but unevenly distributed in all the plant phyla except the ferns and mosses, and have been encountered in almost all parts and organs of plants such as leaves, stems, pods, root, bark, and heartwood.

Biosynthesis
One of the remarkable features of quinone biosynthesis in higher plants is that it is derived from a variety of different precursors and by different pathways

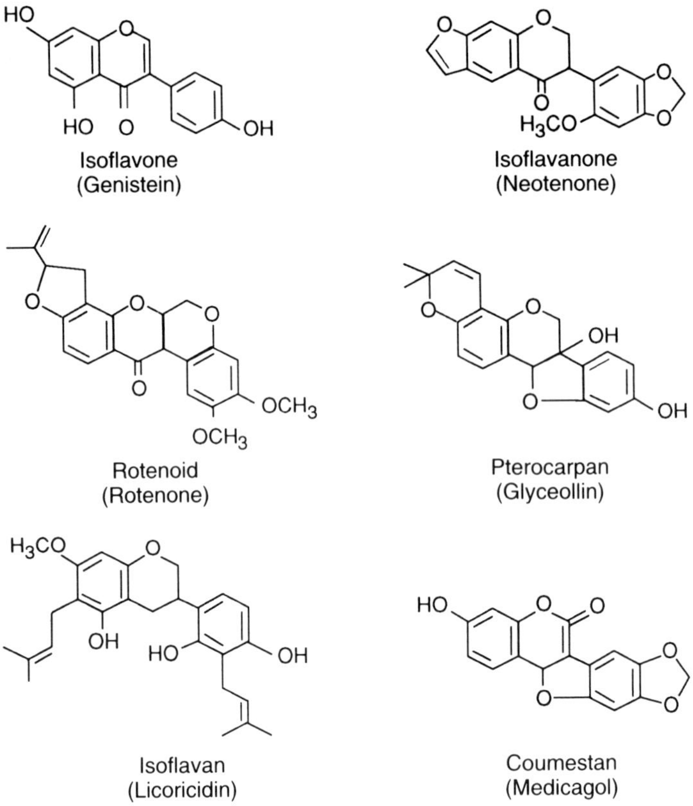

Fig. 3.21. Structure of compounds from six major subgroups of isoflavonoids.

from the acetate–polymalonate pathway, from aromatic amino acids, from mevalonic acid, and from the shikimic acid-*o*-succinoylbenzoic acid pathway.

Antiherbivory functions

In higher plants, quinone pigments seem to be incidental to their more important function as herbivore deterrents. Many quinones are biologically active and are toxic to herbivores. A number of simple benzoquinones are also active in the defense secretions (e.g., pirmin in glandular hairs) against a range of insects.

The simplest naphthoquinone, juglone (5-hydroxy-1,4-naphthoquinone) from the bark of hickory *Carya ovata* can prevent the growth of fungi and bacteria. The extracts of the bark of hickory are antifeedant to the European elm bark beetle *Scolytus multistriatus* (Gilbert and Norris 1968).

Heterogeneous group of secondary compounds

One of the major classes of secondary compounds derived from a diverse biosynthetic process and possessing high toxicity to herbivores is the alkaloids. The alkaloids are a heterogeneous class of natural products that occur in all classes of living organisms. However, most alkaloids so far known have been isolated from higher plants. About 20% of the species of flowering plants are rich in alkaloids and about 10,000 structures are described (Southon and Buckingham 1989).

Alkaloids

The word 'alkaloid' is derived from the term 'vegetable alkali', used earlier to refer to a group of bases of plant origin. No one definition of the term alkaloid is completely satisfactory, but alkaloids generally include basic substances that contain one or more nitrogen atoms, usually in combination as part of a cyclic system.

Chemically and biosynthetically, alkaloids are a very heterogeneous group, ranging from simple monocyclic compounds like coniine to the pentacyclic structure of strychnine. Many alkaloids are terpenoid in structure (steroidal alkaloids of the potato); others are aromatic, with the basic nitrogen in an aromatic ring system (isoquinoline alkaloids) or, less commonly, in a side chain (colchicine).

Because the term alkaloid covers a vast array of unrelated structures, it is more rational to classify alkaloids that are related biogenetically: benzylisoquinoline alkaloids, monoterpene indole alkaloids, simple indole alkaloids, pyrrolizidine alkaloids, quinolizidine alkaloids, and polyhydroxy alkaloids.

Benzylisoquinoline alkaloids
Benzylisoquinoline alkaloids have a limited distribution within the plant families belonging to the superorder Magnoliiflora and Ranunculiflorae, and represent the largest class of alkaloids with about 2500 structures. From early tracer studies, it is well established that the benzylisoquinoline skeleton is derived from two molecules of tyrosine and that reticuline is the common precursor for various alkaloids of this category (morphine, berberine, papaverine, and others). Norcoclaurine is the first benzylisoquinoline of the reticuline pathway.

Monoterpene indole alkaloids
Monoterpene indole alkaloids have been isolated from three plant families of mainly tropical distribution: Loganiaceae, Apocynaceae, and Rubiaceae. These alkaloids are elaborated by the stereospecific condensation of tryptamine (from tryptophan) with the iridoidal monoterpenoid glucoside secologanin (from mevalonate). Such a condensation process leads to formation of

strictosidine, the common biogenetic precursor of the various monoterpenoid indole alkaloids. Some examples of indole alkaloids are strychnine, cantharanthine, vinblastine, and vincristine.

Simple indole alkaloids

Simple indole alkaloids occur widely in various plant families. The majority of these compounds are of low toxicity and are derivatives of tryptamine, part of the molecule from tryptophan. Physostigmine is the principal alkaloid of the Calabar bean seeds *Physostigma venenosum* Balf. (Leguminosae). These beans were formerly used as an ordeal drug in Old Calabar, on the coast of eastern Nigeria. The bases found in the Calabar bean contain a novel heterocyclic system encountered only in this group and in certain *Calycanthus* alkaloids. Biosynthesis of physostigmine is from tryptophan with eseroline as the immediate precursor (Fig. 3.22). On the basis of the structure of this alkaloid, one of the present-day potential synthetic insecticides – the carbamates – is synthesized.

Pyrrolizidine alkaloids

The pyrrolizidine alkaloids (PAs) are frequently found in some genera of the Asteraceae (*Senecio* and *Eupatorium*), in most genera of the Boraginaceae (*Heliotropium*), and in the subtropical genus *Crotalaria* (Leguminosae). Most plants bearing PAs are toxic to domestic animals and to humans. However, some insects have adapted to PAs, not only to cope with these compounds but

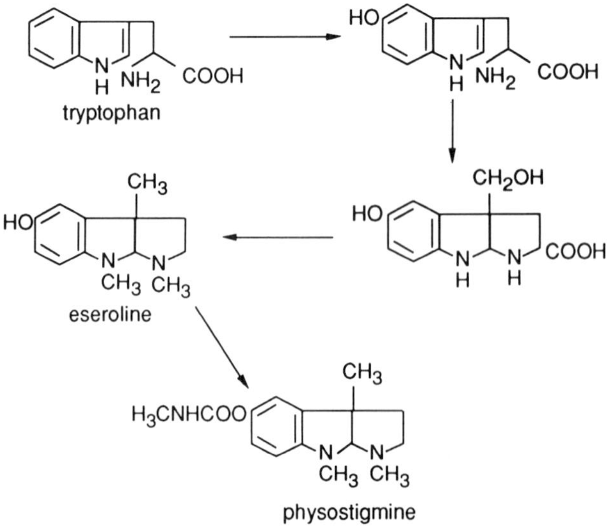

Fig. 3.22. Biosynthetic steps in the production of physostigmine.

also to use them for their own benefits (see Chapter 4).

Tracer studies on the biosynthesis of PAs indicate that the necine moiety (retronecine) is derived from arginine/ornithine by way of putrescine and homospermidine, whereas the necic acid moiety is derived from the branched-chain amino acids isoleucine and leucine as shown below (Herbert 1989):

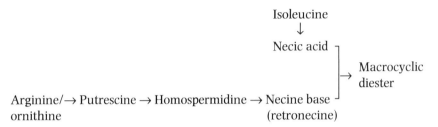

PAs are ester alkaloids formed by the combination of a necine base containing the pyrrolizidine ring systems with one or two necic acids. These ester alkaloids may occur as monoesters, open-chain diesters, or, most frequently, macrocyclic diesters. About 250 PAs have been isolated from natural sources (Mattocks 1986).

Quinolizidine alkaloids

The lysine-derived quinolizidine alkaloids are widespread in the plant genus *Lupinus* and are frequently called lupine alkaloids. The most important alkaloids of this group are lupinine, sparteine, lupanine, anagyrine, cytisine, and matrine. It is well established that lysine and its decarboxylation product 3-cadaverine serve as the only precursors for bi- and tetracyclic quinolizidine alkaloids as shown below:

$$\text{Lysine} \xrightarrow[\substack{\downarrow \\ CO_2}]{\substack{\text{Lysine} \\ \text{decarboxylase}}} \text{3-cadaverine} \rightarrow \text{(intermediate)} \begin{array}{c} \nearrow \text{Lupanine} \\ \searrow \text{Sparteine} \end{array}$$

Polyhydroxy alkaloids

A group of plant alkaloids that has been recently recognized are poly-hydroxylated compounds that structurally mimic sugars. Such compounds interfere with the enzyme glycosidase and receptors of carbohydrate metabolism. They become potential metabolic inhibitors and have been identified as effective antifeedants against various insects (Fellows *et al* 1986).

These polyhydroxy alkaloids have been isolated from five plant families of higher plants (Leguminosae, Polygonaceae, Aspidiceae, Moraceae, and Euphorbiaceae) (Fellows *et al* 1989). Out of the four identified structural types, the 2*R*,5*R*-dihydroxymethyl-3*R*,4*R*-dihydroxypyrrolidine (DMDP), an azafuranose analog of fructose, was isolated from leaves of *Derris elliptica*

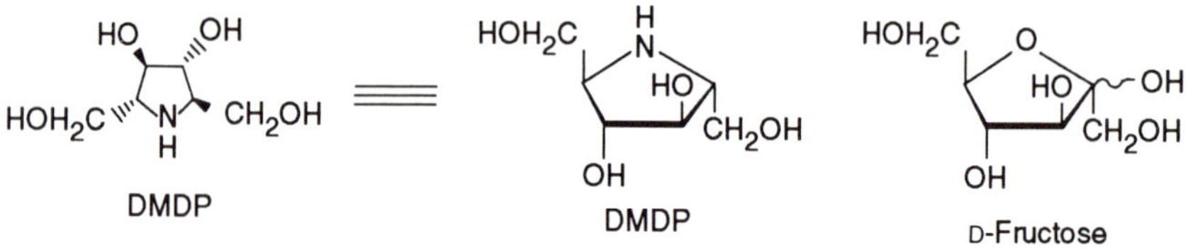

Fig. 3.23. 2R,5R-Dihydroxymethyl-3R,4R-dihydroxypyrrolidine (DMDP), an azafuranose analog of fructose.

(Fabaceae) and many others (Fig. 3.23) (Welter *et al* 1976). In addition to their inhibitory effects on certain glycosidases, the alkaloidal inhibitors also interfere with intracellular glycoprotein biosynthesis. Thus these compounds are promising tools in virus, cancer, and immunologic research (Fellows *et al* 1986).

Antiherbivory functions of alkaloids

The main function of alkaloids in plant–herbivore interactions is toxicity. Because of their nitrogenous structure, many alkaloids interfere with the key components of acetylcholine in the nerve system. Several alkaloids have been reported to be toxic or deterrent to insects (Levinson 1976). Nicotine and nornicotine derived from tobacco plants were used as early insecticides against insect pests. A great number of structurally unrelated alkaloids such as pyrrolizidines, quinolizidines, indole alkaloids, benzylisoquinolines, steroid alkaloids, and methylxanthines are feeding deterrents to many insects. The isoquinoline alkaloids and monoterpenoid indole alkaloids are synthesized and stored at specific sites in the producing plants. When herbivores attack these plants, the alkaloids prove to be insecticidal to the herbivores. The polyhydroxy alkaloid DMDP is a potent inhibitor of insect digestive α-glucosidase enzyme. Larvae of the seed beetle *Callosobruchus maculatus* were killed at a concentration of only 0.03% (Evans *et al* 1985). Similarly, the pyrrolidine DMDP is an effective antifeedant for four species of lepidopteran insect pests (Simmonds *et al* 1990). Furthermore, these alkaloids have been shown to interfere with the normal functioning of insect neurons that respond to sugars and ultimately involve inhibition of glucosidase. Glucosidases are known to be present in taste sensilla and have been implicated as the receptor molecules for sugar (Nakashima *et al* 1982). Thus, alkaloids are regarded as part of the plant's constitutive chemical defense. Because alkaloids are often synthesized and stored at strategically important sites or vulnerable plant parts, plants have a greater chance of remaining unaffected by herbivory, microbial attack, or mechanical damage or stress.

Secondary compounds derived from amino acids

Amino acids contain a carboxyl group and an amino or imino group and possess both acidic and basic properties. Most widespread are α-amino acids, with the general formula R-CH(NH$_2$)COOH. The primary metabolic amino acids belong to the L-series. Twenty L-amino acids are encountered in the biosynthesis of proteins, nucleic acids, chlorophyll, growth hormones, and some intermediates in the primary metabolism. These compounds are indispensable to all living organisms.

Nonprotein amino acids

The nonprotein or unusual amino acids are a group of plant toxins. They are found in a number of unrelated plant families, but are particularly character-istic of legume seeds. There are at least 600 nonprotein amino acids isolated and characterized from higher and lower plants (Rosenthal 1991). They are functionally important in seeds as protection against insects and other herbivores. Because legume seeds are rich in nitrogen, they also serve as a storage medium for ammonia nitrogen that can be utilized during the early metabolism of the germinating seedlings (Rosenthal 1982).

Most nonprotein amino acids are structurally related to one or other of the protein amino acids and usually exist as a zwitterion (Fig. 3.24). These compounds have molecular sizes and configurations that do not differ too markedly from those of the corresponding protein amino acids. They are

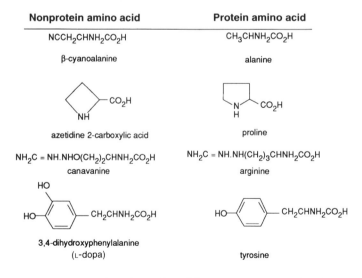

Fig. 3.24. Toxic nonprotein amino acids and their protein amino acid analogs.

antimetabolites because they act as analog molecules. The harmful effects on insects are partly due to the fact that the analog molecule can be incorporated into protein synthesis of the insect. Thus the insect produces unnatural enzymic protein which interferes seriously in its metabolic function, ultimately leading to the death of the organism.

A close approach to isosterism can result from the replacement of one atom (or group of atoms) by another similar in size and polar character, such as canavanine which is an analog of arginine (replacement of $-CH_2-$ by $-O-$). This structural similarity accounts for the ability of arginine tRNA synthetase to activate and attach canavanine to the tRNA that normally carries arginine to the protein assembly site. Replacement of arginine in a protein by the less basic canavanine can affect amino acid R-group interactions. The occurrence of this natural plant product is limited to the plants of the Lotoideae, a subfamily of the Leguminosae. This compound readily elicits pronounced deleterious effects on many insects including the tobacco hornworm *Manduca sexta*. Incorporation of 0.05% (w/v) canavanine into an artificial agar-based diet created drastic growth aberrations both in the pupae and adults of *M. sexta*. When larvae of this insect were given canavanine, they exhibited severe diuresis and often the gut was everted through the anus (Rosenthal 1991).

The toxic effect of another nonprotein amino acid, 3,4-dihydroxy-phenylalanine or L-dopa, on insects is well known. This compound interferes with the activity of tyrosinase, an enzyme essential for the hardening and darkening of the insect cuticle. The seeds of genus *Mucuna* comprise 6–9% L-dopa and are completely free from bruchid infestation. The southern armyworm larvae, *Prodenia eridania*, die upon ingestion of this compound (Janzen 1969).

Cyanogenic glycosides

Cyanogenic glycosides are moderately polar, water-soluble compounds that are normally accumulated in the vacuoles of plant cells. The cyanogenic substances in plants are usually carbohydrate derivatives. They are typically O-β-glycosides of α-hydroxynitriles (cyanohydrins). In the undamaged plants, cyanogenic glycosides are metabolized to amino acids. These glycosides are capable of releasing hydrogen cyanide upon damage to the plant tissues and also release β-glycosidases and hydroxynitrile lyases which commonly occur with the glycosides. The β-glycosidases cleave sugars from the glycosides, and the hydroxynitrile lyases decompose the resulting cyanohydrin thereby liberating hydrogen cyanide spontaneously (Fig. 3.25). Although the enzymes and substrates often occur in the same cyanophoric plants, free cyanide is seldom, if ever, accumulated in the plant.

The ability to produce and accumulate cyanogenic or cyanophoric compounds is found in all major vascular plant groups. About 60 cyanogenic compounds have been characterized from higher plants. Cyanogenins have

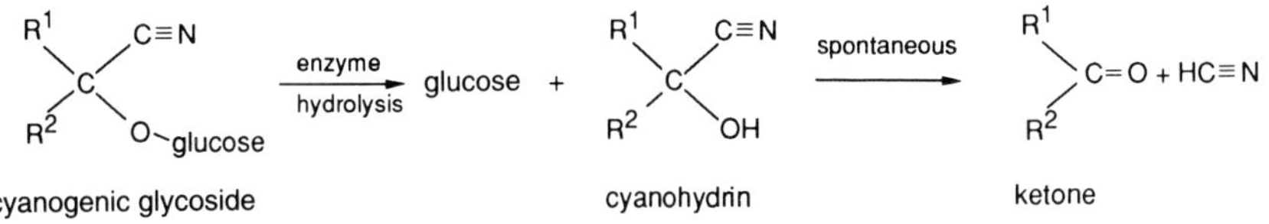

Fig. 3.25. Enzymatic degradation of cyanogenic glycoside.

been reported from at least 2650 plants belonging to more than 550 genera and 130 families (Seigler 1991). Biosynthetically, all these compounds appear to be derived from the L-amino acids phenylalanine, tyrosine, valine, isoleucine, and leucine, the nonprotein amino acid (2-cyclopentenyl) glycine, and nicotinic acid. Glucose is the sugar directly attached to the aglycone of most cyanogenic glycosides. A few occur as free cyanohydrins, which do not require the interaction of β-glycosidases for production of cyanide. Some of the common cyanogenic glycosides are given in Table 3.2.

A group of the major compounds related to cyanogenic glycosides are cyanolipids. The four known cyanolipids possess a long-chain fatty acid (C_{18} or C_{20}) attached to an α-hydroxynitrile. These compounds occur in the seeds of the plant families Sapindaceae and Hippocastanaceae (Mikolajczak *et al* 1984).

Biosynthesis
The biosynthetic pathway of cyanogenic glycosides is closely linked to the production of glucosinolates, organic nitro compounds, and possibly nitrile glycosides in plants (Halkier and Moller 1989). Amino acids are the precursors of cyanogenic glycosides. In plants, the amino acids are converted to aldoximes, which then form cyanohydrins via nitriles or hydroxyaldomines. Oxidation is effected by molecular oxygen. Glucosides are then formed from cyanohydrins by β-glucosyltransferases (Fig. 3.26).

Table 3.2. Common cyanogenic glycosides.

Amino acid precursor	Sugar	Glycoside	Occurrence
Valine	Glucose	Linamarin	*Phaseolus lunatus*
Isoleucine	Glucose	Lotaustralin	*Trifolium repens*
Leucine	Glucose	Proacacipetalin	*Acacia sieberiana* var. *Woodii*
Phenylalanine	Gentiobiose	Amygdalin	Rosaceae
Phenylalanine	Vicianose	Vicianin	*Vicia* sp.
Tyrosine	Glucose	Dhurrin	*Sorghum* sp.

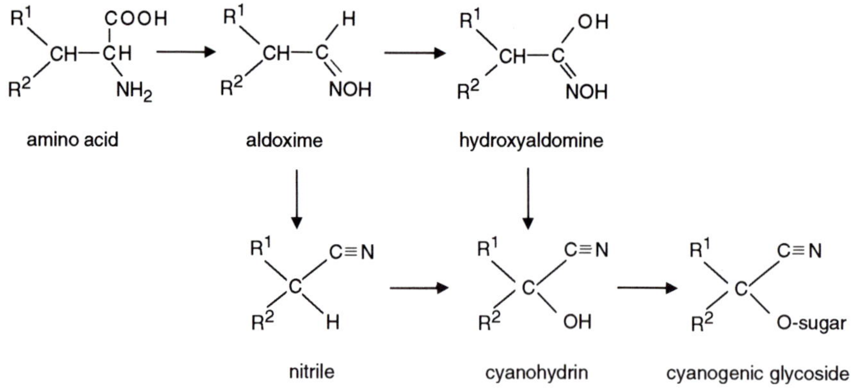

Fig. 3.26. Biosynthetic pathway of cyanogenic glycoside.

Antiherbivory functions

Both cyanide and the aglycones resulting from the hydrolysis of cyanogenic glycosides are toxic to nonadapted herbivores. Cyanide, released by dhurrin from field-grown sorghum, is a potential plant defense chemical against acridid grasshoppers in West Africa and India (Woodhead and Bernays 1977). Since this highly toxic compound is released as a result of chewing by the grasshopper, plant toxicity does not affect the other biotic components. Other studies have produced statistically sound data establishing cyanogenic glycosides as effective deterrents to herbivory (Hruska 1988). Cyanolipids are also highly toxic to many insects, including stored grain bruchids, the European corn borer, and *Lepticoris* sp. (Mikolajczak *et al* 1984). For specialist insects, cyanide and cyanogenic compounds can serve as kairomones. For example, the highly toxic compound dhurrin is a phagostimulant for *Peregrinis maidis* (Bernays 1983b).

Glucosinolates

Glucosinolates are sulfur- and nitrogen-containing secondary compounds. They are abundant in crucifer plants. In plants they are degraded by a thioglucosidase enzyme myrosinase to give D-glucose, sulfate, and iso-thiocyanate (Fig. 3.27). The isothiocyanates are called mustard oil. Glucosi-nolates are the precursors of mustard oil and can be either aliphatic or aromatic. These compounds co-occur with enzymes that catalyze their degradation to yield mustard oil. The acrid flavor of mustard oil is released when the fresh crucifer tissue is crushed during insect feeding. Both glucosinolates and mustard oil act as allelochemicals in mediating plant–

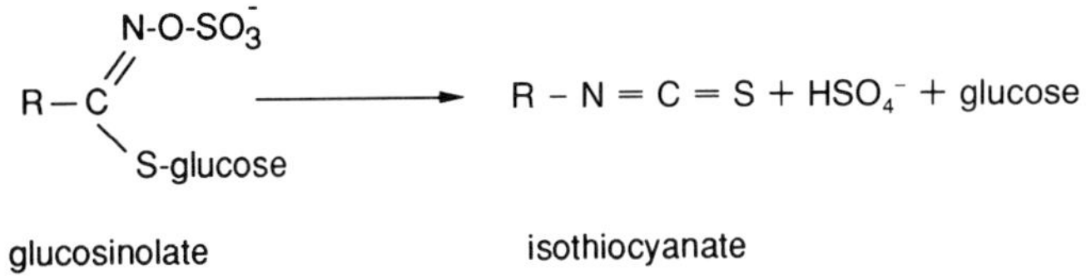

$$R-C \overset{\displaystyle \overset{N\text{-O-SO}_3^-}{\|}}{\underset{S\text{-glucose}}{}} \longrightarrow R-N=C=S + HSO_4^- + glucose$$

glucosinolate   isothiocyanate

Fig. 3.27. Hydryolysis of glucosinolates.

herbivore interactions (Louda and Mole 1991).

Because of their glucose moiety and ionic forms, glucosinolates are hydrophilic and nonvolatile compounds. On the other hand, isothiocyanates are generally volatile and are chemically very active. All the known 100 or so glucosinolates have a similar general structure.

Biosynthesis

Glucosinolates are generally formed from amino acids and, when necessary, these are chain-lengthened by the addition of acetate units. Like cyanogenic glycosides, the glucosinolates are biosynthesized through the intermediate formation of aldoximes. The first well-defined sulfur-containing intermediates are thiohydroxamates, the sulfur being derived from methionine or cysteine. In a further biosynthetic step, desulfoglucosinolate is formed by glucosylation of thiohydroxamate via glucosyltransferase, which occurs in many Cruciferae species. By O-sulfonation, glucosinolates are formed with the interaction of sulfotransferases (Fig. 3.28).

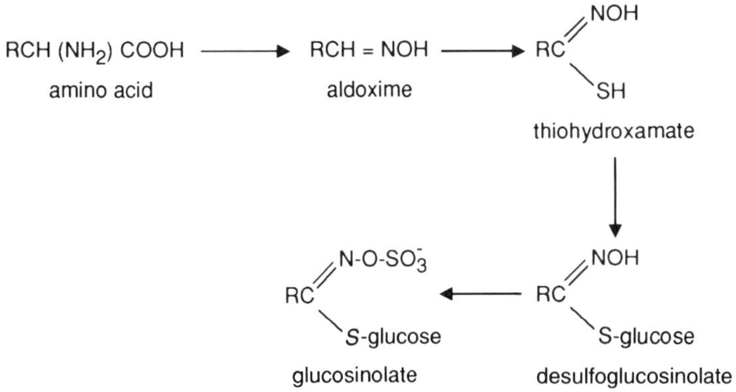

Fig. 3.28. Biosynthetic pathway of glucosinolates.

Antiherbivory functions

Glucosinolates have generally been considered to possess defense functions (Louda and Mole 1991). These secondary compounds interfere with the insect's host plant selection at two levels: the preingestive phase and the postingestive phase. In the former phase, the insects are repelled by the pungent odor of mustard oil; in the latter phase, the insects are confronted with the toxic component of the mustard oil glycoside. Glucosinolates such as isothiocyanate, sinigrin, allylisothiocyanate, gluconappin, and glucotropaeolin act as plant defense chemicals against generalized insects including aphids, grasshoppers, and many noncrucifer-feeding lepidopterans. These insects show reduced survival when induced to feed on crucifers (Bernays 1983b). Glucosinolates can act synergistically with cardenolides and can protect plants against glucosinolate-adapted herbivores (Nielsen *et al* 1979b).

Lectins

Many plant seed extracts can clump or agglutinate animal erythrocytes. Such plant compounds were designated as phytohemagglutinins. Later on, it was discovered that these phytohemagglutinins exhibit a high degree of specificity to human red blood cells of various blood groups. It was this high degree of specificity that led Boyd and Shapleigh (1954) to coin the word lectin (Latin *legere*, 'to pick or choose') for these and other antibody-like substances.

Lectins are proteins produced by living organisms that react with sugar residues. Most lectins are composed of either two or four subunits, each of which contains a specific sugar-binding site. This feature of their multivalency makes the lectins agglutinate cells or precipitate glycoproteins or large polysaccharide polymers. Lectins are also mitogenic, i.e., they induce the transformation of lymphocytes from small resting cells to large actively growing cells that ultimately undergo mitotic division.

Distribution

Although originally applied to plant agglutinins, the term lectin now encompasses a broader range of substances. Lectins are present in a wide variety of life forms from microbes to mammals. They are distributed universally throughout the plant kingdom, where they constitute 6–11% of the total proteins (Liener *et al* 1986). More than 600 species of the family Leguminosae have been shown to contain lectins. They are found mostly in the cotyledons of the seeds, mainly in the protein bodies. However, in cereal grains such as wheat, rye, and barley, the lectins are found largely in the embryo but are absent in the endosperms (Etzler 1985).

Antiherbivory functions

Lectins exhibit a wide variety of interesting biological functions including antiherbivory (Liener 1991). A variety of evidence suggests that lectins are involved in the defense of plants against insects and pathogenic micro-organisms. For example, the lectins from the kidney bean (*Phaseolus vulgaris*) cause lethality of the larvae of the bruchid beetle *Callosobruchus maculatus*, a major storage pest of many legumes (Gatehouse *et al* 1984). When lectins are ingested by bruchid larvae, they bind to the epithelial cells of the larval midgut. The epithelial cell binding not only interferes with the absorption of nutrients, but also results in absorption of potentially toxic substances into the midgut of the insect. A highly purified preparation of the soybean lectin has also been shown to inhibit the larval growth of a leaf-defoliating insect, *Manduca sexta* (Shukle and Murdock 1983).

Chitin-binding proteins with lectins

Plants synthesize a wide array of proteins capable of reversibly binding to affinity matrices composed of chitin, a β-1,4-linked bipolymer of N-acetylglucosamine. Chitin, a natural ligand, has never been detected in higher plants. Instead, it is a vital structural component of the cell walls of fungi and of the exoskeleton of many invertebrates including insects and nematodes. These chitin-binding proteins usually have lectin properties.

Antiherbivory functions. There is considerable circumstantial evidence that one of the major functions of the chitin-binding proteins is plant defense against herbivores (Raikhel *et al* 1993). The chitin-binding complex can be fused to a number of structurally unrelated complexes. Such fusions may recombine with other polypeptides of the plant to create fusion proteins with novel properties. These hybrid proteins can be targeted more specifically to chitin-containing surfaces, such as the insect midgut or the fungal cell wall. Induction of chitin-binding protein upon wounding indicates that these proteins can also act as a part of the induced plant defense mechanism.

The effect of several chitin-binding proteins on insect larvae has been examined using artificial diet containing seed extracts. Chitin-binding proteins of wheat germ agglutinin (WGA) appear to be among the most potent of lectins studied so far. WGA resulted in a 50% mortality of two major corn pests, the European corn borer (*Ostrinia nubilalis*) and the southern corn rootworm (*Diabrotica undecimpunctata*) when incorporated in their diet at 0.59 mg g^{-1} and 3 mg g^{-1} bodyweight, respectively (Czapla and Lang 1991). Similar deleterious effects on the cowpea weevil *Callosobruchus maculatus* F. have been observed with rice lectin and stinging nettle lectins (Huesing *et al* 1991). Every chitin-binding protein tested exhibits either insecticidal or antifungal activities or both. The results obtained with transgenic plants expressing class I chitinases or Gramineae lectins indicate that these proteins can impart resistance to fungal diseases or insect attack (Raikhel *et al* 1993).

Amylase inhibitors

Most cultivars of the common bean *Phaseolus vulgaris* L. have a proteinous
α-amylase inhibitor, which inhibits the activity of insect α-amylases. This
inhibitor is considered to be an important plant defense chemical against
herbivores. Ishimoto and Kitamura (1989) confirmed the presence of
proteinous α-amylase inhibitor (αAI-1) in the common bean that inhibited
the activity of the larval midgut α-amylases of the bean weevil (*Callosobruchus
chinensis* L.) and the cowpea weevil. Some accessions of the wild bean seeds
contain a novel α-amylase inhibitor, designated as αAI-2. It appears that
αAI-2 exhibits a broader spectrum of the inhibitory effect on the growth of
bruchid pests than αAI-1 (Suzuki *et al* 1993).

4 INSECT–PLANT INTERACTIONS

The evolution of plants since the Devonian period and the subsequent development of flowering plants in the early Cretaceous period have transformed the terrestrial environment into a highly valuable resource for the herbivore community. About one-half of all known species of insects are more or less dependent on vascular plants. The dominance of angiosperms, with their diversified plant surface characteristics, plant architecture, and plant chemical factors, has provided the three factors needed for insect proliferation: shelter, oviposition sites, and food. The plant surface, with its enormous variety of new niches, not only enabled the insects to move away from the soil surface, but also offered an ideal habitat for their life cycles (Southwood 1986). The living space of plants is vast, not only in quantity but even more so in quality.

Lawton and Schroeder (1977) first showed the relationship between the structural diversity of plants and the richness of their insect fauna. Insects live in a large and varied world of plant surfaces. For an insect, the host plant is not merely something to feed on but something to live on (Kennedy 1953). A host plant is here defined as 'one on which the insect completes its growth and development'. The fact that these two living systems – plant and insect – have evolved together over hundreds of millions of years and now coexist shows clearly their most 'intricate interdependence and interaction'. This interdependency is broadly depicted in two contrasting ways: some insects are flower pollinators and others are herbivores.

Insect Pollination

The intimate association between flowering plants and their pollinators is testimony to the long periods of well-defined mutualism of these organisms. The most common pollinators of flowering plants are bees, wasps, and butterflies. The distinctive colors and scents of flowers make them conspicuous and easily recognizable to the pollen vectors. Furthermore, many flowers produce nectar which serves as an additional attractant for pollinators. The floral parts are so arranged that pollen collection from the stamens and its transfer to the stigmas are likely to occur during a single visit of the vector. Indeed it has been suggested that the dramatic evolutionary success of the angiosperms can be attributed, at least in part, to the effective cross-pollination of plants by insects.

Insect Herbivory

Another important aspect of insect–plant interaction, herbivory, has received little attention until recently. One reason is that much of the evolution of insect herbivory on plants has resulted from complex physical, chemical, and environmental factors. Insects evolved rapidly during the early Cretaceous period, giving rise to several extant phytophagous families that adapted to their host plants. As a result, insects have developed different patterns of host plant associations coupled with different life histories and feeding strategies necessary for the exploitation of their hosts. Even in closely related groups of insects, the host plant relationships are different. For example, the psyllids (plant lice) are restricted almost exclusively to dicotyledons, whereas the aphids also occur commonly on monocotyledons and conifers as well.

The modes of feeding of phytophagous insects include chewing, sap sucking, mining, and galling. Insects may be monophagous (feed on plants of a single species or, at the most, a few closely related species), oligophagous (feed on plants within one family or members of closely related families), and polyphagous (feed on plants from a number of taxonomically distantly related plants) (Dethier 1953). Most species of herbivorous insects exhibit associations with a narrow range of host plants; that is, they are monophagous or oligophagous. Polyphagy is generally viewed as the more primitive stage in the evolution of insect–plant interactions (Dethier 1954), and oligophagy now seems to be the rule amongst most phytophagous insects, except in the groups of Orthoptera. Occasionally, within a predominantly monophagous group, a species may exhibit wide host preferences; for example the peach–potato aphid *Myzus persicae*, has been recorded on over 50 different plant families. Insects have also frequently been found to switch to a new host plant from a distantly related plant group (Powell 1980).

These fascinating relationships demonstrate that insect–plant interactions are complex and highly adaptable. How plants with relatively long generation times and low recombination rates, as compared with their herbivores, have survived in the evolutionary race is a paradox (Whitham 1981). Green plants, however, dominate the landscape. The survival of plants through evolutionary time is due largely to their own defense mechanisms, deterring the feeding of herbivores and toxicity in the widest sense. Moreover, plants in general have the capacity to either tolerate or escape attack. On the other hand, the selective ability of insects, such as detoxification, sequestration, and avoidance, enables them to overcome these plant defense mechanisms and allows limited feeding.

Interaction Systems

An understanding of insect–plant interactions requires a basic knowledge of the evolution of insects and plants and the factors that promote the feeding behaviors of insects as well as the defense mechanisms of plants. The interaction is a dynamic one, subject to continual variation and change. In any one situation, the insect or plant may appear to have the advantage. Both partners in the interaction adapt themselves in different ways to the changing conditions. However, environmental pressures have an important impact on the interaction system (Fig. 4.1).

Phytophagous insects live in an environment full of all sorts of plant chemicals. Plant chemicals released in the air or soil may affect insects before or just after their contact with the plant substrate. Other phytochemicals present in the plant tissues act upon the insects when they start feeding on the plant. The responses may be behavioral, physiologic, or ecological. The behavioral responses related to the insect's host plant selection are dealt with

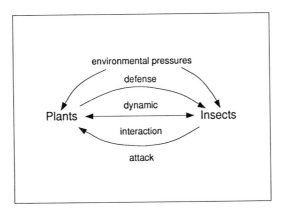

Fig. 4.1. Model of the interaction between insects, plants, and the environment.

in Chapter 5. The physiologic responses involve plant chemicals that influence the physiology and metabolism of phytophagous insects in various ways, including growth and development. Knowledge about the complexity of these interactions can give answers to these questions: how do plants defend themselves and how do insects overcome the defense mechanisms?

How do Plants Defend Themselves?

From a phytochemical standpoint, plants are producers of chemicals (primary and secondary metabolites) and insects are consumers. Plants display physical and'chemical defensive adaptations against insect attack. Obviously, no insect utilizes every plant species as food, and, conversely, no plant species is susceptible to attack by every species of plant-feeding insects. The multitude of species in the class Insecta do not differ greatly in their fundamental qualitative requirements for biochemical nutrients, although some differences and specializations occur. However, the nutritional quality (dietary propor- tions) of plant tissues has influenced their use by insects in some instances (House 1969).

The emerging subject of 'nutritional ecology' is concerned with under- standing the intricacies of invertebrate nutrition and the interaction with non-nutritional substances (Slansky and Rodriguez 1987). Development of an effective host plant resistance strategy requires an understanding of the links between food attributes (nutritional and allelochemicals), food utiliza- tion, and insect abundance on crop plants. Thus, the principles of insect nutritional ecology provide a logical basis for research on selective breeding, biotechnology, and cultural practices to influence the behavior and physiol- ogy of the target pests.

Plant nutritional quality

If a plant is to serve as a host, it must be to able to provide holistic nutrients that can support the growth, development, and reproduction of insects. This concept can properly be referred to as insect dietetics (Beck 1972). Under this concept, the feeding insect must ingest food that not only meets its nutritional requirements but is also capable of being assimilated and converted into energy and structural substances for normal activity and development. Plant species vary in their adequacy as hosts for even the most polyphagous insects, such as cutworms and grasshoppers. Obviously, host plant utilization depends upon the concentration of proximate nutrients, especially proteins (Slansky and Rodriguez 1987), and upon the concentration of defensive chemicals in space and time (Rosenthal 1982). The key role of host plant variation – plant morphology, nutrients, water content, and chemistry – in the regulation of herbivore populations is now well·recognized. Host plants are heterogeneous and insects are therefore faced with locating and utilizing suitable resources within a complex mosaic of less favorable and unacceptable ones.

Host plants are often nutritionally suboptimal *per se* (Southwood 1973). In most Spermatophyta tissues, the quantities of protein (on dry or wet weight basis) are markedly lower than in those of insects. All insects appear to need small quantities of sterols in their diets (Clayton 1970). The ability to utilize different sterols of plant origin is a major dietary hurdle. Furthermore, some of the plant sterols appear to act as neurohormone 'mimics', and pose an additional hurdle for insects in utilizing plant nutrients.

The different parts of the plant, and the same plant parts under different conditions and during different seasons, all vary in biochemical composition. Therefore, it seems reasonable to assume that plant variability changes the nutritional quality of the host plant, which is often far from being optimal food for insects (Schultz 1983a). Using leaf water as a general index of plant growth is surprisingly valid despite seasonal and daily variations (Slansky and Rodriguez 1987). Though water is not a nutrient, it can change the nutritional quality of the plant under water-stress conditions. Similarly, plant allelochemicals (secondary plant metabolites) which differ with plant variability, are another important factor affecting the dietetic nature of plant nutrients for insects (Schultz 1983a). Allelochemicals can function as allomones, which benefit the producing organism, as in the case of the

Table 4.1. Principal classes of allelochemicals and corresponding behavioral or physiologic effects on insects (modified from Whittaker and Feeny 1971, Kogan 1982).

Allelochemical factors	Behavioral or physiologic effects
Allelomones	Give adaptive advantages to the producing organisms
Repellents	Orient insects away from plant
Locomotor excitants	Start or speed up movement
Suppressants	Inhibit biting or piercing
Deterrents	Prevent feeding or oviposition
Toxins	Produce chronic or acute physiologic disorders
Digestibility reducing	Interfere with normal processes of utilization of food
Kairomones	Give adaptive advantages to the receiving organisms
Attractants	Orient insects toward host plant
Arrestants	Slow down or stop movement
Feeding or oviposition	Elicit biting, piercing, or oviposition
Feeding stimulants	Promote continuation of feeding

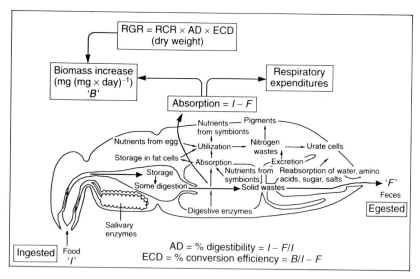

Fig. 4.2. Model of the quantitative nutritional approach for a generalized insect. The relative growth rate (RGR) of the insect larva is a product of the relative consumption rate (RCR), the approximate digestibility (AD), and the efficiency of conversion of digested biomass (ECD). The overall efficiency of conversion is a product of AD and ECD. Reproduced from Daly *et al* (1978) with permission of the author.

resistant host plant, or kairomones, which benefit the recipient insect (Table 4.1) (Whittaker and Feeny 1971).

Therefore, it seems that total organic nitrogen (N), water content, and allelochemicals of plant species exert a strong influence on the 'bioavailability' of nutrients to phytophagous insects. The effect of plant tissues and known allelochemicals on the nutritional physiology of insects has been studied, utilizing three of the most useful nutritional indices: relative consumption rate (RCR), approximate digestibility (AD), and efficiency of conversion of digested food (ECD) (Fig. 4.2). For further details of these indices, the reader should consult the detailed review by Scriber and Slanksy (1981). If growth is inhibited, then it must be reflected either in the amount eaten or in one or more of these indices, or both.

The nitrogen nutritional hurdle

The N in a plant is only one of the many plant nutrients that are vital to herbivores. It is a critical element in the growth of all organisms, but it is scarce and limiting for many herbivores. The large differences in the N content of plant and insect tissues may be the major reason why less than one-third of the insect orders and higher taxa of terrestrial arthropods have achieved the ability to feed on seed plants (Southwood 1973). The protein content of insects and mites is 7–14% N, whereas that of plants or plant parts rarely reaches 7% N and is generally much lower (Mattson 1980). Phytophagous insects are faced with low levels of plant nutrients; hence, their success in growing and reproducing depends upon their ability to efficiently ingest, digest, and metabolize plant N with optimum levels of leaf water (75–95%) (Scriber and Slansky 1981). The water content of the food has a strong effect on the efficiency of conversion of ingested food (ECI) for insects. Sap-feeding insects have a higher ECI than chewing arthropods.

The abundance of N varies during the seasonal growth cycle of plants. The quantitative levels are highest during the early growth stages, but decline with age as N-poor tissues are added to the plant structures. The N content of the phloem and xylem saps of plants also exhibit high seasonal and between-plant variation. The qualitative variations of N are more important from the consumer's point of view. Normally, N occurs in just two forms: protein N (PN), and nonprotein N (NPN). The NPN includes an array of compounds such as inorganic N (nitrates and ammonia) and organic N (amines, amides, amino acids, chlorophyll and other pigments, vitamins, and a large number of secondary compounds) (McKee 1962). Seasonal and ontogenetic cycles also occur in PN and NPN compounds. In general, young moisture-rich tissues have much higher levels of NPN, soluble PN, and N-based secondary compounds than mature tissues, except for senescent tissues where proteins are being hydrolyzed (McKey 1974).

Besides differences due to seasonal and ontogenetic changes, other variations are also superimposed on factors such as temperature–moisture stress, tissue damage by abiotic and biotic agents, and inorganic fertilization. Moisture stress, for example, has a tremendous impact on both carbon and N metabolism. Water-stressed plants show increased levels of amino acids, particularly an increase in proline and high relative concentrations of soluble carbohydrates (Fukutoku and Yamada 1982). Concentrations of qualitative secondary compounds, especially alkaloids and other N-based allelochemicals (cyanogenic glycosides, glucosinolates, various nonprotein amino acids, and peptides), in stressed plant tissues generally increase, whereas concentrations of quantitative secondary compounds (tannins, resins, and essential oils) decrease (Rhoades 1979). Since some of these qualitative secondary compounds are apparently in a state of continuous recycling in plants they may be quickly turned over to primary metabolism. Hence, water-stressed plants conditioned with enhanced nutrients, together with less effective qualitative defensive chemicals, can be more attractive to herbivores and suffer from increased levels of feeding.

Inorganic fertilization can bring about substantial changes in both N quantity and quality. Fertilization increases the levels of soluble N compounds, amino acids, and amides in plants, resulting in a significant increase in digestibility for the concerned insects (Slansky and Rodriguez 1987). There is a vast literature about the effects of soil fertility on the incidence of plant pests. Many pests including several species of aphids have higher rates of reproduction with increased N fertilization. Van Emden *et al* (1969) reviewed such a relationship between various host plants and the aphid *Myzus persicae*. The incidence of most rice insect pests such as stem borers, gall midges, leaffolders, leafhoppers, and planthoppers was raised under conditions of high N application (Singh and Agarwal 1983).

Nitrogen deficiency as another form of plant stress may lead to higher concentrations of starch in C-4 plants (Gallaher and Brown 1977), resulting in an increase of chlorogenic acids in the sunflower (*Helianthus annus* L.) and other plants. The importance of these and other changes in secondary chemicals (tannins, lignin, silica, waxes, resins, etc.) caused by plant stresses are likely to become one of the highest priorities of research in insect–plant interactions in relation to herbivore population dynamics. Such a variation in the chemistry of host plants is a kind of protective system, preventing herbivores from evolving 'fine-tuned' offenses against host plants (Denno and McClure 1983).

Nutrient–allelochemical interactions

Allomones reduce insect feeding and/or fecundity by toxicity; some are also antifeedants or repellents. Some allelochemicals (antivitamins, factors reducing digestibility, and chelators of essential minerals) also have chronic effects on the development or reproduction of insects. Understanding all these aspects is crucial in determining the mode of insect–plant interaction, which ultimately confers insect resistance in a given plant. Insects feeding on host plants containing vast mixtures of allelochemicals usually have problems utilizing the food efficiently. Toxins often affect the nutritional indices (AD and ECD), which in turn affect the growth and reproduction of insects (Reese 1983).

Tannins are examples of substances that can block the availability of proteins by forming less digestible complexes (Feeny 1969). During the growth season, the protein content of oak leaves declines and the tannin content increases; the interaction of the two brings about a marked and accelerated shortfall in the nutritive value of oak leaves. Oak leaf tannins markedly inhibit the growth of the winter moth larvae [*Operophtera brumata* (L.)] because of the formation of relatively indigestible complexes with the available proteins, thus reducing the rate of assimilation of dietary N (Feeny 1969). It is also assumed that many digestive enzymes may further complex and reduce the rate of assimilation across the gut wall. The studies of Bernays (1981), however, showed no reduction in digestion by the polyphagous Acridoidea, even with high levels of tannins. The tree locust *Anacridium melanorhodon* was found to have an increased AD and ECD when fed lettuce plus tannic acid compared to lettuce alone. Even in the graminivorous

Acridoidea, neither the condensed tannin, nor the hydrolyzable tannic acid, adversely affected digestibility in any species (Bernays *et al* 1980). Since tannic acid is hydrolyzed in the gut of these insects, it may be the gallic acid which is beneficial rather than the tannic acid itself. This suggests that the ability of tannins to inhibit growth involves something other than a reduction in assimilation. Whatever may be the mechanism, tannins may serve as formidable defense against many grass-feeding acridids and lepidopterous insects (Zucker 1983).

There are a few examples of reduced predation on food plants having allomones. The glucosinolates in the leaf tissues of the Cruciferae interfere with assimilation of biomass and/or N, because glucosinolates are known to be gut irritants in mammals and to reduce digestibility in non-adapted insects (Erickson and Feeny 1974). The best known nutrient–allelochemical interaction in plants is the case of maize varieties resistant to the European corn borer *Ostrinia nubilalis*. The feeding and growth inhibition of *O. nubilalis* seems to be mediated partly by a mixture of cyclic hydroxamates (Cuevas *et al* 1990). It is likely that a combination of nutritional and allelochemical factors has mediated leaf-feeding and stalk-boring resistance to *O. nubilalis* both in laboratory and field situations. Chapter 6 should be referred to for further details.

A number of plant nonprotein amino acids are usually incorporated into normal insect proteins (Bell 1976). For example, the structure of 3,4-dihydroxyphenylalanine (L-dopa) is similar to that of tyrosine. L-Dopa has been found in relatively high concentrations in a number of legumes. It interferes with the activity of tyrosinase, an enzyme essential for the hardening and darkening of the insect cuticle. It may induce favism and inhibit growth, assimilation of food, the ECI, and normal development in some insects (Reese 1983). L-Canavanine is a structural analog of L-arginine, which is readily incorporated into proteins of most insects. It acts as a competitive inhibitor of arginine metabolism (Rosenthal 1977) and consequently inhibits insect metamorphosis and reproduction. In addition to forming defective proteins, amino acid analogs may also inhibit protein synthesis, block enzymatic reactions, and compete with protein amino acids for transport sites.

Vitamins, like amino acids, also have structural analogs, which are called antivitamins. These structures include any compound that 'diminishes' the effect of the vitamin in a specific way. A well-known vitamin antagonist is dicoumarol in sweet clover (Van Soest 1982). This compound blocks vitamin K synthesis and blocks the clotting reaction of mammals. A number of such compounds in plants may antagonize, destroy, or block the utilization of vitamins by herbivores.

Mechanisms other than the complexing of proteins may prevent nutrients from passing across the gut wall. Protease inhibitors decrease the availability of nutrients and prevent the breakdown of proteins into their component amino acids. The possible involvement of protease inhibitors in plant protection received considerable support in the early 1970s (Green and Ryan 1972) from the finding that Colorado potato beetle infestation induced rapid accumulation of protease inhibitors in the leaves of potato and tomato plants, thus making the host plants less useful as food.

The preceding examples suggest that allelochemicals can interact with nutrients in a number of ways that are deleterious to the survival of herbivores. But, in reality, vast arrays of essential nutrients and allelochemicals occur in plants, and understanding the full range of possible interactions between the two groups is almost impossible. Since some of the allomones can be used by herbivores as feeding stimulants or as kairomones or nutrients, these plant components are situation-dependent, and as such are not characteristics of the molecules. The subtle nature of these compounds is still far from offering a clear-cut answer. The specific area of nutrient–allelochemical interactions deserves attention from researchers involved in basic and applied areas of ecology (Slansky 1992).

The dual role of secondary metabolites

The plant defense chemicals perform allelochemical functions as allomones or as kairomones. Several classifications were proposed to correlate the role of allelochemicals with behavioral or physiologic (growth, development, survival, fecundity, and fertility) effects on insects. Table 4.1 shows a consolidated classification of the two allelochemical factors of plants and the insect response.

The antiherbivory theory suggests that most secondary plant metabolites have evolved to perform allomonal functions in a defense against herbivores including insects (Pasteels 1977). The reciprocal adaptation of specialist herbivores has led to the concurrent evolution of mechanisms that detoxify allomones, avoid plant tissues that contain toxic chemicals, and preferentially attack plants that are in a vulnerable condition due to stress. The specialist herbivores could, in fact, use allomones in the form of kairomones or as cues for host finding, feeding, or oviposition excitation. Consequently, many secondary metabolites act as defense allomones for the majority of sympatric herbivore fauna as well as kairomones for a few oligophagous insects (Norris and Kogan 1980).

Thus, the same plant allelochemical compound has the dual role of repellent as well as attractant to different insects. For example, glucosinolates and their hydrolysis compounds are highly toxic to the unadapted lepidopteran, the swallowtail butterfly *Papilio polyxenes*, but are a feeding stimulant and provide host plant recognition clues for the adapted insect *Pieris brassicae* (Blau *et al* 1978). Such a phenomenon suggests that a chemical messenger can, therefore, be a 'double agent' (Norris and Kogan 1980). This possible principle of chemical ecology means that each chemical molecule can readily assume its role of electronic excitation within a characteristic electrochemical range (Norris 1977). The same plant chemical attracts some species and repels others (Table 4.2). Hence, classification of a chemical as a repellent, deterrent, feeding suppressant, toxin, or digestibility reducer, may be situation- and dose-dependent; it may involve chemical, behavioral, and physiologic feedback systems (Blau *et al* 1978).

Allelochemicals do not necessarily possess dual functions (Schoonhoven 1981). Some compounds have considerable allomonal activity but no obvious kairomonal effect, and some common kairomones have no apparent allomonal properties. The alkaloids are the most prominent elements in the list of insect-feeding deterrents and poisons, and they are rarely feeding stimulants (Smith 1966). The alkaloid α-tomatine, a product of *Solanum* and *Lycopersicon*, interferes with the growth and development of nymphs of *Melanoplus bivittatus* and inhibits feeding of *Pieris brassicae*, but in the presence of sinigrin, α-tomatine becomes a phagostimulant to *P. brassicae* (Ma 1972).

It is even more difficult to find absolute kairomones in green plants because such compounds would not be preserved in the plant's evolutionary process. For example, many attracting plants function as decoys, causing mortality or reduced fecundity, because of the presence of toxins or because of the absence, deficiency, or imbalance of certain nutritional materials. An excellent example concerns the Colorado potato beetle *Leptinotarsa decemlineata*, which is attracted by several members of the Solanaceae in addition to

Table 4.2. Secondary compounds in plants and insect responses.

	Insect reaction	
Secondary compounds	Stimulated	Deterred
Lignin	*Bootettix argentalus*	*Ligurotettix coquilletti*
Tannin	*Anacridium melanorhodon*	*Helicoverpa zea*
Cucurbitacin	*Diabrotica undecimpunctata*	*Epilachna tredecimnotata*
Cucurbitacin E and I	*Diabroticites*	*Phyllotreta nemorum*
Glucosinolate	*Pieris rapae*	*Papilio polyxenes*
Gossypol	*Anthonomus grandis*	*Helicoverpa zea*
Furanocoumarins	*Papilio polyxenes*	*Spodoptera exempta, S. litura, S. eridania*
Tomatine	*Pieris brassicae*	*Leptinotarsa decemlineata, Empoasca fabae*
Cyanogenic glycoside	*Epilachna varivestris*	Many phytophagous insects
Iridoid	*Euphydryas editha*	*Locusta migratoria*
Lupanin (quinolizidine alkaloid)	*Macrosiphon albifons*	*Acyrthosiphon pisum*

its host plant, the potato. In experiments in which 'decoy' foliage was placed in equal amounts with potato foliage, a large proportion of the egg masses were laid on nonhost species (Hsiao and Fraenkel 1968a) (Fig. 4.3). Growth response and, presumably, fecundity differed markedly, with one of the most lethal decoy plants (*Solanum nigrum*) drawing the highest percentage of egg masses. The data suggest that coexisting toxic and nontoxic plants with similar attractant chemistry represent a selectional paradox for would-be-host-specific herbivores.

Plant variability as a defensive strategy

Like plant communities that benefit from diverse defensive interplay, plant populations *per se* benefit from genetic diversity as well as individual physiologic heterogeneity. The enormous genetic potential for interspecific and intraspecific variability in plants is evident in world germplasm collections of various crops. Variation in resistance to many important arthropod pests has been repeatedly demonstrated among crop cultivars not intentionally selected for resistance (Kennedy and Barbour 1992). In addition to

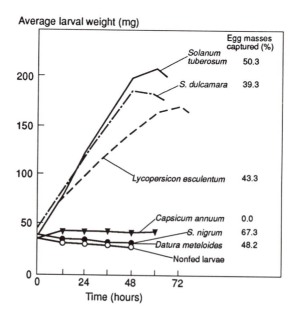

Fig. 4.3. Growth response of the Colorado potato beetle (*Leptinotarsa decemlineata*) on six solanaceous species, contrasted with the percentage of egg masses 'captured' by these plants in the presence of equal amounts of the normal host, *Solanum tuberosum*. Reproduced from Atsatt and O'Dowd (1976) with permission from the American Association for the Advancement of Science, Washington D.C.

variation between cultivars, significant variation has been found to exist within cultivars. For example, sources of resistance to the spotted alfalfa aphid (*Therioaphis maculata* Buckton), pea aphid (*Acyrthosiphon pisum* Harris), and alfalfa blue aphid (*A. kondoi* Shinji), used in the development of resistant cultivars of alfalfa, were found within commercial cultivars (Nielson and Lehman 1980).

Therefore, in the context of the high genotypic and phenotypic variability of most plant populations and individual variability of many plant defenses within plants, the risk of insect attack is spread out among and within individuals. The cost of defense of such variable resources is obviously lower than that of a more homogeneous resource. In such a heterogeneous environment of plants there may be little or no choice of feeding for herbivores, which, ultimately, would reduce selection pressure for the development of specialist herbivore biotypes (Gould 1986).

Optimizing the role of plant defense chemicals

Optimizing the distribution of plant defense chemicals within the plant is another important strategy against herbivores. The limited supply of defensive compounds should be available in those regions of the plant that will result in the greatest defense. Which plant parts need chemical defense depends upon two factors: (1) the *value* or the cost of herbivory to the plant part in question, and (2) the *vulnerability* or the probability that the plant part in question would be successfully attacked by herbivores in the absence of defensive chemicals. Using these criteria of value and vulnerability, plants metabolize their chemical defenses economically in different plant parts. For example, plant surfaces need greater defense than inner layers of cells in the same region, because surfaces are more likely to come in contact with potential herbivores. It is, therefore, expected that selection has favored a greater concentration of defense chemicals in external tissues. Large numbers of alkaloids, saponins, and other defense chemicals are accumulated for protection against herbivores at the appropriate stage of plant growth. As the young tips of roots, shoots, and plant reproductive parts are more vulnerable than mature regions, the young leaves and the fruiting bodies are especially rich in alkaloids on a dry or fresh weight basis (McKey 1979). In the potato plant, solanine occurs wherever growth and metabolism are most intensive. The apical and root tips of the tomato plant have a high tomatine content. The concentration of HCN glycosides (on a wet weight basis) is greater in young leaves than in mature leaves of the acacias (*Acacia chiapensis* and *A. farnesian*) that do not harbor ants. Information on other kinds of plant-produced toxins indicates that this pattern may have some generality (Schultz 1983a).

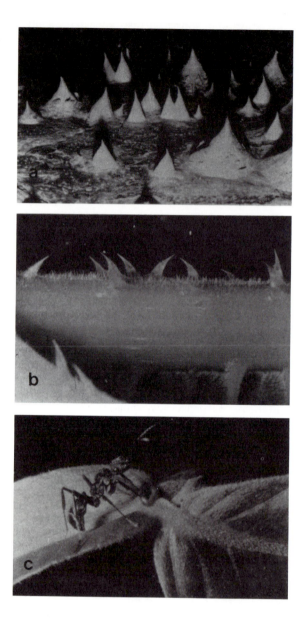

Fig. 4.4. Plant spines as defense mechanisms. (a) Spines on a *Bombacopsis* tree trunk, which may deter large climbing mammals seeking the highly edible flowers. (b) Urticating spines of *Urera baccifera* for protecting ants against ant predators. (c) An *Ectatomma* ant guarding an extrafloral nectary and ready to prey on any small passing insects. Reproduced from Janzen (1975) with permission of the author.

Ant—acacia relationship

Another mode of investing in defensive chemicals is exemplified in the fascinating relationship between ants and acacias. Some species of *Acacia* are protected by ants living on them, which fiercely attack potential herbivores. This is not a casual association between the plant and ant, but a well-evolved symbiosis in which the plant produces food for the ants in the form of extrafloral nectar and protein- and lipid-rich bodies on the leaves, and provides hollow stipular thorns that are the exclusive nesting site of the ants. The ants are unusually hostile to intruders and protect their hosts not only against herbivores but also against vines and competing plants (Janzen 1975) (Fig. 4.4).

These acacia plants lack such chemical defenses as cyanogenic glycosides common on acacia species without ants. If the ants are removed from an acacia plant protected by them, the tree virtually never reaches reproductive maturity unless occupied by another ant colony. If the tree dies it is directly due to severe herbivory by many species of insects. In effect, the ants are analogous to a general purpose chemical defense.

How do Insects Overcome Defense Mechanisms?

In spite of host plant variability, suboptimal plant resources, and the vast array of allelochemicals present in the plant world, 'there is scarcely a plant that does not harbor some insect pest' (Frost 1942). Surprisingly, plant genera (*Chrysanthemum*, *Derris*, and *Nicotiana*) containing well-known broad-spectrum insecticidal chemicals are utilized as host plants by various insect species (Brues 1946). The notion that some plant taxa are better protected than others, presumably by their chemical constituency, often appears to be erroneous. For instance, the statement that ferns are relatively free from insect attack compared with angiosperms is repeatedly found in the literature but probably needs revision (Cooper-Driver 1978).

It follows that phytophagous insects have evolved mechanisms to overcome multiple hurdles posed by host plants. Their behavioral, bio-chemical, and physiologic activities are regulated through sensory inputs. In addition, their natural physiologic feedback mechanisms and environmental factors enable them to adapt to changing situations (Slansky and Rodriguez 1987). Basically, the plant defense hurdles comprise: (1) a nutritional hurdle, (2) a physiologic hurdle, and (3) an ecological hurdle. For each known defensive adaptation in producer organisms (host plants), there are com-plementary offensive adaptations in consumers (insect herbivores) and vice versa. Known defensive adaptations of plants against herbivores and their complementary and possible offensive adaptations in insect herbivores are presented in Fig. 4.5.

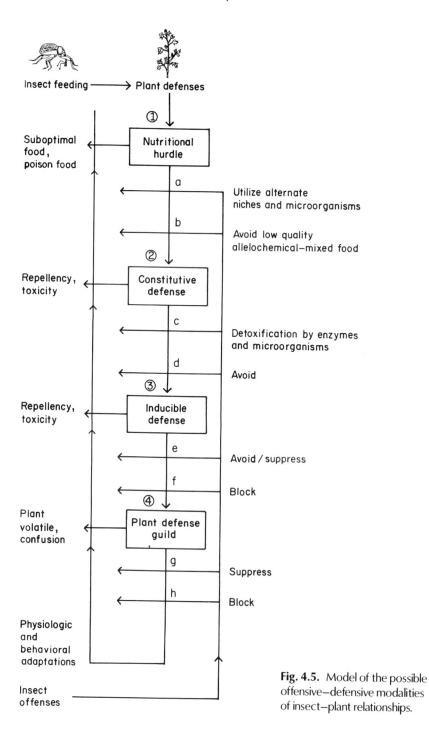

Fig. 4.5. Model of the possible offensive–defensive modalities of insect–plant relationships.

Overcoming the nutritional hurdle

The nutritional status of the plant may be barely adequate or suboptimal (Fig. 4.5(1)), and the herbivore must be able to select plant resources from the available mixture, rather than a specific plant *per se*. Insect herbivore adaptations are mechanisms for adjusting to resource heterogeneity and include: (1) utilizing an alternative niche(s), (2) increasing the consumption rate, (3) modifying the nutritive quality of the host plant tissues, and (4) establishing associations with microorganisms.

Utilizing alternative niches

A switch from old to new leaves – as in winter moths feeding on oak leaves – because of late season tannin–protein complexes (Feeny 1970), or from tree leaves to herbs could result in improved feeding efficiency and growth rate (Fig. 4.5a). Sometimes the insect herbivores avoid feeding and/or ovipositing on unsuitable plant substrates (Fig. 4.5b). In order to find a suitable plant substrate, the insect may move laterally (towards higher leaf water) or upwards (higher N) on the plant surface. In general, feeding, growth, and reproduction of insect herbivores can occur only when the host plant is actively growing. Natural selection, therefore, acts to ensure a high degree of temporal synchrony between insect and plant life cycles. This type of response by some insects to nutritionally variable food is the phenomenon of 'self-selection', in which individuals consume the most nutritionally suitable tissue or a combination of tissues in order to obtain a balanced diet (Waldbauer and Friedman 1991). For example, the last instar of *Helicoverpa zea* caterpillars feed preferentially on the germ of maize kernels (Cohen *et al* 1988).

If arthropods could choose plant parts with optimal levels of water and nutrients, they would be expected to prefer young leaves to old leaves. But the fact is that the generalist insects avoid young leaves that are rich in toxins and deterrents, while specialist herbivores prefer them because the specialists have tolerance for the toxins produced by their specific hosts (Cates 1980).

Some intriguing host shifts have occurred among seed beetles of the family Bruchidae (Southgate 1982). The groundnut seed beetle *Caryedon serratus* was introduced in Africa along with its original host the tamarind *Tamarindus indica*. From the time it reached Africa, the beetle has colonized several indigenous trees (*Acacia, Cassia*, etc.), and populations have reached epidemic proportions on the groundnut *Arachis hypogaea*. Interestingly, the seed chemistry of *A. hypogaea* is entirely different from that of the tree species previously attacked by this beetle. This dramatic example of host shift is an intriguing case of insect adaptation under extreme situations.

Increasing the consumption rate

The consumption and utilization of food by insects are dynamic processes that exhibit compensatory responses to change in the dietary quality. Several species of insects have been shown to increase consumption of food composed of low nutrient levels (Slansky and Wheeler 1989). Increase in food consumption and physiologic changes associated with food utilization may allow an insect to maintain its growth rate on par with that on a richer diet. For example, when larvae of the butterfly *Pieris rapae* are fed on plants with different N levels, they are able to adjust their food intake and assimilation rate to stabilize the rate of N accumulation. On plants with low levels of available N, they eat more and assimilate N more efficiently than on plants with higher N levels (Slansky and Feeny 1977).

Modifying the nutritive quality of host plant tissues

The relationship between the gall-forming insects and their host plant is a remarkable example of insect–plant interaction. Flower galls of the cruciferous plant *Diplotaxis muralis* are formed from the saliva of the gall midge *Paragephyraulus diplotaxis* Solinas. The flower galls are nutritive tissues. They provide food, not only for the larvae of the gall midge, but also for two other insects: thrips and another species of gall midge that colonize some of the developing galls (Feeny 1982).

Some Hemipteran insects, while feeding in considerable numbers, inject their saliva into the plant. In the process, cell tissues are broken down and the nutrient flow is diverted to the infested leaves, thus improving their nutritional quality (Way and Cammell 1970). One group of aphids (*Aphis fabae* Scop.) has become so intimately involved with the physiology of the host plant that it responds to changes in the composition of the host and can influence the flow of nutrients within the plant, thus acting as a 'sink' (Kennedy and Fosbrooke 1973).

Establishing associations with microorganisms

Many insect herbivores have specific requirements in terms of quantitative nutritional balance, specific feeding cues, and, in some cases, qualitatively specific nutrients. However, most qualitative requirements are similar to those of other heterotrophs, with the notable exception of their inability to synthesize sterols (Dadd 1973). All the essential amino acids, vitamins, minerals, and sterols must be obtained from the resource, together with N, lipids, water, and an energy source. With such basic nutritional requirements, the insects are not likely to meet their supply of appropriate nutrients without the aid of mutualistic organisms like microorganisms. This multispecies complexity recently has been receiving increasing attention (Barbosa *et al*

1991). The microbial–plant and microbial–insect associations are likely to be advantageous for the metabolic enhancement of the plant or the insect, increased microbial dispersion efficiency, and increased niche availability for all three parties. Some insects, across a broad taxonomic range, have an obligate association with prokaryotic and/or eukaryotic microorganisms, which are harbored extracellularly or intracellularly (Campbell 1989). By far, the most widely studied extracellular symbionts are the protozoa that inhabit the hindgut of termites.

If plants are nutritionally deficient, these associations should be prevalent because microorganisms modify the resource prior to ingestion by the insects. The occurrence of a microbial–insect 'symbiosis' is very much a function of an intrinsically nutritionally inadequate food source (Buchner 1965). Many termites have a wide host plant range but always have symbionts. The gut flora of termites is capable of producing cellulase for digesting cellulose food. Studies have revealed that cellulase, carboxymethylcellulase, or cellobiase activities in termites are produced either by protozoan flagellates or bacteria inhabiting the hindgut (Hogan *et al* 1988). Some termites are capable of producing their own cellulolytic enzymes in addition to those supplied by bacterial symbionts (Hogan *et al* 1988). If food is barely adequate, the aphids *Myzus persicae* and *Brevicoryne brassicae* can obtain many amino acids via their gut floras (Southwood 1973).

As herbivores cannot normally choose the best resources available, then both specialists and generalists should benefit from facultative microbial associations. Specialists would benefit from resource enhancement while 'solving' their own specific allelochemical problems; generalists would benefit from microbial detoxifications, thereby compensating for the suboptimal nutritional resources.

Utilization of microorganisms by insects helps in overcoming their nutritional constraints and improves the insect–plant nutritional relationships. For example, microorganisms have a wide range of properties that are beneficial for insect growth, development, survival, and fecundity. It has become evident that insects have equally well-established symbiotic relationships with microorganisms that live outside the insect. In such cases, the symbionts live either on the exoskeleton of the insect or somewhere within its habitat (Batra 1979). In certain instances, vegetative mycelia of a particular fungus are 'cultivated' by attine ants. The progeny of these ants are further multiplied in the cultivated fungus garden. Similar to their internal counterparts, these external symbionts provide the host with enzymes, nutrients, metabolic assistance, and even synthesis of semiochemicals such as the aggregation pheromone in Scolytidae (Blum 1987).

Overcoming allelochemicals

Phytophagous insects are continuously challenged by harmful plant chemicals for most, if not all, plants are well endowed with defenses (Fig. 4.5(2)). Plants deploy their chemical defenses in two ways (Levin 1976): (1) a baseline defense or constitutive defense is maintained at all times to repel herbivores through direct toxicity or by reducing the digestibility of plant tissues, and (2) a facultative defense where plants mobilize or produce allelochemicals *de novo* in response to tissue damage from herbivores (Ryan 1983). The combination of the constitutive and inducible (facultative) defenses of a particular plant, with a set of species-specific chemicals, should deter and toxify all herbivores except those few 'adapted specialists' with an appropriate suite of counter-adaptations (Rhoades 1985).

The evolution of defensive metabolites in plants has led to the concurrent development of herbivores with an ability to detect these chemicals, avoid plant tissues with a high allomone content, and preferentially attack plants that are in a vulnerable condition due to stress. However, when highly toxic plant compounds are consumed by specialist insect herbivores, these invertebrates must be able to process these natural products for their benefit. Indeed, this virtuosity of insects in exploiting plant compounds is so remarkable that it can best be described as 'better insect living through plant chemistry' (Blum 1992). These phytophages have broken through the plants' allelochemical defenses by using diverse detoxification mechanisms that also fortify their host plants. Some of the toxicants are sequestered by insect herbivores (Bowers 1990), and once this has been done an allelochemical may be utilized in a variety of ways, such as for production of the insect's own pheromones or for their own defense. Some specialist herbivores metabolize the characteristic allelochemicals of their host and subsequently utilize them for nutrition and kairomones. Examples are given in subsequent sections of this chapter.

In other situations, an increased level of dietary nutrient often mitigates the deleterious impact of allelochemicals on insect growth (Table 1, Slansky 1992). For example, the deleterious effect of the soybean trypsin inhibitor is very much reduced when *Helicoverpa zea* is fed with a protein-rich diet (Broadway and Duffey 1986).

Detoxification

Insects possess a rich and diversified assortment of enzymes that, taken together, constitute a very powerful defense against chemical toxicants (both qualitative and quantitative) of plants (Fig. 4.5c). Qualitative secondary plant metabolites are compounds with a low molecular weight, which are toxic and occur in low concentrations in seasonal plants. These compounds, however, can be costly to maintain owing to high turnover rates (Coley *et al* 1985). Examples of this group of plant compounds are alkaloids, glucosinolates, phenolics, cyanogens, and nonprotein amino acids. Quantitative secondary compounds are tannins and lignins of high molecular weight, and digestibility-reducing compounds. These compounds are cheap to maintain because of low turnover rates, especially in perennial trees.

Detoxification of qualitative defense chemicals

Detoxification usually involves oxidation, reduction, hydrolysis, or conjugation of molecules. Four enzyme categories (multifunction oxidase, epoxide hydratase, phenol β-glucosyl transferase, and aldehyde and ketone reductase) are of major importance in the metabolism of lipophilic foreign compounds. These lipophilic compounds tend to be most hazardous to insects because of their abundance in plants and the way in which they penetrate lipid-containing cell membranes and accumulate in lipid-rich tissues such as the nervous system.

However, enzymes called the 'multi-function oxidases' (MFOs) in the microsomes of the insect midgut and fat bodies constitute probably the most important detoxification system in insects. The MFO enzymes (or more specifically the microsomal cytochrome P-450 system) possess three major characteristics that make them ideally suited as a biochemical waste-disposal mechanism (Brattsten 1979): (1) they catalyze numerous oxidative reactions that produce more polar, and hence more excretable, compounds, (2) they are nonspecific in that a wide range of chemicals are acceptable substrates, and (3) they can rapidly adjust to the presence of allelochemicals or synthetic insecticides via induction. The induction of MFO proceeds rapidly and in response to a small quantity of secondary substances. For example, a diversity of natural plant products induces cytochrome P-450 in the southern armyworm *Spodoptera eridania* (Kramer) (Brattsten *et al* 1977). Compounds as structurally disparate as (+)-α-pinene and sinigrin rapidly induce two- to three-fold enzymatic increases in the larvae. The rise in cytochrome P-450 activity is immediate and its activity proceeds rapidly during the first few hours; however, its activity varies among different insect species as well as among individuals of the same species.

It has been found that host plant allelochemical induction of MFO activity may mediate resistance to synthetic insecticides as well (Yu *et al* 1979). There can also be cross-resistance between pesticides and plant defenses (Gould *et al* 1982), e.g., herbivore populations that have evolved resistance to pesticides in response to chemical control programs should be able to survive on some host plants that are toxic to susceptible populations. This poses a problem in pest management through the integration of plant resistance and chemical control of crop pests.

Detoxification of quantitative defense chemicals

Some phytophagous insects are also physiologically adapted to other types of plant allomones such as tannins and other polyphenols as well as lignins, resins, and silica (quantitative defenses). It appears that tannins are prevalent in many herbaceous plants and woody angiosperms, as well as in many forest trees. Condensed tannins, however, have not yet been found to stimulate feeding because the molecules are generally too large to penetrate the peritrophic membrane of herbivorous insects. As tannins complex freely with enzymes, they can immobilize gut enzymes via nonspecific blockage of active sites and can disrupt insect digestion. Insects have mechanisms to circumvent such a situation by dissociating the tannin–protein complex at more alkaline pH levels (Fox and Macauley 1977). Data from Berenbaum (1980) suggest that most of the phytophagous Lepidoptera feeding on tanniniferous leaves have a mean gut pH of 8.76, which is significantly higher than the gut pH of caterpillars feeding on host plants lacking tannins. It is perhaps this high alkalinity of gut contents and/or presence of surfactants in the gut fluid that have allowed phytophagous insects to overcome the 'quantitative defenses' (Martin and Martin 1984) of tanniniferous trees and shrubs.

Detoxification of inducible defense chemicals

The existence of an inducible response in plants can induce a selection for herbivores to 'suppress' the induced response (Fig. 4.5(3)). For example, the bark beetle *Dendroctonus ponderosae* (Hopkins) utilizes monoterpene alcohols, derived from hydrocarbons produced by its host conifers, as aggregation pheromones. The mature female bark beetles convert α-pinene into *trans*- and *cis*-verbenol as well as myrtenol, which are components of this scolytid's aggregation pheromone (Hunt and Borden 1989). The combined attack caused by the attracted beetles finally exhausts the defensive reactions of host conifers. In some conifers, exudation pressure of preformed resin (constitutive defense) and local synthesis of resin at the site of attack (induced defense) are so reduced by the mass attack of beetles that the tree ultimately succumbs (Raffa and Berryman 1983). Thus, through concerted attack, bark beetles can suppress the effectiveness of host defenses far below that which would result from a single attacking beetle (Fig. 4.5e).

During oviposition, the wood wasps (*Sirex* sp.) inject a toxin and spores of a symbiotic fungus *Amylostereum areolatum* (Fries) into the stems of *Pinus radiata*. The toxin blocks translocation of photosynthates from the leaves, curtailing synthesis of polyphenols and resins in the attacked lesions. Exudation of induced resin into the lesions is also blocked. The fungus then rapidly invades the conductive tissues of the plant to induce transpirational stress. Consequently, a suitable substrate is provided for the developing *Sirex* larvae (Madden 1977) (Fig. 4.5f).

Carroll and Hoffman (1980) have presented experimental evidence for a rapid induced resistance in the host plant and, simultaneously, a behavioral counterresponse by the squash beetle *Epilachna tredecimnotata*, feeding on the leaves of the squash *Cucurbita moschata* (Cucurbitaceae). The beetles have a characteristic solitary feeding habit. They cut a circular trench in a squash

Fig. 4.6. *Epilachna borealis* spins a circular trench in a cucurbit leaf before feeding on the entrenched tissues. The arrow points to the trench. Reproduced from Tallamy (1985) with permission from the Ecological Society of America, College Park, MD.

leaf and feed on the encircled materials (see Fig. 4.6). Trenching takes approximately 10 minutes, whereas complete feeding on the leaf disk takes 1–2 hours. After a morning feeding period, the beetle crawls away from the damaged leaf and does not feed until the following morning. Experimental results show that the induced chemical (cucurbitacin) is rapidly mobilized to the damaged leaf tissues of *C. moschata*. The circular trenching behavior of the *Epilachna* prior to feeding is, therefore, an effective adaptation to counter the mobilization of the feeding inhibitor, the oxygenated tetracyclic triterpenoids of cucurbits. According to Tallamy (1985), the hypothesis has three components: (1) mandibulate feeding damage induces significant increases in the cucurbitacin concentration of squash leaves, (2) within 3 hours of tissue damage, the cucurbitacin content doubles in concentration at the site of injury, and (3) beetles reared on leaves with induced cucurbitacins showed a significant loss in fitness. Individuals that developed mechanisms to avoid ingesting induced plant defense chemicals became genetically established and selection against those chemicals may have been reduced or eliminated. Feeding adaptations in this species block exposure to mobile cucurbitacins so completely that the beetle has not developed effective physiologic resistance mechanisms.

Sequestering plant allomones

Instead of detoxifying plant poisons, several species of insects sequester and deploy the poisons for their own pheromone system and defense. These herbivores are so intimately adapted to the host plants that they can use these compounds to their own advantage. Duffey (1980) provides an excellent review of the phenomenon of sequestration in insects.

Use as pheromones

Insects can detect traces of plant volatile substances in the air and use them to locate their host species. They can also store such substances in either their original or a chemically altered form, and use them as signals or pheromones for other insects. These substances are important in regulating the behavior, growth, and reproduction of some insect populations. For example, the female of the western pine beetle *Dendroctonus brevicornis*, a serious pest of the ponderosa pine in North America, is attracted to its host by the terpene mixture of pine (oleoresin). The female attracts the male by a pheromone, one of whose chemical constituents, myrcene, has been sequestered from the tree's oleoresin. The female can apparently control the number of males attracted, and eventually the local population's sex ratio, by altering the proportion of myrcene in the pheromone mixture. When the insect population reaches an optimum size, the female ceases production of this pheromone and instead produces verbemone, a male-repellent substance that prevents additional males from settling (Edwards and Wratten 1980).

Males of some species of butterflies (Nymphalidae) and moths (Arctiidae) convert plant-produced pyrrolizidine alkaloids into sex pheromones that are especially critical during courtship (Boppré 1990). These pheromones are externalized on secondary male sexual structures such as androconia or hair pencils (coremata). The biosynthesis of the male pheromone (hydroxydanaidal) is related to the occurrence of pyrrolizidine alkaloids in the insect's food. If the larvae have access to plants containing adequate pyrrolizidine alkaloid, full development of the coremata is assured since sufficient hydroxydanaidal is synthesized from pyrrolizidine alkaloid precursors.

Use in defense

The majority of the species that sequester natural plant products store these compounds or their metabolites in a variety of tissues. The sequestered compounds can provide the sequestrators with considerable protection against predators, because these compounds often are unpalatable or emetic. This biomagnification of plant allelochemicals belonging to a variety of chemical classes has been documented for many species of insects as testimony to the virtuosity of these invertebrates as sequestrators (Rothschild 1973). Insects may selectively sequester only some of the allomones belonging to a single chemical class while eliminating related compounds rapidly (Blum 1983).

Larvae of the monarch butterfly *Danaus plexippus* L. (Danainae) utilize a large volume of gut fluid as a critical factor in the processing of cardenolides ingested from the milkweed *Asclepias humistrata* (Asclepiadaceae) host plants. This fluid is fortified with polar cardenolides. During pupal and adult development, these steroids are rapidly exchanged with the hemolymph, providing the insect with a readily available source of cardiac glycosides (Nishio 1980). Larvae can regurgitate gut fluid readily, giving them a cardenolide-rich shield that can be thrust at predators. The cardenolides are even channeled to the cuticle where they can constitute the first line of defense. The prepupal molting fluid is also rich in cardenolides, providing protection to the developing insect. Subsequently, these bitter-tasting steroids reappear in the teneral adult and are concentrated in the wing scales and hemolymph, providing the imago with a readily available defense against avian predators such as the blue jay *Cyanocitta cristata*. In one study, soon after eating the monarch butterfly (about 12 minutes), the bird predator became violently ill and vomited as many as nine times over a period of 30 minutes (Brower 1969) (Fig. 4.7). The jay rapidly learned to refuse these larvae. Thus, the sequestration of milkweed steroids by the monarch larvae provides all developmental stages with a formidable chemical defense. Furthermore, these toxic plant chemicals are often sequestered in the oocytes developing in the abdomen of the female monarch butterfly (Nishio 1980).

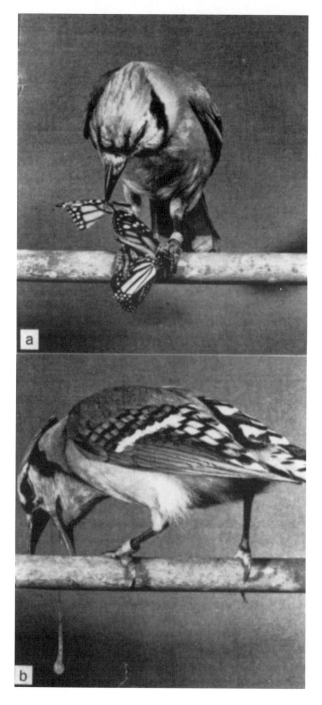

Fig. 4.7. Reaction of a blue jay to unpalatable monarch butterflies. (a) The blue jay attacks an unpalatable butterfly, eats it, and soon vomits it out (b). Reproduced from Brower (1969).

The sequestration of different classes of allelochemicals in insect eggs is widespread in aposematic species that develop on toxic plants (Brown 1984). Sequestration of plant defense compounds is becoming increasingly well documented and understood, for example, insects that feed on plants containing iridoid glycosides (Gardner and Stermitz 1988), pyrrolizidines (Boppré 1990), and cucurbitacins (Nishida and Fukami 1990). Variations in host plant chemistry, as well as in the ability of the insects to sequester these compounds, will also be critical in determining the amount of chemical protection conferred by a particular group of chemical compounds.

Plant allomones acting as nutrients or kairomones

Some specialist phytophages thoroughly exploit their host plants by utilizing not only primary nutrients for growth but allelochemicals as well. For example, canavanine in seeds of the leguminous vine *Dioclea megacarpa* is a harmful nonprotein amino acid. The bruchid beetle *Caryedes brasiliensis* Thunberg is a highly adapted predator of the L-canavanine-laden *D. megacarpa* (Fig. 4.8). The insect has very positive arginase and urease activity, which converts much of the stored N of canavanine to another toxic amino acid, canaline, as well as urea. High levels of urease convert the urea into ammonia for fixation into organic compounds. Additional utilizable N is generated from the detoxification of canaline, which yields large amounts of ammonia. In addition, homoserine, a readily metabolized amino acid, is produced from canaline, thereby conserving the carbon skeleton of canavanine. Thus, the degradation of canavanine by *C. brasiliensis* larvae provides ammonia for N metabolism and an amino acid that can be readily metabolized. The poison canavanine becomes an important nutrient for this particular insect (Rosenthal 1983).

Another example is found in two species of chrysomeline larvae, which ingest salicin from their host plants, *Salix* or *Populus* spp. (Salicaceae). The hydrolyzed compound salicylaldehyde is an effective defense chemical against ants. Ultimately, the chrysomelid larvae convert the defensive allomone to glucose and metabolize this as larval nutrition (Pasteels *et al* 1983). Thus, salicin may be regarded as an allelochemical nutrient.

Some toxic plant compounds act as essential feeding stimulants for many species of specialist insects. Such pharmacophagous species frequently sequester these phagostimulatory allelochemicals, whose bitter or emetic qualities render the insects unpalatable. For example, the allelochemicals (clerodendrins) of the verbenaceous plant *Clerodendrum trichotomum* are powerful phagostimulants for the turnip sawfly *Athalia rosae* (L.) as well as providing a cuticular set of 'armor' to protect against aggressive predators (Nishida and Fukami 1990).

Fig. 4.8. An adult bruchid beetle *Caryedes brasiliensis* that recently emerged from a *Dioclea* seed. The adult's exit hole and the small entrance hole of the first stadium larva can be seen. Developing bruchid beetle larvae feed upon the *Dioclea* seed in spite of the presence of appreciable levels of ʟ-canavanine. Reproduced from Rosenthal (1977) with permission from University of Chicago Press, Chicago.

Avoidance

There is growing evidence that phytophagous insects exhibit an array of avoidance behaviors, such as minimizing direct exposure to plant allelochemicals (Fig. 4.5d) and, in certain instances, blocking inducible pheromonal communication and avoiding physical defenses (Tallamy 1986).

Avoiding plant allelochemicals

The mode of insect feeding, for example, can greatly reduce exposure of the insect to plant allelochemicals. Furanocoumarins in the plant family, Apiaceae, are localized in seeds. These compounds are located within storage structures called vittae (Zangerl *et al* 1989) as well as in vegetative parts known as companion canals (Bicchi *et al* 1990). Insects that feed on these host plants can avoid coming into contact with the toxic compounds. Thus, the tarnished plant bug *Lygus lineolaris* sucks fluids from seeds without contacting the vittae (Flemion and McNear 1951). Even the foliage-feeding insects like the cabbage looper *Trichoplusia ni* can avoid much of the allelochemical content by skeletonizing the foliage but not the major veins. This mode of insect feeding can avoid encountering the furanocoumarins localized in the companion canals adjacent to the veins (Zangerl 1990).

Avoiding pheromonal communication between plants

Red alder trees *Alnus rubra* Bong attacked by the western tent caterpillar *Salix sitchensis* Sanson exhibit a change in foliage quality. Tent caterpillars that were fed with unattacked leaves from attacked trees grew more slowly, died at a faster rate, and produced fewer egg masses than those fed on leaves from unattacked trees. It is postulated that plants could receive pheromonal signals emitted by nearby attacked leaves (Schultz 1983a). Such signals are highly advantageous to plants as a form of advance warning of herbivore attack. Some herbivores could evolve adaptations to suppress pheromonal communication between damaged plants or plants grown in mixed cropping (plant defense guild) (Fig. 4.5(4)). This communication suppression could occur as an indirect result of suppression of the recognition of attack or through herbivore secretions that block the release of communication substances from the wound or from mixed cropping. Some herbivores may themselves emit pheromonal countersignals to confuse, mimic, or otherwise interfere with the signals generated by plants (Fig. 4.5g, h).

Avoiding physical defense

Insects have also evolved many ways of avoiding physical defenses such as leaf spines, hairs, and trichomes (Pillemer and Tingey 1978). Caterpillars of an ithomiid butterfly, *Mechanitis isthmia*, provide an interesting example. They have circumvented trichomes by spinning a silk scaffolding upon which the caterpillars can crawl over safely to the spineless edges of leaves, where they can feed normally. In another example, the caterpillars of the noctuid moth *Pardasena* sp. could mow the trichomes on the leaves of *Solanum coccineum* Jacq. before eating the mown strip (Hulley 1988).

Plant defense–herbivore–parasite interactions

Until recently, most hypotheses developed in the study of insect–plant interactions were limited to the direct insect–plant interacting system. In such a two-component system, insect-resistant crop varieties are developed with increasing levels of allelochemicals responsible for antibiosis. Such a focus has resulted in limited attention being given to the development of natural biological control agents (Campbell and Duffey 1979).

Price *et al* (1980) have argued that insect–plant interactions must be considered along with the third trophic level as a part of a plant's defensive repertoire against herbivores. A holistic approach to understand insect–plant interactions should be based at least on three trophic levels of 'plants, herbivores, and natural enemies of herbivores.' The interactions between insects and plants have been described in the preceding sections, yet the chemical basis of plant resistance cannot be fully appreciated without an

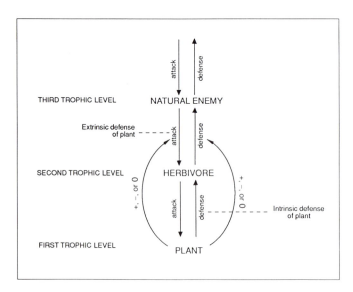

Fig. 4.9. Direct and indirect types of interactions in a trophic food web, showing the relationships between intrinsic and extrinsic defense of plants and the trade-offs between them. +, −, and 0 indicate the positive, negative, or absence of effects on the plant, which may affect the herbivory–natural enemy relationship. Reproduced from Price (1986) with permission from Ellis Horwood, Chichester, UK.

understanding of the modalities of how insects feeding on varied host plants affect the fitness of parasitoids. Between each consecutive trophic level there is a dynamic interaction of defense and offense. An important feature in this trophic system is that members of alternate trophic levels may act in a mutualistic manner (Fig. 4.9) (Price 1986). In addition to defense chemicals, natural enemies of herbivores are beneficial in enhancing the fitness of plants. Therefore, it is a reasonable assumption that plants can defend themselves against herbivores by means of: (1) *intrinsic defense*, where the plant alone produces defense through biosynthesis of phytochemicals, and (2) *extrinsic defense*, where the plant benefits from the natural enemies of herbivores.

A combination of these two approaches has impact on the 'ecological resistance' of the plant, where the intrinsic defense may have a positive or negative influence on the third trophic level (Bergman and Tingey 1979, Price *et al* 1980). To understand the total trophic system, it is necessary to integrate knowledge of each trophic level with that of the different factors that mediate interactions. In this regard, intraspecific plant variation may directly or indirectly affect the evolution and ecology of predators and parasitoids of herbivores (Boethel and Eikenbary 1986). Host plants are likely to vary qualitatively or quantitatively in the production of allelochemicals. Such variation may affect host location by parasitoids and fitness of the parasitoids that feed on the prey. Plant surfaces or other physical characteristics may influence the searching behavior of predators and parasitoids (Hare 1992). Thus, the possible interactions are:

1 synomone-mediated interactions (synomones are semiochemicals released by one organism that evoke a reaction in an individual of another species that is beneficial to both) (Nordlund 1981);
2 chemically mediated interactions involving nutritional and resistance factors;
3 physically mediated interactions involving plant structures and plant architecture.

Synomone-mediated interactions

Plant odors are mediated through volatile chemicals that are often plant defenses against herbivores. Generalist insects are usually repelled, but specialized insects may be adapted to such compounds. The enemies of herbivores may use the plant odor to find the plant first and then its prey on the plant. For example, the aphid *Brevicoryne brassicae* utilizes sinigrin (mustard oil) of crucifer plants as a cue to find host plants. Similarly, its parasitoid, *Diaeretiella rapae*, uses a related mustard oil (allyl isothiocyanate) to find the plant and then the prey aphid (Read *et al* 1970) (Fig. 4.10). On the other hand, gossypol in the glanded cotton varieties that imparts resistance to *Heliothis* spp., acts as an attractant to the *Heliothis* parasitoid *Campoletis*

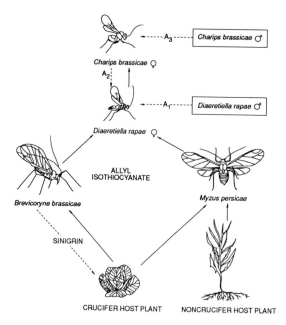

ALLYL
ISOTHIOCYANATE

SINIGRIN

CRUCIFER HOST PLANT   NONCRUCIFER HOST PLANT

Fig. 4.10. Many plants are recognized by herbivores and parasitoids by volatile chemicals. The integrity of the community is maintained by chemical links. Sinigrin from the host plant attracts the specialized aphid *Brevicoryne brassicae,* whereas a closely related compound, allyl isothiocyanate, attracts the primary parasitoid *Diaeretiella rapae.* Reproduced from Price (1986) with permission from Ellis Horwood, Chichester, UK.

sonorensis (Cameron) (Elzen *et al* 1983). The terpenoids released by corn injured by herbivores are used as cues by the parasitic wasp *Cotesia marginiventris*, although they may be produced in defense against herbivores (Turlings *et al* 1990). Thus, in an evolutionary context, it may be difficult to separate the direct benefits of releasing volatile chemicals on plant fitness by reducing insect pest damage from the indirect benefits of the plant being more attractive to the pest's natural enemies (Hare 1992).

Other compounds (kairomones) synthesized by the plants are incorporated into the body odor of herbivores. This body odor is utilized as a searching cue by herbivore enemies. For example, the corn *Zea mays* plant contains a chemical tricosane, which is incorporated into the egg of the corn earworm *Helicoverpa zea*. This token stimulus chemical then is used as a kairomone by the egg parasitoid *Trichogramma evanescens* to find its prey (Lewis *et al* 1972).

In another situation, the volatile kairomones (linalool, guaiacol, and octanone) from the larval frass of the soybean looper *Pseudoplusia includens* (Walker) feeding on the foliage of soybean cultivars have provided excellent cues to the looper parasitoid *Microplitis demolitor*. The interesting feature of this tritrophic relationship is that the volatile kairomones in the host frass include not only host plant compounds but also compounds (i.e., guaiacol) that are larval metabolites of the plant precursors. Thus, there is an evolutionary advantage for a host-specific natural enemy, like *M. demolitor*, to utilize volatile cues associated specifically with the frass of its host, in addition to those associated with the host plant of its prey (Ramachandran *et al* 1991).

Other chemically mediated interactions

Host plants provide food to the natural enemies of herbivores: it may do so directly in the form of floral nectar, pollen, and extrafloral nectar, which may be eaten extensively by parasitoids (Smiley 1978); or indirectly in the form of honeydew or other sugar-rich secretions of plant-sucking insects, which are an important carbohydrate source for natural enemies (Zhody 1976). These food sources are essential in sustaining the natural enemy populations.

The specific mechanism(s) by which the quality of the host's diet is modified for the parasitoid is unknown at the biochemical–physiologic level. Studies on the role of host nutrition indicate that the addition or subtraction of specific chemicals from artificial diets provides a useful tool in understanding the quality of food affecting the tritrophic interrelationship. The plant defense chemicals that impart resistance to field crops against herbivores may not be compatible with parasitoid activity. The use of tomatine (glycoalkaloid) as the chemical basis of resistance in the tomato crop against *Helicoverpa zea* or *Spodoptera exigua* (Hubner) adversely affected the development of the parasitoid *Hyposoter exiguae* and even caused severe developmental abnormalities and/or death (Campbell and Duffey 1979, Duffey *et al* 1986). Laboratory experiments were conducted to quantify the effects of resistant and susceptible host plants of *Cucurbita* on the preimaginal (egg to adult) and adult biologies of the coreid bug *Anasa tristis* and its egg parasitoid *Gryon pennsylvanicum*. The results showed that the preimaginal development of the pest was similar in all the test varieties, but the survival to the adult stage was significantly lower in resistant varieties. Likewise, the lifespan of the adult parasitoid was significantly reduced when its larvae developed in eggs of prey that were reared on the resistant cultivar (Vogt and Nechols 1993).

One interesting example of the biological relationship between plant resistance and insect pathogens has been reported (Chiang and Holdaway 1960). When the European corn borer *Ostrinia nubilalis* larvae fed on resistant corn, the protozoan *Perezia pyraustae* parasite, which the larvae sometimes carry, was eliminated. The few larvae that survived on the resistant plants were healthier and hibernated more successfully than the numerically larger number of larvae that thrive on susceptible plants (Beck 1957).

As a consequence, the increased intrinsic defense in plants, for which selection may be done in a plant-breeding program, is likely to reduce the extrinsic defense, with a net deleterious effect on crop protection. This is especially a problem with specialized herbivores that can tolerate high dosages of toxins, but which are highly detrimental to their natural enemies. Estimates of such effects should be considered when formulating plant-breeding programs.

Physically mediated interactions

Physical effects on the tritrophic interactions are equally important and are as diverse as chemically mediated interactions. Many physical traits (leaf toughness, trichomes, cuticle thickness, etc.) of host plants are likely to influence the herbivore's access to its enemy-searching zone. Plant trichomes play a prominent role as the basis of resistance against a number of small insects and mites (Norris and Kogan 1980, Stipanovic 1983), but they may act negatively or positively to natural enemies. The combination of glandular trichomes in hybrids of *Solanum tuberosum* × *S. berthaultii* and the action of aphid predators and parasitoids (coccinelids, chrysopids, and aphids) was more effective in reducing aphid populations than were either enemies or trichomes alone (Obrycki *et al* 1983). In other crops, however, trichomes have not proved compatible with parasitoids. Glandular trichomes on tobacco impede searching by the small parasitoids *Telenomus sphingis* Ashmead and *Trichogramma minutum* Riley and result in less parasitization of *Manduca sexta* eggs on tobacco than on other solanaceous crops (Rabb and Bradley 1968). In cucumber, the efficacy of searching by the parasitoid *Encarsia formosa* Gaham for its host, the white fly *Trialeurodes vaporarium*, is greatly reduced by the presence of stiff foliar hairs coated with sticky aphid honeydew (van Lenteren *et al* 1980). Exceptions to this generalization are known. Leaf pubescence is an effective resistance mechanism for the cereal leaf beetle *Oulema melanopus* (L.), in that beetles oviposit less on pubescent genotypes of wheat *Triticum aestivum* L. Yet pubescence apparently has no adverse effect on parasitization by the parasites of the cereal leaf beetle (Lampert *et al* 1983).

On the other hand, physical defense may aid natural enemies. Any factor that prolongs the larval life of herbivores will make them more susceptible to enemies. Morphological plant structures such as 'frego bracts' in cotton have a significant impact on the efficacy of natural enemies in suppressing the key pests of cotton (Jenkins and Parrott 1971). A physical attribute of the host plant which provides resistance is considered to be compatible with the third trophic level, as in the case of wheat cultivars. Resistant wheat cultivars showed a reduction in aphid *Diuraphis noxia* (Mordvilko) populations and enhanced parasitoid activity because their leaves did not roll like those in the susceptible wheat entry (Reed *et al* 1991).

These examples of plant characters influencing the third trophic level suggest the need to understand the biology and ecology of each trophic level before we can understand the interaction between plant resistance and biological control. It appears that almost any plant character may influence the trophic system; hence, integration of resistant varieties cannot always be deemed desirable in an integrated pest management system without considering their influence on natural enemies. Mechanisms that operate in a detrimental fashion at the second trophic level may carry over to the third level. Any management strategy that tends to decrease the efficiency of natural enemies should be undertaken with caution.

Another aspect that needs to be considered is the genetic implications of the tritrophic interactions. In this regard, it may be argued that genetic variation within plant species may substantially affect the evolution and maintenance of herbivore—natural enemy relationships. For instance, differential susceptibility to natural enemies among plant genotypes may impose natural selection on herbivores for enemy-free space. Through this phenomenon the host insect will selectively oviposit on plant genotypes that will minimize their progeny mortality due to natural enemies. However, the impact of natural enemies of herbivores as effective agents of natural selection on plant genotypes within species has not yet been ascertained for any tritrophic association (Hare 1992).

Coevolution

In the previous sections, the defensive—offensive mechanisms of plants and insects were elucidated in a stepwise fashion. The diversity of plant defense compounds is enormous and the simultaneous adaptations by insect herbivores against plant defenses have affected the evolution of both groups. However, the processes and evolutionary outcomes of these interactions are not clear.

These interacting systems are seemingly interlinked, possessing some feedback mechanisms, and are believed to have coevolved in pairwise interactions. The term coevolution was introduced in the 'coevolutionary theory' by Ehrlich and Raven (1964) to describe what is believed to have occurred between plants and phytophagous insects. It was further developed by many others and is the most cited theory of evolution in specific insect—plant interactions (Futuyma and Slatkin 1983). In this classic coevolutionary theory it has been hypothesized that the great diversity of plant allelochemicals evolved in response to attack by herbivores, particularly insects, and that many insects evolved in response to changes in their host plants.

However, the coevolution theory has also been criticized because it is apparent that such close reciprocal evolution is only likely to operate under restricted conditions (Thompson 1982). A number of workers have increasingly questioned the validity of coevolution in structuring phytophagous insect communities (Futuyma 1983). Rey *et al* (1981) stress 'that many associations between insects and plants can occur without much evolution, and much evolution can occur without any coevolution in their plants or insects.' In fact, so far no cases are known in which a plant species became measurably resistant to an insect species as a result of its regular attack. However, certain plant defense compounds experience opposing selection pressure by different enemies. For example, glucosinolates in crucifers and hypercin in *Hypericum* are toxic to 'nonadapted' insects but are attractive and/ or feeding stimulants to some insects that specialize on these plants (Chew 1988). Presumably, there are few quantitative data at the population level. On the whole, the insects appeared to respond to different attractant or deterrent characters, and this appears to be the prevailing pattern in field studies of genetic variation due to attack by multispecies assemblages of insects (Maddox and Root 1990). This may imply that major defensive barriers evolve in response to a diverse assemblage of insects that collectively may impose strong selection, but not individually. Several research data also suggest that the evolutionary impact of herbivore damage on plant chemistry may often be confounded by responses to damage from invertebrates, vertebrates, pathogens, drought, temperature stress, exposure to ultraviolet light, and by the effects of primary metabolic demands and nutrient availability (Tallamy and Krischik 1989). Thus, the property of resistance is contingent on the physical or biotic environment. In such instances, an insect cannot adapt to one or more features of a plant with which it has no evolutionary experience (Futuyma and Keese 1992).

Coevolution that occurs between arrays of populations, each generating selective pressures as a group, is more likely to be diffuse than pairwise (Janzen 1980, Strong *et al* 1984). The term diffuse coevolution should not be confused with terms such as interaction, symbiosis, mutualism, and insect–plant interaction. A bee has not necessarily coevolved with the flower it pollinates; a caterpillar has not necessarily coevolved with its sole species of host plant; and the bruchid beetle has not necessarily coevolved with the protease-rich legume seeds that it preys on (Janzen 1980). Hence, adaptation to a new host plant cannot be regarded as an evolutionary 'evasion' caused by interspecific competition.

Sequential evolution

Jermy (1984) has examined the evolutionary aspects of all recent types (monophagous, oligophagous, and polyphagous) of insect–host plant relationships. He presupposes evolution in a 'sequential' way. The evolution of phytophagous insects followed that of plants without major evolutionary feedback, i.e., without affecting plant evolution. The extensive radiation of plant-eating insect groups was made possible by the evolution of angiosperms into very biochemically different taxa and this immense biochemical diversity served as a base for the evolution of phytophagous insects. High evolutionary and speciation rates are characteristics of insects in general (Price 1986).

It may be pointed out that host plant selection is primarily a behavioral process governed mainly by the insect's chemosensory system; hence, it can be assumed that the emergence of new insect–plant relationships results mostly from evolutionary changes in chemosensory responses. Chapman (1980) pointed out 'that a reduction in numbers of sensilla in the mouthparts, associated with the development of labelled lines, could have been a critical factor in the evolution of Hemipteroidea and Endopterygota by facilitating their diversification onto new host plants.' Therefore, specific insect–host plant relationships seem to be the result of evolutionary changes in the insects' chemosensory systems rather than being an adaptation to the nutritional quality of the new host plant. It is hardly surprising that there are to date no definite studies of a chemically mediated pairwise coevolutionary interaction, even in a highly simplified system (Berenbaum and Zangerl 1992a).

5 HOST PLANT SELECTION

In almost every agroecosystem the standing crop of seed plants provides the greatest proportion of the stored energy for terrestrial arthropods and insects, yet only nine orders of insects can effectively utilize such resources (Southwood 1973). This is due to the evolutionary hurdles that insects have to overcome to live on spermatophyta. These hurdles concern: (1) nutrition; (2) shelter; and (3) host finding. The insects can overcome these hurdles by means of physiologic or behavioral adaptations. Overcoming the first two hurdles is elucidated in Chapter 4, whereas host finding and the host selection process are the subject of this chapter.

Host plant selection is the process by which an insect detects a resource-furnishing plant, within an environment of a multitude of diversified plant species. In the course of evolution, most phytophagous insects exhibit specialized feeding habits, and they become adapted to specific plant taxa or to a group of related plant families to the exclusion of others. Other insects may feed on many plant species belonging to different families. Insects with different host-finding abilities have survived well over the eons and are now well established in agroecosystems. This implies that the host plant searching behavior is an active process by which phytophagous insects must recognize and select suitable substrates for food, mates, oviposition, nesting sites, and refuge. The diversity of host plant selection is overwhelming as each insect species shows a series of adaptations to its host plants. Insect adaptations to their host plant may involve the type of mouth parts as well as physiologic and behavioral strategies for meeting the physical and chemical defenses of the host plants (Bernays and Chapman 1994).

The evolution of insect–plant associations has been guided to a large extent by plant chemistry in some, but not all, insect groups (Powell 1980). Related species of butterfly often use plants that are chemically similar, albeit taxonomically distant (Scott 1986). Such a relationship implies that the agroenvironment is charged with a diversity of chemical molecules (Fig. 5.1), in which insect herbivores orient themselves towards plants according to secondary compounds. The phenomenon of host plant selection is a behavioral sequence by which an insect distinguishes between host and non- (unsuitable) host plants and chooses some plants and not others within its host range. Such distinct food preferences are, to a large extent, attributable to the insect's ability to identify preferred host plants on the basis of allelochemicals acting as attractants and/or stimulants (Schoonhoven 1987). Although in some instances optical factors or tactile properties of the plants are involved in the host plant selection process, basically, the chemical stimuli of the plant are perceived by the sensory organs of the insect and elicit differential behavioral responses.

Thus sensory perception represents the link between plants and insects. Plant chemicals involved in host plant selection by phytophagous insects are of two broad categories: (1) those influencing the behavioral process; and (2) those that influence the physiologic processes. The chemicals that affect insect behavior in host plant selection are categorized according to the sensory response that they elicit, i.e., attractants, arrestants or stimulants (positive response) and repellents or deterrents (negative response). Plant chemicals affecting physiologic processes may be classified as antifeedants, physiologic inhibitors, or toxicants. Such diverse effects of plant chemicals imply complicated relationships between insects and plants.

The host plant selection process by phytophagous insects can be usefully considered under the following areas.

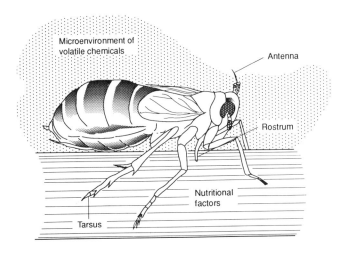

Fig. 5.1. Brown planthopper in the microenvironment of its host plant. Courtesy of R. C. Saxena.

1 *Theories of host plant selection.* Several theories of host plant selection have been proposed and have contributed greatly to our understanding of insect–plant relationships.

2 *Optimal discrimination.* Each insect species has evolved a unique set of sensory apparatus, designed to distinguish hosts from nonhosts with reasonable reliability. It is necessary to understand the insect sensory systems and their sensitivity variations.

3 *Finding a potential host plant.* This is the outcome of processing appropriate external stimuli and the insect's own internal events. Therefore, a critical appraisal of the mechanisms of host plant selection for feeding and oviposition is crucial. Similarly the behavioral responses to plant stimuli with varying proportions of visual, mechanical, olfactory, and gustatory characteristics are of importance.

Theories of Host Plant Selection

In the process of evolution, seed plants have developed insect resistance mechanisms. As early as 1888, the botanist Stahl appears to have been the first to suggest that a variety of botanical compounds probably provided plant defenses against herbivores. The earliest experimental suggestion that plant chemistry plays a role in the feeding behavior of butterflies was provided by Verschaffelt (1911). His experiments revealed that although the food plants of larvae of the cabbage white butterflies *Pieris brassicae* and *P. rapae* belong to different plant families, they share the presence of mustard oils and precursor mustard glucosides (glucosinolates) in their plant tissues. Plants lacking these compounds were only nibbled by *Pieris* larvae. As the larvae did not taste other plants, it was clear that plant odor played a role in host specificity. Verschaffelt's significant discovery undoubtedly provided a major clue to the role of secondary plant compounds in determining the host plant selection of phytophagous insects.

Several theories of host plant selection have been put forward by different workers. Schoonhoven (1990) presented a historical review of the development of research on insect–plant relationships in the last 100 years. Surprisingly, many fundamental theories and discoveries put forward by earlier workers over 100 years ago are still valid today. For instance, De Candolle (1804) used his knowledge to distinguish between monophagous and oligophagous insect herbivores. He described monophagy (in pierid butterflies) as an extreme case and stated that many species of oligophagous insects feed on specific plant genera or families. De Candolle realized that plant-specific compounds were important in host plant specificity. He knew that *Pieris* butterflies were specialized herbivores of the crucifers.

Botanical instinct theory

Brues (1920) was the first to hypothesize that insects are guided by specific chemical cues from green plants in selecting suitable host plants for feeding, oviposition, and other life-history activities. He suggested that food plants are selected on the basis of specific plant odors by larvae and adult insects, and he designated this phenomenon as botanical instinct. He believed that insect species are attracted to plants because of their natural instinct for botanical stimuli. The findings of Verschaffelt and Brues have led to further search for positive cues of individual phytochemical compounds as catalysts of insect behavior. Later works of David and Gardiner (1962) showed that females of *Pieris brassicae* would lay eggs readily on pieces of green paper that had been treated with aqueous solutions of allylglucosinolate.

Token (positive) stimuli theory

Fraenkel (1959) elaborated the role of odd plant substances in host plant selection in the plant families Cruciferae, Umbelliferae, Leguminosae, Moraceae, and Gramineae. He found that the insects were influenced not only by the plant's morphology or primary nutrient substances but also by their chemistry. He strongly argued that the odd chemical substances in which plants differ are glucosides, phenols, alkaloids, terpenoids, saponins, and others, which he considered solely responsible for guiding phytophagous insects to their preferred food plants and as stimuli to induce feeding. Fraenkel proposed the phenomenon as token stimuli for eliciting behavioral responses from insects.

Fraenkel (1969) updated his theory to include the possible effects of secondary chemicals on the developmental and physiologic processes as well as behavioral patterns of insects. He extended the theory to include the concepts that: (1) insects evolve from polyphagy to monophagy to overcome the adverse effects of secondary plant substances, and (2) specialist insects utilize these compounds as cues to locate the host plants.

Dual discrimination theory

The token stimuli hypothesis is an extreme and one-sided view. It seems strange that plant nutrient constituents do not affect the palatability of food for many insects. Indeed, one of the substances ubiquitous in plants, namely sucrose, has long been known to be very acceptable to numerous phytophagous and other insect species (Beck 1965). The concepts of plant 'token stimuli' and 'nutritive materials' are not mutually exclusive and examples are now known in which either of the two groups of compounds or even a combination of them plays a crucial role. The addition of mustard oil glucosides to the artificial diet of some Cruciferae insect feeders has considerably increased their acceptability (David and Gardiner 1966). On the other hand, primary nutrients such as carbohydrates, amino acids, and some vitamins (e.g. ascorbic acid) are phagostimulants (Beck 1972). In the case of the Colorado potato beetle, there are as yet no known secondary plant substances that act as phagostimulants, and the host selection depends mainly on taste. In fact, the feeding stimulants for this beetle are primary nutrients such as sugars, amino acids, phospholipids, and potassium salts (Hsiao and Fraenkel 1968b). Thus, it follows that host plant selection is based on the insect's response to two types of stimuli: (1) the *flavor stimuli*, originating from botanically specific biochemicals; and (2) the *nutrient stimuli*, which are apparently feeding stimulants.

This is the dual discrimination theory propounded by Kennedy and Booth (1951). They studied host alternation in *Aphis fabae* Scop.; these aphids responded behaviorally to at least two main classes of leaf properties: one associated with the age of the leaf and the other with the kind of plant. Among leaves from the same kind of plant, the aphids preferred to feed, and reproduced faster, on young and early senescent leaves rather than on mature ones. Allowing for age differences among the leaves, the aphids settled and reproduced better on spindle leaves than on beet leaves. It is suggested that the shifting patterns of aphid distribution among leaves and plants may depend on the nutrient content of leaves as seasons of active growth and senescence alternate.

Nutritional imbalance theory

Although insect-feeding habits are not determined primarily by the insect's specific nutritional requirements, it must be recognized that not all dietary substrates are equally nutritious. Plant parts and plant constituents vary with their developmental stages, physiologic conditions, and plant genotype. These variations have considerable impact on both the behavior and developmental success of phytophagous insects. Furthermore, the importance of the dietary proportions of essential nutrients in host plants seem to be of greater importance than their absolute quantities (House 1969). The results of Febvay *et al* (1988) indicated that the amino acid balance is one of the factors causing resistance in some cultivars of lucerne *Medicago sativa* L. to the pea aphid *Acyrthosiphon pisum* (Harris).

Negative stimuli theory

Much work has been done to find the specific plant substances evoking feeding responses, e.g., mustard oil glycosides for *Plutella maculipennis* (Thorsteinson 1953). In some cases, such as the silkworm *Bombyx mori*, the feeding stimulants were chemically identified (Hamamura 1959). However, considerably less is known about the role of feeding inhibitors, probably because host selection by oligophagous insects was long considered to be governed mainly by the botanical distribution of specific feeding stimulants.

Research in the 1960s began to emphasize the role of deterrents and other negative stimuli in the food preference of oligophagous insect species. The investigations of Jermy (1961) on host selection by oligophagous insects like the Colorado potato beetle showed that feeding stimuli are not markedly specific, and that host selection is based largely upon negative reactions to repellent stimuli. Further findings of Jermy (1966) on host selection of several chewing phytophagous insects revealed that both polyphagous and mono-phagous species are very sensitive to deterrents, but the former tolerate more deterrent substances than the latter. It seems that sensitivity to deterrents is more important in determining the host range than adaptation to specific phagostimulants. It can be assumed that the more the chemoreceptors are specialized to feeding stimulants, the more they are sensitive to feeding inhibitors (Dethier 1980). This two-way specialization of chemoreceptors appears to be a characteristic feature of food preference by phytophagous insects. Jermy (1966) reported that the narrow negative specialization of the chemoreceptors to deterrent chemicals makes it possible to inhibit feeding by an oligophagous insect and even more by a monophagous one.

Chemical basis of host selection theory

Behavioral and electrophysiologic studies indicate that the host–selection activities of phytophagous insects at the various stages of their life cycle are diverse because of the interaction between the plant allelochemicals and sensory systems. But host plant selection by ovipositing females as a regulator of oviposition is less eludicated than is host plant selection by juvenile forms of insects. Kogan (1977) has forwarded six distinct models of host selection strategies for both adult and larval stages of insect groups. The models range from the more generalized to the more complex and specialized group of insects (Table 5.1).

These six models of host plant selection by phytophagous insects provide an understanding of the chemical basis of plant selection by the ovipositing female and the larval forms (Kogan 1977). Although many other instances may not fit these models, Kogan (1977) has focused on the critical role of allelochemicals in the initiation and maintenance of oviposition and feeding.

Sensory Systems

The preceding discussion of the theories of host plant selection indicates that specific stimuli of a botanical origin or a configuration of stimuli originating in the external environment matches a model in the insect's sensory system, which then evokes a specific olfactory and/or gustatory behavior. To

Table 5.1. Host-selection strategies in six insect–plant interactive systems (modified from Kogan 1977).

Model	Oviposition	Host finding	Larval-feeding stimulants	Host plant range	Insect groups involved
I	Nonselective	Contact chemoreception, active search	General stimulants	A = L	Highly polyphagous species—Acrididae
II	After landing on plant (green, yellow, orange color) at random and then test probing	Wide variety of host plant, passive search	Nonspecific attractants/deterrents	A = L	Polyphagous aphids, whiteflies, thrips
III	Nonselective, eggs laid on nonbotanical substrates	Larvae adapt active random searching (root feeders)	Specific for arrestant and feeding excitant of cucurbitacin, followed by aggregation pheromone	A ≠ L	Diabroticites
IV	Selective on specific host plants	Larvae less discriminatory, passive searching, oligophagous species	Feeding imposed by nature of feeding deterrents rather than excitants	A = L	Chrysomelidae
V	Highly selective by characteristic isothiocyanates	Larval feeding selective, Cruciferae-associated species	Feeding responses by feeding excitants (sinigrin)	A ≠ L	Cruciferae-associated species
VI	Highly selective by visual, contact, chemoreception and learning by *Heliconius* sp.	Larval feeding involves high degree of discrimination on the growth of *Passiflora* sp.	Specific excitants and deterrents	A ≠ L	*Heliconius* sp.

A, adults; L, larvae, =, similar; ≠, dissimilar.

accomplish the holistic recognition of a plant (i.e., form, color, texture, water content, and hundreds of chemical compounds), several types of the insect's interacting sensors are used to detect and identify plants as hosts via the insect's central nervous system. Insects appear to use inputs from different sensors such as the compound eyes, ocelli, and mechanoreceptive and chemoreceptive sensilla. How inputs are integrated by the central nervous system to enable the insect to recognize suitable plants is not well understood. Although recognition involves the sensory system of insects, inconsistencies in sensory input to the central nervous system have been indicated in a number of reports, including that of Dethier and Crnjar (1982). Therefore, the sensory system alone is not sufficient for host recognition. Additional information is required about the insect's developmental stage, feeding experiences, and/or physiologic state, which influences the reception of and response to the incoming information. The following sections deal with these aspects.

Insect sensilla

Snodgrass (1935) provided reviews of the earlier literature and McIndoo (1931) provided a morphological and behavioral approach in studying the sensilla of individual insect forms. With the application of electrophysiologic monitoring techniques and electron microscopy during the 1940s to 1960s, physiologists contributed a rather precise knowledge of the function of individual cells and the type of sensilla in different insect forms.

Morphologically there are two broad categories of insect sensory nerve cells. Type I neurons are bipolar with a dendrite that has a ciliary structure which is unbranched (Fig. 5.2A). Type II neurons are bipolar or multipolar with dendrites that arborize into many fine branches along their length and

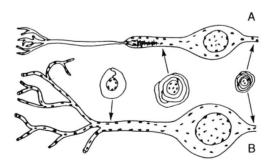

Fig. 5.2. Sensory nerve cells of epidermis. A, Type I neuron with a terminally branched dendritic ciliary segment; B, bipolar Type II neuron with a small part of its terminal dendritic branches. Typical cross-section views of the dendritic and axonal cell sheaths are shown. Modified from Zawarzin (1912).

typically do not have a ciliary structure (Fig. 5.2B). Type II neurons are generally associated with the epidermis and the epithelium of the alimentary tract or its muscles; they may also be associated with muscles of other organs and the peripheral nerves. Both types are ensheathed by glial cells to varying extents. Type I neurons have been associated with at least two other cells of epidermal origin. The origin of Type II neurons has not yet been determined.

The sense organs, that we normally term sensilla in insects, typically contain Type I neurons. Thus, an insect sensillum is defined as a sense organ that has one or more bipolar Type I neurons associated with cuticular regions, with the dendrites of neurons enveloped by at least two associated cells that form the cuticular parts or have other functions (Zacharuk 1980). Type I neurons and the cells giving rise to the cuticular structures usually originate from the same parent epidermal cell. All sensilla seem to be homologous and have been derived from hair or seta.

Types of sensilla

The surface of the insect body is richly supplied with sensilla of a variety of shapes and densities yet, regardless of their sensory modalities, their structures are basically the same. Most types of insect sensilla have either direct or indirect association with the cuticle. They are identified on the basis of the form of their cuticular parts and their positions on, within, or under the cuticle. The classification of nine basic types by Snodgrass (1935) is still in use today (Fig. 5.3). The functions of specific sensilla were later identified

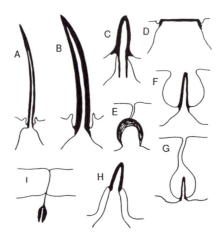

Fig. 5.3. Primary cuticular features of the various types of sensilla. A, trichodea; B, chaetica; C, basiconica; D, placodea; E, campaniformia; F, coeloconica; G, ampullacea; H, styloconica; I, scolopophora. Modified from Snodgrass (1935).

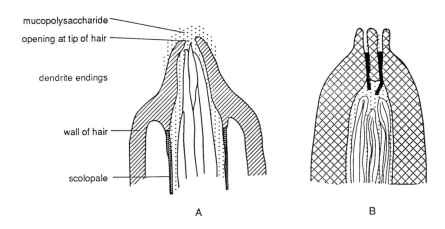

Fig. 5.4. Dendritic terminations and pore structure in two types of uniporous sensilla.
(A) Two-chambered hair with a simple pore opening; (B) one-chambered peg with
finger-like sculptures around a slitted pore opening. Reproduced from Zacharuk (1980)
with permission from Annual Review of Entomology, Annual Reviews Inc., Palo Alto, USA.

electrophysiologically. Additional details on types of sensilla are given by
McIver (1985).

Based on external form and mode of insertion on or in the body wall,
insect sensilla can also be categorized as thick- or thin-walled, according to
the relative thickness and, porosity of the cuticular covering (Slifer 1970).
Thick-walled sensilla have a common opening at one point (uniporous) in the
cuticle through which chemical communication can occur between the
dendrites and the external environment (Fig. 5.4). The uniporous sensilla
have either a simple pit pore (Fig. 5.4a) or a sculptured porous point (Fig.
5.4b) through the cuticle. Internally there is a channel through the body
cuticle and into uniporous papillae, pegs, and hairs. The dendrites of the
sensory neurons extend through this channel towards the porous point
within a tubular dendritic sheath. The dendritic sheath is open opposite the
porous point and fused with the wall around it. The dendritic sheath may
allow certain ions to diffuse between the fluids. The uniporous sensilla
generally function as contact or gustatory chemoreceptors; they also respond
to odors before contact. However, these sensilla are mostly sensors by contact:
hairs on tarsi (Ma and Schoonhoven 1973); and hairs and pegs on the tip of
the labellum, terminal segments of antennae, maxillary and labial palps, pegs,
papillae, or plates on the ovipositor (Zacharuk 1980).

Multiporous chemosensilla are also found on antennae and are often
distributed in large numbers and with specific patterns (Slifer 1970). Many of
them occur in protected locations – sunk in shallow pits or in deep pits with

very small external orifices. They also occur in small numbers on the ovipositor, maxillary palps, and labial palps of different insects.

Sensory modalities

Phytophagous insects, in their food choice, range from diversified hosts to host specificity. They respond to plant color, touch, and chemical cues by utilizing one or several of their sensory modalities including vision, mechanoreception, olfaction, gustation, and other forms of signals (Schneider 1987, Ramaswamy 1988). The operation of the four major senses depends on the insect's distance from the plant source. Modalities like vision and olfaction are thought to operate within a close range from the source, whereas mechanoreception and gustation require direct contact with the substrate (Finch 1980, Prokopy and Owens 1983). Furthermore, insects with varied life-cycle patterns (e.g. Apterygota, Hemimetabola, and Holometabola) utilize their sensory modalities differently from each other (Mitchell 1981).

Vision

Vision is defined as the ability to perceive spatial patterns and true image formation. During their mobile life stages, all insects use photoreception, be it with simple ocelli or highly developed compound eyes. Effective vision is possible with the insect's compound eyes, whereas ocelli are poor image formers (Prokopy and Owens 1983). Recognition of complex and even subtle forms is greatly improved by the multifaceted compound eyes and their highly developed ganglia in the adult insects. The key components of the compound eye are the monopolar neurons (rhabdomeres) that absorb radiant energy and generate action potentials in the transparent areas of the cuticle and internal lenses. Together with clear cellular bodies, the crystalline cones constitute a lens system, known as the 'dioptric apparatus' of the eye. The compound eye is a collection of sensilla called ommatidia. Each ommatidium consists of a corneal lens, a crystalline cone, a number of primary neurons (rhabdoms), and enveloping cells. There are three groups of rhabdoms based on their morphological features: open, fused continuous, and fused layered. The rhabdom type determines the ability of the animal to perceive polarized light. For structural details and functions of the compound eye, see Snodgrass (1935) and textbooks of insect physiology.

The spatial patterns of photo flux differ according to the visual color cues: *brightness* (intensity of perceived reflected light), *hue* (dominant wavelength of reflected light), and *saturation* or *tint* (spectral purity of reflected light). The spatial distribution of photo flux provides information on shape, size, distance, and motion. The foliage color of different plants may serve as visual cues to foraging insects. According to Moericke (1969), the foliage of green plants does not differ much in hue, just oscillating around yellow and green.

Similarly, the spectral reflectance–transmission curves of foliage under diffuse light conditions are remarkably consistent over a wide range of plant species (Gates 1980). This is because the absorption properties of plant chlorophyll have a reflectance ranging between 500 and 600 nm. Such a range of reflectance is also the area of the insect's visible spectrum where green-yellow leaves reflect their peak energy. For example, greenhouse whiteflies can perceive the green-yellow color of leaves and orient themselves towards their host plant (Vaishampayan *et al* 1975). As the foliar color differs in tint and brightness, Moericke (1969) postulated that phytophagous insects may show a preference for a specific tint or a specific brightness of color in their preferred plant. The mealy plum aphid *Hyalopterus pruni*, when leaving the plum and looking for a prospective summer host, alights on the *Phragmitis communis* in response to the unsaturated green (grey-green) color of the leaves rather than on beet plants which have a saturated green color. The whiteflies avoid settling in the presence of short wavelength illumination (400 nm), but will land on green light (550 nm) (Coombe 1982). In fact, most diurnal insects are attracted to yellow, discriminating foliage-like hues from nonfoliage-like substrates. This phenomenon suggests a specific spectral response, but not necessarily through color vision. However, color frequently exerts an influence on the early stages of insect orientation to its host plant (Beck 1965).

Visual stimulation also mediates several other types of behavior, i.e., feeding, oviposition, learning, communication, and sexual behavior. Moths of the African armyworm *Spodoptera exempta* possess a four-color visual system that reaches from ultraviolet to deep red and even functions in moonlight (Langer *et al* 1979). By means of such a visual sensory system, the armyworm adults can select appropriate vegetation for landing, mating, and egg laying.

Among plant species, dimensions and patterns are far more variable than the diffuse spectral quality of foliage. Even within a plant, there may be a wide variation in leaf shape and form during successive growth stages. Studies on the pipe vine swallowtail butterfly *Battus philenor* have shown how females recognize leaves of a particular form for oviposition. After recognizing the leaf form, the butterfly confirms the host plant by its contact receptors. When the host plant grows in dense natural vegetation or when it is late in the season, the height and density of the surrounding vegetation increases, thereby impairing the butterfly's visual ability (Papaj and Rausher 1983).

In spite of a large body of information on the mechanisms of photoreception, our knowledge on the visual performance of phytophagous insects is meager. The ability to detect host plants or plant structures seems to be a comparatively minor factor in shaping insect visual acuity, because most host plant searching activity occurs at very close range. There are instances in which insects utilize a combination of olfaction coupled with close-range responses to visual stimuli for detecting preferred sites for feeding and oviposition (Feeny 1982). The host-finding behavior of the cabbage rootfly *Delia radicum* (L.) shows that initially the fly moves randomly until it encounters the odor of allylisothiocyanate, a volatile mustard oil released from cruciferous plants. This initial step is followed by close-range visual perception by the fly of the host plant (Feeny 1982). Other examples exist where visual and chemical stimuli operate sequentially or simultaneously in locating the host plants by a variety of insects. For example, maximum oviposition by the caged gravid onion fly, *Delia antiqua*, is encouraged by the synergistic interplay of onion-produced alkyl sulfides and a smooth yellow cylinder positioned vertically near the plants (Harris and Miller 1982). Chemicals without the yellow cylinders or cylinders without the chemicals did not stimulate oviposition. The mechanisms behind synergistic effects of combined chemical and visual/structural cues in stimulating oviposition of the onion fly are not known. The presence of olfactory stimuli might cause the fly to respond more strongly to visual stimuli.

Information relating to differences in visual capacities among different insect species and how these insects perceive their environment is still lacking. Data on the optical properties of leaves, such as hue, saturation, reflectance transparency, and others are very sketchy. Obviously the gap is still wide in our knowledge of the visual orientation of insects to plants by a central perception mechanism.

Mechanoreceptors

Mechanoreception is the perception of distortion of the body caused by mechanical energy, which may originate either externally or internally (McIver 1985). The external stimuli may involve touch or air- or water-borne vibration. The internal pressure on muscular activities may also provide stimuli. Mechanical senses are involved in more behavioral activities (such as locomotion, primary orientation, feeding, mating, oviposition, and hearing) than the chemical and visual senses. The rigid exoskeleton of insects is well equipped for detection of mechanical stimuli. This has resulted in the development of a number of effective mechanical sense organs with varying degrees of specialization. These sense organs occur both inside and outside the body, singly, in small groups, or in large groups that function as specialized organs. The sensilla for mechanoreception may be classified according to function as tactile, stretch, auditory, and position receptors, or according to structure. A single structural type may have more than one function. For example, a hair sensillum is usually a tactile receptor, but may also perceive air-borne vibrations. The tactile organs respond during deformation but they may continue to respond throughout the period of stimulus. The latter type of tactile receptors function as proprioceptors. The basis of sensory transduction of mechanoreception is the conversion of the mechanical energy of the stimulus into the electrical energy of the nerve impulse.

Hair sensilla

Hair sensilla are the most abundant, widespread, and extensively investigated type of cuticular mechanoreceptors. The external part of the hair sensilla is either hair shaped or clearly a derivative of a hair such as scale, filament, or peg. The hair-shaped types are commonly called sensilla chaetica, Bohm's bristles, or sensilla trichodea. Hair sensilla may function solely as mechanoreceptors or as chemoreceptors as well. Those that are solely mechanosensitive typically bear no pores or openings in the hair wall and are innervated by one neuron (Fig. 5.5).

The hair is attached to the socket by an articulating (joint-region) membrane, which usually consists of the elastic protein called resilin. The external stimulus for a hair sensillum bends the hair into a certain sector, with the stimulus strength increasing from 0° to about 80° bending angle. The hair bending transmits a mechanical force to the tip of the dendrite, which leads to motion and results in excitation.

Proprioceptors

Proprioceptors are sense organs capable of continuous response to deformations and stress in the body. They provide the kind of information necessary for the animal to maintain certain relations between different parts of its body and of the body as a whole with respect to gravity. Proprioceptors are assisted by other receptors such as photoreceptors and tactile receptors. Five kinds of sensory structures are known to play a role in proprioception: hair plates, campaniform sensilla, stretch receptors, chordotonal organs, and statocyst-like organs. For detailed information on these structures, see Dethier (1963).

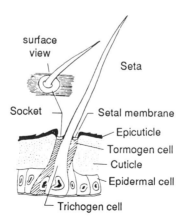

Fig. 5.5. A hair sensillum and its socket. Modified from Snodgrass (1935).

Role of mechanoreceptors

The role of mechanoreceptors on the behavior of insects has been documented earlier by Dethier (1963). The input signals received from various types of mechanoreceptors are integrated and they subsequently influence the behavior of insects. For example, mechanoreceptors including hair, campaniform sensilla, and chordotonal sensilla are involved in the initiation, control, and maintenance of flight. These sensilla are strategically situated on the surfaces of wings and the veins. In locusts, the in-flight function of the sensilla on the antennae, wings, legs, and head has been fully analyzed (Knyazeva 1970). The halteres of Diptera are specialized organs for balancing. They are profusely equipped with all types of mechanoreceptors equivalent to those on the wings of nondipterous insects and they are capable of controlling yaw, pitch, and roll (Pringle 1957). Response to gravity is critical in the lives of many insects; in the social Hymenoptera there are hair plates located between the body regions which function as gravity receptors enabling bees and ants to determine direction, using gravity for orientation (Markl 1966).

Mechanoreceptors of plant-feeding insects are probably involved in discriminating plant surface properties, i.e., leaf geometry, consistency of the leaf or portions of it (Heinrichs 1971), and possibly providing information on distance from and contact with the leaf. The mechanoreceptive sensilla on the ovipositor are necessary for appropriate positioning of eggs in *Bombyx mori* as well as in other ovipositing insects (Ramaswamy 1988). Many taste receptors functioning as contact chemoreceptors contain mechanosensitive neurons, which indicates a link between these two senses. However, processing of the mechanoreceptor message in the central nervous system is not yet well understood (Schneider 1987).

Chemoreceptors

The receptor cells sensitive to chemicals (chemoreceptors) are among the most important components of an insect's sensory system. These chemoreceptors and the physiologic processes that occur in these cells due to stimuli are termed chemoreception. There are two fundamentally different chemosensory mechanisms in insects: general chemosensitivity and receptor chemosensitivity (Schneider 1969). General chemosensitivity is a slow response, which usually extends to the body surface and some inner organs of the body after exposure to a relatively high concentration of harmful broad-spectrum chemicals. Receptor chemosensitivity is a faster response, in which the receptor cells or the dendritic endings of specific cells are excited by relatively narrow-spectrum chemical compounds. With the onset of the receptor response, a nervous message is formed that is transmitted to the central nervous system via the afferent nerve fibers. Similar receptor systems are found in related groups of organisms, but the chemical responses differ for different organisms owing to qualitative and quantitative differences in their chemosensory systems. It is remarkable that the receptor complement with which insects accomplish chemical detection and discrimination is simpler than those of vertebrates by many orders of magnitude (Dethier 1971). The rabbit, for example, possesses 10^8 olfactory receptors as compared with only 48 in a caterpillar.

The importance of chemoreceptors that trigger a wide variety of behavioral patterns in insects has long been recognized by entomologists. Physiologists have also focused attention on these receptors, because the functioning of insect chemoreceptors provides the background mechanisms of chemoreception on a cellular level. Information gathered from such studies often has relevance towards understanding chemoreception in other invertebrates and in vertebrate animals as well. To sort out appropriate terminology for different types of chemoreceptors, Dethier and Chadwick (1948) have distinguished: (1) olfactory chemoreceptors (olfaction or smell), those mediated by chemical stimuli acting in a gaseous state at relatively low concentrations, and (2) contact chemoreceptors (gustation or taste), those mediated by chemical stimuli acting in a liquid or solution state at relatively high concentrations.

The chemoreceptors are not chemosensitive simply because of fortuitous anatomic location, exposure to the environment, or the possession of permeable coverings, but because they are inherently highly specialized and specific in nature (Dethier 1963). Observations on insect behavioral responses to plant substrates for food, mating, and oviposition have established that the principal sites of chemoreceptors are the antennae, maxillae, labial palpi or their homologues, legs, and ovipositors. Olfactory receptors are restricted to the antennae, but in several cases they are also present on the maxillary and labial palpi (Wensler and Filshie 1969).

The olfactory and gustatory receptors of insects are anatomically distinct. Schneider and Kaissling (1957) have identified thin-walled and thick-walled structures as olfactory and gustatory receptors, respectively. Thin-walled chemoreceptors are abundant on the antennae of many species of insects from the Collembola to the Diptera (Dethier 1963). The wall of the sensillum is perforated by large numbers of excessively small pores. Thick-walled receptors with a single pore at the tip act as contact chemoreceptors and occur on almost any part of the body surface. Chemoreception and its role in the ecology of insects have been extensively reviewed by several authorities, notably Visser (1986) and Frazier (1992).

Olfaction

Insects respond to the odors of their surrounding environment and are sensitive to biologically meaningful chemical signals such as those from food, prey, and mates. The system of pheromones is highly developed among the social insects. The colony activities of ants, bees, and termites are very intimately regulated by chemical signs such as the scent of the trail, assembly,

alarm, avoidance, and attraction (Jacobson 1965). The antennal nerves'
response to some olfactory stimuli was recorded first by Boistel and Coraboeuf
(1953). Additional evidence for the flagellum as site of the olfactory sense
organs was also obtained by use of an electroantennogram (EAG) (Schneider
1969).

Olfactory sensilla.   Insect olfactory sensilla have been classified into four
types on the basis of size and shape: sensilla trichodea (e.g., moths), sensilla
basiconica (includes many insects), sensilla placodea (e.g., bees and other
Hymenoptera), and sensilla coeloconica (e.g., Lepidoptera, Hymenoptera, and
many other insect orders) (Schneider and Steinbrecht 1968) (see Fig. 5.3).
Except for the sensilla coeloconica, all the other olfactory sensilla have a pore–
tubule system, which connects the outside medium with the hair lumen and
the receptor dendrite. Sensilla coeloconica are of a different construction and
probably comprise several morphological subtypes having receptor cells that
are sensitive to carbon dioxide, temperature, humidity, or a combination.

   The pore–tubule system of a trichodea sensillum is shown schematically
in Fig. 5.6a. This hair-shaped sensillum is characterized as a thin-walled
cuticular protrusion (10 mm long and 2 mm in diameter). The outer part of

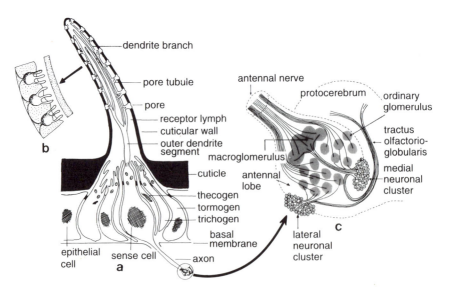

Fig. 5.6. A trichodial sensillum showing: (a) one sensory cell embraced by three auxiliary
cells and the conduction system for the odor molecules (pore and pore tubules), which
make contact with the dendritic membrane; (b) enlargement of the pore and pore
tubules; (c) olfactory pathways in the antennal lobe. Redrawn and modified from
Mustaparta (1984) with permission from Chapman and Hall, London.

the cuticle is penetrated by numerous pores (about 2500) over the entire surface, except at the most proximal region of the hair. Each pore opens to a spherical 'kettle'. From this point, several tubules penetrate the inner parts of the cuticle and reach into the hair lumen nearly as far as the cell membrane of the dendrite (Fig. 5.6b). The hair is innervated by one or two sensory cells, with dendrites penetrating the hair lumen towards the tip. Each dendrite is divided into an inner and outer segment; only the outer dendrite resides within the hair lumen. The outer segment is usually divided longitudinally into branches and has relatively few sensilla trichodea but up to 50 sensilla basiconica. The inner dendrite segment and the cell bodies are surrounded by three auxiliary cells (the thecogen, tormogen, and trichogen cells) that form a sheath around the sensory cells. The auxiliary cells carry out important secretory functions during the ontogenetic development of the sensillum, as well as maintaining the electrical potential between the hemolymph and the receptor lymph (Kaissling and Thorson 1980).

The primary sensory cells receive information from the odor signals and convey it through their own axons to the second-order neurons in the antennal lobe via synapses, which are confined to characteristic glomerular structures (Fig. 5.6c) (Boeckh 1977). After the integration processes in the antennal lobe, involving convergence and connection via interneurons, the information is conveyed further via tractus olfactorio-globularis to higher order neurons in the protocerebrum.

Olfactory mechanism. One of the most striking functions of the olfactory system in insects is to discriminate a limited array of stimuli among a large number of plant volatile chemicals and odors from the external environment. Some insects have been studied with respect to their attraction and orientation to plants (Visser 1986). Their behavior clearly shows that in the majority of cases several odorous components act synergistically. The insect antenna houses a variety of receptor cells that serve various functions such as olfaction, mechanoreception, hygroreception, thermoreception, and contact chemoreception. With such multidimensional features, it is reasonable to assume that odor discrimination is a task fulfilled by the central nervous system. But in reality, the odor image of a plant or any substrate is encoded by a set of peripheral receptor cells called the 'across-fibre-spectrum' cells (Pfaffmann 1941). Peripheral receptor cells then convey the stimulus to the central nervous system.

Olfactory receptors have been studied extensively and their response profiles have been interpreted. It can be postulated that the specificity of cells depends on the specificity of their molecular acceptor system (Kafka 1974). Thus, the physicochemical principle underlying these primary processes of the interaction between odor molecules and acceptors is as follows. The stimulus, i.e., molecules of certain odor compounds, are absorbed on the cuticle of the hair wall (with a layer of electron transport material) and diffuse through the pore openings, which often lead into a fine and short pore channel widening in to a cavity (Fig. 5.7). It is estimated that on a trichoid sensillum of *Bombyx*, any odor molecule on the surface can diffuse to a pore within 2 ms of impaction. Subsequently, the molecule would diffuse in through the pore cavity filled with receptor lymph of the liquor channel, reaching the dendrite membrane in a further 1 ms (Steinbrecht and Kasang 1972). Electron microscopy has established that contact exists between the pore tubules and the dendrite membrane. Such molecular capture is thought to be a binding

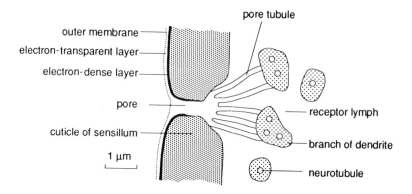

Fig. 5.7. Diagrammatic section through the wall of a basiconica sensillum showing a single pore and connections with the terminal branches of the dendrites. Modified from Steinbrecht and Kasang (1972).

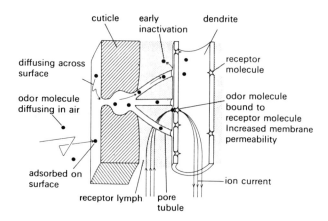

Fig. 5.8. Diagram for the capturing and reception of an odor molecule on an olfactory receptor. Modified from Kaissling (1974).

process with acceptor sites on the dendritic membrane, in a fashion similar to that of enzyme–substrate complexes (Kafka 1974, Kaissling 1974). Only molecules that 'fit' the acceptors of the membrane can elicit a reaction. This fit is brought about by a spatial distribution of atoms and charges in the odor molecule (chain length, location of functional groups, or multiple bonds), which enables the development of weak interactions to certain loci of the acceptor. Such interactions can take place if the molecule reaches a certain position opposite the acceptor. The interactions cause molecular changes in the receptor-cell membrane, which in turn follow a change in the ion permeability of the membrane (Fig. 5.8). This leads to an ionic flux across the membrane and the development of the receptor potential, which spreads proximally along the dendrite and results in generating nerve impulses. The impulses are the signals to the central nervous system.

Contact chemoreceptors
The gustatory sensors are also important for plant-feeding insects, because the process of food selection depends on the taste stimuli. Although a relatively minor difference between taste and smell is suggested, the olfactory capability of gustatory sensilla may not be a general phenomenon. At the most, the gustatory receptor cells can smell at a close range (0.5 mm). Hence, Stadler (1984) proposed the use of the term contact chemoreception to include gustation or taste and close-range olfaction.

Contact or taste hair receptor sensilla have a tip pore and unbranched dendrites with 2–10 cells that reach the tip (Zacharuk 1980). In the insect's thick-walled hair, the sensilla function as contact chemoreceptors, which often contain a mechanosensitive neuron. The mechanoreceptive cell associated with the chemoreceptor cells is mainly for monitoring movements of the sensillum, either the whole or its tips. For instance, the mechanoreceptors of the maxillary palpi of *Manduca sexta* are capable of monitoring the degree to which the palpi can telescope in response to compressive forces (Hanson 1970). The associated cells with these contact sensory neurons and their morphology are described in earlier sections (see Fig. 5.5). For more details, see Zacharuk (1980). Most contact chemoreceptive sensilla are located on the antennae, mouth parts, and tarsi, but they may occur widely on the body surface (Slifer 1970). They have also been identified in the food channels of flies, caterpillars, moths, true bugs, cockroaches, and locusts (Stadler 1984).

Perception of phytochemicals. Many behavioral studies of different insect species have been conducted for the purpose of locating, mapping, determining qualities, and elucidating the nature of receptor action of contact chemoreceptors. These studies were basically of the feeding and ovipos) tional behavior of insects on the basis of acceptance and rejection thresholds (Ma and Schoonhoven 1973, Ma 1977a, Jermy and Szentesi 1978, Roessingh *et al* 1992).

In recent years, electrophysiologic methods have given much impetus to the analysis of the actual factors involved in determining the range of substances that can be perceived by the insect. The first recordings of afferent impulses from individual chemoreceptor cells were obtained from the labellar chemoreceptors of the blowfly *Phormia regina* (Hodgson *et al* 1955). These studies are still the most complete analysis of insect taste receptors. The taste hairs generally contain a sugar-sensitive cell, a salt-sensitive cell, a water-sensitive cell, and a mechanoreceptor. However, the first known example of a contact chemoreceptor cell sensitive to a plant token stimulus was the mustard oil glucoside receptor cell of *Pieris brassicae* discovered by Schoonhoven (1967). These glucosides stimulate biting and synergize feeding stimulation with sucrose at a low concentration (Ma 1972). Inhibitory compounds were at first unexpected in host plants but such examples are known. Stadler and Hanson (1978) isolated fractions from tomato leaves which were deterrent to *Manduca sexta* although tomato is a preferred host plant.

The role of taste in host plant recognition has been well studied with lepidopterous larvae. The styloconica receptors of larvae of a number of lepidopterous species have been recognized in terms of their stimulus. Thus, cells which are stimulated by sugars, inositol, amino acids, or deterrents have been identified (Schoonhoven 1987). In several cases of lepidopteran larvae, a relationship between the concentration of the stimulant and receptor activity has been determined with a fair degree of accuracy (Simmonds *et al* 1990). Once the qualitative and quantitative responses of the receptors are known, sensory input can be correlated with behavioral output as manifested by either food preferences or feeding intensity. A detailed survey of such receptor cells sensitive to specific chemical stimuli and their relevant behavioral reaction to insects has been made by Stadler (1992).

Chemosensilla of lepidopterous larvae. Somewhat more detailed information is available on the taste receptors of lepidopterous larvae. In these caterpillars, the receptor complement is low in number but relatively high in specificity. Such receptors are located in a pair of sensilla styloconica on each maxilla. Each sensillum is innervated by five neurons, one of which is considered to be a mechanoreceptor. The caterpillar of *Pieris brassicae* may illustrate the response spectra of gustatory receptor cells (Schoonhoven 1973). In the larvae of *P. brassicae*, the chemoreceptors involved are located on the antennae and mouth parts (Fig. 5.9). The response spectrum of the individual gustatory receptor cells in the maxillary sensilla styloconica of the larvae of *P. brassicae* is given in Table 5.2.

During the close-range search for an appropriate host plant, the larvae can accomplish odor discrimination using their antennal receptors followed by critical perception using the maxillary sensilla styloconica. Each antenna has 16 olfactory cells that perceive plant odor in overlapping stimuli spectra.

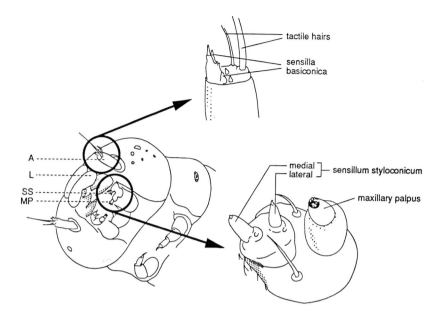

Fig. 5.9. Ventral view of the head of a *Pieris brassicae* larva. The chemosensory sensilla are shown in detail. A, antenna; L, labium; SS, maxillary sensilla styloconica; MP, maxillary palpus. Reproduced from Schoonhoven (1973) with permission from Blackwell Scientific Publications, Oxford, UK.

Table 5.2. The response spectrum of the gustatory receptor cells in the maxillary sensilla styloconica of *Pieris brassicae* larvae.

Maxillary sensilla styloconica	Receptor cells for
Medial sensillum styloconicum	1 Sugars 2 Feeding inhibitors 3 Glucosinolates 4 Salts
Lateral sensillum styloconicum	5 Sugars 6 Glucosinolates 7 Amino acids 8 Anthocyanins
Epipharyngeal sensillum	9 Sugars 10 Feeding inhibitors 11 Salts

The set of maxillary contact chemoreceptors enable the larvae to perceive nutrients such as sugars, salts, and amino acids, as well as secondary plant substances. The feeding-inhibitor-sensitive cell in the medial sensillum styloconicum responds to a variety of alkaloids and steroids.

The gustatory sensilla of the larvae react to glucosinolates of the *Brassica* host plant. Even plant pigments like anthocyanins are perceived as taste by *P. brassicae* larvae. In addition to the two maxillary sensilla styloconica, the larva possesses gustatory sensilla on the maxillary palpi. Eight sensilla basiconicae are located on top of each palpus (Schoonhoven 1973). One pair of epipharyngeal sensilla located in the buccal cavity completes the set of contact chemoreceptors in the larvae of *P. brassicae*; each of these papilla-shaped sensilla contains three sensory cells receptive to sugars, salts, and feeding inhibitors. The taste perception in *P. brassicae* larvae is typical of phytophagous insects, which are able to discriminate a number of compounds including nutrients and specific secondary plant substances (e.g., glucosinolate as a feeding incitant).

Such types of taste cells in insects that are narrowly attuned to unique plant allelochemicals are listed in Table 5.3. However, in the case of *Mamestra brassicae* L. (Lepidoptera: Noctuidae) caterpillars, there are three types of sugar-sensitive cells and inputs from all three correlate with the amount of

Table 5.3. Contact chemosensory cells of phytophagous insects specific for plant allelochemicals.

Insect	Location of sense cell	Plant compound	Reference
Pieris brassicae	Maxillary styloconica	Sinigrin	Ma (1972)
Delia brassicae	Tarsal D	Sinigrin	Stadler (1984)
Chrysolina brunsvicensis	Tarsi	Hypericin	Reese (1969)
Yponomeuta evonymellus	Maxillary styloconica	Prunasin Sorbitol	van Drongelen (1978) van Drongelen and van Loon (1980)
Leptinotarsa decemlineata	Galeal α sensilla	α-Amino-butyric acid L-Alanine	Mitchell (1985)
Spodoptera exempta	Maxillary styloconica	Adenosine	Ma (1977a)
Papilio polyxenes	Tarsi	Sesquiterpenes	Roessingh et al (1991)

sugar-containing diet eaten over a selected concentration range, which means there is an overlap of the sensitivity ranges of individual receptor cells (Blom 1978). In grasshoppers, by contrast, the receptor complement is large in number and low in specificity (Winstanley and Blaney 1978). In fact, the portion of the total sensory system that is required for normal behavioral responses is largely unknown in most phytophagous insects. For detailed information on secondary compounds of different host plants that stimulate or attract insect herbivores, see Stadler (1992, Tables 2–4).

Deterrent chemosensory cells. The description of a bitter or deterrent cell in the maxillary styloconica of the silkworm was first given by Ishikawa (1966) and was subsequently confirmed in many other species (Dethier 1982a, Schoonhoven 1982). Since the 1960s, several plant compounds have been identified that cause a cell to increase firing and also cause decreased consumption of food, confirming the presence of deterrent cells (Chapman and Bernays 1989). In the case of *M. brassicae*, the same cell is sensitive to typical stimulating compounds, as well as to toxic deterrents. Thus, the phytophagous insects choose leaves with a low content of these toxic chemicals.

A special type of deterrent effect was shown by Ma (1977b), in which a sesquiterpene dialdehyde isolated from the warburgia plant blocks the sugar receptor cell of *Spodoptera exempta* for 10–20 minutes after two contacts of 3 minutes each. This inhibitory effect of the sensory neuron has been correlated with a suppression of the feeding response to sucrose.

Sensory coding
In general, the generated spike activity of different sensory cells in response to stimuli is decoded by the central nervous system. To investigate the sensory code of the receptor cells of a sensillum or several sensilla, three types of coding can be distinguished.

1 *Labeled lines* form a system in which the code gives only 'yes' or 'no' outputs in response to receptor cells mediating acceptance (sugar- or host-plant positive stimulus) or rejection (salt, deterrents, or repellents). This code represents an earlier concept derived from the first electrophysiologic studies of the blowfly (Dethier 1974). The characteristics of labeled lines are highly specific receptor cells with narrow-sensitivity spectra and no overlapping with other cells. Classic representatives of such cells are the glucosinolate receptors of insects adapted to the Cruciferae (Stadler 1992).

2 *Temporal patterns* are a type of coding that generate a kind of latency that can be correlated to the deterrency compounds in the olfactory receptor cells of insects. The mechanism of feeding deterrence by glycoalkaloids was elucidated in the Colorado potato beetle (Mitchell and Harrison 1985). Earlier, Dethier and Crnjar (1982) also presented evidence for stimulus-specific time patterns in the response of contact chemoreceptor cells of *Manduca sexta* larvae.

3 *Across-fiber patterns* are a broader and analytical encoding mechanism in the central nervous system. Behavior reactions are usually not bimodal (positive or negative reaction), but represent a kind of graded response. Each of the receptors is sensitive in one way or another to more than one kind of compound. Multiple interactions in taste receptors are particularly striking in caterpillars. Numerous instances of synergism have been recorded. The receptors probably show spectra of sensitivities that overlap but are not congruent. Such complex codes could be achieved by the concept of 'across-fiber patterning.'

The across-fiber patterning mechanism is based on the physiologic principle of Gestalt perception, as postulated some time ago by anethologists (Tinbergen 1951) for the integration of different environmental key stimuli in the central nervous system. Theories and experimental support for across-fiber patterns in insect contact chemoreception have been presented by many workers including Dethier and Crnjar (1982). An important function of across-fiber patterns in lepidopterous larvae is the coding quality. The central nervous system does not merely determine whether the acceptance input dominates the rejection input, or vice versa; it also assesses in a qualitative fashion the sensory impression of a particular plant sap. Across-fiber patterns can assess chemical messages of a complex nature. This conclusion will remain tentative until both the chemistry of natural stimuli influencing insect behavior and the corresponding sensory physiology is better understood.

Sensory variability

The concept of across-fiber patterning is not the only answer to the problems of insect chemoreception. The main difficulty of recording from chemosensory neurons is their variability. The problem of variability in insect chemo-receptors has been discussed by Blaney *et al* (1986). This variability can be due to: (1) the developmental stage, and/or physiologic state of the insect; (2) the age of adult insects; (3) cyclic fluctuations in receptor sensitivity due to changing internal and external factors; and (4) changes in receptor sensitivity related to the feeding history of the insect. For example, the closure of the distal pores seen in palp-tip sensilla of locusts immediately after a meal is correlated with distension of the gut. Furthermore, the influence of hormones, directly or indirectly, on the functioning of insect chemoreceptors has been indicated by a number of studies (Bernays and Chapman 1972). On the basis of the evidence available, it is hypothesized that one or more hormones affect ion-transport processes in the tormogen cells and thus control the ionic composition of the dendrite liquor, which in turn influences receptor sensitivity. Hormones could act directly on the receptors, or indirectly by regulation of dendritic liquor composition or sensillum pore size, or a combination of these mechanisms.

It is evident that the scenario of sensory modulation is still largely *terra incognita*. The chemistry of natural stimuli is not yet fully explored and hence the essential chemical components are unavailable for precise testing of the sensory-coding mechanisms. Furthermore, the existence of strains with different sensitivities for glucosinolates in *Mamestra brassicae* points to genetic determination (Wieczorek 1976). The genetics of chemoreceptive sensitivity as reported by van Drongelen and van Loon (1980) may be a new insight into sensory physiology. These authors crossed two closely related *Yponomeuta* species with different sensitivities for dulcitol, phloridzin, and prunasin. In the crossed insects, the sensitivity to dulcitol and prunasin proved to be dominant, and that for phloridzin had intermediate expression, indicating a simple genetic basis of differences. An analysis of the sensory system involved might elucidate the mechanism underlying the variations in food preference. Electrophysiologic studies on the maxillary responses of lepidopterous larvae to tannins and alkaloids have revealed that more integration occurs in the sensory periphery than was formerly appreciated (Dethier 1982a). The answers to these intricacies may be possible, but it will involve in-depth and multidisciplinary studies on the chemistry of natural stimuli, chemoreception, genetics of behavior, and receptor physiology. Thus, the phenomenon of sensory modulation may open up an interesting new vista on the role of receptors in the control of insect behavior.

Host Selection Processes

The process of host plant selection by phytophagous insects is a lock–key system, in which the lock refers to the host plant and the key stands for the complex sensory pattern of the insect. Only when this pattern corresponds to some innate standard is an appropriate behavioral response triggered (Dethier 1988). The host plant recognition process is not simply a matter of responding to a positive signal (kairomone) or avoiding a negative signal (allomone). The process is the result of an integration of numerous plant chemical and insect internal factors, i.e., a complex multidimensional key and an equally intricate corresponding lock system. Each species of insect has evolved a unique sensory system on the basis of important cues in recognition of its specific range of host plants in its particular biotope.

Sequential behavior

Host selection is a sequence of behavioral events in the plant stimuli–insect response paradigm (Fig. 5.10). Each activity in the sequence brings the insect into a situation in which an appropriate stimulus will lead to the next activity.

Chapter 5

It is a catenary process (Kennedy 1965) involving: dispersal, host finding, host recognition, host acceptance or consuming, and host suitability for food or oviposition (Miller and Strickler 1984). These terms used in the host-selection process are almost similar to categories (search, encounter, pursuit, and handling) used in the optimal foraging theory (Schoener 1971, Stanton 1983).

Such temporal patterns of a behavioral component are associated with an

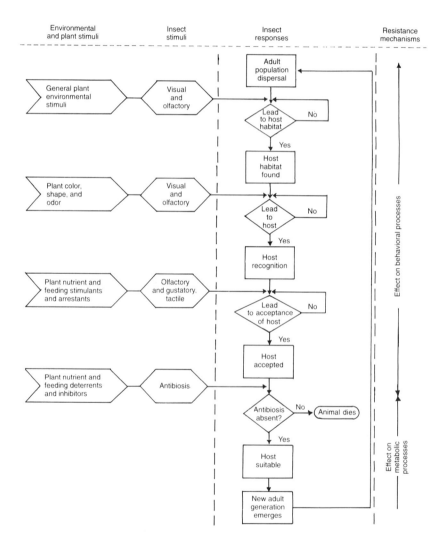

Fig. 5.10. A generalized scheme for the host-selection process in phytophagous insects. Modified from Kogan (1982) with permission from John Wiley and Sons, New York.

internal instinct, such as the urge to feed or oviposit. It may be noted that each chain of behavioral responses is contingent on the perception of stimuli arising as the result of a previous response. Two opposite effects are possible: induction and aversion learning. Induction is the process whereby ingestion of a particular food for 24 hours or longer creates a preference for that type of food over others. Food aversion learning has been demonstrated in polyphagous arctiid larvae. It is sometimes considered to be a kind of associative learning and is defined as an association of an ingested food with postingestional malaise, resulting in subsequent avoidance of ingestion of the untasty food (Dethier 1988).

Hence, any two searching insects may differ not only in the type of response (i.e., taxis versus kinesis; orthokinesis versus klinokinesis; see Note 1, p. 150), but also in the level of response (i.e., velocity of linear movement and rate of turning), or the probability of any response to host plant stimuli. Sometimes individuals of the same species (e.g., bees, butterflies, mosquitoes) exhibit different preferences under identical conditions, and this polymorphism seems to have, at least in part, a genetic basis (Hovanitz 1969).

Furthermore, host plants and herbivorous insects are rarely distributed randomly in a locality or within a geographic region. The nonrandom distribution of plants implies that individual host plants grow in areas of varying floristic texture. In such a type of texture, the plants are aggregated with varying density, dispersion, and species richness. Hence, their distribution in time and space are usually patchy. Secondly, the plant resources utilized by insects vary qualitatively. The strategies or patterns adopted by insects to select their host plant in a heterogeneous environment are addressed in the following sections along with selected case studies of chemical modalities of insect feeding and oviposition.

Dispersal

Before encountering the foraging area, insects pass through the dispersal or migratory phase. During this period, they are primed physiologically to their host plants (Kennedy 1975). Dispersal provides a more homogeneous distribution of a given insect population and scope for invasion into new foraging areas. Dispersal activity is a typical feature of the adult insect, although many neonate larvae disperse from their hatching site and move widely before settling down for feeding on an appropriate substrate. The phenomena of dispersal and search are often difficult to distinguish; they differ in the insect's response to plant stimuli and the insect's internal physiologic drive. For instance, when alate aphids react negatively to green foliage and positively to blue sky, they tend to be oriented to dispersal flight. In contrast, aphids are attracted towards green surfaces during searching behavior. If the plant on which the aphid lands happens to be a nonhost, the aphid promptly switches over to the dispersal and flight behavioral phase.

A similar phenomenon of alternation of dispersal and search behavior is shown by ovipositing butterflies (Stanton 1983).

Host finding

After passing through the dispersal or migratory phase, the phytophagous insects are behaviorally and physiologically tuned towards the next spectrum of foraging behavior, i.e., host finding. The term 'find' has been used (Thorsteinson 1960, Beck 1965) to describe the phenomenon of insects arriving near or on a resource. The process of finding is purposeful search involving behavior that enables the finder to establish and maintain proximity with its desired habitat and consequently locate its appropriate host. Insect habitats are almost invariably heterogeneous in their spatial distribution, hence the foraging insects have to identify initially the habitat, followed by the patch, and then the food (Hassell and Southwood 1978). Patches are spatial subunits of the foraging area (habitat) in which aggregations of food items occur. Patches themselves are not uniformly distributed in a particular habitat, and within them food plants tend to be clustered and represent a third and final level for the insect to overcome in finding its host. To locate host plants, insects utilize a characteristic behavior of trivial movement (dispersal) within the habitat, but adapt migration to move long distances from one habitat to another (Kennedy 1975).

Habitat finding

The first phase of foraging by phytophagous insects is 'habitat finding.' The dispersing adult insect population has to travel a distance towards a general host habitat using mechanisms that include anemotaxis, phototaxis, geotaxis, and probably temperature and humidity preference. These mechanisms have important ecological implications and are of interest in pest management, but they have little effect on plant resistance. Normally most agricultural pests stay within the general area where crops are planted. This phase is, therefore, less important in host selection.

Patch finding

Finding food or an ovipositional site by phytophagous insects probably involves electromagnetic and chemical cues. Both color and form are important; for example, several species of aphids are attracted to yellowish-whitish surfaces (Moericke 1969), and color with high light remission attracts numerous other phytophagous insects (Rausher 1983). Shape may also influence insect orientation to their host plants (Prokopy 1972a).

Host plant finding

Plants have chemical characteristics that are specific for certain taxa (Harborne and Turner 1984). Chemicals appear to play the major role in the orientation of many insects. The botanical specificity of insects restricted to crucifers has long been linked with the presence in host plants of a class of involatile sulfur compounds, the glucosinolates. These compounds induce feeding and/or oviposition by the specialized insects including *Pieris rapae*, *Delia radicum*, *Delia floralis*, and *Psylliodes chrysocephala* (Koritsas *et al* 1991, Traynier and Truscott 1991, Birch *et al* 1992, Roessingh *et al* 1992). Homopterans such as aphids, leafhoppers, and whiteflies are known to respond to olfactory and visual stimuli when at close range or touching the host plants (Nottingham *et al* 1991). Most published information emphasizes the attractiveness of plant volatiles to herbivores (Metcalf 1986b).

The odor-conditioned optomotor anemotactic system appears to be the most powerful mechanism of guidance towards a plant odor source and, on the basis of present evidence, it is probably the sole mechanism within a distance of several meters unless the source is conspicuous enough for a visual response. This type of odor-conditioned positive anemotaxis towards solanaceous host plant odor has been reported for the Colorado potato beetle *Leptinotarsa decemlineata* (Visser and Thiery 1985). In contrast, nonsolanaceous plants are neutral or elicit negative movement. The second type of odor-oriented mechanism involves chemotaxis – klinotaxis and tropotaxis (Kennedy 1977; see Note 2, p. 150). Chemotactic responses occur in the steep concentration gradients very close to local odor sources and along chemical traits on the ground, but there is still no convincing evidence for distant chemotactic guidance for flying insects. In flying insects, however, klinokinetic responses are probably important when close to odor sources in calm air. For many walking insects, plant odors at short range can evoke orthokinesis and klinokinesis as well as klinotaxis and tropotaxis (Harris and Miller 1982).

Olfactory and visual stimuli

The interplay of olfactory and visual stimuli during an insect's search for plant resources has been studied in the case of the apple maggot fly (AMF) (Prokopy *et al* 1987). These workers observed that the gravid AMF tends to move upwind under moderate wind speed conditions, irrespective of the response to the presence or absence of host fruit odor in the area. Later on, in the presence of host fruit odor, an AMF appears to move among trees at a faster rate. At a closer range, both host fruit odor and tree visual stimuli appear to stimulate a greater degree of positive response from an AMF. After arrival on a host tree, an AMF locates individual fruit within the canopy, primarily on the basis of the fruit's visual characteristics (shape). Thus, host finding in apple flies is apparently mediated by a combination of olfactory and visual responses. Similarly, yellow discs or crosses at ground level baited with allylisothiocyanate, effectively attracted the female cabbage root fly *Delia radicum* (Tuttle *et al* 1988).

Variability in searching

Searching at the right time is as important as searching in the right place, because certain resources are available periodically rather than continuously. In addition, searching can be based on stimuli emanating from resources or the searching rhythm and physiologic state of the insect. Thus, an insect usually will not search for food when it is satiated or during its normal inactive period, even if it perceives resource-specific cues (Southwood 1978a). Searching rhythms (circadian, lunar, or circannual) seem to have evolved either in response to competition for similar food sources or to predation. The physiologic state associated with the searching behavior of the blowfly has revealed that neural and endocrine circuitry affects the blowfly's initiation and cessation of search and feeding behavior (Dethier 1982b).

External environmental factors can alter searching behavior directly by means of acting upon environmentally sensitive physiologic processes. An insect can perceive changes in the environmental factors and, on this basis, may alter its behavior. The searching behavior can be affected by abiotic factors such as temperature, humidity, and solar radiation. For example, gravid females of *Pieris rapae* refrain from ovipositing during overcast weather (Gossard and Jones 1977), but on the first fair day they search for host plants and start egg laying.

Biotic factors affecting search patterns include resource distribution and availability, resource quality, as well as the influence of other individuals within their biosphere. The number of resources an insect finds is generally proportional to the number available in time and space. Resource texture is an important component affecting the host plant searching behavior of phytophagous insects. Insects can respond in various ways to resource texture, depending on their search strategy and their perceptual mechanisms. For example, a specialist herbivore used to host-specific cues, might be confused or even repelled by a nearby nonhost plant species in a mixed cropping situation (Stanton 1983); whereas a polyphagous herbivore in a general cropping habitat may perceive diverse plant mixtures without any adverse effect on its host-selection process.

Host recognition

During the host-recognition (examination) process, insects are able to come close to an odor source, where the concentration gradient becomes much steeper and in principle allows chemotactic orientation (Kennedy 1977). The sensing system at this stage is maximally engaged so critical decisions can be made about utilizing potential host plants for sustained feeding or oviposition.

At close range, the insect may obtain additional olfactory information with respect to differences between host and nonhost, as well as between individual host plants. An important component of the recognizing phase is *scanning*. It is a set of mechanisms by which insects engage their sensory apparatus to capture stimuli from multiple points in space. Scanning is vital, because it combines sensory inputs and locomotion; it is involved in predator avoidance or vigilance and in locating resources (Bell 1990). Tobacco hornworm moths use such mechanisms to land on a tobacco plant and almost always avoid landing on a nonhost. Similarly, some individual plants in a cabbage field with a higher content of volatile allyl nitriles are more attractive for oviposition of *Pieris brassicae* than plants with lesser amounts of the volatiles (Mitchell 1977).

Once the insect has made physical contact with the plant, contact chemoreceptors located on the antennae, mouth parts, tarsi, or ovipositor perceive stimuli from the plant surface. Specific movements by the insect may intensify the chemical stimulation. Surface testing by touching, biting, probing, or piercing with sensory organs can assess the host's degree of fitness as a substrate for feeding or oviposition. This phase of host recognition is critical for both the insect and the plant because the latter does not suffer damage until the surface is penetrated; on the other hand, insects investigate carefully before utilizing the plant as a host. Chapman and Bernays (1989) have reviewed different behaviors displayed by insects on or near the plant surface. These include the fluttering of butterflies close to the plant before landing, and palpating with labial and maxillary palpi by larval and adult beetles and moths. Several grasshoppers (Chapman *et al* 1988) and the beetle *Leptinotarsa decemlineata* (Harrison 1987) explore the surface with their palpi at successive intervals. Following palpation, grasshoppers sometimes make a test bite without removing tissues, whereas the beetle compresses the leaf with its mandibles. Palpi and antennae often bear olfactory sensilla and contact chemoreceptors (Frazier 1992). Thus, many different compounds could be perceived with the same organs. Adult butterflies have long been known to drum the leaf surface of nonhosts and hosts before ovipositing. On the ventral side of the prothoracic tarsi of *Papilio polyxenes*, contact chemo-receptor sensilla have been identified and these are sensitive to compounds assumed to be present on the leaf surface (Roessingh *et al* 1991).

A new role of leaf cuticular compounds in insect behavior has been observed (Espelie *et al* 1991). The cuticular lipids of different insects were shown to be dependent upon the composition of the cuticle of their respective host plants. The similarities between the components of the insect cuticle and the insect's host plant could influence the behavior of the insect herbivores as well as that of its parasitoids. Both may act as stimulants or deterrents.

Host acceptance

When the chain of host-selection activities is not interrupted, the insects remain on the host plants for feeding and oviposition. In the host-acceptance process, insect-feeding behavior undergoes three phases: initiation, continuation, and cessation of feeding at the point of satiation (Beck 1965). During feeding, the chemical composition of food is continuously monitored. After the initiation of feeding, the continuation of feeding depends on the olfactory and gustatory feeding stimulants, the intensity of repellents and deterrents, the metabolic state of the insect, and learning acquired as a result of previous feeding experience. On a given resource, the quantity of food ingested is not merely a function of the stimulus/response characteristics of the relevant sense organs nor of their rates of adaptation, but pre- and postingestive factors play an equally important role (Dethier 1982b). The continuation of feeding based on any stimulatory compound also depends on the other compounds present in the natural food mixture, i.e., alkaloids with sucrose and many others. Thus feeding on a particular food plant is a function of the ratio of positive to negative factors and the interaction of excitory and internal inhibitory inputs.

Feeding
An important consideration in host selection is that the degree of association with host plants depends on the life-cycle patterns of insects, which can broadly be categorized as: orthopteran, holometabolous, and homopteran. Each lifestyle may be a specific response to a specific set of selective forces. For example, most of the Orthoptera are active and free-ranging forms. Each time these insects feed, they make individual choices among the available hosts. In holometabolous herbivores (Lepidoptera, Coleoptera, Diptera, and Hymenoptera), the larval food is decided by the ovipositing female; consequently, oviposition is based on cues that correlate with the prospects of larval survival. The oviposition site is a critical factor for the young larvae seeking out hosts because the slow-moving larvae find it difficult to reach the appropriate host plant. The neonate larvae usually suffer very high mortality if forced to leave their original host (Schoonhoven 1973).

Among the homopteran sucking insects (Thysanoptera, Homoptera), the females feed in the same way and often on the same hosts as their nymphs. Hence, the cues that identify nutritional quality of food for the mother can also stimulate oviposition. Some homopteran nymphs are active and feed on many host individuals, but young females of other homopterans settle at one site, lay eggs, and have their brood develop into adults all at the same site (Hinton 1977b). In such species the fecundity of the mother and her brood are affected by the same resource at one site. This means that there is a great need for females to settle at sites of high nutritional quality. Thus, the feeding and oviposition behavior of these three groups of insects are different from each other.

Orthopteran group. In general, grasshoppers and locusts are oriented towards the feeding site by odor activation, but odors do not appear to be used in host recognition. However, responses to lines, shadows, shape, and position of leaves, time of day, and sunlight determine the situations under which a grasshopper orients toward the habitat (Bernays and Chapman 1970). Within its habitat, a hungry grasshopper is an indiscriminate biter and can bite any suitably moist leaf-like object. Grasshoppers will readily bite a very broad spectrum of plants. Rejection *after* tasting is attributable to the tastes of alkaloids and monoterpenoids (Bernays and Chapman 1977). After small bites, materials are taken into the buccal cavity and thoroughly chewed. If the food possesses phagostimulants and is devoid of feeding deterrents, then the swallowing and feeding bout continues. The meal size is determined by the emptiness of the foregut, but is modulated by chemical inhibitors (Bernays and Chapman 1977).

Holometabolous group. The most extensive analysis of the gustatory senses in phytophagous insects has been done in lepidopterous larvae (Hanson and Dethier 1973, Schoonhoven 1973). It has been shown that the larvae of Lepidoptera have a battery of sensilla that may have as much discriminating capacity as those of adults (Dethier 1982a). In the larvae of the cabbage moth *Pieris brassicae*, allylglucosinolate (sinigrin) serves as an incitant and promotes biting activity (Ma and Schoonhoven 1973). However, the response of insects associated with the Cruciferae to glucosinolates seems to be relatively less uniform than believed earlier. Nielsen *et al* (1979a) showed that flavonol glycosides, together with the glucosinolates, stimulate feeding in the flea beetle *Phyllotreta armoraciae*.

Feeding studies using electrophysiologic techniques have been reported in the Colorado potato beetle. This beetle is categorized as oligophagous, feeding on a range of hosts that includes several solanaceous crops, notably potato. The larvae bite substrates that emit odors, until the epidermis is penetrated and they gain direct contact with leaf sap (Visser 1979). Continuation of feeding occurs in the presence of one or more stimulative nutrient chemicals including sugars, amino acids, phospholipids, lecithin, and several other compounds common to most plants (Hsiao and Fraenkel 1968b). To a large extent, the feeding specificity of the beetle is determined by the presence or absence of several alkaloids, notably leptine and α-tomatine occurring in tomato. The alkaloid α-tomatine has a marked inhibitory effect on chemosensory responses to amino acids and to sucrose (Mitchell 1987). This suggests that amino acids and sucrose stimulate the same cell in the α-sensillum of the beetle, which might indirectly inhibit feeding on plants rich in tomatine or other glycoalkaloids (Barbour and Kennedy 1991).

Homopteran group. Aphids are highly specific and have a narrow host range, but exceptions to this rule are not uncommon, as in the case of *Myzus* *persicae* (Eastop 1973a). Aphids as a group choose their food sites using the presence of sapid nutrients rather than any secondary chemical cues. Thus the annual cycles of aphids (e.g., *Aphis craccivora*) are synchronized with the nutritional suitability of their hosts in several ways. Furthermore, aphids can tolerate several of the secondary chemicals (Schoonhoven and Derksen-Koppers 1976).

The probing behavior of sucking insects can now be studied using the direct current electrical penetration graph technique (Tjallingii 1990). This technique is especially suitable for studying possible resistance mechanisms because it provides information about both the behavior of the aphid and the stylet-tip positions in the plant during penetration. Different feeding activities and the location of the stylet tip produce specific voltage changes (patterns), which can be correlated with the probing behavior sequence. The importance of different probing patterns relating to specific stylet movements and locations within the root have been demonstrated in varying resistance levels in lettuce against the lettuce root aphid *Pemphigus bursarius* (Cole *et al* 1993). On susceptible varieties, aphids performed recognizable feeding, but probed rarely on resistant varieties.

Feeding specialization.   On the basis of food preference, herbivorous insects are categorized as monophagous, oligophagous, or polyphagous. All these terms have shortcomings as they represent arbitrary divisions within a continuum – food-selection behavior is subject to change based on the insect's previous association with food plants (habituation, induction preference, and food aversion) (Jermy 1987). There are also differences in feeding specificities of the larval and adult stages of the insect. For example, the northern corn rootworm was considered an example of extreme monophagy, but this is only applicable to the larvae as the adults are polyphagous. There is also a phenological dimension involved in host plant specificity. Different insect species utilize a given plant, or plant tissues, at different times of the year or at different developmental stages of the plant (Feeny 1970).

There is now evidence to show the occurrence of monophagy even in polyphagous insects. This kind of monophagy is recognized as 'ecological monophagy' and 'coevolved monophagy' (Gilbert 1979). The former type involves geographic restriction, i.e, it results when only one plant species is available within the geographic range of the insect. Coevolved monophagy is exemplified by insects that use only one species as a host plant, although closely related, congeneric plants are readily available. This type of specialized feeding habit may be very widespread (Fox and Morrow 1981). Nonetheless, Bernays and Graham (1988) argued that local host plant abundance alone is more of a prerequisite for specialization than a cause of it. They further argued that monophagy is more advantageous and it is more common among herbivores. Whether it is more advantageous to be monophagous or polyphagous may be no more fruitful a question than whether it is more advantageous to be a grass feeder or a dicot feeder, or a leaf miner or a leaf galler (Barbosa 1988b). Alternatively, the distinction between polyphagous and monophagous species (on the basis of plant taxonomy and insect feeding) may be artificial. In many situations, although taxonomic distances are great, few other differences of importance to the herbivore exist. Thus, larvae of the cabbage butterfly feed on plants in two separate orders, but only those containing mustard oil glucosides (Slansky 1976).

Sometimes the most suitable host plant is not found. Under such circumstances, the insects adapt a strategy of utilizing less suitable but more findable host plants. Such feeding strategies are called 'optimal foraging' (Pyke *et al* 1977). Dietary self-selection by insects (see Chapter 4), a newly discovered dimension of insect-feeding behavior, is central to the understanding of how insects interact with their host plants (Waldbauer and Friedman 1991). However, the behavioral and physiologic bases of self-selection in insect feeding are not well understood.

Oviposition

Finding a host plant of appropriate quality for the neonate larvae of phytophagous insects greatly depends on the gravid female's ability to be a 'phytochemist' (van Emden 1972) in selecting a suitable plant substrate for oviposition. Results of Singer *et al* (1988) suggest that oviposition preference and larval performance may be correlated within populations and may vary among individuals such that females prefer the plant species on which their larvae should have the greatest chance of surviving during their first 10 days of growth. Whether this correlation results from a genetic correlation

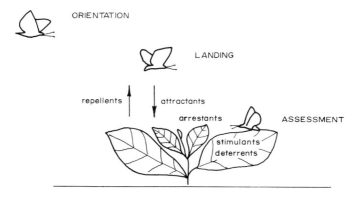

Fig. 5.11. Behavioral events and chemical cues during the process of oviposition. Reproduced from Renwick (1989) with permission from Birkhauster Verlag, Basel, Switzerland.

between traits, pleiotropic effects, or maternal effects has not yet been determined.

The oviposition behavior of several insects has been analyzed in some detail, especially in the case of lepidopterous insects (Feeny *et al* 1983, Ramaswamy 1988, Thompson and Pellmyr 1991). While searching for plants on which to lay their eggs, female insects undergo a sequence of behavior, which falls into three phases: orientation, landing and assessment (Renwick 1989) (Fig. 5.11). During the first two phases of oviposition behavior, the gravid females orient towards an appropriate habitat and search within it for potentially suitable larval food plants by means of vision and odor stimuli. Vision appears to play the dominant role during the first two phases of oviposition behavior, both for assessing gross habitat variables such as light intensity and for guiding the insect towards plants of a particular color or shape (Papaj and Rausher 1983). Plant odor at a closer range, whether specific or mixed, is equally important to orientation and landing on the substrate by female insects. The assessment of a suitable host plant site for oviposition is not uniquely based on some key stimuli, but rather on a large variety of stimulatory and inhibitory plant chemicals acting together (Schoni *et al* 1987). Other factors, such as the physiologic state of the gravid insect, abundance and distribution of suitable plants, environmental conditions, previous experience, or the presence of conspecific eggs, can alter the balance either in favor of or against acceptance of a particular oviposition site (Miller and Strickler 1984).

Olfaction assessment. The oviposition behavior of phytophagous insects involving sensory factors varies widely between species, especially between specialist and generalist insects. Though a large number of volatile compounds emanating from plants seem to play a major role in the orientation and landing of insects on their hosts, few examples of specific chemicals are involved in attracting gravid females to their oviposition sites.

A substantial body of data has been acquired about the oviposition preference of the black swallowtail butterfly *Papilio polyxenes* for the umbelliferous plant, carrot *Daucus carota*. Two secondary plant compounds, luteolin 7-O-(6'-O-malonyl)-β-D-glucopyranoside and *trans*-chlorogenic acid, known to be present in the carrot host plant act as oviposition stimulants to *P. polyxenes* (Feeny *et al* 1988). Roessingh *et al* (1991) have further confirmed the neural responses of the tarsal sensilla of *P. polyxenes* to both of these compounds.

Another well-known compound, glucosinolate, found in the Cruciferae has provided the basis for many investigations on the oviposition and feeding behavior of crucifer-feeding insects. The hydrolyzed products of glucosinolates (allylisothiocyanate) stimulate insects such as the cabbage root fly *Delia radicum* (Roessingh *et al* 1992), flea beetles *Psylliodes chrysocephala* (Koritsas *et al* 1991), the diamondback moth *Plutella xylostella*, and the white butterfly

Pieris brassicae (Ma and Schoonhoven 1973).

Glucosinolates are doubtless important oviposition stimuli for many crucifer-feeding insects (Chew 1988, Louda and Mole 1991, Stadler 1992). However, researchers have warned against generalizations which are too sweeping, and concluded that glucosinolates and their hydrolyzed products are only in some cases responsible for host plant specificity. The results with *Delia radicum* (Roessingh *et al* 1992) indicate that the glucosinolates can be detected at extremely low concentrations, as indicated by the oviposition and electrophysiologic thresholds. The actual stimulatory power in terms of increased oviposition is not very strong, except for the indole glucosinolate glucobrassicin. Since the most stimulatory fraction of the surface extract contained no glucosinolates, it was concluded that other compounds, in addition to glucosinolates, played an important role in the stimulation of oviposition. The results of other experiments also confirmed that the indole glucosinolate glucobrassicin is an important chemical characteristic of the *Brassica*–insect relationship (Renwick *et al* 1992).

Contact assessment. After alighting on plants, the gravid insect assesses the chemical stimuli on the plant surface by means of various contact

Fig. 5.12. Oviposition following contact assessment by tarsal drumming by the cabbage butterfly *Pieris brassicae*. Reproduced from Renwick (1989) with permission from Birkhauster Verlag, Basel, Switzerland.

chemoreceptors. In the case of female butterflies, such as *Pieris brassicae*, it has been observed that the insects move their forelegs vigorously and exhibit a drumming reaction before ovipositing. Most of the time this behavior occurs immediately after landing but sometimes continues during abdomen curving and touching (Fig. 5.12), which are the next two steps in the behavioral chain (Klijnstra 1982) (Table 5.4). In this process, the terminal tarsal segments of the first pair of legs are brought sharply into contact with the leaf and scrape along the surface at the end of each stroke. Diverse explanations have been advanced for the selection of the proper oviposition site on the plants. Ma and Schoonhoven (1973) found that the foretarsi of the female butterflies are richly endowed with chemoreceptors (trichoid sensilla) that are sensitive to nonvolatile chemicals of the plant saps brought out during the drumming. Combined electrophysiologic and behavioral experiments conducted by these workers on the role of this phenomenon have suggested that this kind of behavior is a way of sampling the physical and chemical properties of the

Table 5.4. Qualitative description (ethogram) of the egg-laying behavior of *Pieris brassicae* females in the laboratory (modified from Klijnstra 1982).

Steps in oviposition process	Oviposition behavior of the gravid female moth	Senses involved
Approach	Female flies in rather straight line to the leaf, sometimes interrupted by 'turning', i.e., a sudden change in flight direction (away from the leaf)	Vision Olfaction
Landing	The first contact with the leaf by the tarsi, often immediately followed by drumming	Tarsal taste hairs (olfaction)
Drumming	Alternate tapping movements on upper leaf surface with fore tarsi	Tarsal taste hairs (olfaction)
Curving	Female bends her abdomen around the edge of the leaf without touching lower surface	Taste hairs on tarsi and ovipositor
Touching	Touching lower leaf surface with the ovipositor	Taste hairs on tarsi and ovipositor, mechanoreceptors on ovipositor
Oviposition	Deposition of the eggs, one by one in a batch	Mechanoreceptors and taste hairs on ovipositor

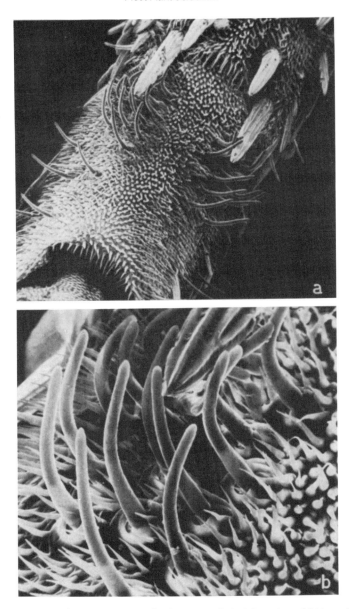

Fig. 5.13. Scanning electron micrograph of a tarsus of *Pieris brassicae*. (a) Ventral side of the fifth tarsomere and the lower part of the fourth tarsomere of a foreleg, showing the distribution of the B-type hairs on the fifth and some large spines on the fourth tarsomere (×331). (b) A closer view of a group of B-type hair sensilla amidst the field of microtrichia. The tip of a large spine protrudes at the top of the electromicrograph (×1102). Reproduced from Ma and Schoonhoven (1973) with permission from Kluwer Academic Publishers, Dordrecht, The Netherlands.

plant. This was supported by the finding that the tarsal chemoreceptors involving the oviposition behavior of *P. brassicae* have one type of trichoid sensilla (designated as 'B-type' hairs), which are larger in the female than in the male. The B-type hairs are located on the fifth tarsomere of the forelegs. Scanning and transmission electron microscopical studies of the fine structure of these B-type hairs have shown that each hair is associated with four contact chemoreceptor cells as well as one mechanoreceptor cell (Ma and Schoonhoven 1973) (Fig. 5.13). At least one of the contact chemoreceptor cells responds to the application of solutions of mustard oil glucosides at the sensillum tip. This suggests a dual function of the B-type hairs: the discrimination of salt solutions from water and the detection of the chemical inducing oviposition. Even among pierids, there is reason to doubt that glucosinolates are the whole answer. Renwick and Radke (1983) have found that several fractions of extracts from cruciferous plants, including a relatively nonpolar fraction, have a stimulating effect on the contact–oviposition response by *P. rapae*.

Orientation of the cabbage root fly to its host plants, wild and cultivated crucifers, has been thoroughly studied by several authors (see review by Nottingham 1988). The oviposition behavior of the cabbage fly was earlier studied by Zohren (1968). The gravid females orient anemotactically to the odor of *Brassica* or allylisothiocyanate in a wind tunnel and in the field. But vision and olfaction are less important for host recognition than contact chemoreception. After landing on the leaf surface, chemo- and mechanoreception are used for the detection of oviposition stimulants and selection of the oviposition site. The insects have been observed to perform an orientation run, such as a leaf-surface run. Subsequently, the insects perform oviposition runs over the plant stem and move down to the base by moving sidewise and around the stem. Females then check the soil near the stem base and perform short walks (Stadler 1978). Thereafter, they return to the stem and climb; then they climb down the stem and again probe the soil before eggs are laid. During these runs or walks, the females are likely to perceive effective olfactory stimuli via the antennae, gustatory stimulation via the taste receptors, and tactile stimuli of the soil texture via the mechanoreceptors.

Based on Zohren's (1968) stepwise observation on the fly's ovipositional process, Stadler and Schoni (1990), developed a scheme for observing behavior patterns (Fig. 5.14): A, short visits (1–30 seconds) with an exploration of the leaf surface; B, resting, usually interrupted by cleaning of the head, tarsi, and wings; C, running over the leaf surface with repeated proboscis contact (exploration); from this pattern the flies return to pattern B. After some time, the exploration activity begins again, i.e., the flies return to pattern C or fly away. Some of the stimulated flies underwent the whole sequence of preoviposition (patterns B–C), oviposition, and postoviposition behavior (patterns D–H). This sequence could be interrupted by departure at any of the steps described. In fact, the flies could repeat some steps of this

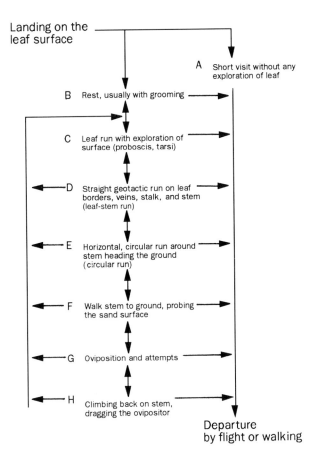

Landing on the leaf surface

A Short visit without any exploration of leaf

B Rest, usually with grooming

C Leaf run with exploration of surface (proboscis, tarsi)

D Straight geotactic run on leaf borders, veins, stalk, and stem (leaf-stem run)

E Horizontal, circular run around stem heading the ground (circular run)

F Walk stem to ground, probing the sand surface

G Oviposition and attempts

H Climbing back on stem, dragging the ovipositor

Departure by flight or walking

Fig. 5.14. Ethogram of the oviposition behavior of the cabbage rootfly *Delia radicum* (L.). Redrawn from Stadler and Schoni (1990) with permission from Plenum Publishing Co., New York.

sequence. The present results of Stadler and Schoni (1990) are in full agreement with those of Zohren (1968). The contact of females with the leaf surface and the characteristic leaf and stem runs are essential for the final host plant recognition and oviposition.

The investigations of Stadler (1978) on the chemosensory basis of oviposition by the female *Delia radicum* revealed that a receptor cell sensitive to glucosinolate and sinigrin has been identified in the tarsal sensory hairs designated as types D, B, and A (Fig. 5.15). The D sensilla on segments 3 and 4 of the tarsus of *D. radicum* females were shown to contain a sensitive receptor cell for glucosinolates. In contrast, the receptor cells of the D sensilla

Fig. 5.15. Scanning electron micrograph of tip of the prothoracic tarsus of a female *Delia radicum*, showing the distribution of chemoreceptive hairs on the prothoracic tarsus. The receptor hairs sensitive to glucosinolate are marked D and A. These are different from the hairs marked B which are sensitive to water. Reproduced from Stadler (1978), with permission from Kluwer Academic Publishers, Dordrecht, The Netherlands.

of the other segments did not respond in a dose-dependent way to these compounds (Roessingh *et al* 1992). The glucosinolate receptors were found to be especially sensitive to glucobrassicin, gluconasturtiin, and glucobrassica-napin with thresholds of about 10^{-8} M to 10^{-9} M. Large differences were observed among different glucosinolates. Hence, optimal release of oviposition stimuli depends upon glucosinolate patterns and other compounds.

Spacing assessment. The concept of host plant chemicals affecting the spacing of phytophagous insects was introduced by Cirio (1971). Female olive fruit flies *Dacus oleae* (Tryptidae) spread olive juice over the fruit surface after egg laying, which provides a signal to deter repeated oviposition on the same fruit. This type of chemical spacing mechanism is also observed in the oviposition behavior of the cabbage looper *Trichoplusia ni* (Hubner). In this case the feeding larvae can deter oviposition by gravid females because the larval frass contains some biologically active material (Renwick and Radke 1980). Even a simple disruption of the plant tissue by feeding is sufficient to release the oviposition deterrent against *T. ni*. Although it is difficult to elucidate the chemistry of oviposition deterrence, a Swiss group has identified the marking pheromone of the European cherry fruit fly *Rhagoletis cerasi* (Hurter *et al* 1987).

Chemical cues and contact chemoreception are responsible for signalling the presence of conspecific eggs or larvae. Convincing evidence has been obtained from several insect pheromone substances and the respective behavioral features leading to the avoidance of oviposition or feeding. The ovipositing females of the apple maggot fly *Rhagoletis pomenella* (Tephritidae) smear a pheromonal substance on the surface of the fruit, which deters further oviposition by apple maggot flies (Prokopy 1972b).

In some cases, oviposition may be enhanced by previous larval infestations. Some of the chemical exudations from maize plants infested by *Chilo partellus* moths are likely to provide a signal for oviposition by *C. partellus* (Kumar 1986). Larval frass of the naval orangeworm *Amyelois transitella* (Pyralidae) also appears to stimulate oviposition by the adults of this insect on green almond fruits (Curtis and Clark 1979).

In a few cases, interactions between plants and pathogens result in the release of chemicals that affect ovipositing insects. For example, onion flies are highly attracted to onion roots infested with bacteria. The microbially produced synergists of the alkyl sulfides have been identified as ethyl acetate and tetramethyl pyrazine (Ikeshoji *et al* 1980).

Oviposition variability

Variability in the oviposition preference pattern seems to be normal (Fox and Morrow 1981). The variation may be due to genetic or nongenetic causes. If it is under genetic control, then the preference pattern tends to remain consistent within an individual or population, as evidenced in several insect species (Tabashnik *et al* 1981, Wilkund 1981). But nongenetic factors, which include the environment, the physiologic state of the insect or plant, phenology, competition among insects, and previous experience, are likely to modify the ovipositional preference. In fact interactions of the genotype and environment are more prevalent in determining the host plant preference of insects. A detailed treatment of this is given in Chapter 7.

Variability in oviposition preferences encompasses at least two dimensions: (1) differences in the hierarchical order of various plants, and (2) differences in the thresholds of acceptance of various plants (Wilkund 1981). It has been suggested that the hierarchical arrangement of oviposition preference by oligophagous insects may have a dual function, i.e., to ensure that most eggs are laid on optimal host plants as long as they are available, and oviposition only occurs on suboptimal host plants when the optimal plants are not available. Specialist females may restrict their oviposition to most preferred plants because the acceptance threshold for less preferred plants is high, whereas generalist females can oviposit on a number of plants because the difference in acceptance thresholds is relatively small. On the basis of these hypotheses, the results of the ovipositional choice of *Papilio machaon* L. showed that some females adapted a generalist oviposition choice strategy and laid eggs on several plants even when optimal host plants were available; whereas others exhibited a specialist strategy and restricted oviposition to the optimal host plant as long as it was available (Wilkund 1981).

In general, however, the probability of oviposition on novel or less preferred hosts is likely to be increased by changes in ecological conditions such as a decrease in abundance and distribution of the preferred host species. The oviposition behavior of the butterfly *Euphydryas editha* provides an excellent example of a graded response. An individual *E. editha* will become increasingly polyphagous with a continuous decrease in the acceptance threshold, and the degree of polyphagy is directly correlated to the rate at which alternate host plants become more acceptable. Ultimately, some of the adults are likely to inherit genetically based behavioral predisposition to choose the new food plant for oviposition (Singer 1982). In another situation, oviposition on a plant species outside the normal range of acceptable hosts (i.e., ovipositional mistake) provides an opportunity for probing into the critical stimuli that determine how gravid females choose oviposition sites among plants (Singer 1984).

Host suitability

The ultimate goal of the entire process of host plant selection comprising host finding, recognition, and acceptance is to reach the point of utilizing the food plant on which fitness or fecundity of the insects is maximized (Mitchell 1981). The behavioral aspects of host finding through to acceptance are all primary and sensory events. Host acceptance depends on the palatability of the food, which is a function of the ratio of positive to negative sensory factors. Once the food plant is accepted by the insect, the plant is considered suitable.

Interestingly, there are cases of larval growth rates that do not closely follow the norms of host plant acceptance. Smiley (1978) pointed out that ecological factors such as predation or plant abundance, rather than differences in host palatability, can determine host plant suitability for phytophagous insects.

Secondly, host suitability is very much influenced by such plant characteristics as nutritional quality, presence or absence of toxins or digestibility-reducing compounds, and leaf water–nitrogen composition (Scriber 1984a).

Table 5.5. Physiologic efficiency of host plant use by phytophagous insects.

Holometabolous insects (lepidopteran)	Hemimetabolous insects (orthopteran)
Ontogenetic changes in feeding habits, larval and adult food and feeding habits are distinct and different	Polyphagous orthopterans tend to feed on the same food and in the same manner at all life stages
Growth ratios for holometabolous species average 1.52	Growth ratios for hemimetabolous species average 1.27
Increase in body size is 45.2 (from one molt to next) because molting efficiency is greater	Increase in body size is 18.6 because molting efficiency is less
Feeding efficiency and efficiency of conversion of digested food to body mass are higher	Feeding efficiency and efficiency of conversion of digested food to body mass are much lower
Gut permeability enhances nutritional efficiency	Gut impermeability due to peritrophic membrane reduces nutritional efficiency
Cuticle comprises < 4% of total immature body dry weight	Cuticle comprises > 40% of total immature body dry weight
Sensitive to antifeedants	Less sensitive to antifeedants
Higher exposure to allelochemicals	Less exposure to allelochemicals
Inducibility of cytochrome P-450 monooxygenase detoxication enzymes	Lower detoxication enzyme activity

Some plants have compounds that reduce digestibility after being ingested. Various interactions between allelochemicals and nutrients may affect the plant's suitability as insect food as well as its 'bioavailability'. Many examples of interactions that interfere with bioavailability exist with insects (Reese 1979). These aspects and the insect adaptations to such allelochemicals are considered in Chapter 4.

Scriber (1984a) has developed a physiologic efficiency model of insect response to plant chemistry on the basis of the water–nitrogen index. This larval performance model, along with seasonal trends in various plant growth forms, provides an empirical framework for interpreting any physiologic advantage (efficiency of growth) of insect-feeding specialization.

Thirdly, the physiologic capabilities of holometabolous and hemimetabolous insects that influence host suitability for the insect in question have been reviewed by Berenbaum and Isman (1989). A comparative physiologic efficiency in herbivory by these two groups of insects is presented in Table 5.5.

The scenario of the host plant selection process is highly complex and involves a subtle interplay of orientation factors, feeding factors, ovipositional factors, plant nutrients, and allelochemical factors along with environmental factors. Research efforts are directed towards providing a clearer understanding of the dynamic equilibrium between insects and their host plants. The necessary technologies have been developed in the 1980s and 1990s to isolate and identify complex biomolecules and analyze the relevant biological systems. Such holistic investigations however, are likely to focus on how insect pests of economic crop plants can be managed.

Notes

1. Taxes are directed movements to determine the direction of the host plant, and require more complex sensory organs than kineses. Kineses are undirected movements, in which the sensory organs discriminate different intensities of stimulation. In orthokinesis the average speed of insect movement is related to the intensity of stimulation. A more complex form of kinesis is klinokinesis, in which the insect moves in a straight line in a favorable environment and turns aside when in an unfavorable environment.
2. In klinotaxis, the insect compares the intensity of stimulation by alternate movements to the left and right sides. In tropotaxis, the insect compares the intensity of stimulation on both sides simultaneously, by means of symmetrically placed sensory organs.

6

MECHANISMS OF RESISTANCE

The mechanisms of resistance need to be understood before the degree of resistance among plants can be ascertained. Resistance may vary between two extremes – immunity and high susceptibility. An immune plant is a nonhost for herbivores and they are usually outside the host range of insect herbivores. Any degree of host reaction less than full immunity is called resistance. The immunity of nonhost plants is heritable and is due to either the presence of genes that interfere with the herbivore's ability to recognize and utilize the nonhost, or the absence of genes necessary for a herbivore to recognize and utilize the nonhost plants. The degree of host plant resistance relates to an extent, to herbivore damage, but much less so than the definition of a susceptible plant. Definitions of host plant resistance have been put forward by several workers (Beck 1965, Harris 1979, de Ponti 1983). However, the most widely accepted is that by Painter (1951): 'the relative amount of heritable qualities possessed by the plant which influence the ultimate degree of damage done by the insect in the field.' In practical agriculture, resistance represents the ability of a certain variety to produce a larger crop of good quality than other varieties under the same level of insect infestation and comparable environment.

Resistance can be assessed by these four characteristics.

1 Resistance is *heritable* and controlled by one or more genes.
2 Resistance is *relative* and can be measured only by comparison with a susceptible cultivar of the same plant species.
3 Resistance is *measurable*, i.e., its magnitude can be qualitatively determined by analysis of the standard scoring systems, or quantitatively by insect establishment.
4 Resistance is *variable* and is likely to be modified by the biotic and abiotic environments.

The phenomena of resistance are usually based on heritable traits. However, some characteristics are very labile and fluctuate widely under the influence of environmental conditions. The environment may favor the plant or the insect unequally, or may prevent or aggravate damage; therefore, it is likely to affect the expression of resistance. Accordingly, plant resistance may be classified as *genetic*, implying the characteristics that are under the primary control of genetic factors; or *ecological*, implying the characteristics that are under the primary control of environmental factors.

Genetic Resistance

The factors that determine the resistance of host plants to insect establishment include the presence of structural barriers, allelochemicals, and nutritional imbalance. These resistance qualities are heritable and operate in a concerted manner, and tend to render the plant unsuitable for insect utilization. Although various workers have attempted to classify the mechanisms of resistance, the terms defined by Painter (1951) – nonpreference, antibiosis, and tolerance – were widely accepted. However, Kogan and Ortman (1978) proposed that the term nonpreference should be replaced by antixenosis because the former describes a pest reaction and not a plant characteristic. The three types of resistance are described in the context of the functional relationships between the plant and the insect.

Antixenosis

Antixenosis is the resistance mechanism employed by the plant to deter or reduce colonization by insects. Generally, insects orient themselves toward plants for food, oviposition sites, and/or for shelter. However, due to certain characteristics, the plant may not be utilizable and may deter the insects. In certain situations, even though the insects may come in contact with the plant, the antixenotic characteristics of the plant do not allow the insect to colonize. Sometimes, the antixenosis mechanism is so effective that the insects starve and die (Painter 1968).

The deterrent mechanisms influence an insect's behavioral response to the plant. The antixenotic mechanism may be due to biophysical or biochemical factors or a combination of both. Plants that exhibit antixenotic resistance should have a reduced initial number of colonizers early in the season; the size of the insect population should also be reduced after each generation as compared with susceptible plants.

Antibiosis

Antibiosis is the resistance mechanism that operates after the insects have colonized and have started utilizing the plant. When an insect feeds on an antibiotic plant its growth, development, reproduction, and survival are affected. The antibiotic effects may result in a decline in insect size or weight, reduced metabolic processes, increased restlessness, and greater larval or preadult mortality. Indirectly, antibiosis may result in an increased exposure of the insect to its natural enemies. Plants that exhibit antibiosis reduce the rate of population increase by reducing the reproduction rate and survival of insects.

In certain cases antibiosis cannot be clearly separated from antixenosis because of the extreme deterrent chemicals and/or physical factor(s) in the plant cultivar. In other words, the deterrent chemicals and toxins in the plant are sometimes difficult to distinguish. Similarly, some of the morphological characteristics of the plant such as leaf trichomes or tissue toughness, are so critical for the insect to be able to react to their host plant, it is difficult to distinguish between antixenotic and antibiotic mechanisms of resistance. Furthermore, there are often overlaps between the morphological and biochemical bases of resistance. The antibiotic properties of the host plant may be expressed as constitutive or induced resistance against herbivores (Levin 1976).

Tolerance

Tolerance is a genetic trait of a plant that protects it against an insect population which would damage a susceptible host variety, so that there is no economic yield loss or lowering of the quality of the plant's marketable product. Tolerance is often confused with a low level of resistance or moderate resistance. The mechanism of tolerance is distinct from antixenosis and antibiosis (Panda and Heinrichs 1983). Tolerance does not affect the rate of population increase of the target pest but does raise the threshold level. Tolerance is an adaptive mechanism for the survival of the plant, and is more or less independent of the effect upon the insect. This type of host plant resistance refers strictly to resultant effects and not mechanisms.

Not all resistance phenomena can unequivocally be assigned to one or other categories of resistance. These functional categories of resistance do not exclude each other, but may interact, complement, and compensate for each other along with other biotic communities and abiotic factors in reinforcing the expression of resistance. Therefore, different cultivars may possess the same levels of resistance with different mechanisms for resistance and/or levels of resistance components.

Ecological Resistance

Ecological resistance has been categorized as pseudoresistance because it results, not from the genetic characters inherent in the host plant, but from some temporary shifts in the environmental conditions favorable to the otherwise susceptible host plants (Painter 1951). Although some factors contributing to pseudoresistance are fortuitous and unusual, plant varieties that exhibit pseudoresistance are of considerable importance in pest management systems and deserve special consideration. Another category of ecological resistance, induced resistance, occurs in response to damage by pathogens, herbivores, environmental stress, or specific chemical and physical treatment (Rhoades 1979).

Pseudoresistance

Alterations in plant growth patterns that result in asynchronies of insect–plant phenologies constitute the modality of resistance known as pseudoresistance (Painter 1951). Certain crop varieties may overcome the most susceptible stage rapidly and thus avoid insect damage. Early-maturing crop cultivars have been used in agriculture as an effective pest management strategy. However, plants that evade insect attack by this mechanism are likely to be damaged if the pest populations build up early.

Induced resistance

Induced resistance is the qualitative or quantitative enhancement of the plant's defense against invading organisms in response to pest-related injury or extrinsic physical or chemical stimuli. The extrinsic stimuli are known as inducers or elicitors. The postchallenge or injury-dependent responses of plants are components of induced resistance (Kogan and Paxton 1983); induced resistance can also result from an environmental change that may lead to a temporary benefit for the host plant. The application of fertilizers, herbicides, insecticides, growth regulators, and mineral nutrients, or a variation in temperature and day length, or insect and pathogen attack can all change the chemical constituents of plant tissue, and consequently their nutritional value for pests (Karban 1991).

Plant Defense Mechanisms

The modalities of plant resistance against different feeding guilds of insects seem to differ. The plant defense systems evolve incrementally and are likely to be proportional to the kind and amount of injury inflicted by herbivores (Berryman 1988). The most diverse and most specific plant defense systems are employed against insects that are 'intimate' with the plant tissue. As the 'intimates' are imbedded in the tissue, plants have greater potential to biochemically and/or physically disrupt the insect's behavioral/physiologic linkages to the plant. Plants may alter the levels and balance of compounds that serve as insect feeding stimulants and deterrents which results in the intimate associations becoming behaviorally/physiologically unacceptable.

Each plant species has a unique set or collection of defense traits ranging from morphological to phytochemical parameters that have behavioral and physiologic ramifications for a potential herbivore consumer. The phytophagous insects must be able to locate the most suitable nutritional substrates among the multitude of plant species available within its temporal and spatial environment. These behavioral patterns of insects can be adversely affected by antixenotic mechanisms involving physical and biochemical factors of the respective host plants.

Antixenosis

Host plant characteristics including morphological, physical, or structural qualities interfere with insect behavior such as mating, oviposition, feeding, and food ingestion. Pubescence and tissue hardness limit insect mobility by acting as structural barriers (Webster 1975). While selecting their hosts, insects respond to various plant stimuli (Hedin *et al* 1977, Stadler 1986) (see Chapter 5); absence of such stimuli, and the presence of repellents, antifeedants, or feeding deterrents contribute to different antixenosis types of plant resistance. As a mechanism of resistance, antixenosis may represent one or more breaks in the chain of responses leading to oviposition or feeding. These breaks take three forms: (1) the absence of an arrestant or attractant, (2) the presence of a repellent, or (3) an unfavorable balance between an attractant on the one hand and a repellent on the other. The relevant chemistry of the host plant seems to influence the herbivore's acceptance or rejection for oviposition or food (Schultz 1988).

Antixenosis to oviposition

Oviposition is often not a fortuitous act. It involves a series of behavioral activities: searching, landing, and contact over the plant surface (see Chapter 5). The sensory cues that mediate host selection for oviposition include visual, tactile, and chemical information (Thompson and Pellmyr 1991). The initiation and completion of each of these steps may be influenced by an array of plant characteristics, including physical and chemical stimuli from the plant surfaces. Resistance to oviposition may come from plant characteristics that either fail to provide appropriate oviposition-inducing stimuli or provide oviposition-inhibiting stimuli. Studies on the chemical cues used by lepidopterans in choosing host plants for oviposition indicate some of the ways in which preference hierarchies among potential hosts can be established (Renwick 1989). Although plant chemicals are closely linked with insect behavior in host finding and acceptance for oviposition and/or feeding, it is equally important to recognize the critical involvement of physical factors in the host-selection process. Hence, ovipositional preference or antixenosis is discussed on the basis of the plant's biophysical and biochemical traits.

Biophysical factors

After alighting on a plant, a gravid insect assesses the acceptability of that plant for oviposition. Plant surface characteristics provide many examples of structures influenced by evolutionary pressures (Southwood 1986). The multifunctional qualities of trichomes are exemplified in the Malvaceae (Ramayya and Rao 1976). The most primitive condition is represented by the pelate scale, which may reduce transpiratory losses; from this evolved the stellate hair and finally the tufted hair, which 'protects against insect attack'. Plant hairs or pubescence interfere with insect oviposition, attachment to the plant, feeding, and ingestion.

Plant pubescence. In the early literature plant pilosity was frequently postulated as being involved in susceptibility or resistance to insects. The breeding of hairy cottons in Africa and Asia to combat the jassids *Empoasca* spp. constitutes the foremost host plant resistance (HPR) success story in cotton. Parnell *et al* (1949) demonstrated that greater hairiness in both upland cotton (*Gossypium hirsutum*) and Egyptian cotton (*G. barbadense*) was related to jassid resistance. Afterwards, breeding of hairy cultivars became *de rigueur* throughout Africa and has proved to be a successful strategy for the management of jassids (Brettell 1980). A similar relationship was established between pubescence and the leafhopper's ovipositional response in the soybean. Soybean varieties with a dense hairiness of foliage can manifest both antixenosis to oviposition and feeding deterrence against leafhoppers. The simple trichomes deter oviposition and feeding by preventing the insect's ovipositor or proboscis from reaching the plant epidermis (Lee 1983).

The pubescent wheat cultivar Vel exhibits antixenosis to adults and larvae of the Hessian fly *Mayetiola destructor* Say (Roberts *et al* 1979). Similarly, Vel attracted a lower number of aphids *Rhopalosiphum padi* (L.) than did the glabrous-leaved Arthur wheat. Aphids moved slowly on Vel, appeared restless, and often left the plant without ovipositing and probing. *R. padi* is an important vector of the barley yellow dwarf virus, thus the leaf pubescence of common wheat provides a potentially valuable resistance to *R. padi* and protection against the barley yellow dwarf virus (Roberts and Foster 1983).

In contrast, the cotton bollworm *Helicoverpa zea* prefers pubescent strains of cotton *Gossypium* spp. for oviposition and attachment. The ovipositing females use plant hairs as footholds, which aid attachment to plants. The results obtained with the soybean isogenic lines Clark and Harosoy, which differ in pubescence (glabrous, appressed, dense, sparse, seedling sparse, irregular, pure line, curled deciduous), showed that the number of eggs laid as well as pod damage by *H. zea* was highest in dense pubescent varieties and lowest in glabrous ones (Panda and Daugherty 1978). Thus, eventual pod damage in the respective pubescent soybean types was primarily due to the ovipositional preference of *H. zea* moths. Although glabrous soybean varieties are least preferred by the gravid female of *H. zea*, they are associated with hypersensitivity to the tarnished plant bugs, and such infestation leads to delayed maturity and reduced yield (Meredith and Schuster 1979).

Frego bract.   Other morphological features of plants, such as frego bract in cotton, help reduce the number of eggs laid and subsequent damage by boll weevils *Anthonomus grandis* (Boh.) (Jenkins and Parrot 1971). The bracts of frego cotton are narrow, elongated, twisted, and flaring away from the bud or boll (Fig. 6.1). In field experiments, frego-bract cotton showed 50% less damage from oviposition than normal cottons did. The role of the frego bract in reducing damage by the boll weevil appears to be due to some adverse effect on insect behavior.

Boll weevils showed eight times as much plant-to-plant movement, required twice as much time for feeding and oviposition puncture, and spent

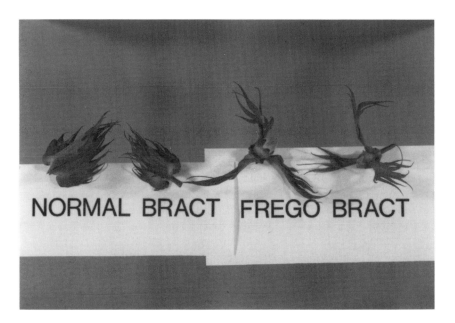

Fig. 6.1. Normal and frego bract cotton. Courtesy of J. N. Jenkins, USDA-ARS, Stoneville, USA.

much less time on squares of frego-bract cotton (Mitchell *et al* 1973). As an indirect result of their restless wandering over the plants, weevils are more likely to come in contact with insecticide residues and exposure to natural enemies. Like glabrousness, however, the frego bract is associated with hypersensitivity to the plant bugs *Lygus* spp. and cotton fleahopper *Pseudato-moscelis seriatus* (Jenkins *et al* 1973).

Visual factors

The role of vision in host finding and acceptance is becoming increasingly appreciated (Prokopy and Owens 1983). Color alone probably does not account for the host specificity of any insect species, but it frequently exerts an influence on the early stages of orientation. In the onion fly *Delia antiqua*, visual stimuli merely supplement the more important chemical cues related to its ovipositional process. Maximal oviposition by caged gravid onion flies is achieved by the synergistic interplay of onion-produced alkyl sulfides and yellow color substrate placed vertically inside the cage (Harris and Miller 1982). Specific color-related resistance, however, does exist. For example, the red and glossy nature of Cruciferae plants was a major factor conferring antixenosis resistance against the cabbage aphid *Brevicoryne brassicae* (Singh and Ellis 1993). These results were based on 39 field, greenhouse, and laboratory trials, and about 950 genotypes have been evaluated during the last 50 years.

Biochemical factors

Chemical cues are involved in all the three phases of host selection behavior: orientation, oviposition, and feeding. Many factors play a role in the process of oviposition by different insects, but long-range orientation of many insects to their host plants is known to be guided by volatile compounds emanating from plants. The eventual acceptance or rejection of a plant as a host usually depends on contact with the plant surface or through probing after landing. These aspects have already been discussed in Chapter 5.

Reviews on the subject have provided lists of insects and the sources of attractant chemicals involved (Hedin *et al* 1977, Stadler 1986). Yet relatively few plant compounds that either attract (Stadler 1986, Visser 1986) or repel female insects that lay eggs on or near food plants are known (Tables 6.1 and 6.2). For example, although the attractive properties of the onion volatile dipropyl disulfide in host finding and oviposition by the onion fly *Delia antiqua* have long been recognized, yet minute quantities of diallyl disulfide in the onion volatiles are antagonistic (Table 6.2) to the stimulating properties of dipropyl disulfide (Harris and Miller 1982).

Other insects, such as the carrot fly *Psila rosae*, are guided to their host plants by a complex combination of compounds with varying degrees of volatility. The involvement of the relatively nonvolatile *trans*-asarone, how-ever, is likely to be critical in the final step of oviposition after the insect alights

Table 6.1. Insect oviposition stimulants.

Host plant	Chemical compound	Insect stimulated	Reference
Cabbage	Allyl nitriles	*Pieris brassicae*	Mitchell (1977)
	Allyl isothiocyanate	*Delia radicum* (cabbage rootfly)	Finch and Spinner (1982)
	Indole glucosinolate	*Pieris rapae* (cabbage butterfly)	Traynier and Truscott (1991)
Crucifers	Sinigrin (allylglucosinolates)	*Pieris brassicae*	David and Gardiner (1962)
	Water-soluble compounds other than glucosinolate	*Pieris rapae*	Renwick and Radke (1988)
	Sinigrin	*Hylemya brassicae*	Nair and McEwen (1976)
Potato	'Green leaf' volatile	*Leptinotarsa decemlineata*	Visser and Ave (1978)
Onion	*n*-dipropyl disulfide and *n*-propyl mercaptan	*Delia antiqua*	Matsumoto (1970)
Cucurbits	Cucurbitacins	Diabroticina	Metcalf *et al* (1982)
		Aulacophorina beetles	
Corn			
Silk	C$_2$–C$_{12}$ alkanols, phenyl acetaldehyde	*Helicoverpa zea*	Flath *et al* (1978)
Kernel	Hexadienal	*Helicoverpa zea*	Centello and Jacobsen (1979)
Husk	Decadienal	*Helicoverpa zea*	Buttery *et al* (1978)
Tassel	Esters: ethyl acetate, ethyl cinnamate; ketones: pentanone, nonanone, octadienone; methylated benzenes and naphthalenes	*Helicoverpa zea*	Buttery *et al* (1980)

Table 6.1. Continued.

Host plant	Chemical compound	Insect stimulated	Reference
Tomato leaves	Nitrogen-containing phenolic glycoside ($C_{17} H_{29} O_{10} N$)	*Manduca sexta*	Yamamoto and Fraenkel (1959)
	Aqueous extracts and steam distillates	*Manduca sexta*	Tichenor and Seigler (1980)
	Ethanolic extracts	*Phthorimaea operculella* (potato tuber moth)	Meisner *et al* (1974)
	Foliar surface chemical	*Keiferia lycopersicella* (potato pinworm)	Burton and Schuster (1981)
Wild tomato *Lycopersicon hirsutum* leaves	Hexane extracts from glandular trichomes – sesquiterpenes ($C_{15} H_{22} O_2$)	*Helicoverpa zea*	Juvik *et al* (1988)
Nicotiana tabacum leaf surface	Combinations of diterpene duvanes and sucrose esters of C_3–C_7 fatty acids	*Heliothis virescens*	Cutler *et al* (1986)
Plantago lanceolata leaves	Iridoid glycoside (catapol)	*Junonia coenia*	Pereyra and Bowers (1988)
Citrus leaves	Flavone glycoside	*Papilio protenor*	Honda (1986)
Carrot *Daucus carota* Leaf surface and whole leaf extracts	Luteolin 7-O-(6"-O-malonyl)-β-D-glucopyranoside and trans-chlorogenic acid	*Papilio polyxenes*	Feeny *et al* (1988)
Cuticular wax of foliage	Falcarindol-propenylbenzene trans-asarone (2,4,5-trimethoxy-1-propenylbenzene)	*Psila rosae* F.	Stadler and Buser (1984)

Tobacco leaves, cuticular chemical components	Duvane diterpenes (α- and β-4,8, 13 duvatrien-1-ols and α- and β-4,8, 13-duvatriene-1,3-diols)	*Heliothis virescens* (F.)	Jackson *et al* (1986)
Citrus plants	Volatiles from lime leaves combined with moisture	*Papilio demoleus*	Saxena and Goyal (1978)
Citrus plant *Citrus unshiu* Marc.	Flavonone glycosides, vicenin-2 6,8-di-C-β-D-gluco-pyranosy-lapigenin mixed with another unidentified component	*Papilio ruthus*	Ohsugi *et al* (1985)
Sour orange *Citrus natsudaidai* epicarp	Flavone glycoside, naringin (naringenin-7β-neohesperidoside)	*Papilio protenor*	Honda (1986)
Pine tree	Pine monoterpenes: β and α pinene	*Panolis flammea*	Leather (1987)
Apple fruit	Isomers of α-farnesene	*Lasperesia pomonella*	Wearing and Hutchins (1973)
Rice leaves	Oryzanone	*Chilo suppressalis*	Munakata and Okamoto (1967)

Table 6.2. Insect oviposition deterrents.

Host plant	Deterrent compound	Insect deterred	Reference
Cabbage tissue	Coumarin and rutin	*Pieris rapae* *Plutella xylostella*	Tabashnik (1987) Tabashnik (1985)
Erysimum cheiranthoides	n-Butanol	*Pieris rapae*	Sachdev-Gupta et al (1990)
Crucifer *Erysimum cheiranthoides*	Specific cardenolides	*Pieris rapae*	Renwick et al (1989)
Sorghum	ODP	*Atherigona soccata*	Raina (1981)
Apple	ODP	*Rhagoletis pomonella*	Averill and Prokopy (1987)
Rice *Oryza sativa* variety TKM6	Steam distillate, oviposition inhibitor pentadecanal	*Chilo suppressalis*	Saxena (1986)
Corn, resistant variety Pioneer X304C, leaves	Aqueous extracts	*Spodoptera frugiperda*	Williams et al (1986)
Cabbage	ODP	*Trichoplusia ni*	Renwick and Radke (1981)
Onion *Allium sativum*	Minute quantities of diallyl disulfide	*Delia antiqua*	Harris and Miller (1982)
Rose-of-Sharon *Hibiscus syriacus* L. calyx	Unsaturated fatty acids and their methyl esters	*Anthonomus grandis* (Boh.)	Bird et al (1987)

ODP, oviposition deterrent pheromone.

on the plant (Stadler and Buser 1984) (Table 6.1). The list of identified compounds or extracts and the cited investigations reveal that a single compound rarely explains completely the ovipositional stimulation by the leaf surface. Even the well-known glucosinolates and the isothiocyanates, which influence the behavior of crucifer-feeding insects so far investigated, cannot fully explain their observed effects on oviposition or food selection. Chemical constituents other than glucosinolates appear to be responsible for the discriminatory behavior of ovipositing butterflies (Renwick and Radke 1988). A combination of several glucosinolates of Cruciferae are necessary to stimulate oviposition by *Delia radicum* and *D. floralis*. An electrophysiologic study shows that specific tarsal-contact chemoreceptors of these insects were stimulated by methanol-soluble fractions, composed of several known glucosinolates, extracted from the leaf surface of crucifer plants (Birch *et al* 1993). As a rule, therefore, it is only mixtures of plant compounds together with nonchemical stimuli that can explain the host plant-selection process (Stadler and Buser 1984). Furthermore, the role of inhibitory stimuli has to be taken into account in the choice of the oviposition site by phytophagous insects (Jermy and Szentesi 1978). The role of chemical deterrents at the leaf surface in preventing oviposition has recently been emphasized. Extracts of nonhost plants are effective in preventing or reducing oviposition on treated host plants in many cases (Renwick 1988). The balance between stimulants and inhibitors is critical in oviposition.

Volatile hydrocarbons and other secondary compounds may act as oviposition deterrents (Table 6.2). Saxena (1986) reviewed the effect of rice plant allelochemicals on the behavior and physiology of *Chilo suppressalis*, *Nilaparvata lugens*, *Sogatella furcifera*, and *Nephotettix virescens*. Steam distillate extracts of the resistant rice variety TKM6 inhibited oviposition, hatching, and larval development of *C. suppressalis*. Extracts of the susceptible variety Rexoro encouraged oviposition. The oviposition inhibitor in TKM6 was identified as pentadecanal.

Antixenosis to feeding

Insects respond to various feeding stimuli when selecting their host plants. The absence of such stimuli and the presence of deterrent compounds presumably contribute to antixenosis types of resistance. The plant surface is embedded with physical and chemical factors responsible for antixenosis to feeding insects (Southwood 1986). Critical observation on a number of phytophagous insect species shows that before feeding on a plant they make some kind of sensory exploration of the plant surface as a prelude to biting. The leaf surface acts as the crucial interface between the insect's battery of chemoreceptors and the plant (Southwood 1986). Hence, antixenosis for feeding in plants is composed of the nonglandular and glandular trichomes, leaf-surface chemicals, tissue toughness, nutrient deficiency and constitutive chemicals (repellents and deterrents).

Nonglandular trichomes

Trichomes are epidermal appendages of diverse forms and structures, such as nonglandular and glandular hairs, scales, or pelate hairs. Nonglandular trichomes frequently impart general antixenosis resistance by providing an effective barrier that prevents small arthropods from landing on the plant surface and prevents movement and feeding (Southwood 1986, Goertzen and Small 1993). Trichomes have basically three types of effects on insect behavior over the leaf surface: (1) simple impedance, (2) physical trapping by hooked hairs, and (3) stickiness caused by exudates from the glandular trichomes. Simple impedance has been demonstrated in pubescent cotton, where Smith *et al* (1975) showed that the rate of travel by the first-instar larvae of the pink bollworm *Pectinophora gossypiella* was more than six times faster on smooth leaves than on those with pubescence. Because of this lack of movement, the larvae were deterred from the plant substrate. A similar phenomenon was encountered by the newly hatched tobacco budworm larvae *Heliothis virescens* on the upper and lower leaf surfaces and petioles of cotton (Ramalho *et al* 1984). The slowing of movement due to plant hairs is likely to starve the larvae leading to increased mortality. Long hairs not only impede movement, but also prevent the insect from reaching the leaf surface to feed on. For example, the 0.2–0.4-mm long proboscis of the leafhopper *Empoasca fabae* can not reach the mesophyll or the vascular bundles of hairy

Fig. 6.2. Fifth-stadium nymph of *Empoasca fabae* unable to penetrate the petiole indumentum of soybean var. Clark, dense pubescent isoline. From Lee (1983), courtesy of M. Kogan, Oregon State University, Corvallis, USA.

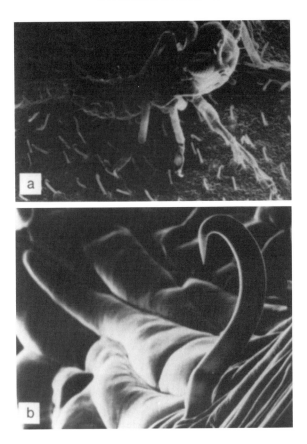

Fig. 6.3. Scanning electron micrograph showing the entrapment of a fifth-instar *Empoasca fabae* nymph by hooked trichomes of *Phaseolus* sp. (a) A nymph of *E. fabae* on the lower leaf surface of *Phaseolus* is entrapped (×82). (b) A hooked trichome characteristic of *Phaseolus* sp. (×886). Reproduced from Pillemer and Tingey (1978) with permission of Kluwer Academic Publishers, Dordrecht, The Netherlands.

soybean leaves (Lee 1983) (Fig. 6.2). Phloem and xylem feeders must insert their stylets deeper into the plant tissue, and trichomes may impede this insertion. When trichomes were removed experimentally from the pods of mustard *Brassica hirta* var. Gisilba, feeding damage from the flea beetle *Phyllotreta cruciferae* increased (Lamb 1980). Hence, the sucking insects avoid utilizing the plant sap because of the presence of trichomes – a case of antixenosis resistance.

Insects may not merely be impeded by trichomes, but when they come in contact with the plant they can be trapped by hooked trichomes. These trichomes are common in climbing plants where they assist the maintenance of the shoot's position among other plants. The hooks also penetrate the cuticle of some insects which are ultimately trapped and get killed. Notable observations on the effects of hooked trichomes of the French bean *Phaseolus* sp. on the behavior of the potato leafhopper *Empoasca fabae* have been made by Pillemer and Tingey (1978). Leafhoppers were impaled by these epidermal appendages, leading to wounding and eventual death (Fig. 6.3).

Glandular trichomes

Glandular trichomes are widely distributed in vascular plants that exude gummy, sticky, or polymerizing chemical mixtures that severely impede the insect's ability to move, feed, and/or survive. The defensive role of glandular trichomes of certain members of the Solanaceae plants against herbivorous insects has recently become an area of intense research at both applied and academic levels. A number of plants of the *Solanum*, *Lycopersicon*, *Nicotiana*, and *Medicago* spp. are particularly adept in producing sticky leaf exudates. In certain wild potato species (*Solanum polyadenium*, *S. berthaultii*, and *S. tariyense*), an exudate is discharged from the four-lobed head of the glandular hairs when aphids *Myzus persicae* or *Macrosiphum euphorbiae* mechanically rupture the cell wall (Gibson 1971). On contact with atmospheric oxygen, the clear, water-souble exudate is changed into an insoluble black substance that hardens around the aphid's tarsi (Fig. 6.4) and seriously impedes its movement. Further accumulation of glandular material sticks the aphid firmly to the plant and starvation leads to death. Polyphenoloxidase and peroxidase activities are exhibited by the glandular trichomes of *S. berthaultii* for oxidation of the phenolic compounds in glandular exudates (Ryan *et al* 1982). Some examples of glandular trichome-mediated resistance are given in Table 6.3.

Leaf-surface chemicals

Plant epicuticular waxes affect the feeding behavior of insects, particularly the settling of probing insects, acting as phagostimulants or feeding deterrents. Evidence that insects respond to chemicals on the leaf surface is obtained from experiments using surface extracts of leaves or pure chemicals that are known to occur on leaf surfaces. Examples of plant-surface waxes which act as feeding attractants and/or deterrents are given in Table 6.4.

Alkanes are amongst the commonest constituents of all plant waxes. Klingauf and his associates (1971) have demonstrated the importance of alkanes in settling and feeding by *Acyrthosiphon pisum* on *Vicia faba*. A specific alkane $C_{32}H_{66}$ common in the wax of *V. faba* caused the insect to probe for longer periods into parafilm sachets (Klingauf *et al* 1971), whereas the alkane fraction of a nonhost deterred feeding (Klingauf *et al* 1978). The apple aphids *Aphis pomi* and *Rhopalosiphum insertum* were stimulated to probe by the glucoside phloridzin, common in apple wax (Klingauf 1971). In the case of rice, an application of epicuticular wax from the brown planthopper-resistant

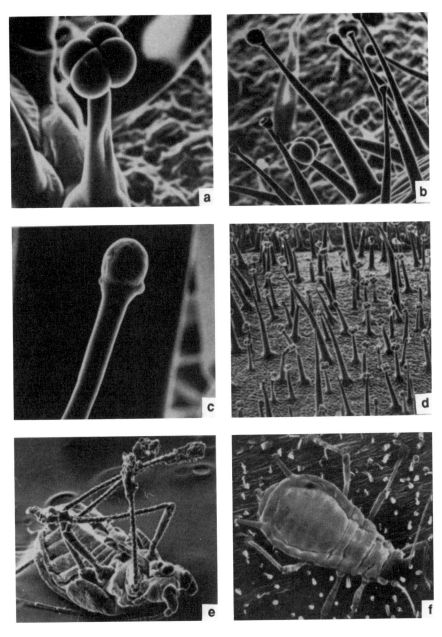

Fig. 6.4. Scanning electron micrographs: (a) glandular hairs on *Solanum berthaultii* consisting of a four-celled head mounted on a stalk (×392); (b) glandular hairs on *S. berthaultii* showing a four-celled head and other types (×157); (c) glandular vesicle of a hair of *S. berthaultii* (×735); (d) upper surface of leaf of *S. polyadenium* showing glandular hairs (×49); (e) glandular material on the tarsi of an aphid (×49); (f) aphid stuck to a stem of *S. berthaultii* (×34). Reproduced from Gibson (1971) with permission from the Association of Applied Biologists, Wellesbourne, UK.

Table 6.3. Examples of insect resistance conferred by glandular trichomes.

Plant with glandular trichomes	Insect	Reference
Solanum berthaultii, S. polyadenium, S. tarijense	Green peach aphid, potato aphid, Colorado potato beetle, two-spotted spider mite	Gibson (1971); Gibson and Turner (1977); Tingey et al (1982); Ave et al (1987)
Solanum berthaultii, S. polyadenium	Potato leafhopper	Tingey and Gibson (1978)
Solanum berthaultii	Flea beetle	Gibson (1978)
	Tarsonemid mite	Gibson and Valencia (1978)
Solanum pennellii	Potato aphid	Gentile and Stoner (1968)
Lycopersicon spp.	Potato aphid	Gentile and Stoner (1968)
	Spider mites	Stoner et al (1968), Aina et al (1972)
	Greenhouse whitefly	Gentile et al (1968)
	Tomato fruitworm	Dimock and Kennedy (1983)
	Tomato fruitworm, tobacco hornworm	Kennedy and Yamamoto (1979), Dimock (1981), Kennedy and Dimock (1983)
Medicago spp.	Alfalfa weevil	Shade et al (1975)
	Alfalfa weevil	Johnson et al (1980)
	Alfalfa seed chalcid	Brewer et al (1986), Danielson et al (1987)
	Potato leafhopper	Shade et al (1979)
	Potato aphid	Shade and Kitch (1983)
Nicotiana spp.	Green peach aphid	Thurston et al (1966)
	Tobacco hornworm	Thurston (1970)
	Two-spotted spider mite	Patterson et al (1974)
Helianthus spp.	Sunflower moth	Rogers et al (1987)
Pelargonium xhortorum	Mite	Gerhold et al (1984), Craig et al (1986)

Table 6.4. Plant-surface waxes which act as insect feeding attractants and/or deterrents.

Plant	Insect	Chemical identity of wax	Insect behavioral response	Reference
Sorghum (immature plants)	Locusta migratoria	Surface wax: nonpolar compound (fractions containing n-alkaline, esters, and p-hydroxybenzaldehyde)	Deterrent	Woodhead (1982)
Avena sativa	Oscinella frit	Hydroxy-β-diketones	Attractant	Hamilton et al (1979)
Vicia faba	Acyrthosiphon pisum	Surface wax: alkanes C_{32}	Attractant/stimulant	Klingauf et al (1971)
Brassica spp.	Acyrthosiphon pisum	Surface wax: alkanes C_{29} fatty acid	Deterrent	Klingauf et al (1978)
Apple leaves	Aphis pomi Rhopalosiphum insertum	Aqueous surface extract: phenolic glucoside, i.e., phloridzin	Attractant/stimulant Attractant/stimulant	Klingauf (1971) Klingauf (1971)
	Aphis pisum Myzus persicae, Amphorophora agathonica		Deterrent Deterrent	Montgomery and Arn (1974)
Rice (resistant variety)	Nilaparvata lugens	Epicuticular wax: higher proportion of short-chain hydrocarbons and carbonyl compounds	Deterrent	Woodhead and Padgham (1988)
Mulberry leaves	Bombyx mori	Surface waxes: C_{26} and C_{28} alcohols	Stimulant	Mori (1982)

Table 6.4. Continued.

Plant	Insect	Chemical identity of wax	Insect behavioral response	Reference
White spruce and balsam fir	*Choristoneura fumiferana* (spruce budworm)	Extracts of epicuticular wax	Attractant/stimulant	Maloney et al (1988)
Cotton	*Anthonomus grandis*	Esters from surface wax of cotton buds and anthers (geranylgeraniol with C_{22} and phytol with C_{12} acid moiety)	Attractant/stimulant	McKibben et al (1985)
Cotton *G. hirsutum* variety Acala SJ2	*Spodoptera littoralis*	Phylloplane alkalinity from cotton leaf surface	Deterrent	Naven et al (1988)

rice variety IR46 to the surface of susceptible IR22 increased restlessness and deterred probing (Woodhead and Padgham 1988).

Leaf-surface extracts containing the nonpolar compounds of two varieties of 14-day-old sorghum plants were feeding deterrents to *Locusta migratoria*, but extracts from older plants had no effect (Woodhead 1983). Seedling sorghum is particularly distasteful to *L. migratoria*, but when the surface wax was removed from the seedling only 10% of the insects rejected eating (Woodhead 1983). Analysis and bioassay of the surface wax of sorghum seedlings (variety Botswana) indicated that both the alkane and ester fractions of the wax are unpalatable and are feeding deterrents to *L. migratoria*. In contrast, wax components serve as feeding stimulants to the larvae of the silkworm *Bombyx mori* on mulberry (Mori 1982), the spruce budworm *Choristoneura fumiferana* on white spruce and balsam fir (Maloney *et al* 1988), and the adult boll weevil *Anthonomus grandis* on cotton buds and anthers (McKibben *et al* 1985).

Wax may physically prevent the movement of an insect across a leaf surface. Mustard beetles adhere much better to *Brassica oleracea* cultivars that do not have a heavy wax bloom than to those with a bloom (Stork 1980). The insect's adhesive setae are much more effective on a smooth surface than on a glaucous surface, where they become covered with debris of the leaf wax. Similarly, movement of first-instar larvae of the stem borer *Chilo partellus* is considerably prevented by wax on the culms of sorghum. In the 'bloom' cultivars, the culm is heavily waxed and the neonate larvae experience considerable difficulty in climbing; their prolegs get stuck in the wax and the larvae never reach the feeding site (Bernays *et al* 1983).

Leaf-surface chemicals undoubtedly affect several aspects of insect behavior. Insects possess the sensory apparatus to detect these chemicals by contact or olfaction. The detectable chemicals include not only secondary plant compounds but also the ubiquitous aliphatic components of plant waxes. There are also suggestions that the nature of the surface may be 'recognized' by insects as indicative of the internal constituents of a plant (Chapman and Bernays 1989). In-depth studies are needed to rationalize the effects of plant-wax chemistry on insect behavior.

Tissue toughness

Various physical resistance factors in plants such as solidness of stems, thickening of tissues, anatomic adaptations, and protective structures affect the utilizability of a plant as a host by phytophagous insects.

Solid-stemmed varieties of wheat proved resistant to some wheat insects. Stem solidness – due to the development of pith inside the stem – gives wheat resistance to *Oulema melanopus* and *Cephus cinctus* (Wallace *et al* 1974). The resistance of sweetcorn lines to the corn earworm *Helicoverpa zea* has been attributed to an increased number of husk leaves and the thickness of the husk. These plant characteristics have been recognized for years as important resistance factors. Silk balling and long tight husks beyond the tip of the cob are effective physical barriers to the corn earworm larvae, preventing them from eating the corn plants. A highly significant positive correlation between resistance to corn earworm injury and husk tightness has been observed (Widstrom *et al* 1973).

Thickening of plant tissues confers an antixenosis type of resistance to insects in a number of field crops. Such physical characteristics of plants will not allow insects to utilize the plant tissues. Extensive testing of rice varieties for relative resistance to the Asiatic rice stem borer *Chilo suppressalis* (Walker) led Patanakamjorn and Pathak (1967) to conclude that resistance tended to be associated with a hairy upper lamina, tight leaf-sheath wrapping, small stems with ridged surfaces, and thick hypodermal layers.

Sorghum varieties resistant to shootfly are characterized by distinct lignification and thickness of the cell walls enclosing the vascular bundle sheaths within the central whorl of the young leaves (Fig. 6.5). Cells with lignified thick walls lie along the circumference of each differentiation bundle, forming a chain-like structure along the section (Blum 1968).

In eggplants the antixenosis mechanism of resistance to the shoot borer *Leucinodes orbonalis* (Guen) is attributed to compact vascular bundles in a

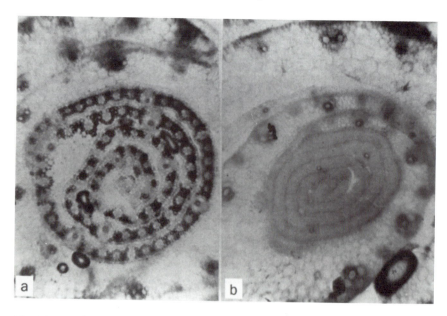

Fig. 6.5. Lignification of cell walls enclosing the vascular bundles within the central whorl of young leaves in a resistant variety (a) compared with a susceptible variety (b). Reproduced from Blum (1968) with permission of the Crop Science Society of America, Madison, Wisconsin, USA.

thick layer with lignified cells and low pith area, resulting in nonpenetration of the apical shoot by the neonate borer (Panda *et al* 1971). The more compact, narrower, and less succulent second and third internodes of resistant soybean varieties apparently constitute a relatively unsatisfactory host plant environment for agromyzid beanfly larvae. Furthermore, the production of the secondary xylem, which specifically reduces the diameter of the pith cavity, and the differentiation and development of lignified xylem fibers are an overall physical hindrance preventing the beanfly *Melanagromyza sojae* larvae from feeding on the pith of resistant soybean plants (Chiang and Norris 1983).

In the case of sap-feeding insects, tissue hardness and stems with small cross-sectional areas have been associated with a restriction in the penetration of stylets and feeding. The highly lignified tissues of *Medicago* clones are associated with resistance to the potato leafhopper (Brewer *et al* 1986). These physical barriers protect host plants from sucking insects.

Nutrient deficiency

An excellent case of nutrient deficiency is exemplified in Canby, an aphid-resistant raspberry. When feeding on the resistant variety, the aphids *Amphorophora agathonica* (Hottes) showed delayed development, significantly reduced size, lower fecundity, and extensive mortality. Mortality on Canby occurred in the first two instars, indicating that the aphids experienced difficulty in colonization. As Canby is virtually immune to infestation by *A. agathonica* under field conditions, Kennedy and Schaefers (1975) investigated the role of aphid nutrition in the immunity of the red raspberry. When the aphid arrived on a Canby raspberry, it probed and fed, imbibing an appreciable volume of sap. However, the sap is nutritionally inadequate for the aphids' normal growth and development. After feeding for 24 hours the aphids suffered from semistarvation, became restless, and abandoned the plant. The nutrient deficiency in Canby is related to the levels of both solids (largely sugars) and nitrogenous compounds in the ingestate of *A. agathonica*. Thus, plant nutrition plays an important role in Canby's resistance to aphid damage. North American raspberry breeders are fortunate that Lloyd George, widely used as a donor for fruit size and quality, is also resistant to *A. agathonica*.

Biochemical factors

Several chemical constituents of plants that serve as olfactory and gustatory stimuli, but differ qualitatively and quantitatively have been discussed in Chapters 4 and 5. These may be nutrients, i.e., sugars, amino acids, phospholipids, etc., or non-nutritive constituents, i. e., glycosides, alkaloids, terpenoids, etc. Such stimuli are specific and are crucial in evoking the behavioral response of insects (preference/antixenosis) to plants. A series of tables listing insect-feeding stimulants and deterrents have been published in recent years (Hedin *et al* 1977, Panda 1979, Smith 1989). Feeding inhibitors and antifeeding compounds have also been identified and their role in host plant resistance has been elucidated (Norris 1986, Jermy 1990).

Antixenosis involving plant biochemical factors may arise through interruptions of the normal feeding pattern at different points (Beck 1965). The insect may be repelled by plant volatile compounds without coming in contact with the plant or, having made contact, feeding may be suppressed; or having bitten the leaf, the insect may be deterred from further feeding. The chemicals inhibiting feeding behavior at these various points are known as repellents, suppressants, and deterrents, respectively. Such types of antixenotic chemical compounds are referred to as feeding inhibitors (Chapman 1974). Other workers use antifeedant and feeding deterrents synonymously for substances that, when perceived, prevent or reduce feeding (Schoonhoven 1982). Plant defense compounds that prevent or reduce contact between the insect and the substrate are known as repellents (Dethier *et al* 1960).

It is widely accepted that all green plants at some stages of their life history contain one or more chemicals that are repellent, antifeedant, and/or toxic to at least some insects (Schultz 1988). Plant allelochemicals involved in resistance have been classified into two broad categories: those that confer general resistance and those that confer specific resistance (Levin 1976). Compounds showing general resistance can repel, deter, or are weakly toxic to most herbivores; they are present in several plant species and sometimes in different families; they are not tissue-specific and their concentration increases as the tissue matures. Examples are volatile hydrocarbons, acids, phenols, terpenes, and alcohol from green plants, as well as chlorogenic acid, quercetin, and tannins. These compounds repel, deter, and even inhibit the growth of insects. Compounds exhibiting specific resistance are extremely toxic to a small group of specialized pathogens or herbivores. Each compound is present in a few species and is often tissue-specific. The compounds tend to reach their highest concentration in young leaves and fruits and decrease as the crop matures. Examples are sinigrin, tomatine, solanine, gossypol, and 2,4-dihydroxy-7-methoxy-1,4-benzoxazin-3-one (DIMBOA). They are toxic in nature and impair either the behavior and/or metabolic functions of insects (Norris and Kogan 1980). The influence of toxic antifeedants will be discussed in the section on antibiosis below.

Repellents. Plant defense compounds that prevent or reduce contact between the insect and the substrate are known as repellents. A variety of insects have been reported to be sensitive to general green plant volatiles and volatiles that are specific to their host plants (Visser 1986). Understanding the olfactory sensitivity of insects to plant volatiles may help in the identification of semiochemicals responsible for attraction or repulsion to an insect.

Some examples of insect-repellent compounds are given in Table 6.5.

Table 6.5. Some examples of insect repellents.

Plant	Repellent compound	Insect repelled	Reference
Rice	Steam distillates of resistant varieties	*Nilaparvata lugens* (brown planthopper)	Saxena and Okech (1985)
Rice	Steam distillates of resistant varieties	*Nephotettix virescens* (green leafhopper)	Khan and Saxena (1985b)
Volatiles of green plants	Thymol (aromatic) and carvacrol (monoterpene alcohol)	*Cnaphalocrocis medinalis* (rice leaffolder)	Ramachandran et al (1990)
Soybean	Steam distillates of resistant varieties	*Trichoplusia ni* (cabbage looper)	Khan et al (1987)
Pinus silvestris	α-Pinene, 3-carene	*Blastophagous piniperda* (pine beetle)	Oksanen et al (1970)
Grand fir tree	Resin vapors (monoterpenes)	*Scolytus ventralis* (fir engraver beetle)	Bordasch and Berryman (1977)
Tomato	Tomatine	*Leptinotarsa decemlineata* (Colorado potato beetle)	Schreiber (1958)
Pepper	Capsaicin	Colorado potato beetle	Schreiber (1958)
Tobacco	Nicotine	Colorado potato beetle	Trouvelot (1958)
Corn	Essential oil	*Helicoverpa zea* (cornworm)	Starks et al (1965)
Nicandria sp.	Alcohol $C_{22}H_{27}O$	*Manduca sexta* (tobacco hornworm)	Fraenkel et al (1960)
Cashew	Anacardiac acid	*Reticulitermes* sp. (termite)	Wolcott (1958)
Alfalfa	Tannic acid	*Hypera postica* (alfalfa weevil)	Bennett (1965)

Steam distillate extracts of resistant rice varieties and nonhost barnyard grass
were found to repel *Nilaparvata lugens*. When applied topically, the extracts
caused high mortality even at low doses (Saxena and Okech 1985). Similar
results with green leafhopper *Nephotettix virescens* feeding on susceptible TN1
rice sprayed with steam distillate extracts of resistant ASD7 plants were
obtained with the use of an electronic monitoring device and lignin-specific
dye (Khan and Saxena 1985b). Phloem feeding by the insects was significantly reduced on the extract-treated plants as indicated by a significant
increase in probing frequency and in duration of salivation and xylem feeding.
Volatile chemicals of plant origin such as thymol (aromatic) and carvacrol
(monoterpene alcohol) exhibited repellent properties to two sympatric species
of rice leaffolders: *Cnaphalocrocis medinalis* (Guenee) and *Marasmia patnalis*
Bradley (Ramachandran *et al* 1990).

Cells of glandular trichomes secrete and accumulate a large variety of
terpene oils and other essential oils that generally act as insect repellents in
plants. More than 100 mono-, di-, and sesquiterpenes that have been isolated
from glandular trichomes (Kelsey *et al* 1984) are clearly the first line of plant
defense against herbivores and pathogens. Many volatile compounds in the
glandular trichomes of potato and tomato repel phytophagous insects (Duffey
1986). The exudate from the glandular trichomes of *Solanum berthaultii*
contain volatile substances including sesquiterpenes. These volatiles repel the
aphid *Myzus persicae* and thus fewer aphids settle on the feeding areas (Ave
et al 1987). The vastly improved techniques for the identification of volatile
compounds, i.e., steam distillation and extraction of plant tissues, integrated
gas chromatography, mass spectrometry, and highly sensitive nuclear magnetic resonance spectrometry, should expedite research on repellents.

Antibiosis

The disruption of the normal metabolic process affecting the biology of an
insect is the basic manifestation of antibiosis. Antibiotic resistance involves
biophysical and biochemical plant defenses, as well as nutritional factors.
Resistance is antibiotic when, upon feeding on a resistant plant, an insect
suffers one or more of the following physiologically adverse effects, which can
be mild to lethal.

1 *Death of early instars.* The early instar larvae die because of the antibiotic
properties of the host plant.
2 *Physiologic disturbances.* There is a decline in size and weight of the larvae
or nymphs, prolongation of the larval period, and reduction in adult
fecundity. The adults that emerge are short-lived and the time before the
female insects mate and lay eggs is limited.
3 *Morphogenetic disturbances.* Precocious larvae and pupae are formed, but ultimately fail to pupate or eclose, respectively. The insect population is reduced.

4 *Continuity of nutrient-poor food reserves.* This affects the survival ability of overwintering or even aestivating insects.

5 *Occurrence of various behavioral and physiologic abnormalities.* Certain dermal glands produce extra secretions that lead to increased sensitivity (restlessness) to stimuli and regurgitation in many insects.

Modalities of plant resistance to insects are complex. For example, the resistance of the alfalfa variety Cody is attributed to the combined effect of antixenosis, antibiosis, and tolerance components (Jones *et al* 1968). The unsuitability of resistant soybean varieties for the Mexican bean beetle and cabbage looper may possibly be due to: (1) physiologic inhibitors as in cultivar PI 227687 (antibiosis); (2) deterrents (antixenosis); and/or (3) nutritional disproportionality (antibiosis) (Norris *et al* 1988). The shootfly-resistant sorghum cultivars IS 2146, IS 3962, and IS 5613 possess a complexity of resistance factors such as a high density of trichomes on the abaxial leaf surface, deterrent compounds (the alkaloid hordenine, the cyanogenic glucoside dhurrin), and an absence of amino acid lysine in the plant sap (Singh and Rana 1986). Therefore, antibiosis cannot be attributed to any one factor – physical barrier, secondary metabolite, or nutritional deficiency.

The following sections discuss some of the factors possibly implicated in antibiosis.

Presence of toxins

Many nonhost plants contain chemicals such as nicotine, pyrethrum, and rotenone which are toxic to insects and may have value as insecticides (Norris 1986). Table 6.6 lists some nonhost plant species that contain toxic compounds capable of killing insects.

The leptine glycoalkaloids in the wild potato *Solanum chalcoence* are toxic to the Colorado potato beetle (Sinden *et al* 1986). The glandular pubescence of potato is a defense against a broad complex of pest species including aphids, leafhoppers, flea beetles, the Colorado potato beetle, potato tuber moth complex, and spider mites (Tingey and Sinden 1982). Pure extract of 2-tridecanone from defensive trichomes (type VI) of wild tomato are a potent feeding deterrent and fumigant, and also neurotoxic to the tobacco hornworm *Manduca sexta*, tomato fruitworm *Helicoverpa zea*, aphid *Aphis gossypii*, and Colorado potato beetle *Leptinotarsa decimlineata*. (Kennedy and Dimock 1983). In addition to 2-tridecanone, the trichomes of tomato PI 134417 also contain several other ketones, including 2-undecanone, which are at such low levels that they cannot be acutely toxic to *H. zea* larvae (Farrar and Kennedy 1987). Since *H. zea* has the ability to detoxify moderate dosages of 2-tridecanone, selection for populations of *H. zea* tolerant of this chemical could be rapid, but the effects of 2-undecanone-mediated resistance would be manifested only in the reduced number of moths produced in the tomato crop.

Table 6.6. Insect antifeeding compounds in nonhost plants.

Plant	Antifeeding compounds	Insect	Reference
Seeds of neem tree *Azadirachta indica, Melia azedarach*	Terpenoid, azadirachtin	*Locusta migratoria*	Butterworth and Morgan (1971)
Extracts of bark of hickory *Carya ovata*	Aglycone 5-hydroxy-1,4-naphthoquinone (juglone) β-Benzoquinone	*Scolytus multistriatus* (small European elm bark beetle)	Gilbert and Norris (1968)
		Scolytus multistriatus	Norris (1970)
Juvenile needles of pine tree *Pinus banksiana*	13-Keto-8(14)-podocarpen-18-oic acid (terpenoid derivative)	*Neodiprion rugifrons* (pine sawflies)	Ikeda et al (1977)
Bark of *Warburgia ugandensis*	Warburganal	*Spodoptera exempta* (nutgrass armyworm)	Kubo and Nakanishi (1977)
Legume *Hymenaea courbaril*	Several sesquiterpene hydrocarbons, caryophyllene, α-selinene, β-selinene, and β-copaene	*Spodoptera exigua* (beet armyworm)	Stipanovic (1983)
Vernonia spp. (Compositae)	Germacranolide-type sesquiterpene lactone, glaucolide A	*Spodoptera eridania, S. frugiperda* (fall armyworm)	Mabry et al (1977)
Parthenium hysterophorus	Sesquiterpene lactone, parthenin	*Dysdercus koenigi* (pyrrhocorid bug), *Tribolium castaneum* (red flour beetle), *Phthormea operculella*	Stipanovic (1983)

Source	Chemicals	Insect species	Reference
Leaves of *Parabenzoin trilobum* ('Shiromoji' in Japanese)	Germacrane sesquiterpenes, shiromodiol monoacetate, and shiromodiol diacetate	*Spodoptera litura* (tobacco cutworm)	Munakata (1977)
Glandular trichomes on the anther tips of wild sunflower *Helianthus* sp.	Sesquiterpene lactone, maximilin C	*Spodoptera eridania* (southern armyworm), *Melanoplus sanguinipes* (migratory grasshopper), and *Homoeosoma electellum* (sunflower moth)	Gershenzon *et al* (1985)
Clerodendrum spp. (Verbenaceae)	Several clerodane diterpenes including clerodendrin A, clerodendrin B	*Spodoptera litura, Euproctis subflava, Ostrinia nubilalis*	Munakata (1977)
	Grayanoid diterpenes, grayanotoxin III and kalmitoxin I and II	*Lymantria dispar* (gypsy moth)	Munakata (1977)
Peruvian plant *Alchornea triplinervia* (Euphorbiaceae)	Anthranilic acid, gentisic acid, senecioic acid, *trans*-cinnamic acid, *trans*-cinnamaldehyde, and camphor	*Anthonomus grandis, Heliothis virescens* (tobacco budworm)	Miles *et al* (1985)
East African medicinal plant *Bersama abyssinica* (Melianthaceae)	Four new bufadienolide steroids: abyssinin, abyssinol A, B, and C	*Helicoverpa zea*	Kubo and Matsumoto (1985)
Nonhost plant *Cocculus trilobus*	Isoboldine alkaloid	*Spodoptera littoralis*	Wada and Munakata (1968)

Table 6.6 Continued.

Plant	Antifeeding compounds	Insect	Reference
Solanum berthaulii, woody plants	Anthraquinone, 2-methyl anthraquinones, 2-hydroxymethyl anthraquinone, and 2-formyl anthraquinone	Termites	Norris (1986)
Malus pumila	Phloretin (flavonoids)	*Scolytus multistriatus*	Norris (1977)
Quercus macrocarpa	Quercetin	*Scolytus multistriatus*	Norris (1977)
Wild rice *Oryza officinalis*	Steam distillates	*Nilaparvata lugens* (rice brown planthopper)	Velusamy *et al* (1990a)
	Steam distillates	*Cnaphalocrocis medinalis* (rice leaffolder)	Velusamy *et al* (1990b)

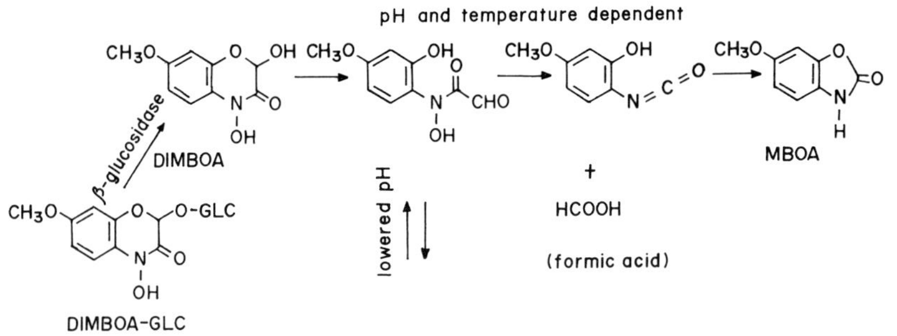

Fig. 6.6. Chemical degradation of DIMBOA-GLC. Modified from Bravo and Niemeyer (1986).

One outstanding piece of evidence of the chemical nature of resistance is seen in the resistance of certain corn lines to the European corn borer. The chemical compound resistance factor A (RFA), identified as MBOA (6-methoxy-2-benzoxazolinone), appears to function as a repellent and/or feeding deterrent (Klun *et al* 1967). MBOA does not exist in uninjured corn tissues, but DIMBOA (2,4-dihydroxy-7-methoxy-1,4-benzoxazin-3-one) in its glucoside form is present in uninjured corn leaves (Beck 1965). Injury to the plant tissue due to larval feeding results in the enzymatic conversion of the glucoside to the aglycone DIMBOA (Fig. 6.6). DIMBOA decomposes in a reaction, via intermediates, to yield MBOA and formic acid (Bravo and Niemeyer 1986). The yield of MBOA from the decomposition of DIMBOA increases with increase in pH of the midgut of most lepidopteran larvae.

Studies were undertaken to evaluate the toxicokinetics and pharmacokinetics of MBOA and DIMBOA on the European corn borer, and the fate of labeled MBOA in feeding trials and topical applications to the insect (Campos *et al* 1988). In feeding trials with MBOA incorporated in meridic diets, the total duration of European corn borer development is significantly increased with an increase in concentration of MBOA. The mean time to pupation and adult emergence is significantly lengthened at concentrations of 1.5 mg g^{-1} and higher. A decrease in the sex ratio (female:total) and in fecundity is observed at a concentration of 0.5 mg g^{-1} and higher. This suggests that European corn borer females are probably more susceptible to the effects of MBOA during their developmental period, which seems to be a new biological effect of MBOA. Since the adverse metabolic effects of MBOA are observed in the following generation and in larval mortality, it may be concluded that MBOA does have toxic effects on the European corn borer. Even so, the insect successfully minimizes the toxic level in tissues other than the hemolymph and eliminates a large amount of toxic materials in the pupal case.

Another factor in corn toxic to the southwestern corn borer (*Diatraea grandiosella*) has been identified as 2-dihydroxy-4,7-dimethoxy-1,4-benz-oxazin-3-one (N-O-ME-DIMBOA) (Hedin *et al* 1993). It is present in the corn whorl surface waxes at a higher concentration than DIMBOA or MBOA. This compound was identified by mass spectrometry and nuclear magnetic resonance techniques. It is suggested that this compound may have some role in the resistance of corn to this pest because the total surface wax content is higher in resistant lines than in susceptible ones.

Presence of growth inhibitors

A 'lethal silk factor' was reported in the silks of some corn lines resistant to the corn earworm larvae (Walter 1957). Since then, a number of corn genotypes possessing antixenosis or antibiosis (or both) resistance to corn earworm have been reported. Feeding trials showed that the silk of the corn cultivar Zapalote Chico was resistant to feeding by corn earworm larvae both as a natural food source and when incorporated into an artificial diet (Wiseman *et al* 1985). In the laboratory, larvae that fed on excised silk of Zapalote Chico remained significantly smaller than larvae that fed on silk from susceptible maize (see Fig. 8.14). The antibiotic factor of the silk, a compound known as maysin, is a flavone glycoside (Waiss *et al* 1979).

Gossypol, the yellow polyphenolic pigment found in the pigment glands of the genus *Gossypium*, is another insect-growth inhibitor that has received considerable attention as a source of some cotton cultivars resistant to the bollworm, tobacco budworm, pink bollworm, and other tissue borers (Shaver and Lukefahr 1969). Researchers have defined gossypol chemically as a phenolic, cadinene-type, sesquiterpene dimer and have also tested the hypothesis that it is responsible for the reduced growth of the tobacco budworm. Larvae reared in a diet containing 0.1% gossypol were about half the weight of 10-day-old bollworm, tobacco budworm, and pink bollworm reared on the gossypol-free diet (Shaver and Lukefahr 1969). The chrysanthe-min (cyanidin-3-β-glucoside) content of red petals is equally efficient in retarding the growth of cotton borer larvae on glandless (gossypol-low) cotton plants (Hedin *et al* 1983). The results suggest that chrysanthemin contributes to the toxicity of glanded red petals and gossypol to the glanded white petals. Furthermore, the biochemical mode of action has revealed that gossypol is an uncompetitive inhibitor of acetylcholine esterase, which augments the mechanisms of growth inhibition in insects (Ryan and Bryne 1988).

The glycoalkaloid tomatine in an artificial diet is a potent growth inhibitor of larval *Helicoverpa zea* (Isman and Duffey 1982). The larval growth of the insects showed linear dose–response relationships. However, tomatine toxicity was completely alleviated in *H. zea* by the addition of equimolar cholesterol into the diet, but some toxicity was maintained in *Spodoptera exigua* (Bloem *et al* 1989).

A number of pest-resistant crop varieties, in which resistance results from the expression of genes conditioning the presence of chemical or physical attributes, have been developed. Chemical resistance, mainly from anti-feedants in crop plants, is attributed to a great array of secondary metabolites (Swain 1977). Some of these secondary plant compounds in host plants are listed in Table 6.7. Within this diverse group, terpenes, alkaloids, and flavonoids are prominent as antifeedants to one or more insects. Other smaller groups, such as 1,4-benzoquinones and 1,4-naphthoquinones, are also extremely important because of their potent antifeedant effects on a broad array of insect species.

In some cases, antifeedants have been shown to be systemic in action, i.e., azadirachtin against the desert locust (Gill and Lewis 1971), coumarin in grass leaves against *Chorthippus parallelus* (Bernays and Chapman 1975), and sinigrin against aphids (Bernays 1983b). Secondary plant substances with antifeedant properties of a systemic nature may at the same time be toxic to insects; thus they provide additional protection to the host plant. The main advantage of antifeedants lies in their relative specificity. Although many allelochemicals are known to be poisonous to herbivores, several are relatively harmless to higher animals. Since antifeedants are of plant origin, the environmental hazards due to their use are considerably less than those due to synthetic pesticides. We should strive to develop crop varieties with a low level of toxicants but with a high level of antifeedants because they would be more environment friendly (Bernays 1983a). Such compounds are likely to play an important role in integrated pest management systems (see review by Jermy 1990).

Nutritional imbalance

It has been pointed out that the nutrient composition of plant tissue strongly influences the performance parameters (i.e., growth, development, reproduction, survival) associated with the fitness of phytophagous insects (Mattson and Scriber 1987).

Plant sugars seem to be vital to insects. They not only provide feeding stimulation, but are also important for proper growth and survival. Reduction in the sugar content of the plant at the critical stages of insect growth may adversely affect the insect. For example, larvae of the European corn borer require glucose up to the fourth instar and are capable of differentiating between varying sugar concentrations in the host plant tissues. Sugar deficiency until this stage of larval development may cause antibiosis (Beck 1957). Knapp *et al* (1966) also showed that the corn earworm could discriminate between concentrations of sugar, and that the sugar balance was considered an important component of certain corn lines' resistance to this insect.

In sap-ingesting insects, the aphids have a more intimate relationship with their host plants than many other pests and are affected by small

Table 6.7. Insect antifeeding compounds in host plants.

Plant	Antifeeding compounds	Insect	Reference
Cotton plants (*Gossypium* spp.)	Gossypol (dimeric sesquiterpenoid) and related terpenoids	*Heliothis* spp. and *Epicauta* sp. (blister beetle)	Maxwell et al (1965)
		Spodoptera littoralis	Meisner et al (1977a)
		Earias insulana (spiny bollworm)	Meisner et al (1977b)
	Isoquercitrin, quercitrin, quercetin	*Helicoverpa zea, Pectinophora gossypiella* (pink bollworm)	Shaver and Lukefhar (1969)
	Gossypol, gossypol-related triterpenoids, sesquiterpenoid quinones, hemigossypols, and cyanidin-3-β-glucoside (chrysanthemin)	*Heliothis virescens*	Hedin et al (1983)
Alfalfa (*Medicago* spp.)	Triterpenoid saponin	*Acyrthosiphon pisum* (pea aphid)	Pedersen et al (1976)
Cucumis salivus (Cucurbitaceae)	Cucurbitacin	*Tetranychus urticae* (mites)	Da Costa and Jones (1971)
		Phyllotreta nemorum (leaf beetles)	Metcalf et al (1980)
Leaves of bittergourd (*Momordica charantia* Linn)	Triterpenoid momordicine II, 23-O-β-glucopyranoside of 3,7,23-trihydroxycucurbita-5,24-dien-19-al	*Aulocophora foveicollis* (red pumpkin beetle)	Chandravadan (1987)
Plants in the Solanaceae (potato, tomato)	Glycosides of steroidal alkaloids, i.e., demissine, solacauline, tomatine, leptine I and II	*Leptinotarsa decemlineata* (Colorado potato beetle), *Manduca sexta* (tobacco hornworm)	Sturchkow and Low (1961), Sinden et al (1986)

Plant source	Compound	Insect	Reference
Wild tomato (*Lycopersicon hirsutum* f. *glabratum*)	2-Tridecanone (methylketone)	*Helicoverpa zea* (tomato fruitworm)	Dimock and Kennedy (1983)
Tomato (*Lycopersicon esculentum*)	Orthodihydroxy phenolics: rutin, chlorogenic acid, and α-tomatine; new caffeyl derivative of an aldaric acid	*Helicoverpa zea*	Elliger *et al* (1981), Isman and Duffey (1982)
Potato (*Solanum tuberosum*)	Tomatine	*Empoasca fabae* (potato leafhopper)	Raman *et al* (1979)
Artificial diet	Indolizidine alkaloid: castanospermine	*Acyrthosiphon pisum*	Dreyer *et al* (1981)
Lupine	Quinolizidine alkaloids	*Acyrthosiphon pisum*	Wegorek and Krzymanska (1970)
Roots of composite coltsfoot (*Tussilago farfara* L.)	Pyrrolizidine alkaloid: senkirkine	*Choristoneura fumiferana* (Clemens) (spruce budworm)	Bentley *et al* (1984a)
Leaves of *Lupinus polyphyllus* (Lindl)	Lupine alkaloid 13-*trans*-cinnamoyloxy-lupanine, 13-tigloyloxy-lupanine	*Choristoneura fumiferana*	Bentley *et al* (1984b)
Solanum spp.	*Solanum* alkaloids, tomatine, solanidine, α-chaconine	*Choristoneura fumiferana*	Bentley *et al* (1984c)
Barley (*Hordeum* sp.)	Indole alkaloid: gramine	*Schizaphis graminum* (greenbug)	Zuniga *et al* (1985)
Apple	Phloridzin (flavonoids)	*Acyrthosiphon pisum*, *Myzus persicae*, *Amphorophora agathonica*, *Aphis pomi*	Montgomery and Arn (1974)
Wheat (variety Amigo)	Polar phenolic fraction (flavone tricin)	*Schizaphis graminum* biotype C (greenbug)	Dreyer and Jones (1981)

Table 6.7. Continued.

Plant	Antifeeding compounds	Insect	Reference
Young leaves of sorghum (*Sorghum bicolor*)	p-Hydroxybenzaldehyde, dhurrin, and procyanidin	*Schizaphis graminum* (greenbug)	Dreyer et al (1981)
		Atherigona soccata (sorghum shootfly)	Singh and Rana (1986)
Young leaves of sorghum (*Sorghum bicolor*)	Cyanohydrin glucoside, dhurrin, and phenolic acids	*Locusta migratoria* (migratory locust)	Woodhead and Bernays (1978)
Alfalfa (*Medicago* spp.)	2–3% coumarin	*Hypera postica* (alfalfa weevil)	Hsiao (1969)
		Sitona cylindricollis (sweet clover weevil)	Akeson et al (1969)
	Dicoumarol	*Acyrthosiphon pisum* (pea aphid)	Dreyer et al (1987)
Sweet clover (*Melilotus* spp.)	Coumarin	*Listroderes costirustris* (vegetable weevil)	Matsumoto (1962)
Sweet clover (*Melilotus* spp.)	cis-o-HCA glucoside and coumarin	*Epicauta* sp. (blister beetle)	Gorz et al (1972)
Wheat (*Triticum aestivum*), rye (*Secale cereale*)	Hydroxamic acid (DIMBOA)	*Metopolophium dirhodum* (wheat aphid)	Argandona et al (1980)
Artificial diet	Hydroxamic acid (DIMBOA)	*Schizaphis graminum* (greenbug)	Argandona et al (1983)
Corn (*Zea mays*)	Hydroxamic acids	*Rhopalosiphum maidis* (corn leaf aphid)	Beck et al (1983)
Young leaf tissue of *Zea mays*	DIMBOA	*Ostrinia nubilalis* (European corn borer)	Klun et al (1967)
Cruciferae	Mustard oil glycoside: sinigrin	*Pieris brassicae*	David and Gardiner (1966)

DIMBOA, 2,4-dihydroxy-7-methoxy-1, 4-benzoxazin-3-one. HCA, hydroxycinnamic acid.

changes in the nitrogen status of the host (Wensler 1962). Phloem analysis (by the artificial exudation method) and subsequent nutritional experiments using artificial diets based on the sap composition of four clones of lucerne *Medicago sativa* were utilized to understand the mechanisms of the plant's resistance to the pea aphid *Acyrthosiphon pisum* (Febvay *et al* 1988). The results demonstrated that a specific amino acid deficiency was responsible for the resistance exhibited by these cultivars. It was found that the aphid-susceptible variety Resistador lacks no single amino acid; in contrast the resistant variety Lahontan is deficient in essential amino acids such as methionine and lysine. These results suggest that resistance is not the result of a simple nutritional effect, but the amino acid balance/deficiency does contribute to the resistance exhibited by some cultivars.

Another example of the relationship between a sap feeder and the amino acid-deficiency effect was provided by a study of the rice variety Mudgo that is resistant to the brown planthopper (Sogawa and Pathak 1970). The brown planthopper did little feeding on Mudgo as it contained a much lower content of asparagine than the susceptible varieties did. In separate tests, female planthoppers exhibited a strong attraction to asparagine. The low asparagine content of Mudgo, therefore, appears to restrict insect feeding and lower fecundity.

Structural factors

Structural factors can serve as defense mechanisms for plants when herbivores come in contact with them. The most common contact factors that impart antibiosis resistance are tissue toughness, cell-wall composition, proliferation of wounded tissues, and pubescence.

Plant-tissue toughness

Leaf toughness may reduce the suitability of leaves as a food source for herbivores in several ways:

1 Indigestible polymers such as cellulose and lignin in secondary tissues may reduce the rate of leaf consumption by herbivores.
2 Indigestible materials in tough leaves may be less suitable for herbivore growth, development, and/or survival.
3 Nutrients such as proteins and carbohydrates may be less available in tough leaves because of the hydrogen bonding between these compounds and lignins (Swain 1979).

Tough leaves of *Salix babylonica* and *S. alba* that can resist tearing, erode the cutting surface of the incisors of the leaf beetle *Plagiodera versicolora* Laich (Fig. 6.7). Consequently, leaf consumption and egg production of the beetles are seriously affected (Raupp 1985). Tough leaves generally contain lower

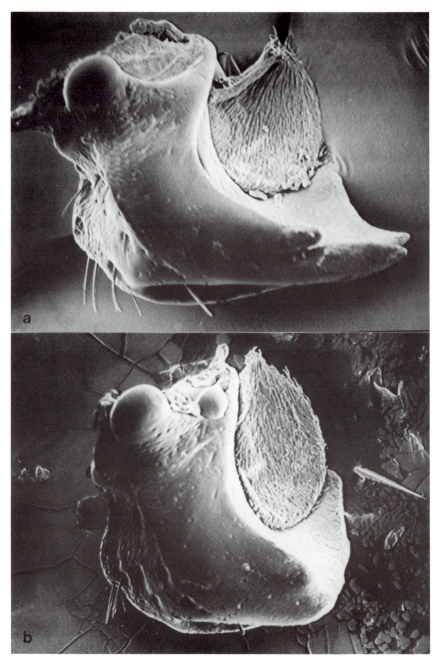

Fig. 6.7. Scanning electron micrographs (×320) of the mandibular incisors of the adult leaf beetle *Plagiodera versicolora* feeding on willow leaves. (a) The left mandible showing sharp incisors when beetles are fed tender leaves; (b) the same mandible showing wornout incisors when beetles are fed tough leaves. Reproduced from Raupp (1985) with permission from Blackwell Scientific Publications, Oxford, UK.

levels of nitrogen and water and this acts as a potent defense affecting morphology, feeding behavior, and ultimately spatial and temporal patterns of herbivores (Raupp and Denno 1983). Although toughness of plant tissues or any other physical characteristic is an efficient defense mechanism, such traits are often eliminated during breeding of improved varieties, especially in those crops that are consumed as leaves. Moreover, such morphological characteristics seem to vary with environmental conditions in different cropping regions.

Cell-wall composition

The presence of neutral detergent fiber (NDF), lignin, and biogenic silica in the cell walls of plants can affect insect feeding at both nutritional and physical levels. Plants high in cell-wall structural components are not desirable for herbivores (Scriber and Slansky 1981). Elevated levels of indigestible fiber and silica may increase the bulk density of the diet to the extent that insects are unable to ingest sufficient quantities of nutrients and water.

The second-generation European corn borers feed initially on leaf sheaths and collars, and then bore into stalks, often causing breakage (Guthrie *et al* 1970). Resistance is apparently associated with factors found in the leaf sheath and collar tissues, but not with DIMBOA content (Klun and Robinson 1969). Rojanaridpiched *et al* (1984) found that the silica and lignin contents of some maize germplasm increase with age and are dominant factors promoting resistance at later stages of plant development. In the maize lines examined in their study, silica and lignin were significantly correlated with the leaf-feeding resistance.

Proliferation of wounded tissues

Antibiosis may also involve the proliferation of cells triggered by insect injury or increased secretion of plant substances known to cause the death of eggs or young larvae inside the damaged plant. Experimental results demonstrated that the larvae of young pink bollworms were crushed or drowned by proliferating cells of the injured tissues in certain corn lines (Adkisson *et al* 1962). Similarly, mustard plants with a hypersensitive reaction to insect attack produced a necrotized zone around the base of the eggs of cabbage worm *Artogeia rapae* L., causing them to desiccate (Shapiro and DeVay 1987).

Pubescence

Pubescence may also interfere with ingestion of food by small mandibulate larvae and adults. The first-instar larvae of the cereal leaf beetle, *Oulema melanopus*, were seriously affected by the pubescence of certain wheat varieties (Schillinger and Gallun 1968). High mortality was possibly due to the fact that the first-instar larvae usually had to bite through the hairs three times before they could reach leaf cellular tissue. In doing so they had to ingest the non-nutritive bulk (cellulose and lignin) of the plant material. These bulk plant materials pass through the insect gut without being digested (Smith and Kreitner 1983), and the young larvae died from the unbalanced diet of fibrous materials.

Tolerance

The basic triad of resistance mechanisms – antixenosis, antibiosis, and tolerance – has been found to result from independent genetic characteristics, which are, however, interrelated in their effects (Painter 1951). Tolerance represents a plant's ability to suffer a smaller yield loss than that manifested by susceptible cultivars by dint of its innate capacity to grow and reproduce itself or to repair injury to a marked degree. The basic difference between tolerance and the two other resistance mechanisms is that the former stems from the plant's response to insect attack, whereas antixenosis and antibiosis are expressed in the insect's reaction to certain specific characteristics of the host plant (Horber 1980). Therefore, tolerance is a basis of resistance due to the plant's ability to withstand insect damage. In natural systems, plant species that can tolerate or compensate for herbivore feeding have obvious selective advantages that lead to genotype maintenance (Trumble *et al* 1993).

Tolerance is useful in pest management programs because of certain inherent advantages.

1 Although not resistant, tolerant varieties have a higher economic threshold than susceptible varieties; they require less insecticide application and enhance biocontrol.
2 Tolerant plants that can support the insect load do not impose selection pressure on the insect population, and thus are useful in preventing the selection of insect biotypes (Horber 1972).
3 In varieties with a combination of the three forms of resistance, tolerance increases yield stability by providing at least a moderate level of resistance when the vertical genes providing a high level of resistance through antixenosis and antibiosis are rendered ineffective by a new biotype (Mac-Kenzie 1980).

In this way, a tolerant variety would 'buy time' between the 'breakdown' of an old variety and the release of a new variety.

Compensatory reactions

Because of the unique nature of tolerance involving the plant's response to insect attack, it is necessary to understand how insects injure plants, how plants repair this injury, and whether tolerance is likely to be influenced by the adaptation of a crop variety to climatic conditions. In agricultural crops, reports of plant compensation are mostly concerned with yield rather than fitness. The hypothesis that herbivory is either detrimental or beneficial to plants, by causing an overcompensation response, has been reported by several workers (Harris 1974). However, the answer is not a biological absolute that favors only one of these contradictory hypotheses. Rather, the compensatory responses vary, and the crop fitness or yield is determined by a variety of endogenous and exogenous factors (Trumble *et al* 1993).

The increase in growth or yield is normally related to carbohydrate production. Sometimes, however, the herbivore damage to foliage may lead to an increase in growth or in yield. The exact mechanisms associated with the partitioning and allocation of photoassimilates in plants are poorly understood. The partitioning of carbon between chloroplasts and the cytoplasm is highly regulated by endogenous (rate of carbon dioxide assimilation, concentration of substrates, products, and effector molecules) and exogenous factors (light and temperature) (Trumble *et al* 1993). Carbohydrate metabolism and their ultimate allocation are regulated by the effector metabolite fructose-2,6-bisphosphate, which is in itself regulated by environmental factors. Thus, a variety of environmental components including arthropod feeding would be expected to alter carbon partitioning and allocation within the plant. Plant compensation for herbivory damage may involve one or more of the following processes.

Sink—source relationships

Sink-limited plants do not undergo yield reduction following leaf loss. Major reductions in leaf area due to leaf mining by *Liriomyza sativae* did not affect tomato yield because there were excess photosynthates (Johnson *et al* 1983). In this case, the insect attack merely decreased the amount of physiologic shedding and delayed the harvest. In contrast, source-limited plants usually suffer marked yield reduction following a decrease in leaf area. The relative effects of sink or source limitation on the yield of agricultural crops are likely to vary with cultivar, growing conditions, and stress.

Photosynthetic enhancement

Herbivory can influence photosynthesis and respiration through a variety of physical and biochemical effects on the plant. Net photosynthesis (P) is the difference between total gross photosynthesis (Pg) and respiration (R) including photorespiration (Ting 1982): $P = Pg - R$. Herbivory injury may decrease P by directly affecting Pg at any level and/or by stimulating an increase in R. In contrast, arthropod damage may increase net photosynthetic activity because leaves often function below maximum capacity (Lovett Doust 1989). Partial defoliation may result in an increased supply of leaf cytokinins or root-derived cytokinins because there is less competition for hormones within the plant. Increased levels of cytokinins have been shown to increase net CO_2 fixation as a result of enhanced assimilate transport and nutrient uptake, to delay senescence, and to decrease intracellular resistance to CO_2 transport. Additionally, the increased availability of nitrogen due to either reduced leaf area or a feeding-induced senescence could enhance protein synthesis (Ting 1982). For example, yields were enhanced or unaffected by partial defoliation of cotton or tobacco by the tobacco budworm (Kolodny-Hirsch *et al* 1986). In a similar study, low levels of simulated defoliation by lepidopteran insects at the preheading or heading stage in cabbage improved yields over those of undamaged plants (Shelton *et al* 1990). Sometimes, the defoliating insects may act as biological pruners, removing superfluous leaf area and an excess respiratory load, decreasing mutual shade, and thus improving yields.

Resource allocation

A substantial body of circumstantial evidence exists for resource reallocation following herbivory. When insect injury occurs early in the life of a wheat crop, the surviving plants grow larger and have greater numbers of normal ear-bearing shoots than usual (Bardner *et al* 1970). Early attack by the sorghum shootfly *Atherigona varia soccata* on the main shoot of sorghum induced the production of a few synchronous tillers that grow rapidly and survive to produce harvestable earheads (Rana *et al* 1985). The Colorado potato beetle may cause almost 100% defoliation within a few weeks of

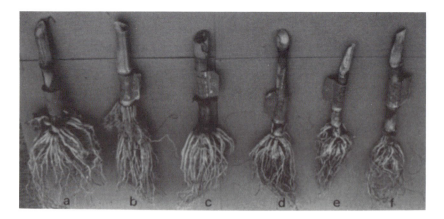

Fig. 6.8. Characteristic root types representing six inbred maize lines. (a) Strong, spreading, deep root system with excellent lodging resistance. (b) Vertical deep roots with weak resistance to lodging under rootworm attack. (c) System intermediate in root numbers and disease resistance. (d) System with relatively high root numbers with the dark lower portion disease infested. (e, f) Relatively poor root system, susceptible to lodging under rootworm attack. Reproduced from Dicke and Guthrie (1988) with permission from the American Society of Agronomy, Madison, USA.

harvest, but tuber growth remains unaffected (Hare 1980).

Inbred strains of corn differ greatly in their ability to recover from the larval attack of the corn rootworm *Diabrotica* spp. Tolerance for larval feeding is attributed to the size of the root system and to the development of secondary roots. This type of well-developed root system is due to the allocation of resources by the plant following rootworm damage (Ortman *et al* 1974). In theory, both tolerant and susceptible maize varieties support larval populations of equal size (Painter 1951), but since a tolerant maize has a larger root system (Fig. 6.8), it is able to withstand the damage better. Figure 6.8 shows high susceptibility in rows **e** and **f** and excellent tolerance in row **a**. Vertical-pull devices have been successfully used to obtain comparative tolerance rating. A highly significant correlation was found among root-pulling resistance, root volume, and root classification, indicating that any one of these three characteristics would be suitable for tolerance evaluation (Penny 1981).

Environmental stress

Some exogenous factors like water and temperature stress can significantly alter plant compensatory capacity, mostly through the alteration of allocation and reallocation of resources and stomatal closure effects on gas exchange and photosynthetic capacity (Benedict and Hatfield 1988, Holtzer *et al* 1988). Air pollution, like water stress, can alter plant compensation for arthropod damage. Although most air pollutants are deleterious to plants, wet deposition of pollutants (acid rain, or acid fog) on foliar surfaces can act as fertilizers (Trumble and Hare 1989) that result in increased photosynthetic activity and subsequent recovery from arthropod damage.

Increasing levels of atmospheric CO_2 alter a variety of physiologic systems and are likely to substantially change plant compensatory responses. Bazzaz (1990) reviewed the possible consequences of increasing global CO_2 levels, such as improved biomass accumulation, improvement in photosynthesis, greater leaf area index, and increased branching. All these plant factors are conducive to tolerance for arthropod defoliation.

Ecological resistance

The environment may favor the plant or insect unequally, or may alleviate or aggravate damage. The resulting type of resistance is ecological resistance. It may be broadly categorized as pseudoresistance and induced resistance.

Pseudoresistance

Pseudoresistance may be the result of transitory characteristics in susceptible host plants. Such plants may evade pest incidence by escaping in space and time. Escape in space appears to be an important component of plant survival in nature, but agricultural plants in monoculture situation are prone to insect depredation. For an oligophagous insect pest, however, not only finding the right host, but also contacting the plant at the appropriate stage of development, are of primary importance. The phenologies of the plant and insect must synchronize for the insect to colonize/infest an appropriate plant structure at the right stage. Host evasion takes place when the plant growth pattern is modified so as to make it asynchronous with the insect phenology.

Examples of such asynchrony include cultivars that will tolerate adjust-ments in the planting time, those which pass through susceptible pheno-logical stages more rapidly than those currently grown, and those whose time to maturity prevents arthropods from reaching pest status. Instances of temporal modifications of phenological characteristics to reduce pests have been reported in a few crops. A study over a period of 16 years (1952–1968) on the incidence of the rice gall midge *Orseolia oryzae* W-M. at the Central Rice Research Institute, Cuttack, India, revealed that rice crops planted in late August and thereafter were worst affected by the pest. The peak period of gall midge infestation was from the second week of October to the second week of November. Thus the time of rice transplanting has directly influenced the prevalence and incidence of this pest (Prakasa Rao *et al* 1971). Early planting could provide the best means of gall midge avoidance. In India the shootfly *Atherigona varia soccata* Rond. of sorghum was almost absent at the beginning of the rainy season, and early plantings in India were completely free from the pest (Vidyabhushanam 1972).

Even though these temporal factors do not constitute true host plant resistance, the fact remains that these approaches are very effective and practical parts of an intergated pest management (IPM) system. Furthermore, as the length of the plant life cycle is under genetic control, the technique of breeding for short- or long-duration crops to enhance asynchronies may be utilized as a component of the resistance program.

Induced resistance

Changes in plants following damage or stress are called 'induced responses.' In some cases, the responses may act as 'induced defense' against herbivore injury or invasion of plant pathogens (Karban and Myers 1989). All these responses need not be defensive, and the more neutral term 'induced resistance' is commonly used in the literature (Haukioja 1991).

Not all induced plant responses increase resistance by making plants less suitable as hosts. In certain situations, plant quality is improved following herbivory (Carroll and Hoffman 1980) or adverse environmental factors (Mattson 1980). However, induced responses act as a defense against herbivores and can regulate herbivore populations under a wide variety of conditions. In conjunction with other regulatory agents (e.g., predators and parasitoids), they can to certain extent depress herbivore populations (Lin and Kogan 1990).

Changes following damage

Injuries to plant tissues cause phytochemical induction in a wide array of plant responses. The nature of the response varies with the plant genotype, ontogeny, phenology, and interactions of the plant with the abiotic and biotic environments (Coleman and Jones 1991). The capacity of a plant to exhibit any induced response will be at least partially under genetic control. Different genotypes may have different thresholds of damage before exhibiting induced responses, and may show different responses once induced. The ability of plant genotypes to express induced responses may be differently affected by the same environmental conditions. Similarly, the physiology, biochemistry, and anatomy of leaves of different ages may have important implications for induction studies. Furthermore, the season of the year, age, and growth habits of the plants are all important variables that can affect the expression and detection of induced resistance.

Many studies of induced responses have indicated changes in the levels of tannins and phenols, which are products of the shikimic acid pathway. The relative activity of the enzyme phenylalanine ammonia lyase (PAL) can determine the production of phenolics, including lignin. Hence, PAL activity is considered an important indicator of induced resistance (Hartley and Lawton 1991). Herbivore damage also affects the concentrations of available nitrogen as well as other important nutrients in foliage.

In some instances, plants synthesize phytoalexins as a result of herbivore attack. The phytoalexin concept was developed by Muller and Borger (1940) from their studies on fungal infection of plants. Phytoalexins are low-molecular-weight, antimicrobial compounds usually present in plants at extremely low concentrations prior to infection (Kuc and Rush 1985). These can be synthesized *de novo* by plants following microbial infection, and their effectiveness is determined by the speed and magnitude at which they are produced and accumulated. There are now an increasing number of examples from several insect–plant systems that exhibit phytoalexin production and accumulation (Table 6.8). The accumulation of phytoalexins in response to insect damage, microbial invasion, or mechanical damage is known to occur in many plant families including Leguminosae, Solanaceae, and Compositae (Bailey and Mansfield 1982). It has recently been demonstrated that attack by insects including the flea beetle *Psylliodes chrysocephala* L. in oilseed rape *Brassica napus* (Koritsas *et al* 1991), and the rootfly *Delia floralis* Fall in Cruciferae plants, can change both the total concentration of glucosinolates in different plant tissues and the relative proportions of aliphatic and aromatic compounds (Birch *et al* 1992). Infestation of oilseed rape by the cabbage stem flea beetle leads to a substantial increase of indole glucosinolates, gluco-brassicin, and neoglucobrassicin (Koritsas *et al* 1991). Damage to the roots of

Table 6.8. Induction of resistance by previous herbivory.

Inducing herbivore	Plant	Induced reaction	Reference
Pea aphid	Alfalfa	↑ Coumestrol	Loper (1968)
European pine sawfly	Pine	↑ Polyphenol	Thielges (1968)
Pine beauty moth	Lodgepole pine	↑ Monoterpene	Leather et al (1987)
Bark beetle	Lodgepole pine	↑ Terpenoid and phenolic compounds	Wong and Berryman (1977)
Sweet potato weevil	Sweet potato	Ipomeamarone (furanoterpenoid phytoalexin)	Uritani et al (1975)
Gypsy moth	Gray birch Black oak Red oak	↑ Nutritional value ↑ Sugar ↑ Phenolics, hydrolyzable tannins	Wallner and Walton (1979) Wratten et al (1984) Rossiter et al (1988)
Lepidopteran larvae	Birch	Phenolics	Haukioja and Niemela (1977)
Lepidopteran larvae	Birch	↑ PAL activity	Hartley and Firn (1989)
Lygus disponsi	Sugarbeet Chinese cabbage	↑ Quinones ↑ Phenolics	Hori (1973) Hori and Atalay (1980)
Cotton bollworm	Cotton	↑ Phenolics	Guerra (1981)
Spider mites	Cotton	↓ Fecundity	Brody and Karban (1989)
Striped cucumber	Squash	↑ Cucurbitacins	Carroll and Hoffman (1980)
Mexican bean beetle	Soybean	↑ Isoflavonoids ↑ Glyceollin ↑ PAL activity	Chiang et al (1987) Fischer et al (1990) Fischer et al (1990)

Soybean looper	Soybean	↑ Glyceollin	Liu *et al* (1993)
Tobacco mosaic virus	Tobacco	↓ Oviposition of *Myzus persicae*	McIntyre *et al* (1981)
Spodoptera littoralis	Tomato	Affects feeding	Edwards *et al* (1985)
Spodoptera exigua	Tomato	↑ Proteinase inhibitor	Broadway *et al* (1986)
Psylliodes chrysocephala	Mustard, kale	↑ Indole glucosinolate	Koritsas *et al* (1991)

↑, increased; ↓, decreased; PAL, phenylalanine ammonia lyase.

Brassica napus by *D. floralis* has led to two- to four-fold increases in indole-base glucosinolate compounds. The largest increase, from four- to 17-fold, for an individual compound after rootfly attack was found for 1-methoxy-3-indolyl-methyl glucosinolate (Birch *et al* 1992). In most of these examples, the induced resistance affects the insects as they feed on the phytoalexins produced by the wounding of the plant.

Whatever may be the adverse effects of phytoalexins on the performance of insect herbivores, reduction in the palatability of foliage following wounding by insects is a widespread phenomenon (Edwards *et al* 1991). The plant may indirectly benefit because insects spend more time foraging for suitable food. In this searching process, mortality risk through predation or falling off the plant will be greater for an insect that must continually move (Schultz 1983b).

The response of plants to herbivore attack can be more extensive than simple modifications of secondary metabolites. For example, spider mites cause widespread changes in the cytology, histology, and physiology of their host plants (Karban and English-Loeb 1988). Herbivores can influence the morphology of their host plants by causing an increase in the density of prickles, spines, and hairs (Myers and Bazely 1991). All these changes influence future herbivore attacks.

Mechanisms of induced resistance

Chemicals seem to be the main components of both constitutive and induced plant defenses against herbivores, and their biosynthesis appears to be a major function of the parenchyma and, probably, other living cells (Lieutier and Berryman 1988). The speed and intensity of biosynthesis may depend on the distance from the site of injury, the amount of damage done, and/or the kind of elicitors involved.

In recent years, studies on the insect-induced proteinase inhibitors in plant leaves and synthesis of phytoalexins in cotyledons and seedlings have provided models for studying the biochemical and molecular mechanisms that underline defense responses induced by elicitors. Over the past 20 years, evidence has been accumulating to show that plant and fungus cell-wall fragments contain oligosaccharins within their polysaccharide structures. The oligosaccharins can be released when cells are damaged by insects and microorganisms, and can subsequently act as signals to activate plant genes that code for enzymes to produce defense chemicals (Duffey and Felton 1989).

Elicitor model. Systematic research on the insect- or wound-induced synthesis of proteinase inhibitors in tomato leaves led to the isolation of a compound called the proteinase-inhibitor-inducing factor (PIIF), which is an active inducer for the synthesis of proteinase inhibitors in excised tomato leaves (Ryan 1983). The possible pathway of these carbohydrate signals in

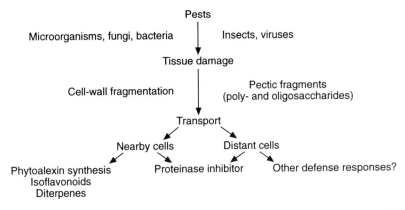

Fig. 6.9. Pathway of induced defense chemicals resulting from insect attack. Modified from Ryan (1983).

response to insect or microorganism attack is presented in Fig. 6.9. The fragments released by insect attack can be simulated *in vitro* by the contact of cell walls with plant endopolygalacturonases (EPGases), released from compartments in the cell by the chewing action of the attacking insects. These fragments are thought to be transported throughout the plants where they can activate defensive genes. Likewise, microorganisms initiate an attack on plants by releasing a variety of enzymes, including polygalacturonides, that degrade the cell wall. The cell-wall fragments are released by the microorganism's EPGases or pecticylases (polygalacturonic acid lyases, PGAlyases) and act as localized signals to elicit the synthesis of phytoalexins. As infection proceeds, the fragments (signals) continue to be released and therefore cause continued phytoalexin synthesis (Davis *et al* 1984).

Using *Spodoptera exigua*, Ryan *et al* (1986) demonstrated that insect herbivory induced a significantly higher accumulation of proteinase inhibitor in the tissue samples from wounded tomato plants than from nonwounded plants (Fig. 6.10). Larvae reared on diets of reconstituted tomato leaves indicated that foliage from wounded plants was a poorer food source than that from nonwounded plants. Broadway and Duffey (1986) showed that chronic exposure to proteinase inhibitors has a detrimental effect on the digestive physiology of *S. exigua* larvae as it induces a pernicious hyperproduction of gut proteases. This kind of adversity, coupled with insufficient dietary availability of sulfur-containing amino acids (i.e., methionine) needed for enzyme synthesis, results in growth inhibition. This evidence further supports the finding that proteinase inhibitors have the potential to contribute towards the defense of plants (solanaceous and leguminous) against herbivorous insects.

A different type of induction occurs with the cucumber beetle *Epilachna tredecimnotata* feeding on squash (*Cucurbita*). Feeding by this beetle results in

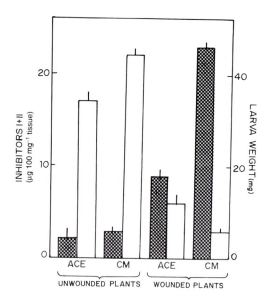

Fig. 6.10. Levels of proteinase inhibitors I and II in leaves of tomato varieties Ace 55 (ACE) and Castlemart (CM). Leaves of unwounded and wounded plants (dark columns) are compared with the growth of *Spodoptera exigua* larvae (open columns) that subsequently fed on the leaves. Reproduced from Ryan *et al* (1986) with permission from Plenum Publishing Co., New York.

an accumulation of cucurbitacins that are strong feeding excitants for the cucumber beetle itself. This kind of relationship between insect–plant feeding improves the quality of the host plant (Carroll and Hoffman 1980).

Even more fascinating is the manipulation of existing or induced plant chemical compounds that regulate the attraction and stimulation of natural enemies. In one example, this is due to an elicitor that mediates a systemic effect for the tritrophic system consisting of lima bean plants *Phaseolus lunatus*, spider mites *Tetranychus urticae*, and predatory mites *Phytoseiulus persimilis* (Dicke *et al* 1990). It was shown that uninfested leaves of lima bean plants, one leaf of which was infested by spider mites, became attractive to the predatory mites. The existence of a water-soluble endogenous elicitor(s) that mediates this systemic effect has recently been suggested (Dicke *et al* 1993). The production of predator-attracting infochemicals was found to occur systemically throughout the spider mite-infested plant. Not only is the predator response correlated with the number of spider mites that infest the leaves, but the predators also prefer the synomone of the noninfested leaves that have been exposed to elicitor solutions obtained from leaves infested by the higher number of spider mites.

Laboratory studies on phytoalexins. Hart *et al* (1983) have done substantial work to prove the antifeeding and adverse qualities of phytoalexins in the laboratory using the Mexican bean beetle. They evaluated the effect of phytoalexins on the feeding preference of the Mexican bean beetle. Soybean seedlings were germinated in the greenhouse and when the cotyledons were fully opened, small disks were cut and irradiated with ultraviolet light to elicit phytoalexin production. Another group of nonirradiated soybean disks were used as controls. It was observed that the phytoalexin-rich disks were probed but not fed upon as shown by fine mandibular markings, whereas sustained feeding occurred on the control disks (Fig. 6.11).

Induction of resistance by bioregulators. Several attempts to use plant bioregulators (PBRs) and plant growth regulators (PGRs) to enhance insect resistance in plants have been reported (Fischer *et al* 1990). A PBR modifies the plant through specific biochemical processes without imparting gross morphological changes, whereas a particular PGR may have both biochemical and growth-modifying effects. Many plant regulators, herbicides, chelating compounds, and fungicides are presently being used experimentally and commercially in the field. Their impact often has been variable – either beneficial or detrimental to plant diseases and insect pests.

One of the earliest deliberate uses of a PGR to manipulate plant protection against insect and mite attack relates to its use on fir trees in order to delay

Fig. 6.11. Feeding pattern of the Mexican bean beetle on cotyledon disks of soybean under dual-choice tests. (a) Feeding ridges observed on the control disks; (b) mandible holes indicating nibbling but no sustained feeding on the disk with phytoalexin. Reproduced from Hart *et al* (1983) with permission from Plenum Publishing Co., New York.

budbreak, and thereby deprive spruce budworms of suitable food on emergence (Eidt and Little 1970). When abscisic acid is applied to balsam fir, bud opening in the spring is delayed and the emerging spruce budworm *Choristoneura fumiferana* larvae are forced to feed on old needles, which are low-quality food.

In several examples, PGRs and herbicides were found to reduce the susceptibility of treated plants to pests (cf. Table 1, Fischer *et al* 1990). Some PGRs, such as CCC (cycocel) and PIX (mepiquat chloride), significantly reduced the susceptibility of sorghum to aphid attack. Treatments with CCC increased the intercellular pectin levels and pectin methoxy content, both of which are major barriers to aphid stylet penetration (Campbell and Dreyer 1985). PIX applied on cotton has induced plant resistance to *Helicoverpa zea* by way of increasing the relative condensed tannin concentration, astringency, and terpenoid concentrations of the plants. Recently Liu *et al* (1993) demonstrated that application of the chemical *p*-chloromercuriphenyl sulfonic acid elicited the production of induced phytoalexin glyceollin in soybean PI 227687 hypocotyls. This phytoalexin is an antifeedant to the soybean looper (*Pseudoplusia includens* (Walker)).

Potential of induced resistance

The research of the past two decades shows that induced resistance in plants remains an exciting and highly viable area for continued investigation. The time period over which resistance can be observed varies from a few days to several months. In the light of 'optimal defense strategies' (Rhoades 1979), a postattack response is a more efficient and economical line of defense than the attack-independent accumulation of costly secondary metabolites. Of more practical interest to applied biologists is the phenomenon of general or cross-protection, whereby attack by one pest or pathogen induces resistance not only to the inducer, but also to other unrelated pests such as a virus (McIntyre *et al* 1981).

Ryan (1983) identified some specific biochemical compounds responsible for the induction of specific defense mechanisms in several plant species. Soon it may be possible to routinely and artificially induce plant defenses through elicitors. The functioning and recognition of these elicitors and the role of oligosaccharides as signals in activating the plant genes for defense are currently receiving much attention (Ryan and Farmer 1991). However, our understanding of how complex carbohydrates in plants may function in signaling and recognition systems is still in its infancy.

Elicitors may be effective in protecting plants for a short period, when they are highly susceptible at a particular stage to a particular pathogen or insect. A temporary disruption in normal metabolism at this stage, as a result of the application of elicitors, may provide protection without affecting yield. The use of specific inducers may be one of the most exciting developments in the fight against insects since the advent of the modern organosynthetic insecticides.

The inducing mechanisms enhancing the ability of the plants to produce their own defenses are more economical and sound than applying potentially hazardous chemicals. Inducible defenses should be more effective as more reliable elicitors become available.

Another area of potential promise is the use of bioregulators that would promote asynchrony of plant growth and insect development. Bioregulators can selectively regulate the expression of key genes by turning on their respective promoters, thereby utilizing the existing genes to greater advantage. Bioregulators may be able to activate (or inhibit) whole gene families and thus control multiple pathways. Once a plant's key resistance mechanism is understood, it may be possible to augment it by manipulating the genes responsible for resistance (Marvel 1985). By and large, the concept of induced resistance may be an alternative or complement to the existing host plant resistance strategies.

Combined Factors of Resistance

Plant resistance mechanisms against herbivores are multidimensional. Chemical defenses are only part of a complex strategy that enables plants to coexist with herbivores, ward off pathogen attack, and successfully compete with other plant species in the same community. Although the antiherbivory theory of coevolution suggests that most secondary plant compounds have evolved to perform allomonal functions in defense against most sympatric herbivorous fauna, yet many secondary compounds are kairomones for a few oligophagous insects on the allelochemical-producing plants (Norris and Kogan 1980). Insects with a restricted host range sometimes shift to unrelated host species, suggesting that behavioral barriers do not simply reflect a need to avoid toxins, and the occurrence of toxins could be a consequence and not a cause of specialization (Bernays and Chapman 1978). Bernays and Graham (1988) argued that chemical coevolution between plants and herbivores has been overemphasized, and that generalist natural enemies, especially predators of the herbivores, may be the dominant factor in pest suppression.

In spite of debate in ecological investigations, it has become clear that advocacy of any single selective force, be it plant chemistry or any of the alternatives proposed by Bernays and Graham (1988), will involve exceptions (Barbosa 1988b). Any argument that relies on a single causative factor like allelochemicals will be less than compelling, because insects are not exposed to secondary chemicals alone, but to whole plants in distinct micro-environments characterized by certain biotic and abiotic factors, constituting a multiple of selective forces. Although the concentration of a given allelochemical may determine the behavioral (antixenosis) and/or physiologic (antibiosis) acceptability or suitability, respectively, of a plant at a given time, the influence of the plant's tolerance qualities, natural enemies, the micro-climate it creates, etc., are likely to determine the herbivore's host range. Other factors such as plant texture and morphology, plant nutrition and availability in time and space, and inter- and intrapopulation variation among herbivores play important roles as well (Reese 1983). The relative importance of allelochemicals *vis-à-vis* nutritive factors on insect performance is hard to determine because the concentrations of some putatively important second-ary constituents correlate negatively with nutritive compounds essential for herbivores.

Almost no attention has been given to the role of microbes in host plant resistance. Microbes also may take an active role against herbivores as they are well known for their ability to produce a diverse array of secondary metabolites (Woodruff 1980). Microbes may produce substances that are not highly toxic but may significantly enhance the activity of plant secondary compounds. Fusaric acid, a secondary metabolite of *Fusarium*, can increase the susceptibility of *Helicoverpa zea* to gossypol, MBOA, and an unidentified saponin by inhibiting nonspecific monooxygenases. The monooxygenases would normally be employed by the *H. zea* larvae to detoxify the secondary compounds (Dowd 1989).

Finally, environmental factors may alter plants in various ways, which in turn alter the modalities of host plant resistance.

Resistance of corn to the European corn borer

Studies on insect resistance in corn, especially to the European corn borer, have been conducted for more than 60 years in the United States. Most of the borer species of corn have several overlapping generations per year or cropping cycles, and corn plants are attacked and damaged from the early seedling stage to near maturity. The plants must therefore possess resistance either for all plant parts or in several key parts (leaf tissue, stalks, sheaths, collars, shanks, ears). It may take a long time to develop genotypes that will withstand multiple attacks by borers (Mihm 1985). The task of finding resistance to multiple generations of the European corn borer continues to be formidable. Early tests on corn resistance to the European corn borer showed differences in larval populations on different cultivars and the ability of some cultivars to tolerate damage (Huber *et al* 1928). With the advent of extensive breeding programs, it was found that the resistance to first-generation borers operates in leaf tissues, and resistance to second-generation borers operates primarily in the sheath–collar tissue (Guthrie *et al* 1978).

The chemical nature of resistance (antibiosis) has been studied for many years. Resistance to first-generation European corn borers was highly correlated with the DIMBOA content in the whorl tissues; in addition, silica and lignin appeared to contribute to this resistance as well. Multiple factor resistance in maize to the European corn borer was reported by

Rojanaaridpiched *et al* (1984). Resistance to the second-generation borer was significantly correlated with silica content in the leaf sheath and collar tissues of some lines; therefore, DIMBOA played a secondary role in imparting resistance to some lines of corn attacked by second-generation borers. DIMBOA was also found to be an important resistance factor against the fungus *Helminthosporium turcicum* in several inbred corn lines (Toldine 1984).

Resistance of cotton to insect pests

In cotton, multiple resistance factors to a number of pests have been identified and summarized by Beck and Maxwell (1976). Plants with glabrous leaves, no nectar and high levels of gossypol have an enhanced resistance to *Heliothis* spp., whereas frego bract, red plant color, increased pubescence, rapid fruit set, and no nectar provide resistance to the boll weevil *Anthonomus grandis*. The resistance of cotton to the pink bollworm *Pectinophora gossypiella* involves the absence of bracts, the presence of glabrous leaves, high cell proliferation rates, high gossypol content, and no nectar (Agarwal *et al* 1976). In addition, short-season cottons have the added advantage of being harvested almost 1 month earlier than indeterminate varieties, well before the environmental conditions that induce diapause in the boll weevil and pink bollworm. It is now possible to develop varieties carrying multiple resistance under varied climatic conditions.

Resistance of solanaceous crops to insect pests

The multiple defense patterns in the Solanaceae are well documented. A first line of defense in the potato is the presence of chemical compounds contained externally in type B glandular trichomes. Small insects like aphids, leafhoppers, and mites are trapped by the exudates of trichomes (Tingey and Gibson 1978). Similarly, glandular trichomes of the wild tomato contain large amounts of the ketohydrocarbon, 2-tridecanone. This compound is a powerful antibiotic against the tomato fruit worm and the tobacco hornworm (Williams *et al* 1980). The flavonal glycoside rutin is also part of the glandular trichome defense system in the tomato; if it is introduced in artificial media it inhibits the growth of *Helicoverpa zea* larvae (Isman and Duffey 1982). About one-third of the total phenolic content in tomato leaves is stored in the trichomes.

The second line of defense in the *Solanum* and *Lycopersicon* genera is the presence of steroidal alkaloids (Gregory 1984). These alkaloids can influence the suitability of a plant for arthropods. A third line of defense has been suggested by Ryan (1983) as an induced mechanism. The wounding of potato or tomato leaves releases a PIIF that translocates within the plant and inhibits further attack by the insect. The antiherbivory role of the accumulation of proteinase inhibitor in tomato has been well documented (Broadway *et al* 1986).

Resistance of Leguminosae to insect pests

Resistance in cowpea *Vigna unguiculata* (L.) to its key insect pests, the cowpea curculio *Chalcodermis aeneus* Boheman, in the southeastern United States operates under three genetically distinct types of mechanism. The antixenotic mechanism inhibits feeding of adults on the cowpea pods. A pod factor inhibits penetration of the pod wall, thereby reducing the colonization of the adult curculio. The antibiotic mechanism causes prolonged development and high mortality of larvae feeding on the seeds (Cuthbert *et al* 1974).

The holistic interpretation of the mechanisms of soybean resistance to insects has been worked out by several workers. Soybean resistance to the Mexican bean beetle operates both at the preingestion (antixenosis) and the postingestion (antibiosis) phases of the host-selection process (Kogan 1986). The antixenosis factors are likely to be cyanogenic glycosides, nonreducing sugars, total nitrogen, sterol profiles, and pinitol. The potential antibiotics caffeic and ferulic acid occurred at higher concentrations in the resistant soybean cultivar PI 227687 than in susceptible ones. In addition to these two types of resistance mechanisms, induced resistance is imparted by iso-flavonoid phytoalexins (coumestrol) in soybean cotyledons as potent feeding deterrents to the bean beetle (Hart *et al* 1983). Thus, in resistant soybean, the Mexican bean beetle first ingests a diet that may be imbalanced in nitrogen/carbon proportions, then the flavonoidal aglycones may act as constitutive resistance, and finally induced resistance, as evidenced by characteristic temporal patterns of PAL activity, may provide protection (Chiang *et al* 1986). In fact, there is considerable genetic diversity among soybean genotypes with regard to their capacity to respond to factors inducing resistance. Inducible defense in soybean is likely to be effective against a broad range of herbivorous insects and pathogens (Kogan and Fischer 1991).

Chemicals identified in soybean against insect pest complexes are repellents, deterrents, antifeedants, antibiotics, and hormone mimics (Norris *et al* 1988). Morphological and anatomic traits of soybean resistance to soybean insects have also been critically worked out (Chiang and Norris 1983).

Other plant-resistance strategies are temporal and spatial escape (asyn-chrony of phenologies), and interaction of various environmental factors like temperature, light intensity, soil moisture, relative humidity, etc. The mech-anisms of resistance to insects are, therefore, seldom simple. The complex interactions between secondary compounds and all elements in the higher trophic levels within the ecological context must be understood in order to realize the full potential of host plant resistance to insect pests.

7

FACTORS AFFECTING EXPRESSION OF RESISTANCE

Genetic resistance that results from inherited characteristics is more desirable and significant than that which stems from ecological conditions. However, even inherited characteristics, especially those involving physiologic attributes, are subject to environmental influence. Environmental factors, both abiotic and biotic, can influence the magnitude and the expression of resistance. Abiotic factors such as climatic, edaphic, and cultural conditions are important modulators of plant and insect interactions. They may operate directly by influencing the plant physiologic processes or by affecting insect populations and degree of insect damage. Plant metabolism is modified under stress conditions, which, in turn, can affect the expression of host plant resistance. Possible biotic factors such as the spatial patterns of plants, competing plant species, plant diseases, variability in insect behavior, and insect biotypes are potentially important in influencing the magnitude of resistance. Analysis of the interactions between the plant and both abiotic and biotic factors is crucial in avoiding any sources of experimental errors in the screening of breeding materials for insect resistance.

Abiotic Factors

The important abiotic factors are the microenvironment in the host plant, edaphic conditions, soil pH, temperature, light intensity, relative humidity, and air pollutants. Agricultural chemicals such as inorganic fertilizers, pesticides, and growth regulators applied to crops influence the level of resistance or susceptibility.

Microenvironment in the host plant

The host plant is a key component of the microenvironment of herbivorous insects. The microenvironmental niche of the host plant is more relevant for insect performance and fitness than the gross environment (Willmer 1982). By virtue of their small size, insects can experience many distinct micro-environments on the plant. The climate of the insect niche is often sharply different from the average climate of the region. An ecological niche, thus, can be defined as the set of resources that provide a species population with its total requirements (shelter, food, oviposition site) for existence and reproduc-tion. Each insect species has its own set of resources essential for life and any deviation from this optimum habitat is likely to limit its growth. Knowledge of the microenvironment of the host plant of a specific insect family is essential for maintaining a similar niche for screening germplasm. For example, optimum temperature is vital for the European corn borer larvae (Beck 1957), whereas a relative humidity of more than 90% is essential for the growth of the rice gall midge (Panda *et al* 1983).

Indirect effects of plant water stress on insect herbivores

The concept of the microenvironment is specifically important in relation to moisture, because water stress experienced by a host plant can have a considerable impact on insect herbivores. Plant morphological responses to water stress consist of leaf shedding, leaf rolling, leaf angle changes, and an increase in the root-to-shoot ratio. Physiologic changes caused by water stress include alterations in leaf cuticular wax thickness, transpiration, respiration, stomatal behavior, photosynthesis, translocation, mineral nutrition, and protein breakdown (Parsons 1982).

Changes in the host plant – its nutritional and plant defense chemicals – are likely to have an impact on the microenvironment of the insect. Water-stressed plants show increased protein breakdown and availability of amino acids. Water-stressed leaves show a higher concentration of free amino acids (particularly proline) as they approach the final stages of senescence (Hanson and Hitz 1983). In addition to the free proline coming directly from the proline in protein that is being degraded, other amino acids are also converted metabolically to proline (Stewart 1981). These changes do not occur merely in the cytoplasm of individual cells, but are also evident in the composition of the phloem sap (Miles *et al* 1982). Water stress is also associated with hydrolysis of starch and a subsequent rise in soluble carbohydrates. Thus, accumulation of proline and soluble carbohydrates in water-stressed plants can serve as a favorable resource for insect herbivory and may permit the herbivore population to grow to the extent of devastating the plants even after the water stress is relieved (Haglund 1980). Depending on the plant species, water deficits lead to increased or decreased concentrations of secondary plant compounds, which may affect the insect herbivores (Gershenzon 1984).

Direct effects of plant water stress on insect herbivores

The total water potential of a plant, U_T, can be approximated as:

$$U_T = U_P + U_S$$

where U_P = turgor potential and U_S = osmotic potential. Typically, U_S is negative and U_P is positive when the plant is under turgor. In a fully turgid plant, U_T is zero and may fall to -2 MPa at wilting (Kramer 1983). The components of water potential vary from cell to cell, so bulk tissue estimates may not represent all individual cells. In the xylem, U_S is very small and U_P is negative. On the other hand, U_S for the phloem can be very strongly negative and U_P positive.

In spite of osmoregulation, U_P in phloem cells decreases to zero under severe water stress. The direct effects of lowered turgor potential are likely to affect aphids, leafhoppers, and other sap-feeding insects. The hydrostatic pressure potential of the plant cell is a key variable determining the physical effort necessary for the sap feeders to obtain food. Aphid populations have been observed to increase, decrease, or remain unaffected on plants grown under moisture stress. The crucifer specialist aphid *Brevicoryne brassicae* depends on sap pressure (turgid cells) for optimum uptake of soluble nutrients (Auclair 1963). On the other hand, sap enrichment during moderate moisture stress is beneficial to the generalist aphid *Myzus persicae* (Wearing 1972). These findings show that the relative importance of turgor pressure depends on the circumstances, i.e., it depends on the aphid species, the age of the leaves being eaten, and the watering regime. Thus, experiments on plant water stress are subject to considerable experimental variation. Precise screening of plant species against sap feeders requires careful assessment from a number of experimental protocols such as laboratory, greenhouse, and field experiments, and computer simulations.

Edaphic conditions

Plant growth is dependent on the chemical properties of the soil. Soil properties are managed by applying fertilizers and irrigation water. Variations in the amount of soil nutrients can create environmental conditions more or less favorable to insects. In a few instances, nutrients themselves are known to alter the level of genetic resistance of some plants to insects. However, parallel tests on resistant and susceptible varieties under different fertility conditions show that the two tend to be affected in the same way. For example, the expression of host plant-resistance characteristics, such as glabrousness and high gossypol levels, appears to depend on optimum soil fertility and moisture (Slosser 1983). As mechanisms of resistance are diverse, this ambivalence should not cause any surprise. The metabolic pathways involved in resistance are not necessarily the same in all plants; hence, a diversity of effects is likely to be observed. Little is known of the intimate metabolic processes that could explain these differences.

The role of mineral nutrients in fundamental plant physiologic processes has been extensively reviewed (Treshow 1970). Several observations relate the patterns of insect incidence on host plants grown in varying levels of soil nitrogen, phosphorus, potassium, and other plant nutrients. Nitrogen is a component of amino acids, amides, and coenzymes as well as of some allelochemicals that affect the feeding behavior of insects. The effects of nitrogen fertilization on insect incidence vary, depending on the insect species, host plant species, soil fertility with regard to nitrogen and other elements, and possibly other environmental factors (Scriber 1984b). When the corn leaf aphid *Rhopalosiphum maidis* (Fitch) was allowed to feed on sorghum plants grown under high and low levels of nitrogen, the aphid population buildup was three times higher in plots with high levels of nitrogen than in those with low levels (Branson and Simpson 1966).

There is a limited literature on the plant-mediated effects of phosphorus on insects. Physiologically, phosphorus deficiency leads to inhibition of protein metabolism and auxin production, and augmentation stimulates root growth compared with shoot growth and enables the plants to withstand insect damage. Resistance to the shootfly in sorghum (Singh and Agarwal 1983) and to the spotted alfalfa aphid in alfalfa (Kindler and Staples 1970a) has been shown to increase with additional phosphorus applications.

Unlike nitrogen, increased levels of potassium fertilizer generally appear to have a negative influence on insect populations. This may be due to a higher proteogenesis in plants, a physiologic phenomenon correlated with the elimination of amino acids and reducing sugars in the sap, which otherwise favor the development of sap feeders (Chaboussou 1972). Application of potassium fertilizers (200–250 kg ha^{-1}) to rice imparted some resistance to the yellow stem borer *Scirpophaga incertulas*, brown planthopper *Nilaparvata lugens*, green leafhopper *Nephotettix* spp., leaffolder *Cnaphalocrocis medinalis*, whorl maggot *Hydrellia philippina* Ferino, and thrips *Stenchaetothrips biformis* Bagnall (Subramanian and Balasubramanian 1976).

While many studies have investigated the effects of soil fertility on host plant resistance to insects in a particular cultivar, relatively few of these have examined the interaction of the fertilizer regime on the resistance reaction of plant lines differing in levels of resistance (Tingey and Singh 1980). However, there are no studies that examine the influence of fertilizer regimes on variation in specific physical or chemical factors mediating the resistance reaction. Barbour *et al* (1991) investigated the effect of the NPK (nitrogen/phosphorus/potassium) fertilizer regime on the level of resistance in the wild tomato *Lycopersicon hirsutum f. glabratum* accession PI 134417 and cultivated tomato *L. esculentum* to three insect pest species of tomato, the tobacco hornworm *Manduca sexta*, the tomato fruitworm *Helicoverpa zea*, and the Colorado potato beetle *Leptinotarsa decemlineata*. The wild tomato accession PI 134417 is highly resistant to *M. sexta* and *L. decemlineata* by virtue of the presence of the methyl ketone, 2-tridecanone, in the glandular tips of the type VI trichomes. Factors associated with the leaf lamella contribute to the resistance in PI 134417 to *H. zea*. The results of this investigation showed that increasing the rate at which NPK fertilizer was applied, from 1.8 to 19.6 g plant^{-1} week^{-1}, reduced the glandular trichome-based resistance of PI 134417 to *M. sexta* and *L. decemlineata* by lowering both the density of type VI glandular trichomes and the amount of 2-tridecanone on the tips of these trichomes. The application of fertilizer also reduced the lamellar-based resistance of PI 134417 to *H. zea*. Even though the resistance of accession PI 134417 is conditioned by several independently segregating recessive genes, yet the resistance is affected by environmental factors such as fertilizer regime. The mechanisms reducing the resistance levels are unknown.

These studies show that variations in soil nutrient levels are likely to affect the level of insect damage as well as modifying the expression of antixenosis and antibiosis factors of resistance. Therefore, nutrient regimes selected for use in the assessment of genetic resistance should be comparable with the recommended dose of nutrients in the cropping system. Excess nitrogen and phosphorus should be avoided in view of their effect on pest performance.

Soil pH

Soil can be acid, neutral, or alkaline. Plants can grow in soils over a pH range of 3.0 to 9.0, but extremes stress the plants. Few studies have investigated the plant-mediated effects of soil pH on insect herbivores. Foliar damage to whorl-stage sorghum by the fall armyworm *Spodoptera frugiperda* was greater in plants growing in acidic soils (pH <5.4) than in plants grown in soils with a pH higher than 6.0 (Gardner and Duncan 1982).

Salinity induces plant responses that can alter the suitability of a plant as a host for insects. Plants exposed to chloride salinity have a greater degree of succulence than plants exposed to sulphate salinity (Rains 1972). Succulence results from the development of larger cells in the spongy mesophyll and the presence of a multilayer palisade tissue. Salinity has varied effects on plant biochemistry, and thus affects the quality of the plant as food for insects.

The effect of salinity stress on rice varietal resistance to the whitebacked planthopper *Sogatella furcifera* (Horvath) was investigated by Salim *et al* (1990). Salinity stress at 1.2 S m^{-1} increased nitrogen, decreased potassium, and reduced the production of allelochemicals in rice-plant leaf sheaths. The salinity-stressed rice plants were conducive to the feeding and survival of the whitebacked planthoppers. The salinity stress significantly increased the insects' growth and development, longevity, fecundity, and population build-up. Insect-resistant cultivar IR2035-117-3 was not affected by salinity, whereas saline-tolerant but insect-susceptible rice varieties Nona Bokra and Pokkali were severely damaged by planthoppers under saline conditions.

Temperature

Temperature is one of the most important physical components of the environment that influence the basic life processes of plants and animals. Temperature-induced stress can limit growth, reproduction, or survival of insect herbivores and their host plants. Plant resistance may be affected by either higher or lower temperature. Tingey and Singh (1980) proposed three possible mechanisms whereby temperature can modify the level and expression of genetic resistance: (1) changing the chemical levels (nutrition quality, allelochemicals) and/or morphological defenses of the host plants influencing their suitability for arthropod pests; (2) influencing directly the behavior and physiology of the insect pest; and (3) influencing directly the plant's physiologic and growth response to insect feeding injury.

Temperature-induced stress affecting plants' nutrition and chemical defenses

The growth and reproductive success of insects depend primarily on their ability to ingest and convert plant nitrogen. Host plant suitability is thus directly linked to the environmental factors related to plant nitrogen status (Whitham 1981). As many allomones and kairomones are nitrogenous compounds or compounds with a nitrogen subunit, any environmental factor affecting plant nitrogen status may affect the quality and/or quantity of secondary plant metabolites.

When plants are exposed to temperature extremes, metabolism changes occur that affect the synthesis of secondary metabolites. Even relatively minor abiotic stresses can enhance allelochemical action. For example, the concentration of ferulic acid, which inhibits growth of grain sorghum, was reduced to half (0.2 μM) in plants grown at 37°C compared with a 0.4 μM ferulic acid threshold at 29°C. Moisture stress caused a similar reduction in ferulic acid. Certain coumarins, phenolic acids, and other allelochemicals increase in plants subjected to water deficits, temperature extremes, abnormal radiations, mineral deficiencies, and herbicides (Einhellig 1989). However, studies on the effect of exposing plant canopies to low ambient temperatures show that it is difficult to derive uniform biosynthetic curves for responses and concentrations of secondary compounds on the basis of temperature effects.

Temperature-induced stress affecting insect behavior and physiology

Insects can survive under a wide range of temperature (0–50°C), although most species have a relatively narrow range of temperature for optimum growth and development. Because of changes in their metabolic functions, which are regulated in part by enzymatic reactions, insects can tolerate temperature-induced stress. In some insects, moderately high temperatures induce production of heat-shock proteins, which protect them from injury due to subsequent exposure to higher temperatures (Stephanou *et al* 1983). Heat and cold stress also induce changes in the insect hemolymph, and alter the metabolic rate and rate of fat consumption. Temperature preconditioning may therefore be a major factor in subsequent physiologic responses (Wigglesworth 1965).

Temperature-induced stress affecting host plant resistance to insect pests

Temperature may affect the stability of resistance indirectly or directly through its influence on plant physiologic processes. Physiologic processes can affect host plant suitability to insects by altering the nutritional levels and allelochemical quality of the host plant. Most examples in the literature on the effects of temperature on host suitability for insects are of cool season insects.

Aphids on alfalfa

Dahms and Painter (1940) were among the first to report that the pea aphid *Macrosiphum pisi* (Harris) exhibited a higher reproductive capacity on resistant alfalfa plants at lower field temperatures. The resistance of alfalfa plants was ineffective at 7°C. Alfalfa resistance to the spotted alfalfa aphid *Therioaphis maculata* (Buckton) was studied by McMurty (1962) under controlled temperature conditions. Two resistant parental clones of the alfalfa variety Lahontan and the susceptible variety Caliverda were studied. The resistant clones showed a lower degree of resistance at low temperatures than at high temperatures. One clone, C-902, appeared to be completely susceptible at temperatures between 10 and 21°C (50–70°F), but became progressively resistant at higher temperatures. The other clone, C-84, maintained at least an intermediate degree of resistance even at 10°C (50°F), and was immune to aphid infestation at temperatures of 20°C (68°F) or higher (Fig. 7.1).

Isaak *et al* (1965) made similar observations with the pea aphid and the spotted alfalfa aphid. The loss or gain in host plant suitability in response to temperature was consistent for all resistant alfalfa clones. McMurty (1962) and Isaak *et al* (1965) concluded that the expression of some unknown aphid-resistance factors in the resistant clones was modified by temperature. McMurty (1962) also suggested that under conditions of fluctuating high and low temperatures, the mean temperature, rather than the extremes of temperature, appeared to be the more important factor affecting expression of resistance under field conditions.

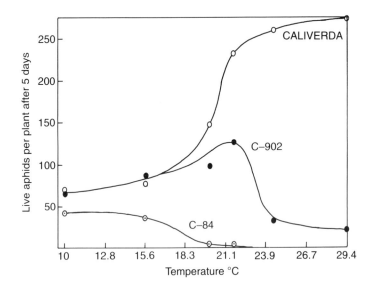

Fig. 7.1. Aphid population levels in relation to alfalfa varietal resistance and temperature interactions. Caliverda (susceptible), C-902, and C-84 (resistant) alfalfa varieties were utilized in the experiment. Reproduced from McMurty (1962).

Kindler and Staples (1970b) studied the influence of fluctuating versus constant temperatures on the resistance of alfalfa to the spotted alfalfa aphid. They noted that differences in the resistance level of susceptible and resistant clones were more apparent at fluctuating temperatures similar to those prevalent under natural conditions. Similar results were recorded by van der Klashorst and Tingey (1979) for potato cultivar resistance to *Empoasca fabae*.

Greenbug on sorghum

A significant loss in the resistance of sorghum cultivars to the greenbug *Schizaphis graminum* was observed at reduced temperatures (Wood and Starks 1972). However, resistance to the three greenbug biotypes, as measured by their reproductive performance on sorghum, differed at different temperature regimes. The optimum temperature for greenbug reproduction was 21–23°C. With an increase in temperature (from 21 to 32°C), greenbug biotype B exhibited a decreasing preference for resistant sorghum cultivar Deer and Piper, compared with the susceptible sorghum OK8. At this temperature less than 15% of the adult greenbugs survived when they were restricted to Deer and Piper. In contrast, temperature did not affect the varietal preference of biotype C. Biotype C seems to be the most adapted to temperature extremes.

Hessian fly on wheat

During the past 30 years, resistant wheat cultivars have been grown widely to control the Hessian fly *Mayetiola destructor* Say in the United States. High temperature regimes have been known to affect the level of resistance of wheat cultivars. In laboratory studies, Cartwright *et al* (1946) showed that at higher temperatures (24–27°C) there was a pronounced increase in the number of infested plants in resistant and susceptible wheat lines; the increase, however, was greater in the resistant lines than in the susceptible lines. Sosa and Foster (1976) confirmed the earlier findings of Cartwright *et al* (1946) that wheat lines with specific genes for Hessian fly resistance responded differently to insect attack at different temperatures. Sosa (1979) showed that when wheat variety Abe was exposed to 1-day-old biotype B larvae at 27°C for 24 hours and the temperature was then altered to 18°C, more than 50% of the plants showed a susceptible reaction. Each additional day of exposure to a temperature of 27°C increased susceptibility until it reached 100% after 7 days.

Critical aspects of the temperature–host plant resistance relationship

Initial assessments of alfalfa, sorghum, and wheat for insect resistance were made in laboratory environments using narrow temperature regimes. But selection of the germplasm resistant to the spotted alfalfa aphid and pea aphid at low temperatures, and to the Hessian fly at high temperatures, might lead to a selection of temperature-sensitive sources of resistance. Germplasm should be assessed over a natural daily range of fluctuating temperatures to increase the chances of selection of temperature-stable sources of resistance.

The interactions of temperature with other factors (i.e., humidity, light intensity, and nutrients) may also affect host plant resistance. Temperature may also interact with humidity and soil fertility to affect the expression of host plant suitability for herbivores (Schweissing and Wilde 1979a). Plant morphology such as the leaf trichome profile that imparts resistance in wheat to the cereal leaf beetle is affected by temperature (Wellso and Hoxie 1982). Trichome profiles differed significantly between years for the same leaf number and between different leaves in the same year. The interactions can also affect the herbivore's developmental biology and its ability to damage the plant. Temperature thus plays a critical role, affecting herbivore and plant success in space and time.

Light intensity

The light changes due to daily day and night rhythm, seasonal changes in day length, different wavelengths, and different intensities affect the physiology of insect pests and their host plants (Beck 1980). Shading, or reduced light intensity, is likely to reduce the photosynthetic activity of the plant. It may also affect its physical integrity, making it vulnerable to insect damage.

The resistance of bread wheat to the wheat stem sawfly *Cephus cinctus* Norton was correlated with 'stem solidness,' which in turn appeared to be affected by the amount of seasonal sunshine. Stem solidness is the major plant factor conditioning resistance to *C. cinctus* (O'Keefe *et al* 1960). This finding became the basis for the development of sawfly-resistant wheat varieties. Stem solidness was generally reduced at decreased light intensities in the resistant, solid-stemmed wheat cultivar Rescue (Roberts and Tyrrell 1961). It was concluded that high light intensity was required for the maximum expression of stem solidness, which, in turn, conferred higher resistance to the wheat stem sawfly. The shaded wheat plants had increased stem sawfly infestation (Holmes 1984).

Shade-induced loss of resistance to *Myzus persicae* has also been reported in resistant sugarbeet and potato genotypes (Lowe 1974). When resistant sugarbeet varieties were grown in the greenhouse under reduced light intensity, the relative expression of resistance as measured by aphid performance was reduced.

There is some evidence that the production of allelochemicals induced by light intensity may mediate interactions between plants and other organisms including insects (Berenbaum 1988). For example, the European corn borer *Ostrinia nubilalis* showed higher feeding rates on both the low DIMBOA WF9 (susceptible) and high DIMBOA B49 (resistant) maize inbreds under low light intensity than under high light intensity (Manuwoto and Scriber 1985). But the reduced larval consumption rates on these varieties under the high light regime were unlikely to be due to the DIMBOA (2,4-dihydroxy-7-methoxy-1,4-benzoxazin-3-one) factor, as its concentrations in the plant leaves under the high light intensity regime were lower in all cases. Leaf nitrogen concentrations were higher under the low light intensity regime and might be at least partially responsible for the greater feeding rates. But there are clear cases of reduction in insect-resistance qualities of crops under reduced light intensity. Resistance to the tobacco hornworm *Manduca sexta* in the wild tomato *Lycopersicon hirsutum f. glabratum* decreased in plants grown under short day-length conditions because of the production of low quantities of 2-tridecanone (Kennedy *et al* 1981). Similarly, the levels of phenolics, which inhibit grasshopper feeding in sorghum, decreased under reduced light intensity (Woodhead 1981).

Soybean cultivar PI 227687, which is resistant to the cabbage looper *Trichoplusia ni* Hubner, showed reduced resistance when exposed to continuous high intensity [24:0 (light:dark) photoperiod] light. When such plants were subsequently placed under a 16:8 photoperiod for 2 weeks, they regained their original level of resistance to the pest (Khan *et al* 1986).

Although some of the examples cited relate to the indirect effects of light on plant–insect interactions, at least one phenomenon is directly involved in determining allelochemical toxicity to insects, i.e., the phenomenon of photoactivation. The linear furanocoumarins, which can form crosslinks with

DNA strands if sufficient ultraviolet (UV) energy is present, are generally toxic to most organisms. At least nine biosynthetically distinct classes of plant chemicals produced by at least a dozen different plant families are known to be phototoxic to insects (see Chapter 3). In the case of furanocoumarins, toxicity to *Helicoverpa zea* (Boddie) is directly proportional to the amount of UV light exposure on the host plant of *H. zea* (Berenbaum and Zangerl 1987).

These findings show the potential involvement of light intensity in the expression of crop plant resistance to phytophagous insects. Researchers should be aware that caged plants and those grown in rearing chambers or in greenhouses during periods of low insolation are subject to shading. In the greenhouse, a range of light intensities equivalent to that of the cropping environment should be maintained by means of metal halide lamps (Duke *et al* 1975) for accurate assessment of resistance. Light and other forms of electromagnetic radiation (EMR) have been largely ignored in insect–plant interactions. It is essential to include EMR along with humidity, temperature, and other environmental factors as a standard parameter in investigations of plant–insect interactions.

Air pollutants

Air pollution refers to the presence in the outdoor atmosphere of one or more contaminants such as dust, fumes, gas, mist, odor, smoke, or vapor in such quantities and duration as to be injurious to human, plant, or animal life or to property (Perkins 1974). Air contaminants can be classified as *primary pollutants*, which are products of combustion and numerous industrial processes, or *secondary pollutants*, which are products of chemical reactions in the atmosphere between primary pollutants and hydrocarbons in the presence of sunshine. Primary pollutants most often affect vegetation in localized areas because they are by-products of particular chemical reactions in local factories. Secondary pollutants, however, are more important in specific regions because they are being produced throughout the air mass. Pollutants considered to be the most important in terms of phytotoxicity include sulfur dioxide (SO_2), fluoride (F), ozone (O_3), nitrogen oxides (N_2O), peroxyacetyl nitrate (PAN), carbon dioxide (CO_2), and some agricultural chemicals (Heck *et al* 1973). In addition, other elements such as metals and hydrocarbons can be important alone or in combinations.

Effect of pollutants on plants

Plant sensitivity to pollutants depends upon the nature of the pollutant, plant species and genotype, growth stage, environmental conditions under which exposure occurs, and other interacting pollutants. These interactions affect the suitability of the plant for insect feeding (Hughes 1988). SO_2 is one of the major air pollutants and is produced in large quantities by the combustion of coal and oil as well as by many industrial ores. It is also emitted from active volcanoes, fumaroles, and vents. The waste products are in the form of sulfides, which combine with oxygen in the air at the higher temperature of the smelter to produce SO_2 which is then released into the atmosphere. Growth and reproduction of plant tissues are severely impaired by SO_2 (Koziol and Whatley 1984); SO_2 is also important in the formation of acid rain.

Fluoride, mostly in the gaseous (hydrogen fluoride) form, comes from varied sources such as volcanoes, dust from F-containing soils, the combustion of coal, and many industrial processes. It is one of the most phytotoxic pollutants which accumulates in plants (Weinstein 1977). Fluoride appears to be an inhibitor of enzymes, and plant susceptibility to injury varies with plant species and conditions (Koziol and Whatley 1984).

Because of fossil fuel consumption and tropical deforestation, global atmospheric CO_2 concentrations are rising. The current atmospheric CO_2 level is 350 ppm, which is expected to reach 700 ppm by the mid to late 21st century (Detwiler and Hall 1988). In addition to potentially altering global climate, an enriched CO_2 atmosphere is expected to influence biotic interactions due to its critical role in photosynthesis (Tangley 1988).

Experimental evidence of interactions

While general observations in the field are often difficult to relate directly to air pollution, a number of well-planned field trials and controlled laboratory experiments have demonstrated the role of air pollution in changing insect–plant interactions. When Mexican bean beetles were fed on soybean, fumigated intermittently with SO_2, the mean number of beetle progeny was 1.5 times greater than that of beetles feeding on the control plants (Hughes *et al* 1983). Endress and Post (1985) have demonstrated that Mexican bean beetles under laboratory conditions preferred to eat soybean leaves prefumigated with ozone-enriched ambient air. Substantial modifications in the form and content of plant nitrogen and sugars have been reported following plant exposure to moderate levels of ozone (Trumble *et al* 1987). Such nutritionally enriched plants are better hosts for insect herbivores, especially aphids (McNeill *et al* 1987).

An increase in atmospheric CO_2 affects herbivore feeding in crop plants. Fajer *et al* (1989) showed that the larvae of *Junonia coenia* raised on foliage of *Plantago lanceolata* which released CO_2 quickly grew more slowly and experienced greater mortality, especially in early instars, than those raised on foliage with a slow release of CO_2.

Air pollution is just one of the many abiotic stresses that affect insect–plant interactions. In nature plants are exposed to multiple stresses and a single change in the environment can trigger many changes in a plant. The problem is complex and an understanding of the actual and potential impact on the ecosystem requires the integration of many factors.

Biotic Factors

The host plant preference of insects is often influenced by several factors. Individual plants are a mosaic of resources that vary temporally and spatially in their suitability as food for herbivores (Whitham 1981). The nutrition, moisture, and allelochemistry of plant tissues vary diurnally and seasonally (Scriber and Slansky 1981). Many recent studies have shown that intra-specific variation in host plants has a significant effect on herbivores that feed on them. Consequently, the populations of insect herbivores may behave quite differently on different individuals of the same plant species (Karban 1992). Therefore, when adopting precise screening techniques for the evaluation of germplasm for host plant resistance, it is necessary to consider variables such as plant density, plant tissue types, tissue age, and herbivore-induced tissue responses (this portion of the interaction is dealt with in Chapter 6). Insects also vary in their ability to exploit individual host plants. Specialists and generalists show preferences for certain tissues in time and space (Rhoades and Cates 1976). Variables within the test insects – age, sex, activity, preassay conditioning, and biotypes – are important considerations when evaluating plant materials for insect resistance. Evaluation studies would be most convincing if they are properly replicated, conducted under controlled conditions, and carried out with at least one complete herbivore generation.

Plant factors

Insects have often shown a feeding preference for specific plant organs or tissues and for foliage and tissues of certain physiologic age. In addition, variations in physical growth conditions such as plant canopy and plant height can influence host selection and subsequent population development.

Physiologic plant age

The effect of age on the suitability of plant tissues as a food resource for herbivores has been treated extensively in several reviews (Raupp and Denno 1983). In general, the concentrations of total nitrogen and moisture decline in older leaves. Leaf age affects the production of defense chemicals; some increase, whereas others decrease (Mattson 1980). Thus, the suitability of leaves as food for defoliators may depend on the age of the leaves. Some insect larvae reared on young leaves survive better, have shorter larval development periods, and grow larger as compared with larvae raised on old leaves, whereas the reverse may be true for other insects. *Myzus persicae*, for example, was relatively more successful on older leaves of Brussels sprouts than was *Brevicoryne brassicae*. On the other hand, *B. brassicae* was relatively more successful in feeding on young leaves than *M. persicae* (Wearing 1972). The greater success of *M. persicae* on older leaves and *B. brassicae* on younger leaves were also apparent in terms of reproduction, survival, and the proportion of the alate progeny.

Depending upon individual insect–plant interactions, varietal screening trials should be conducted with test plants of uniform growth and vigor. Knowledge of the age-dependent expression of resistance is essential in designing appropriate evaluation techniques. For example, maize (*Zea mays*) genotypes resistant to the European corn borer contain high levels of the resistance factor DIMBOA at early growth stages, but the DIMBOA concentration declines rapidly with increasing plant maturity (Klun and Robinson 1969). Hence, the expression of resistance is greatest during the midwhorl stage of development, when the DIMBOA concentration is high and the maize plant is normally infested with first-generation European corn borers. Assessment of plant materials for resistance before or after the normal time of infestation by the first-generation borer larvae would fail to identify the resistant genotypes. Resistance of maize to *Chilo partellus* is affected by plant phenology. In screenhouse studies in Kenya, resistance levels of five maize cultivars differed according to the phenological stages at which they were infested with larvae of *C. partellus* (Kumar and Asino 1993). When infested at 3.5 weeks after germination, there was no difference in dead heart damage between susceptible and resistant maize cultivars. However, when infested at 2 weeks after germination, differences among the cultivars were quite distinct.

The expression of resistance of certain crop plants to insect herbivores can be affected with plant maturity. Studies on soybean cultivars with resistance to defoliating insects has documented the importance of the interaction of plant maturity and insect seasonality (Rowan *et al* 1993). Fourteen (seven resistant and seven susceptible) soybean genotypes of maturity groups V–VIII were evaluated against defoliating insects (velvet bean caterpillar *Anticarsia gemmatalis*, soybean looper *Pseudoplusia includens*, and beet armyworm *Spodoptera exigua*) in four field environments. The results show that the early-maturity genotypes showed greater defoliation than the late-maturity genotypes. This negative association of maturity and defoliation resulted in greater defoliation of resistant maturity group V genotypes than susceptible maturity group VII and VIII genotypes.

Plant growth conditions

Plant growth conditions – plant size and height – may also affect the oviposition and feeding of insect herbivores. If the plant growth stages are not uniform in the population the insects may be confronted with plants of varying suitability. Consequently, they are likely to concentrate egg laying or feeding on plants at the most suitable developmental stages (Rausher and Papaj 1983). Plants, therefore, can gain or lose a certain degree of resistance if the growth stage that confers greatest antixenosis resistance coincides with periods of maximal insect oviposition or feeding requirements (Delobel 1982).

The ovipositional response of *Pieris rapae* depends on the size of the cabbage plant, as indicated by a significant correlation between plant leaf area and number of eggs laid (Latheef and Irwin 1979). As the season advances, the cabbage variety Savoy with the largest plant canopy and tallest plants was most preferred for egg laying by *P. rapae*. Similarly, breeding lines of onion that grew slowly, attaining a basal diameter of 1–2 mm and a height of 90–200 mm after 6 weeks of growth, did not stimulate the oviposition of *Delia antiqua* (Meigen) flies and received very few eggs compared with lines attaining a diameter of 3–4 mm and a height of 200–350 mm (Harris *et al* 1987).

Plant density is likely to affect the expression of resistance to insects. The relationship between plant density and wheat stem sawfly (*Cephus pygmeus*)

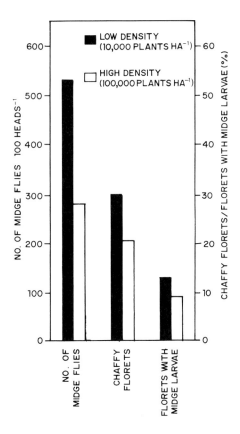

Fig. 7.2. Effect of planting density on adult midge abundance and damage in CSH5 sorghum hybrid (1981–1982 post-rainy season). Reproduced from Sharma *et al* (1988) with permission from the Entomological Society of America, Lanham MD, USA.

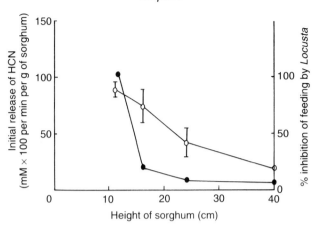

Fig. 7.3. Release of HCN from sorghum at different stages of development in relation to feeding by *Locusta migratoria migratorioides*. Initial rates of HCN release from sorghum as would occur in biting as shown (•). Percentage inhibition of feeding (o) was found by comparing the meal sizes of fifth-instar nymphs on sorghum of different ages with meal sizes on mature, palatable leaves of *Poa annua* used as a control. Standard deviations are indicated by vertical lines. Reproduced from Woodhead and Bernays (1977) with permission from Macmillan Magazines, London.

resistance in Syrian wheats was established at the International Center for Agricultural Research in the Dry Areas (ICARDA), Syria (Miller *et al* 1993). Wheat plants sown at low densities (40 × 40 cm intervals and 2.5 kg seeds ha^{-1}) had a longer interval between plant development stages and had higher stem solidity than those sown at higher densities (10 × 3 cm intervals and 133 kg seeds ha^{-1}). Stem solidness was negatively correlated with the percentage of sawfly-infested stems. These results differed from those of previous studies in North America, where widely spaced plants suffered higher infestations than did closely spaced plants (Wallace *et al* 1974). The results of this study suggest that stand density should be considered while evaluating stem sawfly-resistant wheat cultivars in different wheat-growing locations. In contrast, damage by the sorghum midge *Contarinia sorghicola* (Coquillett) on sorghum hybrid CSH5 was higher in plots with lower plant densities (Sharma *et al* 1988). Planting densities of 10,000 plants ha^{-1} favored a higher number of midge flies and therefore higher levels of midge damage than densities of 100,000 plants ha^{-1} (Fig. 7.2).

Difference in plant height also affects the expression of resistance to insects. For acridids, the plant palatability of sorghum increases with plants up to 60 cm tall. Woodhead and Bernays (1977) showed that young sorghum plants are protected by the release of HCN when bitten by locusts. Sorghum leaves contain the cyanogenic glycoside dhurrin. It is located in the cell vacuole. When the sorghum tissue is crushed by insects, dhurrin is hydrolyzed by an enzyme system probably present in the cytoplasm, resulting in the release of HCN (Saunders *et al* 1977). The release of HCN from crushed leaves was measured quantitatively by monitoring production of *p*-hydroxybenzaldehyde. The hydrolysis of dhurrin yields 1 mol of HCN and 1 mol of *p*-hydroxybenzaldehyde. Concentrations as low as 0.01 μM were sufficient to inhibit insect feeding. The amount of HCN released was very high in young plants and declined with plant age. The initial rate of HCN release was calculated for different plant ages, and its relationship to the amount eaten by *Locusta migratoria migratorioides* in one meal is shown in Fig. 7.3.

Variation within the foliage of individual plants can also affect the resistance to insects. The top two fully expanded leaves nearest the apex of resistant soybean cultivar PI 227687 consistently had higher larval growth rates for the soybean looper *Pseudoplusia includens* (Walker) than the third and lower leaves (Reynolds and Smith 1985). No differences in resistance were observed between leaves below the third leaf. The same trend was observed among leaves of the susceptible soybean cultivar Davis, but larval growth rates were two- to six-fold greater on Davis leaves than on the corresponding leaves of PI 227687.

Plant diseases and insect performance

Another biological factor affecting resistance is the presence or absence of plant pathogens in the host plants. The infection of plants by pathogens can alter the suitability of the plant as a host for an insect herbivore, in some cases by improving host quality and in others by reducing it (Hammond and Hardy 1988). The growth performance of aphids *Rhopalosiphum padi* (L.) and *Sitobion avenae* (F.) was positive on oats infected with barley yellow dwarf virus (BYDV) (Gildow 1980). A greater percentage of winged progeny of aphids was consistently produced on BYDV-infected oats, regardless of the aphid species or the BYDV isolates. The alate progeny was also higher on senescing oat tissues. Differences in nitrogen metabolism, resulting in increased amino acid concentration in diseased or senescing plants, could explain the increased production of alate aphids.

Lee *et al* (1984) reported higher brown planthopper damage on rice varieties infected with sheath blight. On the other hand, tungro-infected rice plants were not favored by the disease vector *Nephotettix virescens* (Distant). Feeding of *N. virescens* on tungro-infected rice plants had significantly reduced adult longevity, fecundity, egg hatchability, and population growth. However, oviposition was not reduced, because the gravid *N. virescens* were attracted towards the characteristic yellowing of tungro-infected rice leaves. Another attribute of the tungro-infected rice leaves was high levels of free sugar, which is an attractant for *N. virescens* (Khan and Saxena 1985a).

Field experiments have demonstrated a positive relationship between the ryegrass *Lolium perenne* L. infected with clavicipitaceous endophytic fungi and resistance to the fall armyworm *Spodoptera furgiperda* (J. E. Smith). Fall armyworms reared on perennial ryegrass infected with an endophytic fungus showed reduced larval weight and delayed development compared with larvae reared on uninfected ryegrass (Hardy *et al* 1985). In preference tests, neonate larvae preferred uninfected to infected ryegrass plants and consumed more fungus-free leaves. The negative response of insects to fungi-induced resistance is due to fungi-produced alkaloids.

Thus, it is necessary to guard against the possibility of negative or positive effects of plant diseases (viral, fungal, or bacterial) on the expression of all the three components of plant resistance. Test plants whether in field or greenhouse experiments meant for screening for insect resistance ought to be kept free from any plant diseases to obtain unbiased results.

Insect factors

Use of uniform insect cultures is one of the most important requirements in screening germplasm for insect resistance. Heterogeneity in the test insects' age, sex, preconditioning, and level of infestation must be avoided to minimize discrepancies in screening results. Furthermore, species of generalists and specialists, as well as populations or strains of the same insect species, may differ in their response to a particular group of chemical compounds in an artificial diet or to chemical compounds within the host plant. Such

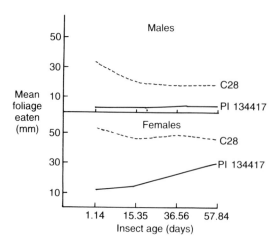

Fig. 7.4. Effect of age and sex on the feeding of adult Colorado potato beetles on C28 (susceptible) and PI 134417 (resistant) tomatoes. Reproduced from Schalk and Stoner (1976) with permission of the authors.

differences may confound attempts to generalize how plant allelochemicals might affect different groups of insects (Puttick and Bowers 1988).

Insect age and sex

Insect age and sex are likely to be factors that affect the preference for host plants with varying levels of resistance. Young (3-day-old) female Mexican bean beetles showed no preference for leaves of susceptible Deane soybean over resistant soybean leaves. In contrast, 14-day-old Mexican bean beetle females showed a marked preference for leaves of susceptible plants. Considering these results, 14-day-old females were used to measure feeding responses to soybean leaf extracts (Smith *et al* 1979). In feeding bioassays, the female beetles demonstrated a distinct preference for leaves of susceptible soybean over the resistant soybean leaves, but males did not. In another instance, there was no difference in the feeding of male Colorado potato beetles of different ages on resistant tomato accession PI 134417, but the females increased their feeding on this accession as they became older (Fig. 7.4) (Schalk and Stoner 1976).

Insect preconditioning on host plants

Rearing insects on certain host plants can influence insect preference for plant materials meant for screening. In other words, preconditioning the test insects on particular plants prior to testing is likely to affect insect behavior and the results of screening for plant resistance. For example, rearing greenbugs on certain alternate host plants can influence the antixenosis resistance of sorghum to this insect. Schweissing and Wilde (1979b) evaluated several small-grain hosts of greenbugs for their predisposition effect on the feeding preference for sorghum. When the greenbugs were cultured on barley, oats, rye, and wheat they did not show a differential preference for sorghum. However, greenbugs reared on susceptible sorghum RS 671 showed a definite preference for sorghum. But antixenotic resistance to greenbugs in resistant sorghum KS 30 was not influenced by prior culturing on RS 671.

The efficacy of screening for resistance was influenced by rearing the aphids on the proper host plant genotypes. Recently, Robinson (1993) showed that conditioning of *Diuraphis noxia* on host genotypes can influence antixenotic as well as antibiotic types of resistance of the test cultivars. *Diuraphis noxia* conditioned on oats showed reduced antixenosis on oats relative to the two barley and wheat genotypes. However, *D. noxia* conditioned on barley and wheat showed increased antixenosis on those genotypes. Conditioning also influences the host plant antibiosis to aphids. *Diuraphis noxia* conditioned on barley produced more eggs on barley (no antibiosis) than on oats; however, *D. noxia* transferred from any of the conditioning host plants to oats contained significantly fewer eggs (presence of antibiosis) than on the other hosts. Therefore, the diet of test insects should be standardized prior to their use for artificial infestation on the test host cultivars. Currently, test insects are conditioned by starving before use in varietal screening trials.

Levels of insect infestation

The level of insect infestation is important in rating plants for resistance, moderate resistance, susceptibility, and tolerance. Host plant resistance is a relative phenomenon, with insect–plant interaction often observable within segregating plant populations. To screen for truly resistant plant materials among a large number of plants in greenhouse or field cages, a susceptible cultivar is first carefully chosen as the experimental standard, and measurements of resistant plants are compared with that standard. This bioassay procedure can only be effective with the optimum level of insect infestation, not with either too low or too high insect infestation levels (Harris 1979). Although known susceptible plants are shown to clump towards the susceptible end of the spectrum in both high and low levels of infestation (Fig. 7.5), the range encountered in light infestation indicates pseudoresistance

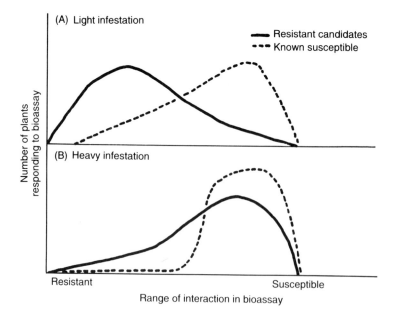

Fig. 7.5. Hypothetical responses of the same population of genetically segregating resistant candidate plants and a population of plants known to be susceptible when exposed to light (A) and heavy (B) insect infestations. Reproduced from Harris (1979).

(Fig. 7.5A) because there is no visible difference between susceptible and resistant lines. In this technique, many candidate plants that are retained for testing will not be reliable sources of resistance. Likewise, under heavy infestation, resistant plants exhibit relatively high levels of damage compared to susceptible plants (Fig. 7.5B), resulting in a narrow difference between resistant and susceptible plant materials. Even highly resistant plants may succumb to high densities of arthropod attack.

Figure 7.6 indicates that the ideal range of infestation within which to test for resistance is within areas II and III. Area I contains some pseudoresistant candidates; under area IV the infestation levels are so high as to eliminate all moderately resistant as well as resistant lines. Harris (1979) noted that abnormally high insect infestations were often as unsatisfactory for resistance testing as were unusually low infestations, and both situations should therefore be avoided. Field- and greenhouse-screening studies can be modified by introducing a known number of insects of known age and sex into the experimental area at a specified time. This establishes a baseline of infestation which can largely eliminate pseudoresistance responses.

In a screening study, Blum (1969) found that the size of the sorghum shootfly *Atherigona varia soccata* population had a pronounced effect on both absolute and relative rates of oviposition in various sorghum genotypes. Data

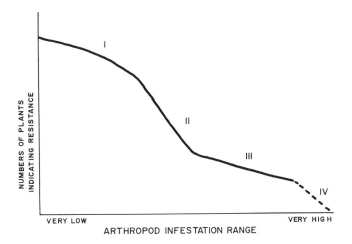

Fig. 7.6. Hypothetical response of a population of plants segregating for resistance to a range of insect populations. Area I indicates a low level of infestation where many susceptible plants escape (pseudoresistance). Area II indicates population pressure, where pseudoresistant plants will be eliminated. Area III represents a population level where tolerant plants will be eliminated. Area IV represents a very large insect population where even resistant plants will be killed. Reproduced from Harris (1979).

on the F_2 sorghum population indicated that resistance (antixenosis) was partially dominant when evaluated under low shootfly populations; when evaluated under high populations, susceptibility appeared dominant.

Biotypes

In 1907, Biffen reported the genetic basis of disease resistance in crop plants for the first time. However, within nine years of this discovery, it was realized that the stability of this resistance was threatened by the problem of variable pests. By 1918, Stakman and associates (Johnson 1961) had discovered physiologic races of the wheat stem rust *Puccinia graminis tritici*. Physiologic races are common in plant pathogens of crop plants, but are comparatively less common in insect pests. Painter (1930) discovered the existence of biological strains of the Hessian fly *Mayetiola destructor* on resistant varieties of winter wheat. Other workers subsequently recognized that races and strains are more common among Hemiptera – particularly aphids, leafhoppers, and planthoppers – than among Lepidoptera, Coleoptera, or Hymenoptera. These insect races or strains, now called biotypes, are primarily distinguishable on the basis of their interaction with different host plant varieties (Eastop 1973b). Some authors (Claridge and Den Hollander 1980, 1983) have voiced objections to the term 'biotype' without suggesting any alternate terminology. This subject is also discussed in Chapter 10.

Insect biotypes constitute an important biotic stress affecting the host plant resistance. Biotypes develop when antibiosis is the major component of resistance. They seldom, if ever, develop when antixenosis or tolerance is the mechanism of resistance (Gallun 1972). With antibiosis, selection pressure is exerted on the insect pest, resulting in severe mortality. The virulent individuals within the insect population survive and multiply to form a new population or biotype. The erstwhile resistant plant variety can be attacked by the new biotype and becomes susceptible. Biotypes, therefore, are a natural survival mechanism for the perpetuation of an insect species, selected on a resistant cultivar widely grown in areas where exposure to such insects is common (Nielson *et al* 1971).

Insect biotypes and host plant resistance

Crop cultivars previously known to be resistant to a particular insect may succumb to a new biotype of the same pest. Such biotypes have been recognized in several crop insect pests belonging to different orders. Painter (1951) reviewed the information on the biotypes of the grape phylloxera, European corn borer, Hessian fly, pea aphid, greenbug, and corn leaf aphid. By 1970, biotypes were known to occur in eight species of insect pests affecting agricultural crops (Pathak 1970). Now, 12 species of the phytophagous insects are known to have biotypes and seven of these are in the

Table 7.1. Some examples of host plant resistance-mediated insect biotypes.

Crop	Insect	Biotypes (no.)	Reference
Wheat	*Mayetiola destructor* (Say) (Hessian fly)	9 (field) 2 (laboratory)	Everson and Gallun (1980), Obanni *et al* (1989)
	Schizaphis graminum (Rondani) (greenbug)	7	Beregovoy and Peters (1993)
Raspberry	*Amphorophora idaei* (Born) (raspberry aphid)	4	Keep *et al* (1969), Keep (1989)
Alfalfa	*Acyrthosiphon pisum* (Harris) (pea aphid)	4	Cartier *et al* (1965), Frazer (1972), Auclair (1978)
	Therioaphis maculata (Buckton) (spotted alfalfa aphid)	6	Nielson and Lehman (1980)
Sorghum	*Schizaphis graminum* (Rondani) (greenbug)	5	Harvey and Hackerott (1969)
Corn	*Rhopalosiphum maidis* (Fitch) (corn leaf aphid)	5	Singh and Painter (1964), Wilde and Feese (1973)
Rice	*Nilaparvata lugens* Stal (brown planthopper)	4	Heinrichs (1986)
	Nephotettix virescens (Distant) (green leafhopper)	3	Heinrichs and Rapusas (1985)
	Orseolia oryzae (Wood–Mason) rice gall midge	4	Heinrichs and Pathak (1981)
Apple	*Eriosama lanigerum* (Hausm) (woolly aphid)	3	Sen Gupta and Miles (1975)

Aphididae (Wilbert 1980). Some important examples of insect biotypes in crop plants are given in Table 7.1, and three of them are discussed here.

Hessian fly

More comprehensive research has been conducted on biotypes of the Hessian fly and greenbug than any other insect. Painter *et al* (1931) first listed six biotypes of the Hessian fly; the number was subsequently raised to eight (Gallun and Hatchett 1968). The characteristics for identifying biotypes involve the capacity to survive on and stunt wheat plants with different genes for resistance. There is a gene-for-gene relationship between the gene for virulence in an insect pest species and the gene for resistance in a host plant. Eleven biotypes of Hessian fly have now been reported (nine field biotypes: Great plain, A, B, C, D, E, J, L, and M; and two biotypes (N and O) created in the laboratory). Nine biotype populations showed differential reactions to wheat genotypes possessing different genes for resistance (Obanni *et al* 1989). Genes in the wheat plant that condition resistance are either completely or partially dominant. On the other hand, the genes in the Hessian fly biotypes that condition virulence to a specific wheat genotype are recessive (Hatchett *et al* 1981). This subject is treated in greater detail in Chapter 10.

Brown planthopper

The first brown planthopper (BPH)-resistant variety of rice, Mudgo, was identified by Pathak *et al* (1969). It was found to be resistant to BPH populations prevalent in the Philippines, Indonesia, Malaysia, Thailand, Vietnam, China, Japan, and Korea, but not to those in the Indian subcontinent (India, Bangladesh, Sri Lanka). Thus, two biotypes of BPH existed before the large-scale introduction of BPH-resistant varieties. The BPH-resistant variety IR26 with the *Bph-1* gene for resistance was released in the Philippines in 1973, and in Indonesia and Vietnam in 1974. It was widely planted in those three countries. However, a biotype capable of damaging IR26 appeared in the Philippines in 1977 and in Indonesia and Vietnam in 1978. It was designated as biotype 2. Biotype 3 was developed in the laboratory by rearing the insects on the resistant variety ASD7, which has the *bph-2* gene for resistance. These biotypes can be differentiated by their reaction on different varieties (see Fig. 10.1).

After the breakdown of the resistance of IR26, varieties IR36 and IR42 with *bph-2* for resistance were released (Khush 1977). They were widely adopted in the Philippines, Indonesia, and Vietnam, but were found to be susceptible to the South Asian biotype. IR42 became susceptible to the BPH in the North Sumatra Province of Indonesia in 1982. In other areas of Indonesia, the Philippines, and Vietnam, IR36 and IR42 continued to be resistant until 1989–1990. IR56 with the *Bph-3* gene for resistance was released in the Philippines in 1982 and in North Sumatra in 1983. Several other varieties with the *Bph-3* gene for resistance – IR60, IR62, IR68, IR70, IR72, and IR74 – have been released and are widely grown in the Philippines, Indonesia, and Vietnam. These varieties are also resistant to the South Asian biotype, also called biotype 4 (Khush 1992). No biotype designation has been given to BPH populations prevalent in the North Sumatra region of Indonesia and Vietnam, but they are presumably similar to biotype 3. Some rice varieties are resistant in the Indian subcontinent but not in Southeast and East Asia. They have either *bph-5*, *Bph-6*, or *bph-7* genes (see Chapter 10 for more details).

Greenbug

Biotypes of the greenbug *Schizaphis graminum* (Rondani) are identifi characteristic damage patterns on wheat entries (Beregovoy and Peters 1993). Seven wheat entries with different genes for resistance were tested for damage by seven greenbug biotypes (B, C, E, F, G, H, and I). There were two kinds of damage: (1) feeding lesions (necrotic or chlorotic) which appeared during the first week, and (2) plant death by the 18th day after infestation. Combinations of both kinds of damage for the seven entries allowed a description of each biotype's unique damage pattern.

Plants were killed only in those cases where the greenbugs made necrotic lesions on the leaves. In nine out of 31 comparisons, when greenbugs made necrotic lesions on leaves, they did not kill the plants; this may be due to low reproductive rates of greenbugs on these wheat entries. Greenbugs which formed only chlorotic spots or no spots on wheat leaves, grew slower and did not kill plants. Variation of patterns found in the two damage characteristics indicated that to overcome resistance determined by each of the five genes for resistance to greenbug in wheat, the greenbug ought to have at least two qualities – one for causing necrotic lesions and another for a high reproductive rate.

Insect biotypes in host plant-resistance programs

The preceding discussion reveals one of the many ways in which insects cope with their environment – through the evolution of biotypes. The study of insect pest biotypes of crops is particularly relevant to programs involving resistance breeding (Gonzalez *et al* 1979). Insect biotypes serve as a tool for understanding the genetic interrelationships between the host plant and the insect pest species. For instance, biotypes of the Hessian fly can distinguish wheat varieties with different genes for resistance and can determine whether a combination of genes occurs in a single plant (Gallun 1972). By maintaining pure line populations of Hessian fly biotypes, genetic sources of resistance in the world collection could be identified, distinctions could be made between different genetic sources without inheritance studies, and donors with a combination of resistance genes could be identified. Biotypes are also useful in screening breeding materials for incorporating different genes for resistance and for pyramiding two or more genes for resistance in a single variety. For example, rice breeding lines at the International Rice Research Institute are screened with BPH biotypes 1 and 2 if the *bph-2* gene is to be incorporated and with biotypes 2 and 3 if the *Bph-3* gene is to be incorporated.

8

SCREENING FOR INSECT RESISTANCE

In recent decades, many national and international organizations have established research programs for developing insect-resistant crop varieties that are suitable for different agroclimatic zones. A prerequisite of a breeding program for host plant resistance is the availability of screening techniques.

The techniques employed for screening plant genotypes for resistance vary with the crop, insect, resistance mechanisms, and site for screening (natural field conditions, field cage, screenhouse, greenhouse, or laboratory). The basic requirements for an insect-resistance screening program are: (1) adequate germplasm resources; (2) adequate supply of target insects; (3) efficient techniques for artificially infesting the test plants with insects; and (4) efficient methods/techniques for evaluating the levels of resistance among test entries. Germplasm resources are dealt with in Chapter 11.

Insect Sources

A reliable and continuous supply of insects is essential for effective screening of crop germplasm. An adequate supply of insects at the desired life stages must be available to infest test plants at their most vulnerable growth stage(s) and thus provide maximum differentiation between resistant and susceptible genotypes. The principal sources of insects for studies of host plant resistance (HPR) are natural field populations, greenhouse colonies, and laboratory-reared colonies. All these sources can be effectively utilized, but each has its inherent limitations.

Natural field populations

The major advantages of field populations are convenience and low cost. In cropping localities where the specific insect pest is endemic, natural field populations of insects may be available in adequate numbers for field screening of germplasm. Some insect species normally occur in high numbers at intervals of several years. Considerable data on host plant resistance have been collected by taking advantage of these outbreaks. Several of the earliest wheat strains resistant to the Hessian fly were discovered in this way. Many local rice cultivars in the gall midge endemic regions of Andhra Pradesh, Assam, Orissa, and Bihar (India) were observed to be highly resistant to field populations of the gall midge, and some of those cultivars were utilized in developing gall midge-resistant varieties (Panda *et al* 1983). Field insects can be collected for artificial infestation of plants by means of insect-collecting devices ranging from the simple sweep net to the motorized portable D-vac sampler (Dietrick 1961) and tractor-powered suction equipment (Stern *et al* 1965). Along with other advantages, the field population of insects represents a gene pool of the natural agricultural environment.

However, the major limitations of natural insect infestation are seasonality, unpredictability, and uneven distribution that make field screening at times unreliable. Seasonal and year-to-year variation in the target insect population levels may be misleading in interpreting field data. Sometimes the target species are infected by pathogenic microorganisms and parasitoids, and represent a weak population for proper infestation of test cultivars. In some instances, plant materials are infested with nontarget insects that cause plant damage symptoms similar to those due to the target pest. In such cases, plant damage assessment is not reliable.

The problems of uncertainty in obtaining a sufficiently high field population of insects can be alleviated through the use of special techniques such as mass collection, trap crops, adjusting planting dates, light traps, chemical attractants or pheromones, and resurgence-causing insecticides and/or cultural practices.

A careful integration of simple and mutually complementary methods described below can lead to a buildup of desirable insect pressures in the screening plots under natural infestation conditions.

Mass collection

The mass collection of target insects of an appropriate age from field populations and then releasing them in field plots forms one of the simplest techniques for infesting plant materials with insects for screening. Egg masses of several lepidopterous insects can be collected from field crops and the emerging larvae utilized in infesting test germplasm for resistance screening (Davis 1985). Diapausing eggs of the western corn rootworm *Diabrotica virgifera* Leconte, egg masses of the rice yellow stem borer *Scirpophaga incertulas*, diapausing larvae of the southwestern corn borer *Diatraea gradiosella* Dyar., adults of the clover head weevil *Hypera meles* (F.), and many others have been collected from their respective field-grown host plants and utilized in field screening experiments. Insect-infested host plants can be transplanted into field plots to provide adequate infestation levels (Tingey and van de Kalshorst 1976).

Trap crops

Susceptible trap crops can be interplanted within field plots to attract, concentrate, and enhance populations of the target insect pests (Stern 1969). Later on, the susceptible trap crop is cut or allowed to senesce, and the target insects are forced to move into the experimental plots. For example, a trap crop of mustard, *Brassica juncea*, highly preferred by *Lygus lineolaris* (Palisot de Beauvois), was planted throughout a cotton nursery in March followed by the cotton crop in May. The mustard was cut when the cotton plants were at the six-leaf stage and the *Lygus* bugs moved from the mustard to the adjacent rows of cotton plants (Laster and Meredith 1974).

Adjusting planting dates

Adjusting planting dates is another technique to increase insect populations in the field. In India, the sorghum shootfly is almost absent at the beginning of the wet season. If planting is delayed, the shootfly population increases and, depending upon the variety, the seedling attack goes up to 100% (Vidyabhushanam 1972). Sharma *et al* (1988) adopted various techniques to increase infestation and improved the efficiency of screening sorghum for sorghum midge resistance. Adjustment of planting dates to synchronize flowering with the period of peak abundance of adult midges, planting infester rows of susceptible cultivars 20 days before test cultivars are sown, spreading midge-damaged sorghum panicles containing diapausing midge larvae on infester rows, and using sprinkler irrigation during flowering in the post-rainy season helped to increase midge abundance in the field. Sorghum midges can also be maintained on continuous ratoons of wild sorghum *Sorghum verticilliflorum* and cultivated susceptible sorghum lines.

The rice crop is most vulnerable to the gall midge at the maximum tillering stage, i.e., 60 days after sowing the dibbled crop or 45 days after transplanting. The test-plant cultivars can, therefore, be planted so as to synchronize the maximum tillering stage with the peak period of gall-midge emergence to provide an adequate midge population in the field (Prakasa Rao 1975).

Light traps

Lights can be effectively used over the test plots to attract phototropic insects such as green leafhoppers, brown planthoppers, stem borer moths, and gall midges of rice. To attract the gall midge, electric lamps are installed 3 m above the ground at the rate of one lamp for 25 m^2 of crop area when the rice crop is 40 days old (Prakasa Rao 1975). The lamps are lighted from 1900 h to 0400 h during rainless nights till the crop is 80 days old. Gall-midge incidence in lighted rice fields may reach 2.5 times of that in unlighted fields.

Baits and attractants

Baits or attractants such as fish meal attract adult sorghum shootflies to test rows of sorghum cultivars (Doggett *et al* 1970). Fish meal spread in the test rows about 10 days after plant emergence attracted flies, and over 90% of the plants in test rows were infested.

Resurgence-causing insecticides

Resurgence-causing insecticides were utilized on rice to increase the populations of the target insect, the brown planthopper. These chemicals eliminated the nontarget pests, predators, or parasites and helped increase the population of the brown planthopper (Chelliah and Heinrichs 1980). Selective insecticides were also used to increase the field populations of the sorghum midge in sorghum-screening trials (Sharma *et al* 1988). Sprays with contact insecticides at complete anthesis (after oviposition) and at the milk stage (before adult emergence) could control sorghum head bugs *Calocoris angustatus* Lethiery, to minimize interspecies competition with the sorghum midge as well as to minimize midge parasitization by *Tetrastichus diplosidis* Crawther. Insecticides with selective properties can be powerful tools in conserving and enhancing pest populations (Shepard *et al* 1977). A number of insecticides with useful properties for selectively managing insect populations are described by Tingey (1986).

Cultural practices

Cultural practices can be manipulated to encourage target pests in field trials. For example, closer spacing (15 × 15 cm or 10 × 10 cm) of hills in transplanted rice was found to result in a higher number of tillers and more foliage per unit area and higher relative humidity (94–98%) at the basal portions of rice plants. These environmental conditions are conducive to the successful hatching of eggs, larval invasion, and higher buildup of the rice gall midge. An absolute increase in gall-midge infestation is achieved by bringing about higher tiller numbers per unit area which, in turn, are obtained through higher levels of nitrogenous fertilizer application (Prakasa Rao 1975). Tillage and irrigation facilities can sometimes be modified to encourage insect pest populations (Hollingsworth and Berry 1982).

Mass rearing of insects

The mass rearing of insects on natural hosts in the greenhouse and/or laboratory or on artificial diets in the laboratory offers a dependable method for obtaining a large and continuous supply of insects for host plant resistance studies. In the last 25 years, entomologists have made excellent progress in mass culturing of insects on natural hosts (Heinrichs *et al* 1985) as well as on artificial diets (Singh and Moore 1985, CIMMYT 1989, Smith *et al* 1994). The feasibility of rearing insects on natural host plants or on artificial diets depends upon the insect, crop, and financial resources.

Natural diets

Many plant-feeding Homoptera, Hemiptera, Diptera, and Lepidoptera can be mass reared on their natural hosts. Starks and Burton (1977b) described techniques for mass rearing of greenbugs in the greenhouse. The host plant resistance studies with the Hessian fly can also be conducted with the flies raised on native host plants in a greenhouse. Heinrichs *et al* (1985) have developed efficient mass-rearing techniques for rearing brown planthoppers, whitebacked planthoppers, green leafhoppers, leaffolders, caseworms, striped borers, and yellow stem borers on rice, which is their natural host. These rearing methodologies permit a timely and adequate supply of insects for greenhouse and field screening experiments.

Meridic diets

Meridic diets are diets with a known chemical composition. Beck *et al* (1949) were the first researchers to develop an adequate meridic diet for the European corn borer. Encouraging progress has been made in the mass rearing of insects, especially phytophagous Lepidoptera, on artificial diets (Singh and Moore 1985). Many standard artificial diets are now available commercially. Rearing equipment and procedures for handling insect cultures to minimize pathogen attacks and to reduce contamination of the diet are some of the factors needed for a successful rearing program (Reinecke 1985).

The mass-rearing program for maize insects at Tifton, Georgia, began more than 20 years ago. Through the years, great strides have been made toward the development of a rearing system designed primarily for the corn earworm (CEW) and fall armyworm (FAW) (Mihm 1983a,b). Diets and rearing procedures were developed as described by Burton and Perkins (1989). The four main aspects of this program are adult manipulation, egg

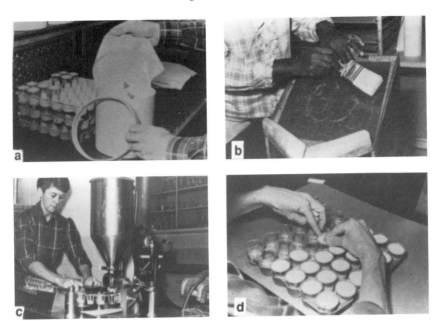

Fig. 8.1. Rearing and handling of the corn earworm and fall armyworm. (a) A food carton used as an oviposition cage; (b) the eggs are moved to a slanted flat surface and brushed; (c) the different diets can be dispensed within 30 minutes using an automatic filling machine; (d) infecting the cups by hand and then covering with a lid. Reproduced from Burton and Perkins (1989).

management, diet formulation and dispensing, and larval rearing. Food cartons (3.8-liter capacity) are used as oviposition cages (Fig. 8.1a). Eggs are collected from day 2 to day 6 by removing the cloth covering the eggs and replacing it with a fresh cloth (Fig. 8.1b). Cheesecloth is used as the oviposition substrate for the CEW and paper toweling for the FAW. A corn–soya-milk diet is selected for the CEW and a modified pinto bean diet for the FAW because of ease in their formulation, low cost, and nutritional adequacy. The diets are mixed in a large blender in 34-liter batches and then dispensed into 30-ml plastic cups using an automatic filling machine (Fig. 8.1c). Food cups are infested by hand with CEW eggs and FAW larvae (Fig. 8.1d). Infested cups are stored under controlled conditions throughout larval development, pupation, and emergence. This rearing system can process up to 10,000 cups per day.

The mass-rearing program for the southwestern corn borer *Diatraea grandiosella* Dyar and the FAW established at the Mississippi State University has developed semiautomatic equipment to prepare and dispense the diet, infest rearing containers with larvae, cap containers, and harvest pupae (Fig.

8.2). The oviposition cages have been improved by changing from small cages to large ones holding approximately 1000 adults per cage. The rearing procedures are effectively automated (Davis 1989).

Since the development of artificial diets with wheat germ as a basic ingredient, a highly efficient laboratory system for rearing the European corn borer (ECB) has evolved at the Corn Insects Laboratory, Agricultural Research Service, US Department of Agriculture (ARS-USDA) at Ankeny, Iowa (Guthrie 1989). The use of wheat germ marked the advent of practical artificial diets for rearing plant-feeding Lepidoptera. Slight modifications in the existing wheat germ diets have been successfully used to rear other species away from their natural host plants. This is the single most significant breakthrough in screening maize for resistance to the ECB. This program has served as a model for ECB rearing by researchers in the private and public sectors in the United States and elsewhere. In 1986, researchers in the United States and several other countries produced 50 million ECB egg masses for host plant resistance research.

Fig. 8.2. Machine used to infest diet-filled cups with neonate larvae mixed in corncob grits. The cups are capped and moved onto a loading tray in a continuous process. Reproduced from Davis (1989).

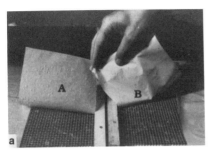

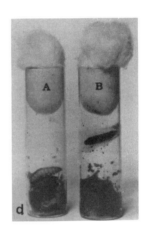

Fig. 8.3. Advances in rearing the European corn borer (ECB). (a) Oviposition cages: (A) egg masses on a waxed paper sheet, and (B) egg masses on a plastic sheet; (b) transferring newly hatched ECB larvae to rearing vials; (c) trays of ECB larvae in the incubator; (d) (A) fifth-instar larvae on diet plug, and (B) ECB pupa. Reproduced from Guthrie (1989).

Guthrie (1989) described the methods and equipment for preparing ECB diets, infestation, rearing, and egg collection. The ECB moths deposit egg masses on waxed paper or plastic sheets (Fig. 8.3a) between openings in hardware cloth. The sheets containing the egg masses are removed and replaced with new sheets each morning. Egg masses laid on waxed paper are placed in small screwtop jars and incubated to hatch. Newly hatched larvae are transferred individually with a small artist's brush to the diet in vials (Fig. 8.3b), which are then plugged with nonabsorbent cotton. Trays of the vials (Fig. 8.3c) are then maintained at 27°C and 75% relative humidity. Under these conditions, larvae reach the fifth instar on the 11th day and pupate on the 14th day (Fig. 8.3d).

At the International Crops Research Institute for the Semi-Arid Tropics (ICRISAT), the sorghum stem borer *Chilo partellus* (Swinhoe) is mass reared on a kabuli bean diet (Taneja and Leuschner 1985). The insects are utilized for screening sorghum germplasm for resistance to borers.

The mass multiplication of these insect pests has evolved to become a model program that has been adopted by many researchers in plant resistance programs. Books on insect rearing, such as *Advances and Challenges in Insect Rearing* (King and Leppla 1984) and *Handbook of Insect Rearing* (Singh and Moore 1985) may be referred to for additional information on mass rearing techniques.

Limitations of using laboratory cultures

The major disadvantage of laboratory cultures is the high cost due to expensive artificial diets, time, and labor-intensive work. Special facilities and equipment are required for the control of temperature, humidity, and illumination as well as for isolation and handling. Futhermore, the continued production of test insect populations on an artificial diet often decreases their genetic diversity (Berenbaum 1986). Laboratory-reared insects are likely to differ from natural populations in genetic, behavioral, and physiologic characteristics and this may limit their usefulness in the assessment of plant resistance (Guthrie *et al* 1974). This situation can be circumvented by periodically infusing insects from the field into the laboratory colony. The effectiveness of ECB cultures is maintained by establishing new cultures of the insect each year with larvae collected from cornfields (Guthrie *et al* 1974). To ensure that quality insects are being produced, periodic evaluation using standard resistant and susceptible lines should be made. When introducing new genetic variability into the existing culture, precautions should be taken to avoid introducing disease pathogens, parasites, and microbial con-taminants along with the field-collected insects (Shapiro 1984).

Insect-infestation Techniques

In addition to controlled mass rearing and production of test insect species, the techniques for artificially infesting plants in the field, screenhouse, and greenhouse are extremely important in any screening program. Artificial infestation is required for successful and sustained progress. Many techniques have been developed for infesting plants with insects and evaluating the resulting insect–plant interactions (Smith *et al* 1994). The techniques vary with the insect and the crop and amongst researchers working on the same insects and crops. Uniformity of insect infestation is critical to screening. While devising infestation techniques, researchers should take the following factors into consideration: (1) stage(s) of the insect to be used; (2) number of insects to be deposited on each plant or released within a given plant area; (3) number of releases required; (4) sites of insect release; and (5) growth stage of the plant. Efforts should be made to ensure that each of the test plants is infested with a uniform number of insects. The level of insect infestation can be determined by experiments using varying densities of insects on the susceptible standard plant genotype. The desired level of infestation is the minimum number of insects required to consistently rate the susceptible check variety in the susceptible category, and the moderately resistant lines as not susceptible. This aspect has been discussed in Chapter 7 (Figs 7.5 and 7.6).

Several examples are available in the literature to illustrate techniques that are being used to infest plants and/or field plots with insects. In the greenhouse, seedlings grown in flats are infested by brushing on them or shaking over them, test insects from the culture plants. This technique has been followed for infesting sorghum lines with greenbugs (Starks and Burton 1977b) and rice seedlings with planthoppers and leafhoppers (Heinrichs *et al* 1985). Field plots can likewise be infested by shaking test insects from host plants.

Infesting with eggs

Artificial infestation with the maize stem borer is done by placing egg masses into the whorl of the plant, or pinning discs with egg masses to leaves or other

Fig. 8.4. Egg masses of the European corn borer are pinned near the midrib on the underside of maize leaves for infestation. Reproduced from Davis (1976).

plant parts (Fig. 8.4) (Davis 1976). Palmer *et al* (1979) have developed a wet method of infestation that involves a suspension of *Diabrotica virgifera* eggs in agar water. Sutter and Branson (1980) developed a procedure to quantitatively and uniformly distribute *D. virgifera* eggs in large-scale field plots. A mechanical, pressurized system dispenses a known quantity of eggs suspended in agar water for use in the field screening experiments. This procedure promotes uniform infestation, which is considered a major breakthrough for research into plant resistance to *Diabrotica*. Wiseman *et al* (1974) used a pressure applicator (handlotion dispenser) for infesting corn silks with *Helicoverpa zea* eggs suspended in 0.2% agar. For labor savings, low cost, and convenience, the pressure applicator method of infesting with eggs has replaced the previously used method of infesting with larvae.

Infesting with larvae

Neonate lepidopterous larvae that used to be deposited on test plants with the aid of a camel hair brush are now applied with a larval dispenser. In this process, newly hatched larvae are mixed with corncob grits and dispensed mechanically onto plant parts by a plastic dispensing device referred to as 'bazooka' (Mihm *et al* 1978). This inoculation method is rapid and efficient in uniformly depositing insects on plants. Davis and Oswalt (1979) developed a modified version of the bazooka – the 'hand-operated inoculator' – for dispensing various insect species including lepidopterous larvae (Fig. 8.5).

A few points are essential to effectively operate the inoculator.

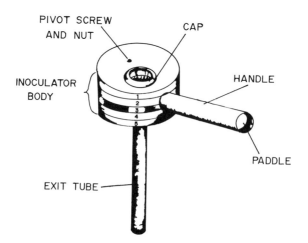

Fig. 8.5. Hand-operated inoculator (bazooka) for infesting insect larvae. Reproduced from Davis and Oswalt (1979).

1 It is important that the corncob grits are sterilized in an autoclave at 120°C for 2 hours to avoid any microbial contamination of the diet. If grits are not readily available, other materials such as corn meal, millet seed, or sorghum meal may be used (Hall *et al* 1980).

2 The next step involves the measurement of larvae for delivery through the outlet of the applicator. The larvae can easily be mixed with corncob grits. The mixture is poured into a supply bottle (1000-ml polythene wash bottle with a 28-mm cap), and additional grits may be added until the desired number of larvae per delivery is obtained. Occasional gentle tumbling of the mixture ensures uniform larval distribution.

3 The inoculator functions by paddle movement. When the paddle is operated, the grits move to the exit hole and down through the exit tube into the diet-filled cups. The amount of each delivery can be regulated by adjusting the paddle hole. After dispensing is completed, the inoculator is sterilized in 10% Chlorox (bleach) solution.

This inoculator is simple to construct, inexpensive, easy and fast to operate, and accurate. It is most effective for depositing larvae onto test plants grown in the greenhouse as well as in field plots. Maize, sorghum, cotton, and rice plants are usually infested at the seedling stage with this inoculator. To use the technique on cotton, the plants are sprayed with water first (Hall *et al* 1980); after rain or heavy dew, it is unnecessary to spray the plants. With experience, maize plants at midwhorl stage can be infested at the rate of 1500 plants per man hour. At the International Wheat and Maize Improvement Center (CIMMYT) approximately 1.5 million FAW larvae could be deposited on about 50,000 plants in 1 day. The inoculator is now commercially available for use not only with lepidopterans, but also with aphids and leafhoppers (Mihm 1989). It has also been used for dispensing insect eggs mixed with corncob grits or similar materials. Reese and Schmidt (1986) have cautioned that neonate larvae may be sensitive to physical handling while mixing with grits as this process may interfere with the insect's oxidative metabolism.

Mass Screening Techniques

With the advent of insect-infestation devices and simple agronomic procedures for growing healthy plants, it is now possible to adopt techniques for mass screening large populations of segregating plant materials. This is an initial step in the screening technology needed to eliminate the majority of susceptible segregants and select the resistant ones. Such large-scale evaluation where insects are offered a free choice of plant materials can be accomplished in the greenhouse, screenhouse, or small field plots. The approach for screening and evaluating resistance will depend upon the insect and the crop under study, and the required insect numbers, as well as the availability of research facilities. If insect damage occurs at more than one stage of plant growth, it is important to evaluate resistance at each of those stages. Some screening and evaluation techniques are presented in the following sections.

Greenhouse screening

Breeding materials can be screened rapidly by infesting plants at the seedling stage, especially during early mass screening cycles, in the greenhouse. Greenhouse screening techniques are economical in space, time, and labor, and have been successfully employed in screening cultivars of several grain and forage crops including rice (Heinrichs *et al* 1985), wheat (Webster and Smith 1983), sorghum (Starks and Burton 1977b), and alfalfa (Sorensen and Horber 1974).

Screening rice for resistance to brown planthoppers

The conventional seedbox screening test is a rapid method for screening large numbers of rice germplasm for qualitative resistance to the brown plant-hopper (Heinrichs *et al* 1985). Seed sowing and infestation are timed according to the hopper-rearing schedule. Seeds are sown in rows in a standard seedbox (60 × 40 × 40 cm). The number of insects per seedling can easily be determined, and 39 test entries can be screened in a box (Fig. 8.6). Twenty-five seeds of each test entry are sown in a 12-cm row. Seven days after sowing (DAS), when the seedlings are at the two-leaf stage, the seedboxes are placed in a water pan inside a screened room. Weeds are removed and the seedlings thinned to about 20 per row. The seedboxes are kept in the pan containing 5-cm water. Brown planthopper nymphs cultured on the suscepti-ble variety are uniformly distributed on the test seedlings by holding the base of the feed plant and lightly tapping the plants and blowing on them. In this way, approximately 10 hopper nymphs are deposited on each seedling. Grading of the entries in each seedbox is done when about 90% of the susceptible check seedlings in that box are dead. The Standard Evaluation System (SES) scale (0–9) for rice (IRRI 1988) is used to grade seedling damage: 0, no damage; 1, very slight damage; 3, first and second leaves of most plants are partially yellow; 5, pronounced yellowing and stunting or about half of the plants wilting or dead; 7, more than half the plants wilting or dead; 9, all plants dead.

Screening soybean for resistance to foliar-feeding insects

Any procedure to screen crops for resistance to foliar-feeding insects should meet the following criteria: (1) accuracy reflecting natural conditions, (2)

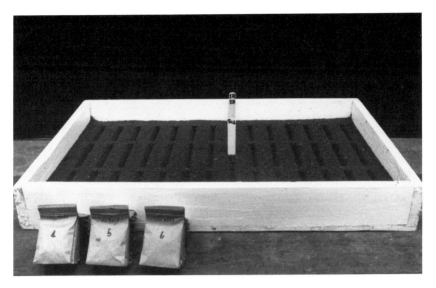

Fig. 8.6. A standard seedbox for sowing seeds of different rice varieties for screening against the brown planthopper. IR29, a resistant check, and TN1, a susceptible check, are planted in the middle row. The remaining 13 rows, each subdivided into three, are planted with test entries, thus accommodating 39 entries in each box. Reproduced from Heinrichs *et al* (1985).

rapid screening cycles, (3) applicable to most major foliar-feeding insects, (4) efficient and simple, and (5) economical.

In the greenhouse screening of soybean, three seeds of test genotypes are planted in each of 0.5-liter polystyrene foam cups filled with a methylbromide fumigated soil and sand mixture. Each cup has three 8-mm-diameter holes punched at the bottom. The cups are arranged in a randomized complete block design with three to five replications in a stainless steel pan (4.9 m × 1.2 m × 8 cm). The plants are thinned to one per cup at the one-leaf stage (Fehr and Caviness 1977). When the plants are 12–16 days old, four neonate larvae are placed on a trifoliate leaf of each plant with a camelhair brush. The trifoliate leaves of different plants touch each other, enabling the larvae to move freely among the plants within a block, but movement among blocks is prevented by maintaining a gap between them. This method can be used for screening soybean against the major soybean defoliating insects. Up to 900 soybean genotypes can be screened by this method at one time (All *et al* 1989). Damage evaluations are made 14 days after infestation. Each plant is examined and the percentage leaf area defoliated is estimated by comparing the feeding damage with photographs of similar leaves with known damage percentage. Leaf area meters are also used to assess the degree of insect defoliation of the cultivars (Kogan and Goeden 1969).

It is assumed that defoliation of test plants by *Helicoverpa zea* in the greenhouse screening system can be correlated with insect defoliation of soybean plants in the field nurseries. In fact, defoliation ratings for seven insect species indicate that the greenhouse procedure is useful for differentiating the resistant from the susceptible genotypes. The greenhouse procedure is a good substitute for the field screening program of soybean for resistance to defoliating insects.

Screenhouse screening

The yellow stem borer (YSB) is a major borer pest of rice and is widely distributed throughout South and Southeast Asia. It develops only on rice and feeds within the stem, causing deadhearts and whiteheads. As YSB egg masses as well as larvae are poorly distributed in field plantings, the test plants must be artificially infested with first-instar larvae. With the technique developed by Medrano and Heinrichs (1985) for rearing the YSB it is now possible to

Fig. 8.7. A rice plant in an oviposition cage showing yellow stem borer moth and egg masses.

Fig. 8.8. Recently hatched larvae of the yellow stem borer on rice inflorescences in a petri dish.

provide a regular supply of YSB larvae for artificial infestation. Potted rice plants are kept in the oviposition cage for egg laying. The egg masses collected from the oviposition cages (Fig. 8.7) are kept in a vial (Fig. 8.8) at room temperature (25–30°C) for hatching and are used for artificial infestation. By this rearing technique, about 80% of all infested tillers produce adult moths.

Screenhouse screening is the best procedure for evaluating rice germ-plasm resistance against the YSB. A screenhouse of $28 \times 22 \times 2.5$ m used at the International Rice Research Institute (IRRI) can accommodate six beds (25×2.5 m) separated by pathways and provide the capacity to evaluate 600 entries in nonreplicated studies.

Fourteen-day-old rice test entries are transplanted in the plant beds of the screenhouse at 20×20-cm spacing, each entry to one row. One row each of the susceptible (Rexoro) and the resistant (IR40) checks are planted for every 20 rows of test entries. One row of each entry is planted for an initial screening test. For retesting, entries are replicated three times in a randomized complete block design. Thirty days after transplanting, the entries are infested by placing newly hatched larvae on the youngest leaf or auricles at the rate of one larva per tiller, using a fine camelhair brush. Deadhearts are counted four weeks after larval infestation. The test is considered valid when the percentage of deadhearts on the susceptible check is at least 50%. The percentage of deadhearts for each entry is computed as:

$$\text{Percent deadhearts} = \frac{\text{Number of deadhearts counted}}{\begin{array}{c}\text{Total number of tillers observed}\\ \text{(healthy + infected + damaged)}\end{array}} \times 100.$$

The percentage of deadhearts (dh) is transformed to a 0–9 scale (0, no dh; 1, 1–10% dh; 3, 11–20% dh; 5, 21–30% dh; 7, 31–60% dh; 9, >60% dh) (IRRI 1988).

Field screening

Most screening for resistance to maize insects at the CIMMYT is conducted throughout the year under field conditions at experimental stations in Mexico. Even the greenhouse or laboratory studies at the CIMMYT nearly always utilize plant materials grown in the field, or at least include field-grown plants as checks (Mihm 1989). The main emphasis in the resistance program is to identify stable resistance in maize to feeding by tropical stem borer larvae.

The preliminary maize screening nurseries at the CIMMYT consist of a large number of exotic entries planted without replication. For initial screening, four-row plots are planted per entry, with two rows being infested and two rows protected with the recommended insecticide. Each 5 m row has 16–30 plants. In the process of screening, standard resistant and susceptible checks of the appropriate adaptation and breeding status are planted at regular intervals in the screening nursery for comparison. For routine screening and selection in the breeding nurseries, plant materials are planted in duplicate (nonrandomized). Promising entries are then replanted in verification trials the following season, with adequate replications in properly designed and randomized experiments. These are usually planted in stan-dardized yield trials, in four- to eight-row plots, where performance under both artificially infested and protected split plots is evaluated. This allows *per se* yield performance in the absence of pest attack and performance under heavy pest attack to be evaluated.

Larval infestation by the bazooka is more efficient than other means of infesting because it eliminates many laborious steps. When studying lepidop-terous borers, egg masses ready to hatch are transported to the field for mixing with grits and for infestation. As larvae of these species are positively phototactic, it is imperative that they are held in complete darkness during final incubation and larval hatching. Measured quantities of corncob grits are added to the dishes containing neonate larvae, and the larvae are uniformly distributed. The mixture is then passed through a No. 12 US Standard brass sieve to remove any unhatched masses, clumps of larvae, or debris. By serial dilution and check counts, the mixture is adjusted to the desired larval concentration of ten larvae per shoot for field infestation. If the air is hot and dry during preparation of the larvae–grit mixture, the addition of 25–50 ml of water per 1000 cm^3 of grit usually enhances larval survival. Infestation is usually done in early morning or late afternoon to evening. Test entries are infested with 30–45 larvae per plant, through three to four successive applications of about 10 larvae per delivery. It is critical that the mixture is agitated by gently swirling the bazooka after infesting a row or two.

Larval feeding sites vary with the plant growth stage. Therefore, infestation sites and techniques for evaluating resistance depend upon plant growth stage at the time of infestation. Infestation for evaluating resistance to first-generation and second-generation ECBs is done by infesting maize at the midwhorl stage and at the silking stage, respectively. Infestation sites are in the leaf axils of the ear leaf and the leaves above and below (Fig. 8.9).

For evaluating resistance to leaf feeding by the first generation of most maize stem borer species, entomologists use a standard scale of 1 to 9 based on the types of feeding lesions. The scale can be divided into three categories:

Fig. 8.9. Larval infestation with an inoculator for evaluating maize lines for resistance to the European corn borer. Plants are infested at the silking stage. Reproduced from Mihm (1989).

Fig. 8.10. Rapid screening of sorghum seedlings in (a) microplots, (b) trays, and (c) infestations with the 'Bazooka' for resistance to *Chilo partellus*. Reproduced from Nwanze and Reddy (1991) with permission from the South Carolina Entomolgoical Society, Clemson, USA.

1–3, resistant; 4–6, moderately resistant; and 7–9, susceptible (Guthrie *et al* 1960).

For evaluating resistance to sheath and collar feeding and stalk boring by second-generation ECBs, the number of 'cavities' per plant or the percentage of stems tunneled is estimated as an index of the establishment of the third- and fourth-instar larvae. Visual sheath and collar feeding ratings (1–9) are used to screen maize genotypes for resistance to second-generation borers (Guthrie *et al* 1978). However, as the number of entries and plants evaluated per season is large, the number of damaged internodes per plant is the easiest and fastest parameter to use. Using this system, an experienced recorder can rate more than 15,000 entries in three or four days.

Microplot screening

A rapid screening method for resistance to the sorghum stem borer *Chilo partellus* (Swinhoe) was developed at ICRISAT, in which sorghum seedlings, sown in microplots or seedlings in trays are evaluated under artificial infestation conditions (Nwanze and Reddy 1991).

Sorghum plants are grown either in 3×1-m field microplots with a plant spacing of 15×10 cm (Fig. 8.10a) or in plastic trays ($40 \times 30 \times 40$ cm) with a spacing of 5×4 cm (50 plants per tray) (Fig. 8.10b). A randomized complete block design with three replications is utilized for all experiments. Test plants are infested with stem borers reared on a kabuli bean diet (Taneja and Leuschner 1985). The larvae are dispensed with a bazooka at the rate of three to four per stroke (Fig. 8.10c).

For screening plant materials in microplots, 9–10-day-old plants are artificially infested by applying one stroke of the bazooka containing larvae and carrier into the leaf whorl of each plant. A second application is done two days later, especially if it rains within two days of the infestation. Leaf-feeding damage is scored seven days after infestation (DAI) on a visual rating scale of 1–9 (1, highly resistant; 9, highly susceptible) and deadhearts are recorded at 14 DAI (ICRISAT 1989). Using microplots, it is possible to plant test materials at three-week intervals at the ICRISAT Center between June 15 and September 15 under rainfed conditions and screen 12,000 entries per year. This method is valuable for the initial screening of germplasm for resistance.

Assessment of Resistance

The degree of insect damage to crop cultivars is due to pest intensity, the characteristic feeding or oviposition behavior of the pest species, and the plant resistance characteristics. A variety of criteria are employed for evaluating resistance. Resistance can be measured as the percentage of damage to the foliage or to fruiting parts, reduction of stand, per cent yield reduction, and general vigor of plants. It can be measured in relation to the insect, i.e., the number of eggs laid, aggregation, food preference, growth rate, food utilization, mortality, and longevity.

Plant damage assessment

The ultimate objective of host plant resistance is to minimize the insect damage to crop yield. The degree of yield loss depends upon the plant organs attacked by the insects (Solomon 1989). The insects may be *direct pests* that attack the plant parts of economic value, thus causing significant yield reduction. Relatively low pest densities of direct pests (e.g., corn earworm, boll weevil in cotton, fruit borers, sugarcane shoot borers, and rice stink bugs) may cause a significant yield loss. *Indirect pests* (e.g., *Epilachna* beetles feeding on the foliage of solanaceous crops, rootworms on corn, and sap feeders) at high pest densities invade the plant parts that may be indirectly related to yield loss.

Assessing plant damage by direct pests

The techniques for evaluating damage usually refer to absolute or relative numbers of damaged units: for example, damaged ears of corn per ten plants, infected bean pods per 5-m row, or affected tomato fruits per ten plants. Such quantitative levels of damage can be assessed and converted to actual yield loss per unit area. The damage caused by the CEW to soybean pods was evaluated by the percentage of infested pods in each line and the data were correlated with the plant resistance factors (Panda 1969). Jackai (1982) utilized an overall plant resistance index for assessing cowpea resistance to the pod-borer *Maruca testulalis* (Geyer) and correlated this with yield loss.

Linear regression, depicting the relationship between the incidence of rice YSB damage and grain yield in rice was established by Gomez and Bernardo (1974), where 2% deadhearts and 2% whiteheads were estimated to cause a 4.4% yield loss in the field at a yield level of 3 tons ha^{-1} and a 6.4% yield loss at a yield level of 4 tons ha^{-1}. Precise measurements of damage by direct pests is difficult because of the complexity of pests. Proper evaluation and statistical analysis of crucial samples are, therefore, necessary to arrive at correct assessments.

Assessing plant damage by indirect pests

These measurements concern such plant growth criteria as leaf area, growth condition, plant weight, ability to recover from damage, and effects on senescence and maturity. Photometric techniques have been used by researchers to measure tissue removal and defoliation of leaf area. The defoliated leaf areas are magnified 100-fold with a photographic enlarger and the excised portion of leaf traced onto a clear acetate sheet with a pointed felt-tip marker pen. The traced areas are darkened and measured with a leaf area meter (Lambda Instrument Corporation, Model LI-3000) (Bristow *et al* 1979). Feeding injury expressed by discoloration of plant tissue (necrosis, premature senescence) can also be measured by using reflectance sensors and infrared or fluorescence photography (Kogan 1972).

Gross damage symptoms caused by sap-feeding hemipterous and homopterous insect pests, whose damage potential is associated with removal of plant photosynthate, may not be obvious until late in the plant's development. However, subtle effects on growth and development can be observed soon after injury and measurement of plant growth condition (such as stunting, yellowing, or burning) can effectively complement or in some cases supplant analysis of losses in yield and quality. For example, brown planthopper damage to rice seedlings results in stunting, yellowing, and loss of dry weight, which can be accounted for as parameters for yield loss.

Assessing plant defense traits

Traits such as physical characteristics (leaf or stem features, leaf pubescence, and tissue strength) and/or specific secondary compounds that are well known to impart resistance to cultivars can be assessed without an insect bioassay. Determining leaf trichome length and density or plant crude fiber and silica content can indirectly indicate the degree of plant resistance and consequently be related to loss of plant quantity and quality.

Since biological methods tend to be laborious and time-consuming, quick biochemical methods of screening germplasm for resistance to different insects are desirable. Cole (1987) developed a simple and rapid technique for the visualization of phenolic acids using an ultraviolet fluorescence microscope. Fluorescence microscopy can be utilized to visually distinguish phenolic levels of susceptible and resistant lettuce and carrot cultivars. Experiments have shown that carrot cultivars known to be resistant to carrot fly attack can be distinguished from susceptible cultivars by the fluorescence of ammonia-treated seedling roots. There was a significant correlation between laboratory measurements using fluorescence microscopy and assessments under natural conditions in the field. This method is simple, relatively cheap, and quick; one person can examine 1000 plants in one day.

Recently, Liu *et al* (1993) reported that the constitutive and inducible resistance factors are associated in the soybean looper-resistant soybean cultivar PI 227687. In this cultivar's hypocotyls, the glyceollin concentration is positively associated with resistance. These workers suggested that for evaluating soybean for insect resistance, the hypocotyl assay is a good alternative to tests with fully developed plants.

Another novel approach for screening sorghum germplasm involves using greenbug toxin in the absence of the insect (J. C. Reese, Kansas State

University, personal communication). Commercial enzyme preparations that mimic the action of the greenbug toxins can be employed in inducing a 'red spot' reaction on the host plant. Techniques have been evaluated to quantify the development of this 'red spot' spectrophotometrically and relate it to germplasm tolerance for the greenbug. The new instrument, the 'Spadmeter', very quickly measures red spot response as it is reflected in a rapid drop in chlorophyll content.

Insect response assessment

Plant damage assessment provides an approximate measurement of qualitative plant loss by insect feeding. However, assessing insect behavior and insect growth can provide critical information for determining the existence of different modalities of resistance in the test plant cultivars. It also provides information supplemental to the plant damage measurements.

Insect behavioral response

Plant resistance is often expressed in terms of the insect response to plant stimuli. Many techniques have been developed to study the behavioral response of insects to host plant recognition, selection, and acceptance. In the 'free-choice tests,' whole plants, plant organs, or plant allomones/kairomones are utilized to assess their positive, negative, or neutral effects on insects. Some typical insect behavioral responses for evaluation include directed movement towards or away from the host, restlessness, acceptance or rejection of the host for oviposition and feeding, and diurnal or phenological shifts in behavior patterns. Insect perception of plant volatiles can be assessed by the olfactometer (Traynier 1967), but the electroantennogram (EAG) has been widely used for more precise assessment (Visser 1979). The electronic measurement system (EMS) greatly facilitates the behavioral study of feeding and probing by piercing and sucking insects (McLean and Weigt 1968).

Insect growth response

Plant resistance to insects can be critically assessed by the trend in growth and development of insect feeding on the test plants. The type of information useful in such studies includes survival and mortality, duration of immature and adult stages, weight of life stages, and life-tables. Information on antibiosis aspects of resistance in feeding experiments can be obtained using nutritional indices (Beck and Reese 1976). Measurements of number, size, or volume of feces (Kasting and McGinnis 1962) or honeydew excreted (Paguia *et al* 1980) are also used in assessing the insect's food utilization and growth response.

Techniques for Determining Mechanisms of Resistance

Greenhouse, microplot, and field screening techniques will probably eliminate the majority of susceptible strains. Following further testing and retesting in replicated trials, resistant entries are identified. Selected entries, such as those being considered as resistant donors in the breeding program or elite breeding lines, are further evaluated to determine the mechanisms of resistance. Such information is essential for developing varieties with effective resistance to insect damage. Different experimental designs and test procedures are necessary to differentiate each of the three mechanisms of resistance: antixenosis, antibiosis, and tolerance.

To assess the antixenosis type of resistance, uniform plant materials are infested with insects of the same age, sex, and growth conditions. The test plant cultivars in the greenhouse are often arranged in a circle, and test insects released in the center of the test plants to offer them free choice to accept or reject the test plants. The insects are left on the plants until the susceptible control cultivar sustains heavy damage or shows population accumulation. These experiments can be conducted with potted whole plants in the greenhouse, or with excised leaves in the laboratory, or small nylon cages in the field. Plant materials possessing antixenosis can be identified in such free choice tests.

Antibiosis can be assessed by 'no-choice tests' in which plant materials are caged and infested separately. Under such conditions, insects have no choice but to feed or not feed on the test plants. Antibiosis tests are designed to determine if the biology of the insects is adversely affected when they feed on the test materials. These may be whole plants, excised plant parts, or leaf discs, artificial diet or membrane filters containing plant extracts. The tests are designed to compare biological parameters such as larval survival, larval weight, days to pupation and/or population buildup on resistant and susceptible genotypes. However, antixenosis and antibiosis are not easily separable, for retarded larval growth may be due to potential deterrent factor(s) responsible for antixenosis or to the presence of growth inhibitors or toxins resulting in antibiosis (Renwick 1983).

Assessment of plant tolerance does not include plant interactions with insect behavior or physiology. Unlike the two other types of resistance, tolerance involves a comparison of the loss of plant biomass under insect-infested and noninfested conditions. Yield differences between infested and noninfested plants can then be used to estimate the per cent yield loss. Tolerance evaluation is conducted with comparable measurements of the insect populations and yield losses of the tolerant versus susceptible cultivars.

Examples of experiments to determine the mechanism of resistance to leaf-feeding and plant-sucking insects are presented in the following sections to illustrate the techniques employed.

Antixenosis tests

Much of the work on mechanisms of resistance has been conducted with larvae-feeding inhibitors, but this is not the first component of antixenosis in host plants. Insects encounter leaf volatiles, surface waxes, and ovipositional deterrents of the host plants before they start feeding. Techniques employed in assessing olfactory stimuli, oviposition preference, and feeding preference are discussed below.

Olfactory bioassay

The EAG is used as a bioassay technique to establish the potential of plant volatiles in olfactory stimulation of an insect (Guerin and Visser 1980). The EAG records the change in the potential for olfaction between the tip of an antenna and its base in response to stimulation by a plant odor.

    The insect is anesthetized with carbon dioxide and its head is removed. The antenna of the insect is inserted into the tip of a glass pipette. The EAG readings are recorded with glass capillary electrodes filled with electrolyte

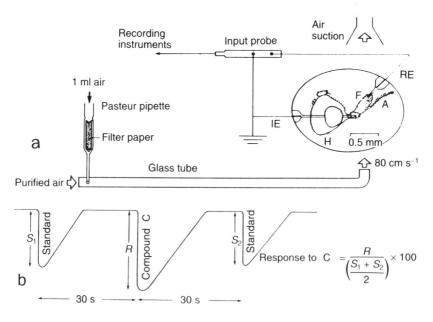

Fig. 8.11. Figure showing the electroantennogram recording technique. (a) Preparation: A, arista; F, funiculus; H, insect head; IE, indifferent electrode; RE, recording electrode. (b) Response evaluation method. Reproduced from Guerin and Visser (1980) with permission from Blackwell Scientific Publications, Oxford, UK.

(NaCl 3.75 g, KCl 0.175 g). The indifferent electrode (IE) is placed in the pedicellus and the recording electrode (RE) is pierced through the ventral tip of the funiculus (Fig. 8.11a). The IE is inserted into the occipital opening and through the scapus into the pedicellus. With this setup, movement artifacts are eliminated and desiccation is prevented, and the dissected antenna has an effective lifespan of 1–2 hours. The electrodes are connected via Ag–AgCl (chlorinated silver wire) to the recording instruments.

The odor delivery system and response evaluation system involves the continuous flow of charcoal-filtered air at a flow rate of 80 cm s^{-1} from a glass tube positioned within 5 mm of the antennal preparation. Test chemicals are dissolved in paraffin oil. Each solution (25 ml) is pipetted onto a piece of filter paper, which is placed in a pasteur pipette and attached to a syringe. The tip of the pipette is inserted through a hole in the glass tube and the syringe plunger is depressed to pass 1 ml of air through the pipette into the air stream. The duration of the pulse is 1 second. The antenna is stimulated at 2-minute intervals with each test chemical, each followed and preceded with a standard and control. The amplitude of the response to the test compound is expressed as a percentage of the mean of the two adjacent standard response amplitudes (Fig. 8.11b). The differences in volatility between the test compounds are relative.

The antennal olfactory receptor system in some phytophagous insects is very sensitive for the detection of odor components of green plants and shows different sensory sensitivity for individual odor components (Visser 1983). In rice, volatiles extracted as steam distillates from rice plants significantly affected the behavior and biology of the brown planthopper. The volatiles of resistant and susceptible rice varieties acted as feeding deterrents and attractants, respectively. In a multichoice test conducted inside plastic cages, more females settled and fed on tillers of a susceptible rice variety than on the resistant variety (Saxena and Okech 1985). Similar behavioral and feeding responses to steam distillates from resistant and susceptible soybean plants were observed in the cabbage looper (Khan *et al* 1987).

Ovipositional bioassay

Laboratory choice experiments may use filter papers soaked in plant extracts, whole plants, or plant parts. Oviposition studies are conducted with a simultaneous choice, and preferences are estimated on the basis of a relative number of eggs laid on the test plants. Reed *et al* (1989) bioassayed chemical stimulants of *Brassica* for ovipositional preference by the diamondback moth *Plutella xylostella*. Bioassays were used to examine the contribution of glucosinolates in the three *Brassica* spp. (i.e., *Brassica napus* L., *B. juncea* (L.) Czerniak, and *Sinapis alba* L.) to the ovipositional preference of the diamond-back moth. Compounds from aerial portions of 4–6-week-old plants were extracted and fractionated using ion-exchange liquid chromatography. The activity of glucosinolates was neutralized by treatment with myrosinase or sulfatase enzymes, which degrade glucosinolates. Bioassays were conducted in clear plastic chambers (30 × 30 × 20 cm) with screened tops. A drawer in the side of the chamber was added for the insertion of an 18.5-cm-diameter Whatman No. 1 filter paper disk which contained the extracts to be bioassayed. The disks were allowed to dry before exposing them to the moths. Twenty-five newly emerged (0–24-hour-old) male and female moths were placed in a bioassay chamber and provided with 10% sucrose solution. The moths were allowed to oviposit for 1–6 days. All bioassays were choice tests with equal-size areas treated with different solutions for comparison. At the end of the experiment, the filter papers were removed and the eggs counted. The data were log transferred and analyzed by two-way ANOVA (analysis of variance). This bioassay experiment demonstrated that the major oviposition stimulants for the diamondback moth found in the three cruciferous species tested were glucosinolates. The results also revealed that the adult diamond-back moth could not discriminate different sidechain structures of glucosino-lates.

Whole plants in screen cages, treated with test solutions under choice and no-choice tests, can also be used for assessing oviposition deterrents to gravid insects (Tabashnik 1987). Field cage experiments were carried out to determine if maize hybrids possess antixenosis to oviposition by the south-western corn borer and FAW (Ng *et al* 1990). When maize plants reached the midwhorl stage of growth, a screen cage consisting of a frame (6 × 6 × 2 m) was placed over each set of plants. Twenty pairs of mated moths were released inside the cage and egg counts on each plant were recorded after 2 days. Female moths of both species laid significantly fewer eggs on the resistant than on the susceptible maize hybrids. Small oviposition cages can also be utilized to examine antixenosis to CEW oviposition on maize cultivars (Wiseman 1989).

Feeding bioassay

Bioassay techniques for insect feeding have been employed to study the role of naturally occurring plant chemicals in the insect's choice of host and to determine the mechanisms of resistance in crop plants. The basic design in studying antixenosis of plant substrates is to present a choice of different substrates to the insects. The substrates may be whole plants, excised plants, leaf discs, or artificial substrates with an incorporated resistance factor. Since the behavioral effect of a chemical may or may not be independent of its nutritional value, these two properties must be experimentally separated before evaluating the chemical as a feeding stimulant or deterrent. This is done by running short-term assays to avoid postingestional effects of food. In the long-term tests, consumption, effect on growth, and digestive efficiency of the compound can be measured separately to assess the antibiosis nature of the phytochemicals. Test plant materials are typically compared with controls or other test substances in dual-choice, multiple-choice, or no-choice tests. The number and arrangement of substrates may vary greatly with species, cage design, and objectives of the experiment.

Chewing insects

The dual-choice arena test was conducted to assess the relative antixenosis mechanism of resistance in cowpea to the legume pod borer *Maruca testulalis* Geyer (Jackai 1991). The basic component of the assay is a round plastic container or 'arena' (18.5 cm diameter × 7.5 cm depth) with a 25-cm-thick piece of styrofoam fitted at the bottom (Fig. 8.12). A filter paper of similar size, moistened to keep the arena slightly humid, is placed on top of the styrofoam. Before the filter paper is installed it is marked with a pencil into four equal parts and pods of two varieties are put in each part to provide a choice for the

Fig. 8.12. Dual-choice arena used in the laboratory to screen cowpea for resistance to the legume pod borer. The pod borers are released at the center of the arena and within 3 hours they select the susceptible pod segments (identified with light-colored pins) and begin to feed on them. After 72 hours they still have not fed on the resistant pods (identified with dark-colored pins). Reproduced from Jackai (1991) with permission from Butterworth-Heinemann, Oxford, UK.

insect. The pod segments are arranged concentrically in an alternating sequence of a susceptible (IT84E-124) and resistant (TVNu 72) varieties. Four third- or fourth-instar larvae of *M. testulalis* reared on artificial medium are introduced at the center of each arena and allowed to choose between the test varieties for 72 hours. At the end of this period, feeding measurements are taken: the feeding ratio (FR, fraction of feeding out of four) and the feeding severity (FS, extent of feeding). Thereafter, a feeding index (FI) to estimate the insect's preference for, and consumption of, the test varieties is computed as FI/(FR × FS); FI varies from 1 to 4. To make comparisons among varieties against the control plant, a preference ratio (PR) is computed: PR = $2(FI_{test\ plant}) - (FI_{control} + FI_{test\ plant})$. The PR has a minimum value of 0 and a maximum value of 2, where PR > 1 indicates a preference for the test plant, PR < 1 indicates a preference for the control plants, and PR = 1 indicates no preference. These laboratory results for measuring antixenosis are comparable with those from field trials.

Sap feeders

An improved electronic measurement system developed by Kawabe *et al* (1981) was utilized in experiments related to the feeding behavior of planthoppers and leafhoppers. The device detects the changing electrical impedance in the insect and the substrate in which the insect is probing. To accomplish this, a small alternating current or direct current voltage is applied across the insect and the substrate, which results in a small flow of charge. When the insect probes into a plant or diet, the electric circuit is completed. A complex signal originating from the insect and the substrate is amplified and the waveform patterns within the signal are then subsequently correlated with the various phases of feeding: probing, salivation, and ingestion.

The electronic monitoring system was used for assessing the feeding behavior of the brown planthopper to determine antixenosis resistance in rice varieties (Velusamy and Heinrichs 1986). The insects were starved but water-satiated for 24 hours. A gold wire (20 μm diameter) was then attached with electroconductive paint to the dorsum of a 1-day-old female brachypterous brown planthopper. The opposite end of the gold wire was attached to a larger wire leading to the input of a current detection amplifier. The insect was then placed on the leaf sheath of a potted rice variety. The final amplifiers were adjusted to 500 mV full-scale output. The speed of the strip-chart recorder was maintained at 2 cm min^{-1}. The sequence of brown planthopper feeding on susceptible and resistant rice cultivars was electronically recorded as waveforms. The insect made more frequent and shorter probes, the salivation period increased, and the ingestion period was shortened on the moderately resistant and resistant varieties compared with those on the susceptible variety (Fig. 8.13). An increase in probing frequency indicated the presence of a deterrent factor that signified an antixenosis type of resistance. The

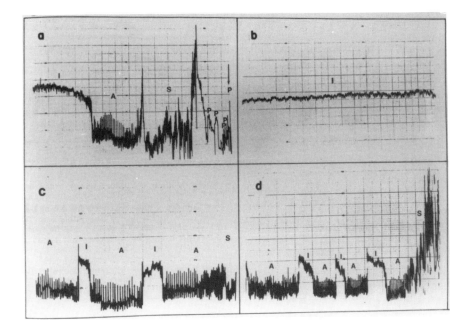

Fig. 8.13. Typical electronically recorded waveforms produced during feeding by brown planthoppers on 40-day-old rice cultivars. Charts should be read from right to left. The arrow denotes the initiation of probing: P, probe; S, salivation; A, waveform; I, ingestion. (a) Waveforms associated with the initiation of feeding on susceptible cultivar TN1. After an initial period of probing (3 minutes), salivation (5 minutes), and waveforms (4 minutes), ingestion begins and continues for up to 110 minutes as shown in (b). In (c) (IR36) and (d) (Utri Rajapan) ingestion duration is short and interrupted by the waveform. Reproduced from Velusamy and Heinrichs (1986) with permission from the Entomological Society of America, Lanham MD, USA.

electronic monitor, therefore, can be used to assess distinct differences in the feeding preference of insects for resistant and susceptible cultivars.

Antibiosis tests

Most of the antibiosis mechanisms of resistance are assessed under no-choice tests, with the insects confined on plants or plant materials inside a cage. Such tests are performed mostly in the greenhouse or in the laboratory, and sometimes under field conditions. Meridic or artificial diets can also be used in antibiosis tests.

Chewing insects

Antibiosis can be assessed with leaf discs, excised leaflets, lyophilized resistant plant materials, or membrane filters with incorporated test plant leaf extracts.

Laboratory antibiosis tests can be made with trifoliate leaves of soybean. When soybean plants are six weeks old, the uppermost fully expanded trifoliate leaves are collected at random from plants of each line for testing against the CEW (Smith and Brim 1979). Three first-stage CEW larvae are placed on a single trifoliate in a 75-ml polystyrene plastic vial with a snap-on cap (30 vials per line). Moistened paper toweling is placed at the bottom of each vial, and the vials are kept in plastic trays in an incubator at constant temperature and humidity. The larvae are thinned to one per vial after 48 hours. They are observed daily for mortality and weighed after 14 days. The pupae are weighed four days after pupation. Antibiosis is determined by estimating percentage larval and pupal mortality, larval and pupal weights, and duration of larval and pupal development. Workers who have conducted similar tests believe that as long as mature leaves are used, the excised leaf method is reliable for screening crop varieties against chewing as well as sucking insects.

The use of an insect bioassay method using the artificial diet for screening corn lines has greatly enhanced the studies on antibiosis and the search for the biochemical basis of maize resistance to the ECB (Wilson and Wissink 1986) and CEW (Wiseman 1989). In laboratory bioassays, the pinto bean diet is used as the base diet for the CEW (Wiseman 1989). Diet ingredients are blended along with distilled water and a preheated agar solution. Fresh or freeze-dried corncob silks of resistant corn variety Zapalote Chico are blended with the diet materials. Silks of a susceptible corn variety as well as the pure bean diet are included as checks. The diet–material mixture is dispensed into plastic diet cups and allowed to solidify at room temperature for 2 hours, after which one neonate CEW larva is put in each cup. Larval weight, days to pupation, pupal weight, and days to adult emergence are recorded to assess the antibiosis type of resistance.

With a much smaller amount of maize plant material, a microtechnique can be employed to assess antibiosis to the CEW (Wiseman *et al* 1986). A bean diet (40 ml diet and 10 ml distilled water) is blended thoroughly with 2 g of finely ground dry silks or other plant tissues of the test plants. The blended mixture is then aspirated into plastic soda straws and allowed to solidify. The soda straws (Fig. 8.14a) are then cut into 2-cm sections, with each end beveled at a 45° angle to expose a maximum diet surface. Two cut sections are placed into each of the 18-ml diet cups. One neonate larva is placed in each cup and the cups are closed with polycoated lids which prevents moisture loss from the diet. The weight of the larvae is recorded at 7 days (Fig. 8.14b). This bioassay technique is particularly useful for the rapid evaluation of resistance of any plant part as well as of chemical extracts of the plant parts.

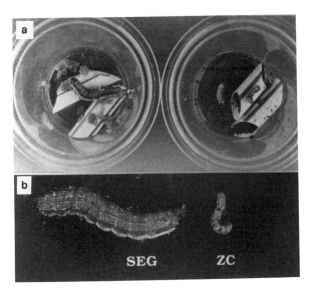

Fig. 8.14. Microbioassay technique for assessing antibiosis in corn to the corn earworm. (a) The petri dish on the left has soda straw segments with beveled ends which contain a bean diet into which powdered silks and other tissues of susceptible corn variety Stowell's Evergreen (SEG) have been mixed. The diet in the soda straws of the petri dish on the right contains powdered silks and other tissues of resistant corn variety Zapalote Chico (ZC). (b) The differences in the larval growth are evident here, as well as in (a). Reproduced from Wiseman (1989).

Hoppers

Several methods have been developed for assessing the antibiosis resistance of rice varieties to the brown planthopper. These methods include hopper feeding, survival of nymphs, and population growth. The feeding activity of hoppers can be determined by: (1) measurement of the amount of honeydew excreted, and (2) electronic monitoring.

Honeydew excretion

Honeydew excreted by the brown planthopper has been used as a criterion for determining the amount of sap ingested by the insect on resistant and susceptible rice cultivars (Sogawa and Pathak 1970). The quantity of honeydew excreted is low on resistant cultivars and high on susceptible cultivars (Paguia *et al* 1980). Hopper-excreted honeydew on filter paper (Fig. 8.15) and in parafilm sachets is collected (Fig. 8.16) (Pathak *et al* 1982) and measured by determining the area of filter paper with honeydew or the weight of the honeydew in the parafilm sachet. The ninhydrin-treated filter paper

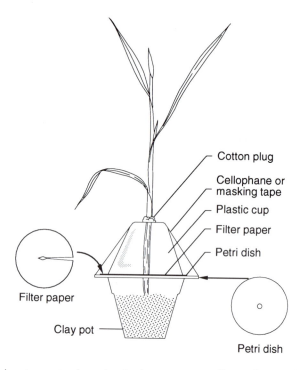

Fig. 8.15. Laboratory setup for estimating hopper-excreted honeydew on filter paper. Reproduced from Heinrichs *et al* (1985).

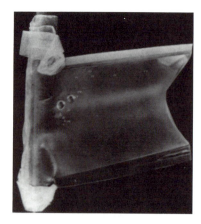

Fig. 8.16. Parafilm sachet cage to collect honeydew from leafhoppers and planthoppers feeding on resistant and/or susceptible rice cultivars. Reproduced from Heinrichs *et al* (1985).

method is useful for the brown planthopper, which is a phloem feeder. The filter paper is sprayed with 0.001% ninhydrin in acetone solution. The measurement of the honeydew spots, which are bluish or purple, indicates the intensity of feeding by the hoppers. Each treatment, including the control, is replicated several times. When using the parafilm sachets, the quantity of honeydew excreted by one insect during a 24-hour period can be used as a parameter in comparing the insect's feeding activity on susceptible and resistant varieties. This technique has proved useful for determining the level of resistance of rice varieties to the brown planthopper, whitebacked planthopper, and green leafhopper. The amount of honeydew excreted is highly correlated with weight gain over the same time period for a given rice variety. Sometimes, however, this assessment is not accurate, hence other techniques need to be adopted.

Electronic monitoring

The electronic feeding monitor can quantify precisely the feeding activity of various species of sap feeders. A direct current system has been developed which provides detailed electrical signals caused by insect stylet penetration activities (Tjallingii 1978). These systems are called electrical penetration graphs (EPGs). Kimmins (1989) used the EPG systems to study the probing of brown planthoppers on rice plants (Fig. 8.17). The correlations between the

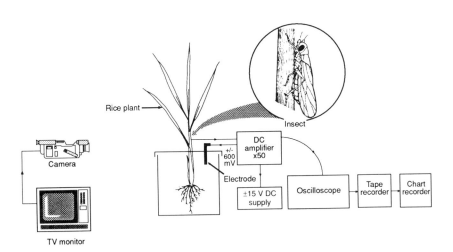

Fig. 8.17. Diagram of video and electronic recording equipment for assessing the feeding activity of the brown planthopper. The electrical penetration graphs (EPGs), using a direct current (DC) system, describe the insects' probing into rice. The correlations between individual patterns and ingestion activities were investigated by recording EPGs and observing honeydew production simultaneously. Reproduced from Kimmins (1989) with permission from Kluwer Academic Publishers, Dordrecht, The Netherlands.

EPGs and ingestion activities were investigated by recording EPGs and observing honeydew production simultaneously. Two insects were placed on separate plant tillers and observed using individual cameras and amplifiers and a screen-splitter and TV monitor. EPG signals were recorded for 8 hours (0500–1300 h) on separate channels of an FM tape. After the observation period, the EPGs were played from the tape onto a chart recorder (bandwidth 75 Hz). Extreme care was taken in interpreting the EPGs and in categorizing correctly the waveforms observed. Honeydew excretion was observed via the video monitor and recorder using a portable computer as a data logger. The times of droplet production were marked on the corresponding section of the EPG charts. The honeydew drops that fell onto the filter paper were estimated by the ninhydrin method described in the previous section. Sap ingestion and honeydew excretion by the brown planthoppers were also assessed in the EPGs by incorporating radioactive phosphorus into the rice plant and monitoring the amount within the insect and its excretion after 24 hours on the plant (Hopkins 1991). The total radioactivity of the insect plus the excreted honeydew increased exponentially with duration of the ingestion pattern. The total amount of label taken up by the insect from the resistant cultivar was significantly less than that from the susceptible cultivar.

Population increase

In addition to the nature of feeding and honeydew excretion by the brown planthopper, population increase and the growth index of the insect can provide additional information on antibiosis type of resistance (Heinrichs *et al* 1985). Thirty-five-day-old, greenhouse-grown, potted rice plants were cleaned at the base and enclosed with a 13-cm diameter, 90-cm high mylar cage (Fig. 8.18). Plants in each cage were infested with 10 1-day-old nymphs with the aid of a mouth aspirator. The generation time of the insect is about 20 days. At 30 days after infestation, the insects produced in each cage were counted. Population buildup varied according to the level of resistance in the different rice varieties.

Differentiating the effects of antixenotic and antibiotic resistance on the insects

It is often not clear whether a secondary plant compound is a feeding inhibitor or has a toxic effect on the test insects. The 'no-choice' feeding tests do not differentiate between the effects of antixenotic and antibiotic resistance. By confining insects in no-choice tests, the behavioral aspects of resistance are minimized. Differences in the insect's reproduction and longevity cannot necessarily be ascribed to antibiotic factors alone. Because of the presence of feeding inhibitors in the test plants, insects may not feed, may lay few eggs, or may die due to starvation (Puttick and Bowers 1988). Thus, antibiotic resistance cannot be ascertained by means of individual life-history parameters of insects.

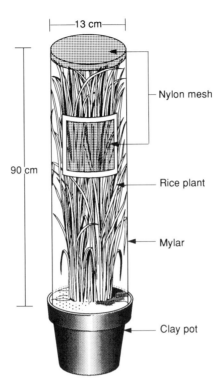

Fig. 8.18. Potted rice plants infested with brown planthoppers and enclosed with a Mylar cage for population build-up studies. Reproduced from Heinrichs *et al* (1985).

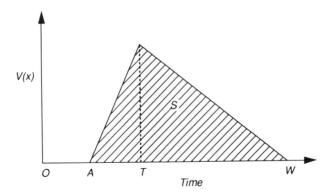

Fig. 8.19. The Lewontin reproductive function model showing an exponential population increase as a simple function, *V*, of age (*x*) where *A* = start of reproduction, *T* = peak reproduction, *W* = end of reproduction, and *S* (area) = total reproduction. Reproduced from Lewontin (1965).

The population model developed by Lewontin (1965) is a useful instrument to compare insect population dynamics on different host plants and to select critical periods of the reproductive cycle to use as criteria for resistance. This model, often referred to as the 'Lewontin reproductive triangle' (Fig. 8.19), describes reproduction during the period of exponential population increase as a simple function, V, of age (x) where A, T, and W are the time reproduction starts, peaks, and ends, respectively. The area, S, of the triangle represents the overall reproduction. From these values, the intrinsic rate of increase (r_m) of the insect population can be calculated for each plant cultivar (Romanow *et al* 1991). Lewontin's model can precisely assess how the insect reproductive function varies among test plant varieties and translates those differences into a single value for comparison of plant varieties possessing antixenotic and antibiotic resistance.

Following Lewontin's model, Romanow *et al* (1991) assessed the resistance of tomato to greenhouse whiteflies. Their tests could differentiate small differences in antibiosis between breeding lines. These differences are valuable for evaluating certain parameters of insect population development for resistance screening. Because the estimation of total reproduction included several parameters of insect reproduction, any differences between antibiosis as well as between antixenosis and antibiosis are likely to be distinguished. Where resistant parent materials possess both antibiosis and antixenosis components of resistance, they can contribute to the durability of resistance.

Tissue culture for evaluating insect resistance

Tissue culture for evaluating insect resistance in crops is a potentially useful technique. Williams *et al* (1983) demonstrated that maize borer larvae fed on calli of maize genotypes with varying levels of insect resistance grew differentially. The mean weights of larvae of the southwestern corn borer, sugarcane borer, and ECB reared on calli of resistant maize hybrids were significantly lower than the mean for larvae reared on calli of susceptible maize hybrids. When larvae were allowed to choose among calli of the resistant and susceptible hybrids, significantly fewer of each insect species chose to feed on the calli of resistant hybrids. Both antibiosis and antixenosis appear to be operating in the maize tissue calli as mechanisms of resistance to the borer insects (Williams *et al* 1987).

Preliminary results of investigations in rice showed that larval development in the YSB, striped stem borer, and rice leaffolder was normal on the callus of susceptible rice variety Rexoro, whereas none of the insect larvae could grow on the callus of resistant *Oryza ridleyi* (Caballero *et al* 1988).

Enzyme activity of insects in identifying resistance

The activities of carboxylesterase, and acid and alkaline phosphatases in the delphacid *Sogatella furcifera* were used to assess the resistance of 15 rice varieties. The activities of these three enzymes in adults and second-generation nymphs were significantly greater when the delphacid fed on susceptible varieties (Wu *et al* 1993).

Tolerance tests

Different plant growth parameters are used for measuring the tolerance of plant cultivars to insect damage. Tolerance was measured by differences in plant height and plant injury caused by greenbugs on infested sorghum plants (Schuster and Starks 1973). Other plant characteristics used for assessment of insect tolerance are the growth rate of plant leaves, stems, petioles, roots, and fruits under insect pressure.

   The conventional seedbox test was modified to identify rice varieties with tolerance for the brown planthopper. Ten-day-old plants were infested with three to five brown planthopper nymphs. Plant damage, assessed 28 days after seeding, was caused by the progeny of these nymphs. Formulae were devised to assess the tolerance of rice cultivars (Panda and Heinrichs 1983). Plant dry weight loss due to brown planthopper feeding was estimated by the functional plant loss index (FPLI)

$$FPLI = 1 - \left[\frac{\text{Dry wt of infested plants}}{\text{Dry wt of uninfested plants}} \right]$$
$$\times \left[1 - \frac{\text{Damage rating}}{9} \right] \times 100.$$

The regression of FPLI (y) on brown planthopper dry weight (x) was computed for each test variety and a pooled regression over all varieties was calculated. The pooled regression estimate for the FPLI on insect dry weight was significant, indicating the possible separation of varieties having various combinations of antibiosis and tolerance (Fig. 8.20). The mean value of the independent variable, brown planthopper dry weight (x), is computed and a vertical line is drawn. The intersection of the mean and the pooled regression line is used to separate the components of resistance into four categories. The mechanisms of resistance involved in each variety are based on the spread of points of individual observations within the four categories. The rice variety Utri Rajapan is considered tolerant of brown planthopper damage in the absence of antibiosis. The susceptible IR26 was classified as lacking antibiosis and tolerance. The regression analysis technique provides a meaningful

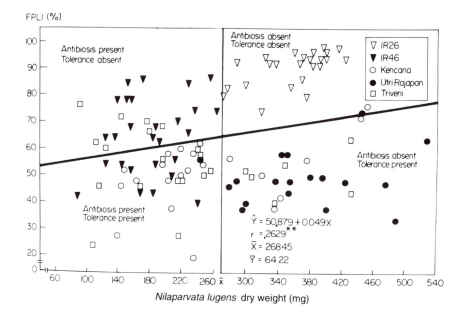

Fig. 8.20. Identification of the components of resistance of rice varieties to the brown planthopper using brown planthopper dry weight as the antibiosis indicator and the functional plant loss index (FPLI) as the tolerance indicator. Reproduced from Panda and Heinrichs (1983) with permission from the Entomological Society of America, Lanham MD, USA.

relationship between tolerance and antibiosis in populations of plant materials evaluated for insect resistance. The tolerance index based on the brown planthopper dry weight on the test varieties is also used for the assessment of tolerance, where:

$$\text{Tolerance index} = \frac{\text{Brown planthopper dry weight on test variety}}{\text{Brown planthopper dry weight on susceptible variety}}$$

Antibiosis index = 1 − tolerance index.

Another parameter used to measure the level of tolerance in the test varieties is the plant dry weight loss per milligram of brown planthopper dry weight produced, based on the formula of Schweissing and Wilde (1978).

9

Plant Resistance and Insect Pest Management

The indiscriminate use of chemicals for insect control in agriculture during the past 40 years has led on numerous occasions to 'ecocatastrophe' (Metcalf 1986c). In general, heavy reliance upon insecticides has resulted in such phenomena as acquired pesticide resistance in insect pests (Georghiou 1986), pest resurgence (Chelliah and Heinrichs 1980), and development of secondary pests (Metcalf 1980). It is now recognized that most insect pests have high reproductive rates, they can eventually adapt to natural and synthetic pesticides, and it is likely that most of them will never be eradicated (Gould 1991). Control measures should therefore aim at containment of insect populations rather than eradication. This thinking has led to the development of the strategy or philosophy of integrated pest management (IPM) (Huffaker and Smith 1980). IPM aims to keep pest populations below the economic threshold levels through an ecologically sound, economically practical, and socially acceptable technology (Metcalf and Luckmann 1982). Host plant resistance (HPR), along with natural, biological, and cultural control measures, is the basic component of the IPM system. Patterns of pest population buildup under different pest control methods should be looked into when developing IPM strategies.

Insect Population Trends under Different Strategies of Pest Control

The trends in insect populations in an agroecological zone vary, depending on the species and the associated biotic and abiotic factors. If all basic conditions are assumed to be constant, the relative trends in insect population dynamics under different pest control strategies can be compared. It is well known that

271

most insect populations can increase at a rapid rate. But interventions such as predation, parasitization, and climatic factors greatly limit the actual rate of increase. The normal intrinsic rate of increase is five-fold. In the absence of natural hazards, the insect population is likely to increase 10-fold or even 20-fold (Knipling 1979).

Insect population trends with the use of pesticides

Undoubtedly, insecticides are effective in reducing the initial insect population. But, even if 90% of the insect population is destroyed by insecticide application, the remaining insects would multiply at a faster rate in the absence of natural enemies, which are also destroyed by insecticide application (Knipling 1979). Consequently, the insect population increases with each generation and the growers need to apply higher dosages of insecticide, but fail to control the target pests. One recent example of insect resurgence caused by synthetic pyrethroids is that of the bollworm *Heliothis armigera* on cotton in India. The bollworm now poses the biggest threat to India's cotton industry (Sharma 1991).

Intensive insecticide use has resulted in an exponential growth in the numbers of insecticide-resistant insect species (Georghiou 1986). In several instances, insecticide-resistant strains have evolved that are adapted to almost all pesticides used. Georghiou and Taylor (1977) have enumerated specific factors that influence the rate of evolution of insecticide resistance in field populations of a pest. All of the factors (genetic, biological, ecological, and operational) relate directly and indirectly to the potential selection intensity per unit of time. Using a computer simulation model, Tabashnik and Croft (1982) studied the evolution of insect resistance to pesticides under the influence of biological and operational factors. The strong interaction between these factors is seen in the variation in the three operational components: dose, spray frequency, and the fraction of the insect lifespan exposed.

Tabashnik and Croft (1982) developed a flowchart decisionmaking scheme for devising strategies to retard the development of insecticide-resistant insect strains. They identified two main strategies – low-dose pesticide use and high-dose pesticide use – and the respective conditions under which each strategy might be effective. The principle behind the low-dose pesticide use strategy is to use pesticides in as small amounts as possible to reduce the rate at which susceptible insects are removed from the population, thereby reducing the rate of development of insecticide-resistant strains. In practical terms, spraying less often is the most efficient and economical way to implement this strategy. This strategy is likely to maximize the potential for biological control.

The results suggest that a high-dose pesticide use is not likely to suppress the buildup of resistance in insect species having high reproductive potential, high survival, or many generations per year. Furthermore, since most insects are not randomly distributed over plant surfaces, insecticide sprays are not likely to hit all the target insects. The insects are mobile and can avoid or seek out areas already treated with insecticides. Thus, insect behavior has the potential to significantly alter the efficacy of insecticides (Sparks *et al* 1989).

Insect population trends with the use of host plant resistance

The most important advance in facilitating IPM has been the breeding of pest-resistant crop cultivars. Host plant resistance has often been deployed alone as an approach to pest management. But the benefits of resistance depend on the modality (antixenosis, antibiosis, or tolerance) and the level of resistance against the specific pests and the crop production system under consideration (Kennedy *et al* 1987). The benefit of a specific type and level of resistance would depend on the pattern of insect invasion of a crop. Many arthropod species such as aphids, whiteflies, and mites, initially invade crops in low

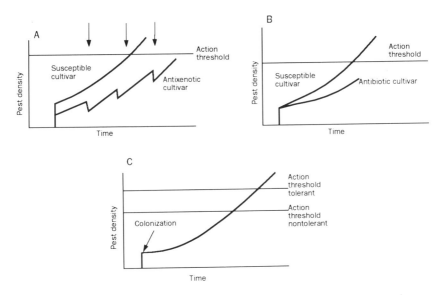

Fig. 9.1. Effects of different types of resistance on the population increase of the target insect species over the course of a single crop season and the relative action threshold levels for initiating pest control measures. (A) Antixenosis reduces the initial number of insect colonizers early in the season and the size of the population in the successive generations. The action threshold level for the antixenotic cultivar is much below the level for the susceptible cultivar. Arrows indicate the start of new insect generations. (B) Antibiosis reduces the rate of population increase by reducing reproduction and survival. The action threshold level is far below the level of the susceptible cultivar. (C) Tolerance does not reduce the rate of population increase but does raise the action threshold level as compared to the nontolerant cultivar. Redrawn from Kennedy *et al* (1987) with permission from the Entomological Society of America, Lanham MD, USA.

numbers, but their populations increase gradually over several generations before reaching damaging levels. For such herbivores, even low to moderate levels of antixenosis and antibiosis would be effective in delaying the time required for the population to reach economic damage levels. Antixenotic resistance is likely to reduce the rate of increase of both the initial and successive insect population buildup (Fig. 9.1A). Antibiosis also reduces the rate of population increase by reducing reproduction and survival, and prolonging generation time (Fig. 9.1B). The plant with tolerance can withstand damage from an insect population which its nontolerant counterpart cannot (Fig. 9.1C).

Antixenosis-based resistance may have the effect of shifting the population of the pest, especially polyphagous pests, to other nearby crops, away from the antixenotic crop. In crops with a single major insect pest, a high level of resistance (antibiosis) can provide complete control of the pest, as has been the case in wheats resistant to the Hessian fly Mayetiola destructor (Say) (Painter 1968). The effect of antibiosis on the reduction of pest development over generations and years has been calculated by Knipling (1979) using theoretical models. Antibiosis can decrease insect populations within a few generations by cumulative effect. With the advent of new genetic engineering techniques, antibiosis is expected to produce cultivars with the capacity to kill over 95% of a pest population. This technique can slow down the pests' growth rate and/or decrease the pests' reproductive capacity (Vaeck et al 1987).

In some cases, even low levels of resistance appear to be very promising. Van Emden (1990) reviewed the synergistic effects of partial or low levels of plant resistance and natural, biological, cultural, and chemical control on the population buildup of insects. He observed that partial resistance is often more evident in field tests than in greenhouse tests because of the synergistic effect of natural enemies in the field. He is of the opinion that such resistance could prevent pest outbreaks in certain situations.

Kennedy et al (1987) used the HELSIM Helicoverpa zea population dynamics computer model to illustrate how simulation models can be utilized to explore the consequences of deploying particular modalities and levels of insect resistance for durability. The results indicated that the antixenotic maize hybrid would not reduce the Helicoverpa zea population and the damage they caused. In another series of simulations, it was observed that antibiotic maize could cause about 50% mortality of first- and second-instar larvae of H. zea. Using the same model, Kennedy et al (1987) explored the genetic response of H. zea to a hypothetical antibiotic-resistance factor in maize silks, which raised mortality from 20 to 80%. Antixenotic resistance reduced the attractiveness of silking to only 10% of that of susceptible maize, but it imparts only minor resistance which does not last long. After approximately 23 generations more than 50% of the H. zea population would be unaffected by antixenosis (Fig. 9.2A). If instead of antixenosis, the antibiosis factor of maize silk is introduced

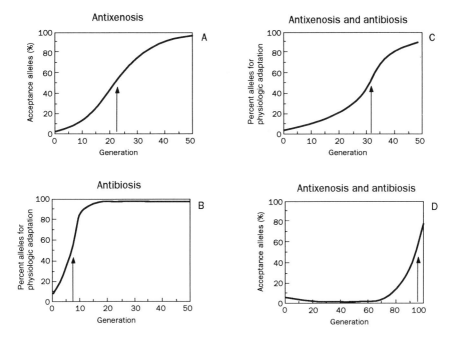

Fig. 9.2. Durability of antixenosis and antibiosis resistance of silking-stage corn to *Helicoverpa zea* when the two types of resistance are used separately (A, B) or together in a single cultivar (C, D). (A) and (D) exhibit the increase in frequency of alleles in *H. zea* for accepting the antixenotic corn (antixenosis lasts for almost 100 generations). (B) and (C) show the increase in the frequency of alleles of *H. zea* for adaptation to antibiosis (antibiosis lasts about 32 generations). Redrawn from Kennedy *et al* (1987) with permission from the Entomological Society of America, Lanham MD, USA.

into the system, there would be high selection pressure on the insect and the resistance would be overcome in only seven generations (Fig. 9.2B).

The combination of antibiosis and antixenosis would be more durable. When the two types of resistance are combined, antibiosis would last for about 32 generations (Fig. 9.2C) and antixenosis for almost 100 generations (Fig. 9.2D). Gould (1984) was also of the opinion that crop cultivars with both antibiotic and antixenotic resistance should last longer and would be less prone to insect adaptation.

Simulated results were obtained with two genetic loci in the insect for overcoming resistance conditioned by plant-produced toxins (Gould 1986). If the simulated insects are exposed exclusively to the toxin-producing plants, a biotype capable of overcoming resistance emerges quickly; adding just 10% susceptible plants delays the development of a biotype for almost 150 generations; and with 30% susceptible plants, a biotype does not appear until after 500 generations. The benefits of reduced selection pressure are smaller but still significant when the ability to overcome resistance in the pest is inherited as a monogenic recessive trait.

Plant Resistance and Suppression of Insect Pest Populations

Plant resistance has six outstanding characteristics that greatly enhance its utility in IPM systems (Kogan 1982).

1 *Specificity*. Plant resistance is specific either to a single key insect species or a complex of pest organisms.
2 *Cumulative effectiveness*. Any reduction in insect pest density due to antibiotic resistance usually is the result of lowered fecundity, growth, and development. Even a limited degree of resistance is sufficient to suppress the insect pest population in successive generations.
3 *Persistence*. Some varieties with durable resistance are likely to maintain their resistance for long periods.
4 *Compatibility*. The unique feature of plant resistance is that it is compatible with most of the other techniques of pest management.
5 *Environmental friendliness*. As no unnatural elements are used, there is no danger of contaminating the environment or endangering humans or wildlife. However, the environmental consequences of genetically engineered crops need to be understood.
6 *Ease of adoption*. Once the resistant varieties are developed, they are easily adopted in normal farm operations at no additional cost. Earlier concerns that resistance is frequently associated with poor grain quality and low yield are no longer valid.

A higher level of resistance may not always be attainable, nor is it called for. The goal should be to achieve some degree of resistance. A resistant plant variety that reduces the insect population by 50% in each generation is ideal for reducing the insect population to below the economic injury level within a few generations (Painter 1958). Such a cumulative and persistent effect of a resistant variety contrasts with the explosive effects of most insecticides. Once a population has been considerably suppressed in a cropping area, the continued use of resistant crop varieties should serve as a major deterrent to the insect's re-establishing its status. One notable example is the near elimination of the Hessian fly by the extensive growing of resistant wheat varieties in the United States. Crops with an adequate level of resistance to the wheat stem fly, chinch bug, cereal leaf beetle, spotted alfalfa aphid, greenbug, cotton boll worm, and weevil have been developed in the United States (Sprague and Dahms 1972). Another notable example of the success of HPR is the control of rice pests such as the rice gall midge and brown planthopper

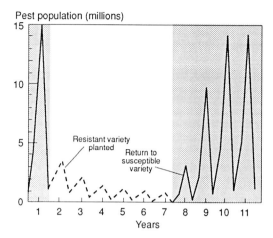

Fig. 9.3. Decline in a hypothetical pest population following the planting of a resistant plant variety that produces 60% mortality in each generation, and the resurgence of the pest population following the return to a susceptible variety. Reproduced from Knipling (1979).

by growing resistant varieties in tropical and subtropical Asia (Heinrichs 1988).

A model for insect pest population suppression by the growing of resistant crop varieties is presented in Fig. 9.3. The trend of pest suppression is rather slow but steady. The figure indicates the pest suppression process by a resistant variety and how the pest tends to rebound to explosive levels when a susceptible variety is re-introduced. The resistant variety is assumed to possess a 60% level of pest suppression by antibiosis factors. If it is grown for 6 years, the overwintered insect population will decline to about 70,000, which is 7% of the original steady-state population level. If a susceptible variety is then planted for successive years, the pest population would rapidly increase and would exceed the original steady-state level, causing higher than normal damage before it restabilizes.

Complementary Functions of Plant Resistance

The use of HPR as the key component of a successful IPM system has even greater potential than any other tactic or strategy for pest suppression. Resistant rice varieties provide an inherent control that involves no environmental pollution problems. They are generally compatible with other insect control methods and are more likely to contribute stability to a pest management system in an annual crop situation. The cultivation of a resistant rice variety is not subject to the vagaries of weather as are chemical and biological control measures. In certain pest endemic situations and with typical monsoon climates, it is the only effective means of pest control. In sustainable farming, farmers are also using resistant cultivars (Mochida 1992). The use of resistant varieties of rice without using pesticides is a very important aspect of the high yields achieved in Indonesia (Gallagher 1992). The rice pest management system developed at the International Rice Research Institute (IRRI), in the Philippines, for example, relies upon the key role of varietal resistance. Numerous rice varieties with multiple resistance to several key pests developed at the IRRI and by national rice improvement programs are now grown on more than 50 million ha in Asia (Khush 1989).

In rainfed agriculture, planting pest-resistant cultivars of sorghum is especially useful under the subsistence farming conditions of the semi-arid tropics. Resistance to the major pests of sorghum is available in diverse genotypes. HPR is now utilized for the management of the sorghum midge, greenbug, mites, aphids, and head caterpillars (Sharma 1993).

Plant resistance with biological control

DeBach (1964) defined biological control as 'the action of parasites, predators, and pathogens in maintaining another organism's density at a level lower than would occur in their absence.' Under this definition, biological control includes both natural biological control agents, which occur without human intervention, and applied biological control measures. Applied biological control requires the purposeful introduction or manipulation of parasites, predators, or pathogens to reduce populations of undesirable organisms. Entomophagous insects (parasitoids and predators) are major components in natural as well as applied biological control of insect pests (Price 1986).

In addition to the direct and indirect effects of plant allelochemicals that impart resistance to insect herbivores, the selection pressure imposed by natural enemies should result in the magnification of plant resistance (van Emden 1990). Plant resistance and biological control are generally considered to be compatible. They are the key components of IPM for field crops and forests (Starks *et al* 1972). The use of these two control tactics brings together unrelated mortality effects, thus reducing the pest population's genetic response to selection pressure from either plant resistance or natural biological control. Acting in concert, they provide density-independent mortality at times of low pest density as well as density-dependent mortality at times of pest increase (Bergman and Tingey 1979).

Utilizing moderate levels of plant resistance and natural enemies has a long-term benefit because the insects do not adapt to such a system as rapidly as they would to a host plant whose high level of resistance is the sole approach to pest suppression. With the use of qualitative and quantitative models, Gould *et al* (1991) showed that the rate of insect adaptation to plant defenses will tend to increase without the involvement of natural enemies. However, if mortality due to natural enemies can decrease the fitness of the insect, the rate of adaptation to plant defenses will decrease. In general, therefore, the rate of insect adaptation on a resistant cultivar is lower at a given level of insect population suppression when that suppression is achieved by the combined action of plant resistance and natural enemies than by a high level of plant resistance alone.

Biological control processes involve interactions between plants and organisms of the third trophic level. It is surprising that, until recently, few studies have examined the effects of plant physiology and plant allelochemicals on the biology of parasitoids and predators (Boethel and Eikenbary 1986). These interactions are not limited to the impact of plant chemistry on herbivores and the influence of herbivore fitness on natural enemies, but include direct effects of plants on the third trophic level. It is very likely that plants would evolve mechanisms to attract natural enemies and reduce herbivore damage. For example, the female parasitoid wasp *Campoletis sonorensis* responds to the volatiles of cotton over a short distance while searching for its prey, *Heliothis* spp. It is easier for the wasp to find the host habitat first and then the prey itself within the vicinity of the cotton plant. This type of strategy adapted by the wasp is compatible with effective biological control (Williams *et al* 1988). In contrast, plant secondary compounds such as nicotine or tomatine in the insect host's diet may affect parasitoids by parasitization. In other cases, changes in host suitability due to the insect host's diet can influence the developmental rate, size, percentage emergence, parasitization success, sex ratio, fecundity, and lifespan of the parasitoids (Vinson and Barbosa 1987). Depending on the situation, plant resistance may or may not be compatible with biological control.

Compatibility of plant resistance with biological control

Insects feeding on resistant host plants commonly experience retarded growth and an extended developmental period. Under field conditions, such poorly developed insect herbivores are more vulnerable to natural enemies for a longer period and the probability of their mortality is higher. Insect herbivores that develop slowly on resistant varieties are more effectively regulated by the predators than those developed robustly on the susceptible varieties. This is because the predator has to consume more small-sized prey to become satiated (Price *et al* 1980).

In certain cases, the plant secondary compounds that impart resistance are compatible with the performance of natural enemies. For example, *Cotesia congregata*, the relatively effective monophagous parasitoid of the specialist insect *Manduca sexta*, shows very few detrimental effects of exposure to nicotine in tobacco when it parasitizes the insect host. Larval and pupal development as well as the size of the adult *C. congregata* are also unaffected by nicotine. In field experiments using low- and high-nicotine cultivars, results paralleled those obtained in the laboratory (Barbosa *et al* 1986). Similarly, the gossypol in the pigmented cotton ingested by *Heliothis* has no adverse effect on its parasitoid *Campoletis sonorensis*. In fact, at very low concentrations of gossypol, larger adults of *C. sonorensis* were produced than from hosts fed on gossypol-free diets (Williams *et al* 1988).

Greenbug–parasitoid interaction
Some in-depth studies of the interaction between resistant varieties of barley and sorghum and the parasitization of the greenbug *Schizaphis graminum* (Rondani) demonstrated the compatible nature of plant resistance and biological control (Starks *et al* 1972). The movement of the greenbugs on the resistant sorghum cultivars exposes them to greater parasitization (Starks and

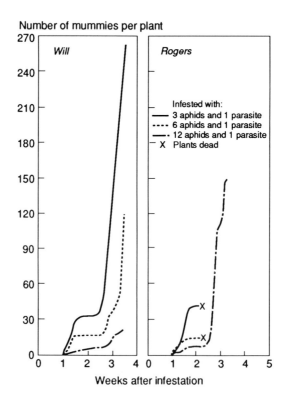

Fig. 9.4. Reduction of *Lysiphlebus testaceipes* mummies on greenbug-resistant (Will) and susceptible (Rogers) barley infested with three levels of greenbug population. Reproduced from Starks *et al* (1972) with permission from the Entomological Society of America, Lanham MD, USA.

No. of pest individuals

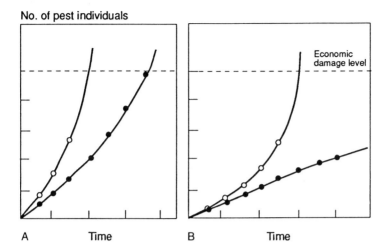

Fig. 9.5. Influence of a low level of plant resistance to pest attack on the effectiveness of natural enemies. o, without predators; •, with predators. Predator activity fails to exert economic control of insect pests on susceptible plants (A), whereas the same degree of predator activity exerts economic control of the insect pest on plants with some degree of resistance (B). Reproduced from van Emden and Wearing (1965) with permission from the Association of Applied Biology, Wellesbourne, UK.

Burton 1977a). The control of greenbugs on barley is very effective when they are infested with the generalist parasitoid *Lysiphlebus testaceipes* (Cresson). This parasitoid was able to keep the biotype C greenbug population nearly static on both susceptible and resistant barley, when the initial population of the aphid was three per plant. But when 12 aphids and one female parasite were introduced per plant, the parasitoid could suppress the aphid population only on the resistant barley (Fig. 9.4). Thus, damage to barley was reduced by the combined effect of varietal resistance and parasites.

A similar trend was seen in the interaction of varietal resistance and biological control agents in controlling greenbugs on sorghum (Teetes 1980). The predator–prey ratio is about equal, but is sometimes greater on resistant sorghum than on susceptible sorghum. A greater predator–prey ratio indicates that varietal resistance complements biological control. Since insecticides are generally not used on resistant cultivars, the natural enemies of the aphid are protected.

Moderate resistance–parasitoid interaction
When varietal resistance is combined with other crop-protection strategies, a high level of resistance is not necessary. On the basis of a simple model, van Emden and Wearing (1965) proposed that the reduced rate of multiplication

of multivoltine aphids on moderately resistant varieties should magnify the plant resistance in the presence of natural enemies (Fig. 9.5). This model is based on an equation by Bombosch (1963), which shows that the voraciousness of a given predator fails to control the aphid population multiplying daily by a factor of 1.2 on a susceptible cultivar (A), but could adequately suppress it if the population growth rate decreases to 1.15 on a resistant cultivar (B). It is apparent that even a slight difference (0.05) in insect population growth is adequate to trigger predator efficiency in pest suppression.

Other positive instances of the interaction between low levels of plant resistance and biological control were also observed in the field with predators of *Brevicoryne brassicae* on Brussels sprouts (Dodd 1973). A moderate level of resistance in rice varieties IR46 and Utri Rajapan effectively increased the predation rate of the spiders *Lycosa* spp. on the brown planthopper (Kartohardjono and Heinrichs 1984).

A number of predators and parasitoids including *Campoletis sonorensis* and *Cardiochiles nigriceps* attack early instars of *Heliothis virescens*, a pest of tobacco and soybean, but generally do not attack bigger larvae. But because of a low level of antibiosis in moderately resistant plants, the larvae of *H. virescens* remain in early instars for longer periods and are more likely to be parasitized (Danks *et al* 1979). The antixenotic factor may decrease or increase pest fitness due to density and/or frequency-dependent predation. In avoiding antixenotic plants or plant parts, young larvae become aggregated at a limited number of feeding sites. This situation is more favorable than normal circumstances for natural enemies seeking their prey (Gould *et al* 1991). In a recent review, Sharma (1993) highlighted the positive role of moderate levels of resistance and natural enemies in protecting sorghum against shootfly, stem borers, head bugs, and armyworm damage. A synergism between plant resistance and biological control of the kind observed above is likely to be a particularly valuable phenomenon in the development of practical insect pest management.

Incompatibility of plant resistance with biological control

Sometimes, plant morphological characteristics and plant defense chemicals have adverse effects on the predators. Foliar pubescence offers an effective plant defense mechanism against arthropods, but it can interfere with the mobility of natural enemies (Stipanovic 1983). For example, certain genotypes of tobacco with hooked and/or glandular trichomes have been shown to severely limit the parasitization of the tobacco hornworm eggs by *Telenomus sphingus* and *Trichogramma minutum* (Rabb and Bradley 1968). Similarly, Treacy *et al* (1985) have shown that increased trichome density on cotton leaves reduces the ability of the parasitoids *Trichogramma pretiosum* and *Chrysopa rufilabris* to find and parasitize corn earworm eggs.

When corn earworms were reared on an artificial diet containing plant defense compounds such as nicotine, α-tomatine, and gossypol, the development rate of parasitoids fed on these host insects was impaired (Vinson and Barbosa 1987). Recently, Barbour *et al* (1993) found that methylketone adversely affected the egg predators (*Coleomegilla maculata* and *Geocoris punctispes*) of *Helicoverpa zea* that fed on the foliage of wild tomato PI 134417. However, plant breeders can sometimes manipulate plant characteristics to promote the effectiveness of natural enemies. For example, the hairiness of cucumbers interferes with the biological control of the greenhouse whitefly by the parasite wasp *Encarsia formosa* Gahan. Cucumber hybrids now have half the number of hairs. Consequently, the movement of *E. formosa* is 30% higher and its parasitism is significantly greater on these experimental hybrids (van Lenteren 1991). Reference should be made to Chapter 4 for more examples of tritrophic interactions with biochemical and physical components of plant resistance.

Plant resistance–insect pathogen interaction

The role of plant allelochemicals in the interactions between herbivores and their pathogens has been reviewed by Barbosa (1988a). Based on research data, Schultz (1983b) hypothesized that the effectiveness of insect pathogens may be reduced or improved, depending upon plant chemistry and variability of plant resistance. The effectiveness of *Bacillus thuringiensis* is better on high-tannin-adapted insect pests (with a gut pH of 8.0–9.5), such as those feeding on late-successional or slow-growing tree species, but less effective on early-successional plant species. Interactions among herbivores, their pathogens, and allelochemicals (like nicotine and rutin) can alter the pathogenicity of *B. thuringiensis* on specialist *Manduca sexta* and generalist *Trichoplusia ni* herbivorous insects (Krischik *et al* 1988). The specialist *M. sexta* can tolerate higher levels of toxins such as nicotine and gain protection from entomogenous bacteria. In contrast, when the generalist *T. ni* was reared on diets with increasing concentrations of *B. thuringiensis* the presence of dietary nicotine enhanced mortality. Insect susceptibility to the entomopathogenic fungus can also be influenced by plant defense chemicals. The pathogenicity of the fungus *Nomuraea rileyi* is diminished considerably if corn earworm larvae have ingested α-tomatine of tomato plants (Gallardo *et al* 1990).

The synergistic interaction between maize cultivars resistant to the leaffeeding fallworm *Spodoptera furgiperda* and the nuclear polyhedrosis virus (NPV) disease of fallworm was reported by Hamm and Wiseman (1986). Insects feeding on resistant inbred lines were more susceptible to infection and mortality from NPV. NPV appears to be more effective (with a fallworm gut pH between 4.5 and 8.5) in early successional situations or where tannins are not important plant defenses. Felton and Duffey (1990) reported the possible incompatibility of chlorogenic acid in the resistant cultivars of *Lycopersicon esculentum* with NPV control of *Helicoverpa zea*. The two workers found that both rutin and chlorogenic acid significantly inhibited the infectivity of NPV because chlorogenic acid is oxidized by foliar phenol oxidases to chlor-ogenoquinone, a highly reactive alkylating agent that binds to the occlusion bodies of NPV, thereby decreasing its infectivity against *H. zea*. These studies show that the application of insect pathogens in the IPM system can be successful only if antibiosis factors of host resistance are not antagonistic to the insect pathogens.

Plant resistance with cultural control

Cultural control can be a powerful tool in our efforts to suppress arthropod pests in agroecosystems. This pest control technique involves two basic approaches: (1) to make the environment less favorable to the pest, and (2) to make the environment more favorable to the pest's natural enemies. Cultural control may not by itself reduce the pest population to below economic threshold levels, but may aid in reducing losses due to pests (Glass 1975). Plant resistance in concert with cultural control can effectively reduce the need for pesticides.

Synchrony of plant and herbivore phenologies

An effective control of the boll weevil and pink bollworm through reducing the overwintering insect population has been developed. This involves early uniform planting, early-maturing varieties, defoliation, and stalk destruction of cotton in late August and September before the larvae are forced into diapause by short days and cool nights (Adkisson and Gaines 1960). These cotton insects have been controlled through a combination of plant resistance and short-growth-duration cotton varieties (Walker and Niles 1971). Thus the development of agronomically acceptable early-maturing varieties has received considerable emphasis in cotton-breeding programs. The short-season, rapid-fruiting cotton variety matures about 2–3 weeks earlier than the long-season cotton. The United States cotton industry survived the boll weevil plague of the early 1900s by using early-maturing varieties (Helms 1980). When an early harvest was coupled with areawide stalk destruction before mid-September, the overwintering populations of diapausing insects were drastically reduced by as much as 90%. This approach was further combined with selective insecticide treatment of the sites where the insects were likely to overwinter. Insect population development was slowed down in the following season. Such a system has not only suppressed the pest population but has also restored biological control and significantly reduced insecticide applications to the crops.

Vegetational diversity

Vegetational diversity plays a key role in IPM. Polyculture (i.e., growing more than one crop simultaneously in the same area) is one way of increasing vegetational diversity. Polycultures are ecologically complex because inter-specific and intraspecific plant competition occurs simultaneously with herbivores, insect predators, and insect parasites (van Emden 1965). Elimina-tion of alternate habitats may lead to decreased predator and parasite populations and to increased insect pest populations (Southwood 1975). Polycultures have received considerably more research attention than other forms of vegetational diversity (Andow 1991).

Population densities of herbivorous insects are frequently lower in polyculture habitats (Risch *et al* 1983). Two hypotheses have been proposed to explain this phenomenon: (1) the associational resistance or resource concentration hypothesis (Roots 1973), and (2) the natural enemies hypoth-esis (Russell 1989). The resource concentration hypothesis proposes that the specialist herbivores are generally less abundant in vegetationally diverse habitats because their food resources are less concentrated and natural enemies are more abundant. The natural enemies hypothesis states that a diversity of plant species may provide important resources for natural enemies such as alternative prey, nectar and pollen, or breeding sites.

In diverse plant communities, a specialist insect is less likely to find its host because of the presence of confusing or masking chemical stimuli, physical barriers to movement, and other adverse environmental factors. Consequently, insect survival may be lower (Baliddawa 1985). Letourneau (1986) examined the effect of crop mixtures on squash herbivore density in the tropical lowlands of Mexico. He found that *Diaphania hyalinata* (L.), the most abundant insect in the system, generally had lower population density in polyculture (maize–cowpea–squash) than in monoculture (squash alone) systems. The total crop yields in polycultures were higher when estimated as a land equivalent ratio.

Variety mixtures
The preceding discussion shows that the feeding preference of polyphagous insects can be altered by including genetically different crop cultivars in the cropping area. An interesting potential for integration of plant resistance and polyculture practices was pointed out by Casagrande and Haynes (1976). They compared damage by the cereal leaf beetle *Oulema melanopus* L. in mixed and pure stands of resistant and susceptible wheat varieties. They reported that biological control was more effective in the mixed cropping of beetle-resistant and beetle-susceptible wheat varieties than in a pure stand of either one of those varieties on a regionwide basis.

Wilhoit (1991) developed an aphid model for predicting the rate at which a virulent aphid genotype will replace an avirulent genotype in a variety mixture with different proportions of resistant and susceptible plants in the presence of an aphid predator. The model predicts that the simulated growth of predators of aphids on susceptible plants in variety mixtures slowed the development of virulent aphid genotypes. Sexual reproduction slowed the development of virulence. The action of predators was also effective in slowing down the development of virulence. Thus, the combined effects of varietal mixtures and natural enemies are likely to be effective in suppressing the aphid populations.

Trap cropping

Trap crops are plant stands that are grown to attract insects or other organisms so that the target crop escapes pest attack. Protection is achieved either by preventing the pests from reaching the crop or by concentrating them in a certain part of the field where they can easily be destroyed (Hokkanen 1991). The principle of trap cropping is similar to associational resistance, in which the insect pests show a distinct preference for certain plant species, cultivars, or a crop stage. Crop stands are manipulated in time and space, so that attractive host plants are offered at the critical time of the pest's phenology. Such tactics lead to the concentration of the pests at the desired site, i.e. the trap crop. One major benefit of trap cropping is that the main crop seldom needs to be treated with insecticides; thus, natural control of pests may remain fully operational in most of the field. Trap-cropping techniques may be particularly important to subsistence farming in developing countries. Practical applications of trap cropping in cotton and soybean have been very successful (Newsom *et al* 1980). In cotton/sesame intercrop trials, row strips of sesame, constituting 5% of the total acreage, were used as a trap crop to attract *Heliothis* spp. from the main crop of cotton. Sesame, which is highly attractive to *Heliothis* species from the seedling stage to senescence, attracted large numbers of insects away from the cotton. It also attracted the parasitoid *Campoletis sonorensis*, which ultimately parasitized large numbers of *Heliothis* insects (Pair *et al* 1982).

These examples illustrate the manipulation of cultural practices in crop production systems to either suppress pest abundance in crops or to allow the crop to escape pest damage. Cultural practices cause specific physiologic changes that reduce the suitability of host plants for phytophagous insects (Hare 1983). Most of these practices are compatible with other pest control tactics and have long been associated with subsistence farming as an appropriate approach to low-input cropping systems.

Plant resistance with chemical control

Synthetic insecticides are very effective against many insects and arthropods affecting agriculture, forestry, and the health and comfort of humans and domestic animals. Insecticides are most advantageous for the immediate control of pest outbreaks. However, the broad-spectrum activity of most synthetic insecticides drastically destroys beneficial insects, and thus has adverse consequences in the IPM system. The suppressive action of insecticides is independent of the density of the pest being controlled. The same level of kill can be expected from a given dose whether the pest population is high or low. This characteristic of insecticides is important in applying certain principles of pest population management – i.e., the use of insecticides to reduce populations to levels that can subsequently be controlled by other pest control techniques. The integration of other methods of control with different modes of action to supplement a minimal use of insecticides would best ensure an efficient IPM system.

Antibiosis and chemical control

The most obvious advantage of plant resistance is the containment of pest population pressure. Chemical control measures are geared mostly towards increasing the death rate of an insect population. In contrast, antibiosis aims at indirectly decreasing the birth rate through reduced vigor, thus preventing the insect population from reaching economic threshold levels. Because the toxicity of an insecticide is a function of insect bodyweight, it is expected that a lower concentration is needed to control insects feeding on a resistant variety than those feeding on a susceptible variety. In this regard, nymphs of the wheat grain aphid *Sitobion avenae* reared on resistant wheat variety Altar possessing the antibiosis compound DIMBOA (2,4-dihydroxy-7-methoxy-1,4-benzoxazin-3-one) were significantly more susceptible to the insecticide deltamethrin than nymphs on the susceptible wheat variety Dollarbird. The LD_{50} (median lethal dose) adjusted for weight was reduced by 91% for nymphs reared on the cultivar with a high DIMBOA content (Nicol *et al* 1993). Further investigations by Leszczynski *et al* (1993) revealed that in concentrations as low as one-hundredth of those affecting the survival and reproduction of the aphid, DIMBOA strongly inhibited detoxifying enzymes (peroxidase, polyphenol oxidase, *N*-demethylase, glutathione *S*-transferase and UDP-glucose transferase).

Moderate levels of resistance and chemical control

Muid (1977) compared the susceptibility to malathion of the aphid *Myzus persicae* feeding on two varieties of Brussels sprouts. Although the aphid population on the partially resistant variety was about 85% of that on the susceptible variety, the LD_{50} was only about 55%; i.e., the pesticide requirement was much less on the partially resistant variety. This increased susceptibility to malathion of insects on a partially resistant variety appeared to be due to the interaction of the insecticide with insects of low vitality. It appears that even in the presence of small levels of plant resistance, insecticide concentration can be reduced to one-third of that required on a susceptible variety (van Emden 1990).

Reports from Norway indicate that insecticide use against insect pests was considerably less in vegetable crop varieties with even low levels of pest resistance. A half dose of the insecticide chlorfenvinphos gave equal or better control of the turnip rootfly *Delia floralis* on resistant cultivars of swede (Cruciferae) variety S7790 than the full dose on the susceptible cultivar Ruta. The reduction in indices of fly damage and the increase in percentage of marketable roots were more pronounced in the resistant than in the susceptible cultivar (Taksdal 1993). The soil–insect complex for resistant sweet potato could be controlled with one-half the amount of insecticides required for susceptible cultivars (Cuthbert and Jones 1978). The reduced pesticide use on vegetables for controlling insects not only benefits the agroecosystem and/or natural enemies but also results in lower pesticide residues in the human foodchain.

The susceptibility of the whitebacked planthopper and brown plant-hopper to insecticides is affected by the level of varietal resistance in the rice cultivars on which they are reared. Hoppers reared on moderately resistant rice varieties had lower LD_{50} values when placed on sprayed plants than those reared on susceptible varieties (Heinrichs *et al* 1984). The LD_{50} of the whitebacked planthopper was 9.4 on the susceptible variety TN1 treated with ethylan, but only 2.8 on the moderately resistant N22. The combination of moderate varietal resistance and a low dose of pesticide resulted in effective hopper control. Each combination lessens the resurgence of insect populations and helps conserve natural enemies, preserves environmental quality, slows down the rate of selection for insecticide-resistant insect strains, and increases the profitability of crop production (Adkisson and Dyck 1980).

High levels of resistance and chemical control

McMillian *et al* (1972) investigated the effect of multiple applications of an insecticide on resistant and susceptible varieties of sweet maize exposed to artificial and natural infestations of the corn earworm. The resistant untreated sweet maize hybrid 471-U6 × 81-1 had 48% more undamaged ears under artificial infestation (heavy population pressure) than a susceptible hybrid treated with seven applications of Gardona (2-chloro-1-(2,4,5-trichlorophenyl)vinyl dimethyl phosphate). Under natural field infestation, the susceptible hybrid treated with seven applications of Gardona had only 8% more undamaged free ears than the untreated resistant hybrid. When the resistant hybrid was treated with Gardona and exposed to both natural and artificial infestations, it had 15 and 4% more undamaged ears, respectively, than when untreated. These results demonstrate that resistant maize hybrids require very low levels of insecticides to achieve an equivalent high level of earworm control. Even one low-dose application of insecticide to the resistant hybrid gave an earworm control equal to that achieved with seven applications to the susceptible hybrid (Wiseman *et al* 1973).

One of the most successful examples of drastic reduction in insecticide use with resistant varieties is the large-scale planting of rice gall midge-resistant varieties in India. In Warangal (Andhra Pradesh), Raipur (Madhya Pradesh), and Sambalpur (Orissa), the rice gall midge was an endemic pest causing enormous losses. Since 1976, more than 60% of the area has been planted with resistant rice varieties such as Phalguna, Surekha, Samalai, and Shakti. Prior to the introduction of resistant varieties, 50% to total yield losses used to occur (Kalode *et al* 1986), but since the large-scale introduction of resistant varieties the gall midge has not caused any economic damage. Consequently, no insecticides are used on the resistant varieties.

Antixenosis and chemical control

In the antixenosis type of resistance, repellent chemicals and/or morphological traits may be effective either at the egg laying or feeding stage of the insect. Frego bract provides an excellent example of antixenosis resistance. In frego-bract cotton, the square has a rolled, twisted, and open bract unlike in normal cotton where the bract is flat and encloses the square (see Fig. 6.1). These morphological traits prevent oviposition by the boll weevil on cotton. Jenkins and Parrot (1971) planted frego-bract cotton M-64 in 4–8 ha and assessed resistance to the boll weevil over three to four insect generations. No insecticides were required for weevil control and up to 94% of the weevil population was suppressed and levels of overwintering weevils were very low. A high level of oviposition suppression can be very useful in an eradication program. As boll weevils feed and oviposit on the buds, the exposed buds in the frego-bract cotton can ideally be covered with insecticides sprayed over the plants. When sprayed with methyl parathion, frego-bract buds had seven times more deposits of insecticide residue than the normal bract buds. A combination of the okra-leaf (open types) and frego-bract characteristics improved the efficiency of insecticide that fully covered all plant surfaces (Jones *et al* 1986).

Gould (1984) presented the two-locus model for resistance–avoidance interactions to illustrate that insect resistance can be managed in the field by the use of insecticides having insect-repellent properties. In this model, he considered the repellent quality of pyrethroids in suppressing populations of *Heliothis* spp. It is usually assumed that insecticidal action will reduce pest damage and will at least temporarily decrease the target pest population. Repellency is likely to be effective in limiting pest damage to treated crops and it may also keep the pests away from their suitable resources and therefore cause indirect mortality or lowered fecundity. Repellency is equivalent to using low doses of insecticides along with the repellent properties of the host plant. The model predicts that insecticide formulations containing noninsecticidal compounds with repellent properties (naturally derived or synthetically developed products) could lower the rate of insect adaptation to an insecticide. Insects lack the sensory apparatus to respond to most insecticides, but they have evolved the ability to respond to volatile secondary plant compounds. Therefore, the selective pressure due to insecticide-treated populations of insects on plants without antixenosis is likely to be much stronger than that occurring in treated populations faced with a resistant host variant with antixenosis resistance (Pluthero and Singh 1984).

Plant resistance with virus control

Crop cultivars resistant to insect vectors of plant viruses alter the population size, activity, and probing/feeding behavior of the vectors, thereby influencing the pattern of virus spread. The effect of a vector-resistant cultivar on virus spread depends upon several factors: the type of resistance (antixenosis, antibiosis, or tolerance), the level of resistance, the type of virus, and the effect of virus infection on the vector-resistant plants (Painter 1951). Each combination of these factors may affect virus spread differently. Speculation in the literature is that nonpersistent viruses such as mosaic and yellows in beet may spread more rapidly in varieties resistant to the vector, because the latter may be restless and increase probing. The complexities involved are such that, without a thorough understanding of the ecology of both the virus and the vector and the biology of the vector, it would be impossible to predict

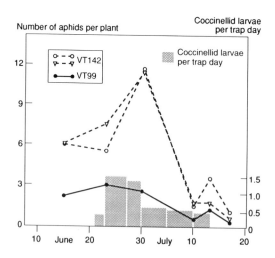

Fig. 9.6. Population development of *Myzus persicae* on aphid-resistant (VT99) and susceptible (VT142) sugarbeet, with the average number of coccinellid larvae in pitfall traps. Modified from Lowe (1975).

the effect of vector resistance on virus spread (Kennedy 1976). Despite these drawbacks, varietal resistance to the vector has considerable impact in reducing the spread of certain types of plant virus diseases.

In large sugarbeet fields in the UK, fewer *Myzus persicae* aphids developed on a resistant (VT99) than on a susceptible (VT142) variety. The early decline of the natural population of the aphids on the resistant variety was attributed to predation by a large population of coccinellid larvae (Fig. 9.6) (Lowe 1975). There was apparent synergism between the resistance of the sugarbeet and the action of natural enemies in decreasing aphid infestation, resulting in the reduced natural spread of virus yellows in plots planted to the aphid-resistant sugarbeet variety. On the other hand, aphids multiplied rapidly on the susceptible beet variety within a short period.

Several virus diseases of rice are transmitted by insect vectors. The green leafhopper *Nephotettix virescens* (Distant) is an efficient vector of rice tungro virus (RTV). RTV causes severe yield losses in many rice-growing countries of Asia. Resistance to the vector provides good protection against the virus. Under field conditions, green leafhopper-resistant rice variety IR28 treated with insecticides had a very low incidence of RTV, whereas susceptible variety IR22 under similar treatment had an extremely high incidence. In contrast, only insecticide-treated plots of the moderately resistant variety IR36 showed lower infection of RTV than untreated plots in which RTV incidence was high (Heinrichs *et al* 1986).

Biotechnology in insect pest management

Until now pest resistance in crops has been achieved primarily through conventional plant-breeding approaches. Plant breeders have relied on crop germplasm including induced mutants and related wild species as resistance donors. Recent advances in biotechnology, particularly cellular and molecular biology, have opened new avenues for developing resistant cultivars. Products of biotechnology are expected to be widely available for the diagnosis and control of pest populations by the end of this century. From this diagnostic perspective, molecular techniques are likely to play an important role in identification, quantification, and genetic monitoring of pest populations (Whitten 1992). The diagnostic information is a necessary prerequisite for implementing rational insect control strategies. To date, morphological, cytological, and biochemical techniques have been employed for this purpose, but the separation of sibling species in a pest complex is beyond the reach of such methods. However, appropriate molecular techniques can be employed to study the species composition of the pest population and to identify strains, races, or biotypes of the same species.

Another important application of molecular diagnostic techniques is for monitoring both the presence and frequency of genes of particular interest. For instance, genes for resistance to a specific class of pesticides and their frequency in a particular region can be assessed. Such information is very useful for designing and implementing rational pest management strategies (Whitten 1992).

Perhaps the most important application of biotechnology in pest management is the introduction of novel genes for resistance into crop cultivars through genetic engineering. A number of toxins that kill insects but have no adverse effects on vertebrates have been identified and isolated from various organisms. The *Bt* genes from *Bacillus thuringiensis*, for example, encode protein toxins that are lethal to the larvae of lepidopteran insects. The *Bt* genes have now been introduced into several crop species (see Table 11.5). Transgenic plants with the *Bt* gene produce their own biocide and are resistant to lepidopteran insects. Transgenic rice plants expressing the *CryIA(b)* gene from *B. thuringiensis* are resistant to the stem borer and leaffolder (Fujimoto *et al* 1993). Transgenic potato plants expressing the *CryIIIA* gene are resistant to the Colorado potato beetle at all stages of the insect's life cycle. The potato tubers conform to standards in terms of agronomic and quality characteristics including taste (Perlak *et al* 1993). Laboratory assays using field-grown transgenic cotton expressing the *Bt* gene indicate that the transgenic strategy has great potential in controlling *Heliothis virescens* larvae under field conditions (Jenkins *et al* 1993).

Many novel genes for insect resistance such as trypsin inhibitors, α-amylase inhibitors, and lectins have been isolated from various plants and are being used in developing transgenic crops for insect resistance (Toenniessen 1991). Further research is needed to control the expression of genes encoding insecticidal properties so that the gene product is delivered at the correct site. The biocide should be produced in the plant organ where the insect normally attacks. If the target insect is a phloem feeder, the toxin should be available in the phloem; if the insect attacks the grain, the gene product should be expressed in the endosperm and should not be toxic to humans.

Very little is known about the stability of transgenes. Control of gene expression to prolong the usefulness of the transgene is a major topic of research. Engineering constitutive expression of transgenes at high levels throughout the life cycle of the crop and its homogeneous plantings over large areas would exert tremendous selection pressure on the insect population to evolve a new biotype. Gould (1988) has suggested that concepts of evolutionary biology and ecology should be kept in mind during the genetic engineering process. Genes producing mild toxins may last longer than those with strong toxins. Combining genes encoding a toxin and genes encoding a repellant may offer longer lasting resistance than either approach alone (Toenniessen 1991).

Modeling and Systems Analysis

Advances in computer science have contributed to the development of insect population growth models that can be utilized in IPM systems. A model is a representative of a system; it mimics the essential features of a particular system and is taken as a limited part of reality. Four main levels of systems in agriculture have been distinguished: pathosystems, cropping systems, farming systems, and agroecosystems (Rabbinge and de Wit 1989). Most models in IPM deal with the pathosystem and the cropping system because at those levels the amount of available information makes the model less speculative. Models can mimic an existing system and most models require a certain level of mathematical skill.

Insect pest management is obviously a complex subject. Even separate components of IPM present diverse and complicated interactions. For instance, if the interactions of resistant plant cultivars and natural enemies and the effects of cultural practices are included in the IPM system, it would be difficult to assess the role and effect of each interaction. To deal with such complexity, the whole system must be systematized. If a system is systematized as a conceptual model, the complexity of interactions can be studied in a holistic perspective.

Models define the objectives, system boundaries, and interrelations of the different components of a system (Southwood 1978b). The major components of the system need to be identified and the key processes described. For instance, the relative effects of the different resistance modalities and individual insect life-history components must be considered as potential resistance factors. From this, it should be possible to formulate key questions that relate to the ecological, socioeconomic, and technical information required for the management of the pests. A conceptual model of the system can then be produced to address the problem. The complexity of models increases if modules of more crops and more insects, their natural enemies, and all other strategies as well as insecticides are included. However, each model and system will be different and the boundaries of the system and the complexity of the model will exclusively be determined by the objectives of the study. The model should neither be too complex nor too simplistic, but should describe the key processes and adequately reflect the system.

The next step is to identify the route(s) by which the objective of pest control may be achieved and then to formulate measures of performance by which the model can be assessed. These may be based on the degree to which the model agrees with the accepted theory, actual events, or economic criteria. A model is a complex hypothesis, and its validity needs to be tested experimentally. Once it is conditionally validated, it can be implemented and its results put to use. Strategic models are the most common in IPM. They attempt to provide general guidelines and elucidate principles that are applicable over a wide range of circumstances. In this way, the entire pathosystem or pest system of a specific agricultural system and its pest management can be simulated. Such studies can substitute for the already existing experimental results and can identify results that are expected to provide useful information for effective implementation of the IPM program. For further details on modeling and systems analysis, the reader is referred to Gutierrez (1987).

The model of the population dynamics of the groundnut leafminer *Aproaerema modicella* (Deventer) in India can be cited as an example (Dudley *et al* 1989). This model was built in the early stages of an IPM research project in which mortality due to natural enemies, HPR, and insecticides to control the leafminer were evaluated. It was verified with experimental and farm management data. The model consists of a submodel with a single difference equation to simulate pest density, yield, cost and return functions to calculate net returns from a groundnut crop. The simulation submodel is linked to a dynamic programming submodel to optimize farmers' insecticidal application decisions. The modeling results indicated that the optimal number of sprays is insensitive to changes in groundnut price. The sensitivity of the optimal number of sprays on resistant plants, or mortality from natural enemies increases when the chemical control efficacy is reduced. The model emphasizes the importance of increased HPR and natural enemies in implementing IPM in the groundnut.

Limitations

Certain limitations and major problems will always beset any insect control program, and HPR is no exception. The time involved in the search and development of a new plant variety resistant to a target pest is often considerable. It took 18 years to develop the Hessian fly-resistant wheat variety Kawvale. In contrast, only 3–5 years was needed to develop the spotted alfalfa aphid-resistant Cody, Moapa, and Lahontan varieties. Developing insect-resistant crop varieties requires a great deal of expertise and resources. It is usually necessary to organize a well-planned multidisciplinary team of entomologists and plant breeders. Commitment of relatively long-term funding is often a critical factor in the ultimate success of plant-resistance programs.

Some mechanisms of plant resistance may involve the diversion of some resources by the plant to extra structures, or the production of defense chemicals at the expense of other physiologic processes including those contributing to yield potential (Mooney *et al* 1983). Mitra and Bhatia (1982) pointed out that although concentration of defense chemicals responsible for resistance is low in plant tissues, the total amount per hectare may be high. They indicated that the production cost of 34 kg of gossypol in terms of glucose equivalent will be 70.7 kg of glucose ha^{-1} (or 1,212,850 kJ of energy).

The genetic, biochemical, and physiologic mechanisms of resistance are also likely to involve the initial cost of resistance as well as the extent to which they can be modified by selection. Some mechanisms are less costly whereas others involve high costs that cannot be modified. More information is needed regarding the genetic regulation of resistance traits, their biochemical pathways, and physiologic effects (Simms 1992).

In HPR programs, one might expect a negative correlation between the potential yield of a cultivar and its level of resistance to pests. This is illustrated by the failure to evolve insect-resistant varieties of soybean with substantial yield. Although HPR promises to contribute greater effectiveness and stability to pest management systems for soybean than any other tactic of pest population suppression, progress towards developing acceptable resistant commercial varieties has been slow, mainly because of the lower yield of resistant varieties (Newsom *et al* 1980). Yields of promising soybean lines at the advanced stages of testing usually fall short of the performance of commercial varieties by about 2 bushels per acre (134 kg ha^{-1}). The growers are not willing to plant a resistant variety unless it can yield about as much without insecticide treatment as susceptible varieties with insecticide treatment. Although the fundamental objective of breeding for insect resistance in crop plants is to produce an acceptable level of sustainable resistance, such resistance should at the same time be compatible with optimum crop yield and quality.

Despite many dramatic successes in HPR, there are still cases where host resistance to one insect led to increased susceptibility to another insect or microorganism. Cotton varieties high in gossypol content are resistant to *Heliothis* spp. and blister beetles, but gossypol at concentrations found in most cotton plants attracts boll weevils (Bell *et al* 1987).

It is well known that plant chemicals have negative effects on the fitness of many herbivores. It is equally evident that many herbivores possess remarkable potential for utilizing toxic plant chemicals; no defense chemical is sacrosanct. Thus insects can evolve biotypes to overcome antibiosis resistance. However, the partially resistant varieties would probably last longer in the field than those with strong resistance (Lamberti *et al* 1983).

The current theories on plant chemical defensive strategies do not adequately take into account the complexity of tritrophic systems. Plant resistance based on antibiosis may not always be compatible with biological control (Boethel and Eikenbary 1986). Elucidation of these interactions can further our understanding and provide greater potential for the manipulation of these systems in IPM suited to specific crop species and varieties (Whitman 1988). The possibility of using naturally occurring toxic or attractive compounds from plants to reduce herbivore damage and increase the effectiveness of biological control agents is attractive. Ideally, plant resistance should strive to reduce substances attractive to herbivores while increasing substances attractive to natural enemies.

The chemical bases of resistance in plants can sometimes modify the toxicity of insecticides to insects. The 2-tridecanone (a neurotoxin) in the glandular trichomes of the wild tomato *Lycopersicon hirsutum f. glabratum* is toxic to corn earworm. But in the presence of 2-tridecanone, the efficacy of carbaryl (a carbonate insecticide) against corn earworm is greatly reduced and the larvae of the insect become more resistant to the insecticide (Brattsten 1988).

Some plant defense chemicals are toxic or interfere with the processing of the crop for use by humans and animals, and interact with the biosphere. Gossypol and related polyphenolic compounds in the cotton plant confer resistance to the cotton bollworm and other pests, but are toxic to non-ruminant vertebrates at pest-active levels, unless removed during the processing of cotton seed oil and meal (Lambou *et al* 1966). Natural products like rutin, chlorogenic acid, tomatine, and phenolics in fruits and vegetables exert acute and/or chronic effects upon humans. Some of these compounds are carcinogenic and mutagenic (Ames *et al* 1990). The potato variety Lenape is resistant to the Colorado potato beetle because it has a high level of glycoaldehyde. It was taken off the market because glycoaldehyde affected its palatability (Day 1992). Insect-resistant tomato plants with elevated levels of allelochemicals, if plowed into the field after each harvest, may lead to the buildup of allelopathic agents in the soil and subsequently may show detrimental effects on the growth of successive crops. The accumulation of allelochemicals may alter the nature of the rhizosphere. All these intricacies related to substantial amounts of toxic secondary compounds in resistant crop plants must be kept in mind during the process of developing resistant varieties.

10 GENETICS OF RESISTANCE TO INSECTS

The ultimate objective of host plant resistance programs is to develop insect-resistant varieties for on-farm production to minimize crop losses from herbivore attack. To this end, germplasm collections are necessary. These may be farmer varieties in use, landraces, primitive or weedy types, and wild species (see Chapter 11). Using the techniques described in Chapter 8, germplasm collections are screened and donors for resistance are identified. These donors are used in hybridization programs to develop resistant varieties with other desirable attributes.

Information on the inheritance of resistance is useful to the breeder in deciding what breeding methodology and breeding strategies to adopt. Diverse genes for resistance are needed to cope with the development of new biotypes, to develop multiline varieties, and to attain regional deployment of genes. For this reason, genetic studies are carried out to determine the mode of inheritance of resistance and to identify diverse genes for resistance.

Types of Resistance

The functional definitions of resistance are given in Chapter 6. When defined on a genetic basis, resistance is of two types – vertical and horizontal. These definitions were originally proposed by Van der Plank (1963, 1968) to describe plant–disease interactions but are equally applicable to plant–insect interactions (Gallun and Khush 1980).

Vertical resistance

If a series of different cultivars of a crop show differential reactions when infested with different insect biotypes (Fig. 10.1), resistance is vertical. In other words, when infested with the same insect biotype, some cultivars show a resistant reaction while others show a susceptible reaction. It is also referred to as qualitative or biotype-specific resistance. Vertical resistance is generally, but not always, of a high level and is controlled by major genes or oligogenes; it is considered less stable.

The major genes for resistance may be dominant if the F_1 progeny of a resistant and a susceptible parent are resistant; they may be recessive if the F_1 hybrid is susceptible. Sometimes two or more dominant genes for resistance may be present in a cultivar, or a cultivar may have a dominant and a recessive gene for resistance. Polygenes quite often reinforce the effect of oligogenes and are referred to as 'modifiers.'

Horizontal resistance

Horizontal resistance describes the situation where a series of different cultivars of a crop show no differential interaction when infested with different biotypes of an insect. All resistant cultivars show similar levels of resistance to all biotypes. This type of resistance is called biotype-nonspecific resistance, general resistance, or quantitative resistance. Generally, horizontal resistance is controlled by several polygenes or minor genes, each with a small contribution to the resistance trait. Horizontal resistance is moderate, does not exert a high selection pressure on the insect, and is thus more stable or durable.

Parasitic Ability

Parasitic ability refers to the ability of an insect to survive at the expense of its host. The term has the same meaning for insects as pathogenicity, which is the term used to describe the ability of a disease organism (fungal, bacterial, or viral) to attack its host. Parasitic ability is determined by the genotype of the parasite and that of its host. The insect's genetic constitution determines its ability to parasitize its host. If the insect is able to feed and multiply on its host, it is virulent and has one or more genes for virulence. However, if the parasite is unable to feed on the host, it is avirulent and has a gene or genes for avirulence (Day 1974).

Fig. 10.1. Reaction of different rice varieties to three biotypes of the brown planthopper. IR8, with no gene for resistance, is susceptible to all biotypes. IR26, with *Bph-1*, is resistant to biotypes 1 and 3. IR36 with *bph-2* is resistant to biotypes 1 and 2. IR62, with *Bph-3*, is resistant to all biotypes.

Biotypes

A biotype is defined as a population or an individual distinguished from other populations or individuals of its species by some nonmorphological trait, usually adaptation and development to a particular host, host preference for feeding or oviposition, or both. This definition does not imply any reproductive barrier, morphological, cytological, or biochemical differentiation, or geographic isolation between such populations.

The term biotype has been loosely used to describe insect populations that differ in diurnal or seasonal activity pattern, size, shape, color, insecticide resistance, migration or dispersal tendency, pheromone response, or disease vector capacity (Eastop 1973b, Diehl and Bush 1984). For example, insect biotypes have been distinguished on the basis of their response to insecticides. Biotype D of the greenbug differs from the four other biotypes in its ability to resist certain insecticides. It has low mortality to disulfon and about 30-fold resistance to other organophosphate insecticides (Starks *et al* 1983). Biotypes have also been differentiated on the basis of their differential virus transmission ability (Thottappilly *et al* 1972). Such usage is unacceptable and the use of the term should be restricted to describe populations that differ in virulence patterns on host varieties. Such populations may or may not differ in morphological or biochemical traits.

Knowledge about the genetic basis of virulence or parasitic ability of the insect population is not absolutely necessary but is desirable before designating them as distinct biotypes. Biotypes, as defined above, have been reported in more than 10 species during the last 50 years. The genetic basis of differential virulence has been investigated only in the Hessian fly of wheat (Hatchett and Gallun 1970), the greenbug of wheat (Puterka and Peters 1989), and the brown planthopper (BPH) of rice (Den Hollander and Pathak 1981). The differences are under monogenic control in the former two and under polygenic control in the latter. Claridge and Den Hollander (1980, 1983) favor the use of biotypes for the Hessian fly but not for the BPH, their main argument being the overlap between virulence patterns of BPH populations. However, as shown in Fig. 10.1, biotypes of BPH show clear differences in virulence patterns and rice entomologists, geneticists, and breeders have found it useful to call such populations biotypes (Khush 1977, Pathak and Khush 1979). Diehl and Bush (1984) agree with this usage.

Genetic Analysis for Resistance

Inheritance of resistance to insects has been investigated in several crop species (Khush and Brar 1991). The various steps involved in genetic analysis are described below.

Plant materials

The resistant varieties to be analyzed, whether farmer varieties or land races, should first be purified. Sometimes a variety is composed of resistant and susceptible individuals and may have off-type plants. It is advisable to purify test varieties on the basis of their reaction to insects, their morphological features, or through single-plant progeny tests. Resistant and susceptible checks should be pure lines. Tests should be done on well-grown and vigorous plants under uniform and disease-free environments. Diseased plants or those suffering from nutritional deficiencies may succumb to insect attack, even if they are resistant.

Insect populations

An insect population of significant density is required to test segregating materials. It is advisable to use laboratory- or greenhouse-raised insects multiplied from a standard susceptible variety. To overcome the problem of variability within the insect population, an insect colony started from a single-pair mating of insects of known virulence is highly desirable. Field-collected populations may contain individuals of different biotypes and may not yield critical results. Insect species which vector the viruses should be free of the virus.

Scoring for resistance

The determination of resistance may be based on plant injury or other symptoms of attack or on the reaction of the insects to the plant. Plant injury may be in the form of defoliated leaves as in the case of the striped cucumber beetle on squash (Nath and Hall 1963), stunted plants as with the Hessian fly on wheat (Gallun 1965), or death of the plant as in 'hopperburn' caused by the BPH in rice (Athwal *et al* 1971). Insect reactions to resistant plants may be death (Hessian fly and wheat stem sawfly), reduced fecundity (spotted alfalfa and green bug), loss in weight (cereal leaf beetle), restlessness (aphids), a longer period between stages in the life cycle of the insect, or avoidance of plant feeding. A detailed discussion of the procedures for assessing plant resistance is given in Chapter 8.

    For genetic analysis, plant resistance is generally assessed on the basis of the reaction of the plant to the insect and, less often, by the reaction of the insect to the plant. A scale to classify plant reaction to the insect on the basis of degree of damage has to be devised. Hybrid populations from crosses of resistant and susceptible parents are classified using the scale. Resistant and susceptible checks are planted along with the hybrid populations. The hybrid populations are eventually classified into resistant, segregating, and susceptible and genetic ratios are calculated. For quantitatively inherited resistance, various biometric genetic techniques are used.

Inheritance of resistance

In the initial stages, resistant parents are crossed with a susceptible parent. It is advisable to use a highly susceptible parent lacking any modifiers. A portion of the F_1 seeds are planted to test the reaction of the F_1 progenies to the insect and the rest are planted to obtain F_2 seeds. F_2 seeds from each F_1 plant are harvested separately. Part of the F_2 seeds are planted to grow F_2 progenies and the rest are used to determine the insect reaction of F_2 plants. F_2 populations from two to three F_1 plants are tested for insect reaction. A random sample of 150–200 plants from the field-grown F_2 populations is harvested to obtain F_3 seeds. F_3 progenies are evaluated for resistance to verify the F_2 results.

The F_1 and F_3 progenies are scored on a row basis and F_2 progenies on a plant basis. Thus all the F_1 plants should have similar reactions. If resistant, a dominant gene(s) governs the resistance; if susceptible, the gene is recessive. Occasionally, the F_1 progenies may be intermediate in reaction between the two parents and the gene for resistance is incompletely dominant. Each F_2 plant is classified either as resistant, moderately resistant, susceptible, or highly susceptible. Resistant and moderately resistant plants are combined into a resistant category and susceptible and highly susceptible into a susceptible category. A 3:1 ratio of resistant to susceptible plants indicates a single dominant gene for resistance and 1 resistant:3 susceptible indicates a recessive gene. In case of two dominant genes, the ratio is modified to 15:1; it is 7:9 if two recessive genes convey resistance. If a dominant and a recessive gene for resistance are present, a 13:3 ratio is obtained.

It is advisable to confirm the results of F_2 analysis from the F_3 progeny tests. Some of the resistant F_2 plants may be killed because of soil-borne pathogens or extremely high populations of the insect. Conversely, some susceptible plants may escape insect injury and may thus be classified as resistant. It is not uncommon to find a few dead seedlings in the resistant checks. For these reasons, it is important to confirm the F_2 results from F_3 analysis. Approximately 25–30 plants of each F_3 progeny are tested and the progenies are classified as resistant, segregating, or susceptible. The F_3 progenies segregate in a ratio of 1 resistant:2 segregating:1 susceptible in the case of monogenic resistance, and 7 resistant:8 segregating:1 susceptible in the case of digenic resistance.

Allele tests

After completing the studies on inheritance of resistance for a number of varieties, they are crossed with each other to determine if they have a common gene for resistance or if they have different genes. If two parents have recessive genes and their F_1 hybrid is resistant, the recessive genes of the two parents are allelic. A susceptible reaction of the F_1 hybrid indicates that the two parents have different genes for resistance. If the parents have dominant genes, the F_1 progenies would be resistant and the F_2 and F_3 progenies will segregate for susceptibility in case of nonallelic genes, but there is no segregation if the genes are allelic. The F_2 and F_3 data can also be analyzed to determine if the nonallelic genes are independent or linked. For example, a 7 resistant:9 susceptible F_2 ratio would indicate that two recessive genes are independent from each other. Deviation from this ratio would indicate linkage. Similarly, an F_2 population from a cross between two parents having independent dominant genes for resistance would segregate into 15 resistant:1 susceptible in the F_2 progeny and into 7 resistant:8 segregating:1 susceptible in the F_3 families. Deviations from these ratios would indicate linkage.

Once a number of genes for resistance are identified, the varieties with those genes are used as testers in the genetic analysis of additional resistant varieties. The resistant varieties to be analyzed are crossed with a susceptible parent as well as with parents with known genes for resistance. The susceptible and resistant parents with known genes are called testers. It is advisable to use the same susceptible and resistant testers throughout the genetic analysis. The crosses with susceptible testers provide information about the mode of inheritance and gene action and the crosses with resistant testers yield information about allelic relationships. If a number of biotypes of the insect are available an estimate of the genetic constitution can be made from the differential reaction to biotypes before embarking on genetic analysis.

Genetic analysis for major gene resistance

There are numerous reports about major genes for insect resistance in crop plants. Resistance to the BPH (*Nilaparvata lugens*) of rice is a classic example of major gene resistance. The BPH is a serious pest of rice and outbreaks of the pest have been reported since 697 AD (Dyck and Thomas 1979). However, the pest attained major importance after the widescale adoption of high-yielding varieties and improved cultural practices. Breeding programs to develop BPH-resistant varieties were initiated in 1969. Varietal resistance to the BPH was first reported by Pathak *et al* (1969). Athwal *et al* (1971) carried out genetic analysis of four BPH-resistant varieties using a mass screening test and their results are summarized in Table 10.1. F_2 populations from the crosses of susceptible TN1 with resistant Mudgo, MTU15, and CO 22 segregated into a ratio of 3 resistant:1 susceptible, indicating that these three varieties have a dominant gene for resistance. The F_2 population from the cross TN1 × ASD7 segregated into 1 resistant:3 susceptible, indicating that ASD7 has a recessive gene for resistance. The monogenic nature of resistance in the four varieties was confirmed from the F_3 data as the F_3 families of the four crosses segregated in a ratio of 1 resistant:2 segregating:1 susceptible.

Table 10.1. Genetic analysis for resistance to the brown planthopper in rice (data from Athwal et al 1971).

		Reaction to brown planthopper						
		F_2 seedlings (no.)			F_3 families (no.)			
Cross[*]	F_1	Res	Sus	χ^2 3:1 or 1:3	Res	Seg	Sus	χ^2 1:2:1
TN1 × Mudgo	R	227	87	1.09	27	54	24	0.26
TN1 × ASD7	S	206	621	0.0004	34	77	40	0.54
TN1 × CO 22	R	579	201	0.21	32	63	22	2.40
TN1 × MTU15	R	227	73	0.04	35	65	30	0.38
Mudgo × ASD7	R	1034	37	–	124	0	0	–
Mudgo × CO 22	R	1023	16	–	127	0	0	–
Mudgo × MTU15	R	1366	4	–	133	0	0	–
ASD7 × CO 22	R	759	8	–	129	0	0	–

R/Res, resistant; Seg, segregating; S/Sus, susceptible.
[*] TN1 is a susceptible parent and other varieties are resistant.

Athwal *et al* (1971) designated the dominant resistance gene of Mudgo as *Bph-1* and the recessive resistance gene of ASD7 as *bph-2*.

Allele tests between Mudgo on the one hand and CO 22 and MTU15 on the other showed that CO 22 and MTU15 have the same gene for resistance. A small proportion of F_2 seedlings from the crosses of Mudgo with CO 22 and MTU15 were killed (Table 10.1). A similar proportion of seedlings of the resistant checks were also killed due to causes not related to insect damage. However, as none of the F_3 families were susceptible, the conclusion about the allelic nature of genes in these three varieties was confirmed. The value of F_3 analysis in the genetics of insect resistance, as pointed out earlier, is evident from these data.

In the F_2 populations of the crosses of ASD7 with Mudgo and CO 22, 3.5% and 1.0% seedlings, respectively, were classified as susceptible. If *Bph-1* and *bph-2* were independent, 18.75% of the seedlings (13 resistant:3 susceptible) would have been susceptible. Moreover, none of the F_3 families of these two crosses were susceptible, indicating that *Bph-1* and *bph-2* are closely linked. The close linkage between these two genes has been confirmed in subsequent studies.

Martinez and Khush (1974) investigated the inheritance of resistance to the BPH in two improved breeding lines, IR747B2-6 and IR1154-243. The former was found to have *Bph-1* and the latter *bph-2*. For further genetic analysis, IR1154-243 was used as a tester for *bph-2* and a dwarf line IR1539-823 (from the cross of Mudgo with *Bph-1* for resistance) was used as a tester for *Bph-1*. Lakshminarayana and Khush (1977) carried out a genetic analysis of 27 BPH-resistant varieties. The reaction of F_1 and F_2 populations from crosses of test varieties with TN1 indicated that 11 varieties had a dominant gene and 16 had a recessive gene for resistance (Table 10.2). To determine the allelic relationships of the genes of these varieties, 11 varieties with dominant genes were crossed with IR1539-823 and 16 varieties with recessive genes were crossed with IR1154-243. The F_1, F_2, and F_3 progenies from these crosses were evaluated for resistance. The F_1 hybrids of IR1539-823 with 11 varieties were resistant and the F_2 populations and F_3 families from 10 crosses did not segregate for susceptibility (Table 10.2). It was concluded that these ten varieties have *Bph-1* for resistance. The F_2 populations from the cross IR1539-823 × Rathu Heenati, segregated in a ratio of 15 resistant:1 susceptible. The F_3 families of this cross did not segregate in the expected ratio of 7 resistant:8 segregating:1 susceptible, but when resistant and segregating families were pooled, a good fit to the 15:1 ratio was obtained. Apparently, some of the segregating families were misclassified. These results show that the single dominant gene of Rathu Heenati is nonallelic to, and independent of, *Bph-1*. This gene was designated as *Bph-3*.

The F_1 hybrids of IR1154-243 with 15 varieties were resistant. Although a few F_2 seedlings of these crosses died, the proportion of dead seedlings was

Table 10.2. Reaction to the brown planthopper of F_1, F_2, and F_3 populations from the crosses of IR1539-823 (*Bph-1*) and IR1154-243 (*bph-2*) with resistant cultivars of rice (data from Lakshminarayana and Khush 1977).

Cross		F_1 reaction	Reaction of F_2 seedlings			Reaction of F_3 lines			
			Res (no.)	Sus (no.)	χ^2 15 : 1 or 7 : 9	Res (no.)	Seg (no.)	Sus (no.)	χ^2 7 : 8 : 1
IR1539-823	× Balamawee	R	351	0		110	0	0	
	× CO10	R	302	0		128	0	0	
	× Heenukkulama	R	314	0		129	0	0	
	× MTU9	R	356	0		130	0	0	
	× Ptb 21	R	419	6		132	0	0	
	× Rathu Heenati	R	374	30	0.95	187	99	21	
	× Sinnakayam	R	346	3		130	0	0	
	× SLO12	R	309	0		130	0	0	
	× Sudhabalawee	R	334	3		130	0	0	
	× Sudurvi 305	R	312	0		130	0	0	
	× Tibiriwewa	R	324	0		128	0	0	

IR1154-243	× Anbaw C7	R	840	32		130	0	0	
	× ASD9	R	309	9		128	0	0	
	× Babawee	S	426	580	0.79	49	71	12	3.39
	× Dikwee 328	R	758	0		132	0	0	
	× Kosatawee	R	647	39		132	0	0	
	× Madayal	R	625	20		132	0	0	
	× Mahadikwee	R	610	18		132	0	0	
	× Malkora	R	544	14		132	0	0	
	× M.I. 329	R	582	26		132	0	0	
R1154-243	× Murungakayan 302	R	972	45		132	0	0	
	× Ovarkarupan	R	809	26		132	0	0	
	× Palasithari 601	R	810	53		132	0	0	
	× PK1	R	628	22		132	0	0	
	× Seruvellai	R	1005	41		132	0	0	
	× Sinna Karuppan	R	667	37		132	0	0	
	× Vellailangan	R	1038	12		132	0	0	

R/Res, resistant; Seg, segregating; S/Sus, susceptible.

similar to that observed in the check rows of IR1154-243. All the F_3 families of these crosses were resistant, indicating that these 15 varieties have *bph-2* for resistance. The F_1 hybrid of IR1154-243 with Babawee was susceptible, indicating that the recessive gene of this variety is nonallelic to *bph-2*. The F_2 and F_3 populations of this cross segregated in a ratio of 7 resistant : 9 susceptible and 7 resistant : 8 segregating : 1 susceptible, respectively. This confirmed the conclusion of the F_1 reaction. This new recessive gene was designated as *bph-4*.

Genetic analysis of 20 additional BPH-resistant varieties was carried out by Sidhu and Khush (1978). From analysis of crosses with TN1, three varieties were found to have two genes each, seven had single dominant genes, and ten had single recessive genes for resistance. Allele tests were conducted with four testers. Varieties with dominant genes were crossed with IR1539-823 and Rathu Heenati. Varieties with recessive genes were crossed with IR1154-243 and Babawee.

Segregation data revealed that the dominant genes in seven varieties were allelic to *Bph-3* and the recessive genes in ten varieties were allelic to *bph-4*. One of the two genes in three varieties was dominant and the other was recessive. No new gene was identified on the basis of analysis of these 20 varieties.

As discussed in Chapter 7, four biotypes of BPH are known. Biotype 1 was prevalent in Southeast and East Asia before the large-scale cultivation of BPH-resistant varieties. Biotype 2 originated in 1976 in the Philippines, and biotype 3 was selected in the laboratory by rearing insects on resistant variety ASD7. *Bph-1* conveys resistance to biotypes 1 and 3, *bph-2* to biotypes 1 and 2, and *Bph-3* and *bph-4* convey resistance to all three biotypes. Studies on the inheritance of resistance discussed up to now were carried out using BPH biotype 1 as test insects. Results of international nurseries indicated that varieties with *Bph-1* or *bph-2* were not resistant in southern Asia (India, Bangladesh, Sri Lanka) but varieties with *Bph-3* and *bph-4* were resistant. In addition, some varieties resistant in the Indian subcontinent were susceptible in Southeast Asia (Table 10.3) (Seshu and Kauffman 1980). Genetic analysis of one such variety, ARC10550, was carried out. Since ARC10550 is susceptible to the three biotypes in the Philippines, it was necessary to use the South Asian biotype (biotype 4) to test the segregating populations. Crosses with TN1 were made at the International Rice Research Institute (IRRI) and F_2 and F_3 seeds were also produced. F_1, F_2, and F_3 seeds were sent to the Bangladesh Rice Research Institute (BRRI) where they were evaluated for resistance using biotype 4. The data showed that ARC10550 has a recessive gene for resistance to biotype 4. This gene was designated *bph-5* (Khush *et al* 1985). To determine the linkage relations of *bph-5* with *Bph-1*, *bph-2*, *Bph-3*, and *bph-4*, F_1, F_2, and F_3 progenies from the crosses of ARC10550 with testers for *Bph-1*, *bph-2*, *Bph-3*, and *bph-4* were evaluated for resistance to biotype 1 at the IRRI, Philippines, and to biotype 4 at the BRRI, Bangladesh. A two-way classification of F_3 families for their reaction to the BPH at the IRRI and the BRRI indicated that *bph-5* is independent of the other four genes. Seventeen additional rice varieties – resistant to biotype 4 but susceptible to biotypes 1, 2, and 3 – were genetically analyzed by Kabir and Khush (1988) using the biotype 4 insects for testing. Seven were found to have single dominant genes which segregated independently of *bph-5*. The dominant gene of Swarnalata was designated *Bph-6*. The remaining 10 cultivars were found to have recessive genes for resistance and eight of them were allelic to *bph-5*. However, recessive genes of two cultivars were nonallelic to *bph-5*. The recessive gene of T12 was designated *bph-7*.

Two Thai varieties, Col. 5 Thailand and Col. 11 Thailand, and Chin Saba from Myanmar were found to have single recessive genes which are allelic to each other but are nonallelic to *bph-2* and *bph-4*. The recessive gene of these three cultivars is nonallelic to *bph-5* and *bph-7* because *bph-5* and *bph-7* do not confer resistance to biotypes 1, 2, and 3 but the new gene does. Therefore, this new recessive gene is different from all the other recessive genes and was designated as *bph-8* (Nemoto *et al* 1989). Three varieties – Kaharamana, Balamawee, and Pokkali – were found to have single dominant genes which are allelic to each other but are different from *Bph-1* and *Bph-3* (Ikeda 1985). This gene was designated as *Bph-9*.

Several genes for resistance to the BPH have been transferred from wild *Oryza* species to cultivated rice through wide hybridization (Jena and Khush 1990). Genetic analysis to determine allelic relationships of these genes with known genes are underway. An introgression line from the cross of cultivated

Table 10.3. Genes for resistance to the brown planthopper in rice and their reaction to different biotypes.

Gene	Chromosome location	Reaction to indicated biotype			
		1	2	3	4
Bph-1	4	R	S	R	S
bph-2	4	R	R	S	S
Bph-3	10	R	R	R	R
bph-4	10	R	R	R	R
bph-5	–	S	S	S	R
Bph-6	–	S	S	S	R
bph-7	–	S	S	S	R
bph-8	–	R	R	R	–
Bph-9	–	R	R	R	–
Bph-10(t)	–	R	R	R	–

R, resistant; S, susceptible.

rice and *O. australiensis* has a dominant gene for BPH resistance which has been tentatively designated as *Bph-10(t)*. It confers resistance to three biotypes of BPH. Thus, on the basis of genetic analysis of about 90 varieties, 10 genes for resistance have been identified. The reaction of the known genes against four biotypes is shown in Table 10.3.

The genes for resistance in rice varieties can be inferred without genetic analysis by determining their reaction to different biotypes. If a variety is resistant to biotypes 1 and 3, it is likely to have *Bph-1*; if it is resistant to biotypes 1 and 2, it has *bph-2*; and if it is resistant to all three biotypes, it may have any of these – *Bph-3*, *bph-4*, *bph-8*, or *Bph-9*. Thus, varieties can be classified into varietal groups on the basis of their reaction to different biotypes before embarking on genetic analysis.

Inheritance of resistance to various insects has been investigated in most of the major cereal crops (rice, wheat, maize, sorghum, barley, and rye), forage and food legumes, vegetables, fruits, and cotton. Numerous cases of monogenic resistance have been reported. The monogenes may be dominant, incompletely dominant, or recessive but are more often dominant. The most thoroughly studied inheritance is that of wheat resistance to the Hessian fly. Twenty-six genes for resistance to this insect have been identified, all of which, except one, are dominant. Table 10.4 summarizes information on the genetics of resistance to major insects of important crops. The review paper of Khush and Brar (1991) should be consulted for original references.

Genetic analysis for polygenic resistance

Polygenic (or quantitative) resistance to insects has been reported in several crops. Unlike major genes for which phenotypes can be placed into discrete classes, levels of resistance among individuals in polygenic inheritance differ by degree and show continuous variation. The inheritance of polygenic traits is complex. The basic assumption is that many genes with small and roughly equal effects govern the trait, and expression of the trait is strongly influenced by the environment. The heritability or genetic component of variation is lower than in monogenic resistance. In spite of this complexity, the genes that determine continuous variation are assumed to behave in the same way as the major genes (Allard 1960).

Besides the small additive effect of many genes, quantitative genetics also presumes that phenotypic expression of a trait is modified by intra-allelic interaction (dominance), interallelic interaction (epistasis), effects of genes on multiple traits (pleiotropy), and by the environment. In its simplest form, the total phenotypic variation of a quantitatively inherited trait can be presented as

$$V_P = V_G + V_E$$

where V_P is the total phenotypic variation, V_G is genotypic variation, and V_E

Table 10.4. Genetics of resistance to major insect pests of crop plants (from Khush and Brar 1991).

Crop species	Insect pest	Gene(s) for resistance
Rice	Brown planthopper	*Bph-1, bph-2, Bph-3, bph-4, bph-5, Bph-6, bph-7, bph-8, Bph-9*
	Whitebacked planthopper	*Wbph-1, Wbph-2, Wbph-3, wbph-4, Wbph-5*
	Green leafhopper	*Glh-1, Glh-2, Glh-3, glh-4, Glh-5, Glh-6, Glh-7, glh-8*
	Zigzag leafhopper	*Zlh-1, Zlh-2, Zlh-3*
	Gall midge	*Gm-1, Gm-2*
	Striped stem borer	Monogenic (antibiosis)
		Polygenic (tolerance)
Wheat	Hessian fly	*H-1, H-2, H-3, h-4, H-5, H-6, H-7, H-8, H-9, H-10, H-11, H-12, H-13, H-14, H-15, H-16, H-17, H-18, H-19, H-20, H-21, H-22, H-23, H-24, H-25, H-26*
	Greenbug	*gb (Gb-1), Gb-2, Gb-3, Gb-4, Gb-5*
	Cereal leaf beetle	2–3 genes, polygenic
	Wheat stem sawfly	1–4 genes (stem solidness)
Maize	European corn borer	Polygenic
	Corn earworm	Polygenic
	Western corn rootworm	Monogenic, polygenic
	Corn leaf aphid	Monogenic, polygenic
	Fall armyworm	Polygenic
	Spotted stem borer	Polygenic
Sorghum	Corn leaf aphid	
	Greenbug	Monogenic, 2–3 genes
	Shootfly	Polygenic, *tr* (trichome on leaf surface)
	Sorghum midge	2–3 genes, polygenic
	Chinch bug	Monogenic, 1–2 dominant genes
	Stem borer	Polygenic
Barley	Hessian fly	*H-f, Hf-1, Hf-2*
	Cereal leaf beetle	Monogenic
	Greenbug	*Grb (Rsg-1a), Rsg-2b*
	Corn leaf aphid	*s-1, s-2* (complementary)
Cotton	Boll weevil	A few genes (pubescence, frego bract, okra leaf)
	Thrips (*Thrips* spp.)	*H-1, H-2, Sm* (hairiness, smooth leaf) nonpreference
	Tobacco budworm	*Gl-2, Gl-3* (nonpreference/antibiosis)
	Jassids (*Empoasca* spp.)	*H, H-1, H-2*, a few modifiers (hairiness)
	Tarnished plant bug	
	Pink bollworm	A few genes, polygenic

Table 10.4. Continued.

Crop species	Insect pest	Gene(s) for resistance
Apple	Rosy leafcurling aphid	*Sd-1, Sd-2, Sd-3, Sd-pr*
	Rosy apple aphid	*Smh*
	Wooly apple aphid	*Er*
Raspberry	Rubus aphid *Amphorophora rubi*	*A-1, A-2, A-3, A-4, A-5, A-6, A-7, A-8, A-9, A-10, A-k4a,*
	Amphorophora agathonica	*A-cor1, A-cor2, Ag-1, Ag-2, Ag-3* (complementary)
Black currant	Leafcurling midge	*Dt*
	Gall mite	*Ce*
Citrus	Red scale	Polygenic
Musk melon	Melon aphid	Monogenic
	Red pumpkin beetle	Monogenic, polygenic
Cucumber	Striped cucumber beetle	*cu*, polygenic
	Two-spotted spider mite	*Bi*, polygenic
Summer squash	Squash bug	Polygenic
Pumpkin	Fruit fly	*Fr* (two complementary dominant genes)
Lettuce	Leaf aphid	*Nr*
	Root aphid	Cytoplasmic and nuclear genes (modifiers)
Okra	Cotton jassid	Polygenic, monogenic (cotyledonary stage)
Tomato	Fruit borer	Polygenic
Potato	Potato tuber moth	A few genes, polygenic
Soybean	Mexican bean beetle	2–3 genes, polygenic
Mungbean	Azuki bean weevil	Monogenic
Common bean	Leafhopper	Polygenic
Cowpea	Cowpea aphid	*Rac, Rac-1, Rac-2*
	Cowpea seed beetle	*rcm-1, rcm-2*
Alfalfa	Spotted alfalfa aphid	A few major genes
	Pea aphid of alfalfa	Monogenic
	Sweet clover aphid	Monogenic
	Potato leafhopper	Polygenic

is environmental variation. Genotypic variation can further be partitioned into additive and nonadditive variation. The nonadditive variation can also be partitioned into dominant (intra-allelic) and epistatic (interallelic) inter-actions. These parameters can be used to estimate the heritability of the trait which, in turn, influences the breeding strategy to be adopted.

Because of the complexity of the inheritance of quantitative traits, biometric genetic procedures are used to infer the existence and properties of genes controlling them. The components of variation can be estimated either from the mean values of the parents, F_1s, and segregating generations (generation mean analysis) or from the variances of parents and F_1s in a series of crosses (diallel analysis).

Generation mean analysis

Different components of variation which can be estimated from the means of parents, F_1s, and segregating generations are:

m = mean of the parental values,
$[d]$ = additive component of variation,
$[h]$ = dominance effect,
$[i]$ = additive × additive interaction,
$[j]$ = additive × dominance interaction,
$[l]$ = dominance × dominance interaction.

These parameters can be estimated from the following relationships, where bars represent mean values:

$$m \ = \tfrac{1}{2}\bar{P}_1 + \tfrac{1}{2}\bar{P}_2 + 4\bar{F}_2 - 2\bar{B}_1 - 2\bar{B}_2,$$
$$[d] = \tfrac{1}{2}\bar{P}_1 - \tfrac{1}{2}\bar{P}_2,$$
$$[h] = 6\bar{B}_1 + 6\bar{B}_2 - 8\bar{F}_2 - \bar{F}_1 - 3/2\bar{P}_1 - 3/2\bar{P}_2,$$
$$[i] = 2\bar{B}_1 + 2\bar{B}_2 - 4\bar{F}_2,$$
$$[j] = 2\bar{B}_1 - \bar{P}_1 - 2\bar{B}_2 + \bar{P}_2,$$
$$[l] = \bar{P}_1 + \bar{P}_2 + 2\bar{F}_1 + 4\bar{F}_2 - 4\bar{B}_1 - 4\bar{B}_2.$$

These relationships are modified if data from more generations become available. For more details see Mather and Jinks (1982).

The ratio $[h]/[d]$ measures the average degree of dominance. If $[h]/[d]$ = 1, there is complete dominance; if it is >1, there is overdominance; and if it is <1, there is partial dominance.

The heritability estimates can also be obtained from generation means.

Diallel analysis

Diallel analysis is one of the most widely used approaches to study the inheritance of quantitatively inherited traits. When a number of inbred lines

are available, these can be crossed in a diallel fashion (all possible combinations) and genetic analysis can be carried out in one generation to estimate
the various components of variance. The Hayman (1954) approach is the
most widely used method to estimate the components of variation from an F_1
diallel. (For a more recent treatment of the subject, Simms and Rausher
(1992) may be consulted.) The various components of variation which can be
estimated using this approach are:

D = variance due to additive effects of genes,
H_1 = variance due to dominance effects of genes,
H_2 = dominance indicating asymmetry of positive (resistant) and negative
       (susceptible) effects of genes,
h^2 = dominance effects over all loci,
F = covariance of additive and dominance effects,
E = environmental variance.

   Several other parameters can be obtained from these estimates. $(H_1/D)^{1/2}$
measures the average degree of dominance. If this value is equal to 1, it means
complete dominance, <1 means partial dominance, and >1 means overdominance.
   $(H_2/4H_1)$ measures the product of frequency of positive (resistant) and
negative (susceptible) genes. This can have a maximum value of 0.25, which
means the frequency of the positive and negative genes is the same at all loci.
The sign of F determines the relative frequency of dominant to recessive alleles
and the magnitude determines the variation in dominance level over all loci.
   $F = 0$ implies that the frequency of positive and negative alleles is equal.
If F is positive, the frequency of dominant alleles is more than that of recessive
alleles, irrespective of whether dominant alleles have increasing or decreasing
effects. Heritability of the trait can be calculated using the formula of
Crumpacker and Allard (1962).

Heritability = $[\frac{1}{2}D/(\frac{1}{2}D + \frac{1}{4}H_1 - \frac{1}{4}F + E)]$.

   Resistance to the European corn borer (ECB) *Ostrinia nubilalis* is an
excellent example of polygenic or quantitative resistance to insects. It has
been studied for more than 60 years. There are two generations (broods) of
ECBs each season. Several studies have been conducted to study the
inheritance of resistance to first- and second-generation ECB using a varying
number of resistant and susceptible lines. The studies of Jennings *et al* (1974)
and Chiang and Hudon (1973) will be discussed here to explain the use of
generation mean analysis and diallel analysis in studying the inheritance of
polygenic resistance to insects.
   Jennings *et al* (1974) studied the inheritance of resistance to second-
brood ECBs. The basis of evaluation was the number of cavities per stalk

Table 10.5. Mean number of cavities per stalk in maize caused by European corn borer feeding in various generations (data from Jennings *et al* 1974).

Generation	B52 × L289	B52 × WF9	B52 × Oh43
P_1	0.76	1.51	1.59
P_2	12.70	13.19	16.69
F_1	3.77	4.63	2.99
F_2	4.54	10.31	6.16
BC_1	2.05	4.18	3.04
BC_2	7.78	8.73	8.52
F_3	4.72	10.32	6.83
BS_1	2.31	6.15	3.46
BS_2	8.82	8.59	8.99

Table 10.6. Estimates of genetic effects and standard error of mean number of cavities per stalk in maize caused by European corn borer feeding (Jennings *et al* 1974).

Parameter	B52 × L289	B52 × WF9	B52 × Oh43
m	−4.60 ± 0.16	8.04 ± 0.27	5.43 ± 0.24
[d]	−5.98 ± 0.42	−3.50 ± 0.70	−5.10 ± 0.81
[h]	−1.52 ± 0.65	−7.07 ± 1.09	−4.31 ± 0.81
[i]	1.50 ± 0.64	−5.02 ± 1.07	1.56 ± 0.97
[j]	0.05 ± 0.50	2.08 ± 0.84	2.35 ± 0.76
[l]	0.40 ± 1.56	−1.57 ± 2.62	−0.68 ± 2.37

See text for meaning of parameters.

caused by ECB feeding as counted 50–60 days after flowering. They crossed resistant inbred line B52 to four susceptible lines – L289, B39, Oh43, and WF9 – and evaluated nine generations (P_1, P_2, F_1, F_2, F_3, BC_1, BC_2, BS_1, and BS_2). Results of some of these crosses are presented in Table 10.5. Scores of the F_1s showed resistance to be dominant over susceptibility. Estimates of components of variation are presented in Table 10.6.

Additive effects were negative and highly significant in all crosses and so were the dominance effects, indicating some degree of dominance in all the crosses. Significant additive × additive effects were also detected in all crosses. Thus, in all cases, additive gene effects were predominant with some degree of dominance in all the crosses. Heritability in the narrow sense was estimated to be 30.7% in cross B52 × Oh43.

Chiang and Hudon (1973) used diallel analysis to study the inheritance of resistance to first-brood ECBs. Four inbred lines of maize (A619, CI31A, Oh43, and Romania T393) selected for resistance and four susceptible lines

Table 10.7. Average visual leaf-feeding damage score by the European corn borer of parental lines and single crosses of maize* (data from Chiang and Hudon 1973).

	A619	C131A	Oh43	Romania T393	France F564	M14	WF9	Hungary 1429
A619	1.10	1.10	1.25	1.35	1.10	1.40	1.90	2.20
C131A		1.10	1.20	1.55	1.80	1.55	1.10	1.80
Oh43			1.20	1.55	2.70	1.60	1.90	2.95
Romania T393				2.20	3.65	3.45	3.95	4.10
France F564					4.00	4.60	5.75	5.15
M14						4.90	4.40	5.00
WF9							6.00	5.10
Hungary 1429								6.25

* Damage score based on a scale of 1–9 (1, most resistant; 9, most susceptible).

Table 10.8. Genetic components with standard errors (SE) and ratios between components in leaf feeding of the European corn borer (from Chiang and Hudon 1973).

Component of variance	Magnitude and SE	Proportional value	
D	$4.0653 \pm 0.9880^{**}$	$(H_1/D)^{1/2}$	0.80
F	$-0.3880 + 1.1614$	$H_2/(4H_1)$	0.24
H_1	2.6058 ± 2.2682	h^2/H_2	0.38
H_2	2.5622 ± 1.9759	$[4(DH_1)^{1/2} + F]/[4(DH_1)^{1/2} - F]$	0.89
h^2	0.9703 ± 1.3248	Heritability	40%
E	$0.9765 \pm 0.3392^{*}$		

*, **, significant at the 5 and 1% levels of probability, respectively. See text for meaning of components.

(France F564, Hungary 1429, M14, and WF9) were crossed in a diallel manner and 28 F_1s and eight parents were evaluated for resistance by artificially infesting the plants with ECB larvae. Insect damage was scored 2 weeks after infestation on the basis of a 1–9 scale (1, most resistant; 9, most susceptible). Results are given in Table 10.7. High levels of significance of the additive component D and the nonsignificance of dominance component H_1 (Table 10.8) suggest that the additive genetic component played the most important role. A relatively high value of $(H_1/D)^{1/2}$ (0.8) indicates partial dominance of resistance. The nonsignificant value of F and higher value of $H_2/4H_1$ (0.24) indicates that both positive (resistant) and negative (susceptible) genes are almost equally distributed among the parents.

The significance of the *E* component suggests considerable environmental influence. Heritability in its narrow sense was estimated at 40%. In both of the above mentioned experiments, additive gene action was more important and heritability was reasonably high, therefore selection for a high degree of resistance to first- and second-brood of ECBs should be effective. This was demonstrated by Penny *et al* (1967) in selecting for resistance to the ECB. After three cycles of recurrent selection for resistance to leaf feeding, five synthetic varieties with borer resistance were selected. It is evident that information from quantitative genetic analysis is useful in selecting the breeding methodology and in predicting genetic advance under selection. As discussed in Chapter 11, the recurrent selection method is the most widely used in breeding for polygenic resistance, whereas the pedigree method is most suitable for incorporating monogenic resistance.

Several cases of polygenic resistance to insects have been reported in crop plants (Table 10.4) and crop varieties with horizontal resistance to insects have been developed.

Mapping genes for resistance

The genes for insect resistance can be located to specific chromosomes by different techniques such as standard linkage tests and trisomic or monosomic analysis through the use of substitution or alien addition lines and translocation stocks. Ikeda and Kaneda (1981) located *Bph-3* and *bph-4* for BPH resistance in rice on chromosome 10 through trisomic analysis. Similarly, Ikeda and Kaneda (1983) located *Bph-1* and *bph-2* on chromosome 4. Genes for greenbug resistance in barley were also located to specific chromosomes through trisomic analysis (Gardenhire *et al* 1973). The chromosomal location of genes for Hessian fly resistance in wheat has been determined primarily through monosomic analysis (Gallun and Patterson 1977). Substitution lines were employed to determine the chromosomal location of genes for cereal leaf beetle resistance in wheat (Smith and Webster 1973) and a gene for resistance to rice BPH was located on chromosome 12 through an alien addition line (Multani *et al* 1994). Genes for corn earworm resistance were located to specific chromosomes through the use of translocation stocks (Robertson and Walter 1963). Similarly, chromosomal translocations have been used for mapping polygenes conferring resistance to the ECB in corn (Scott *et al* 1966, Onukogu *et al* 1978).

After the chromosomal location of a gene is known, it can be mapped with respect to morphological and molecular markers. Ikeda and Kaneda (1983) detected linkage of *bph-2* with a morphological marker d-11 on chromosome 4. A gene for resistance to the BPH, introgressed from the wild species *Oryza australiensis* to cultivated rice and located on chromosome 12, was mapped with a restriction fragment length polymorphism (RFLP) marker. Of the 14 polymorphic probes on chromosome 12, only one (RG457) detected

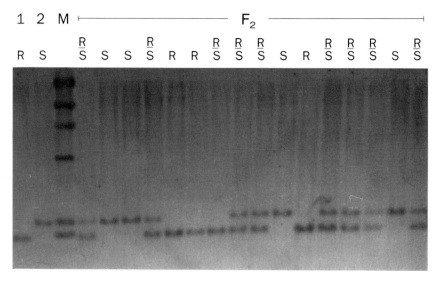

RG457

Fig. 10.2. RFLP pattern in an F_2 population of a cross of a brown planthopper (BPH)-resistant introgression line (1) with the recurrent parent (2). Total DNA was digested with *Eco*R1 and probed with RG457 of chromosome 12. Also indicated are the results of BPH-resistance tests performed on the F_3 progeny of the 16 F_2 plants. The seventh F_2 individual from the marker-lane was the only recombinant. R, true breeding for resistance; R/S = segregating for resistance; S, true breeding for susceptibility. Reproduced from IRRI (1993).

introgression from *O. australiensis*. Cosegregation for BPH resistance and RG457 was studied (Fig. 10.2) and the gene was found to be linked with RG457 with a crossover distance of 3.68 ± 1.29 cM (Ishii *et al* 1994). Similarly, the *Wbph-1* gene for resistance to the whitebacked planthopper *Sogatella furcifera*, another important pest of rice, was found to be closely linked with two RFLP markers (McCouch *et al* 1991). Linkage between molecular markers and three genes (H3, H23, and H24) for Hessian fly resistance in wheat was reported by Ma *et al* (1993). Close linkages between resistance genes and molecular markers are useful in transferring genes from one varietal background to another, following marker-aided selection (see Chapter 11).

Genetic analysis for resistance through associated traits

If resistance to insects is due to alteration in the plant morphological traits – such as the resistance of frego-bract cotton to the boll weevil – the inheritance of resistance can be simply inferred from the inheritance of the morphological trait. Frego bract is a monogenic recessive trait and so is resistance to the cotton boll weevil (Jenkins and Parrot 1971). The presence of trichomes on leaves and stems is often associated with resistance to herbivores and inheritance of resistance can be determined simply by studying the inheritance of trichomes. In tomato *Lycopersicon* spp. type VI glandular trichomes (see Chapter 6) play a role in arthropod resistance (Snyder and Carter 1984). In crosses between *L. esculentum* and *L. pennellii*, the presence of type VI trichomes was found to be controlled by two independent genes, while the density of type VI trichomes appeared to be inherited quantitatively (Lemke and Mutschler 1984).

Genetics of insect resistance can also be inferred from the inheritance of secondary compounds responsible for resistance. For example, the presence of polyphenolic gossypol in cotton contributes resistance to the tobacco budworm (Wilson and Shaver 1973). Three dominant genes (G1, G2, and G3) control the level of gossypol production (McMichael 1960). Cucurbitacins confer antibiosis resistance to spider mites. The absence of cucurbitacins in cucumber (*Cucumus sativus* L.) is controlled by the recessive gene *bi*. Cucumber plants lacking cucurbitacins (*bi bi*) are highly susceptible to mites whereas those possessing cucurbitacins (*Bi Bi* or *Bi bi*) are resistant to mites (Da Costa and Jones 1971). A more comprehensive discussion of inheritance of secondary compounds imparting resistance to insects is given by Berenbaum and Zangerl (1992b).

Genetics of Parasitic Ability in Insects

As discussed earlier, there are numerous studies on genetics of resistance to insects. However, studies on the genetics of parasitic ability in insects are limited. The most well-known studies are on the Hessian fly of wheat.

The Hessian fly larvae feed on susceptible wheat, resulting in stunted and dark green leaves. The new leaves generally fail to form. The resistant plants remain light green, as do the uninfested plants. The larvae that feed on the resistant plants generally die. The small number that survive remain small and do not stunt the plants. Thus, virulent larvae (those that can survive and stunt the plants) can be distinguished from the avirulent larvae by the reaction of the plant to them and by the reaction of the insect to plants of known genotype. In the case of the Hessian fly, biotype distinction is based on the virulence/avirulence of larvae to different wheats having known genes for resistance. One biotype may be virulent on a specific line whereas another biotype may be avirulent. Thus, the terms virulence and avirulence describe the insect's reaction to the plant, whereas resistance and susceptibility describe the plant's reaction to the insect. Biotype can be determined by scoring plant reaction to the larvae of the progeny and the ability of the larvae to survive. Wheat seedlings of known genotype can be grown and infested with a pair of flies. The female lays eggs at random, showing no host preference. The larvae can be scored as dead or alive 15 days after egg laying. At this stage, the length of the living larvae can also be measured.

Using this technique, inheritance of parasitic ability was studied by Hatchett and Gallun (1970). They evaluated the F_1, F_2, and backcross progenies from the crosses of biotype GP and E on wheat variety Monon. Monon is homozygous for a dominant gene H_3 for resistance to the Hessian fly and this gene confers resistance to biotype GP but not to biotype E. All the F_1 flies were avirulent on Monon and had the phenotype of GP, showing that virulence is recessive. The reactions of the F_2 and backcross progeny showed that the difference between biotypes GP and E is governed by a single gene. The Hessian fly has one recessive gene for virulence that enables it to overcome the resistance conferred by the H_3 gene of Monon. This recessive gene for virulence was designated as m.

Five loci for virulence in the Hessian fly are known. They have been designated t, s, m, k, and a. Eight biotypes of Hessian fly were reported by Gallun (1977). However, 11 biotypes are now known (Obanni *et al* 1988).

Gene-for-gene Concept

The gene-for-gene relationship between hosts and parasites was first proposed by Flor (1942). Flor (1956) further elaborated the concept and it was formally defined by Person *et al* (1962). Simply stated, it means that for every major gene for resistance in the host species, there is a corresponding matching gene for virulence in the parasite species. The host plant shows a resistant reaction if it has a resistance gene and the insect has an avirulent gene at the corresponding locus. If the insect has a virulent gene at the corresponding locus, however, the plant is susceptible. This gene-for-gene relationship has been called the matching gene theory.

The gene-for-gene relationship was first shown in flax *Linum usitatissimum* and its rust fungus *Melampsora lini* (Flor 1942). This relationship has been shown to exist in at least 12 pathogen–plant systems (Thompson and Burdon 1992). However, the concept as applied to insect–plant systems has only been explored for wheat and its insect parasite, the Hessian fly. As discussed in the last section, five loci for virulence in the Hessian fly were identified by Gallun (1977) and were designated t, s, m, k, and a; they correspond to five genes for resistance in the wheat plant. The virulence of an insect biotype on a wheat variety is conditioned by homozygosity for a recessive allele for virulence at a specific locus corresponding to a specific dominant gene of wheat that conditions resistance. A Hessian fly biotype can, therefore, be virulent to a specific wheat cultivar only if the biotype is homozygous for recessive virulent alleles at all loci corresponding to loci at

Table 10.9. Genotypes of eight Hessian fly biotypes* (from Gallun 1977).

Biotype	Wheat varieties				
	Turkey	Seneca	Monon	Knox 62	Abe
GP	tt	S-	M-	K-	A-
A	tt	ss	M-	K-	A-
B	tt	ss	mm	K-	A-
C	tt	ss	M-	kk	A-
D	tt	ss	mm	kk	A-
E	tt	S-	mm	K-	A-
F	tt	S-	mm	kk	A-
G	tt	S-	mm	kk	A-

* Hessian fly biotypes homozygous for recessive alleles are virulent to wheat having at least one dominant allele at the corresponding locus that conditions resistance.

which the wheat plant has dominant alleles for resistance.

The wheat variety Turkey is susceptible to all fly biotypes because it has no gene for resistance. The varieties Seneca, Monon, Knox 62, and Abe are resistant to biotype GP because that biotype has at least one dominant allele for avirulence at loci corresponding to the loci in wheat with dominant alleles that condition resistance (Table 10.9). Only when there is homozygosity for the recessive virulent allele at a locus in the biotype is the insect virulent to the wheats that have at least one dominant allele at the corresponding locus for resistance. Thus, biotype A, in addition to being virulent to variety Turkey, is also virulent to Seneca because it is homozygous for the virulent gene *s* which corresponds to the dominant gene for resistance in Seneca. However, it is avirulent to Monon, Knox 62, and Abe because it has dominant avirulent alleles at the loci corresponding to the loci that carry dominant alleles for resistance in those varieties. Biotype D is homozygous recessive for alleles at virulence loci *s*, *m*, and *k* and thus is virulent to varieties Seneca, Monon, and Knox 62, as well as to Turkey. But it is avirulent to Abe because it has a dominant avirulent allele at the *a* locus which corresponds to the resistance gene of Abe.

The gene-for-gene concept as applied to plant–insect systems has only been demonstrated for the Hessian fly–wheat relationship. Diehl and Bush (1984) question the general applicability of the concept to all insect–plant systems. However, Robinson (1991) believes that in all cases of vertical resistance, the gene-for-gene relationship exists. In other words, vertical resistance is the result of a gene-for-gene relationship.

Vertical versus Horizontal Resistance in Host Plant Resistance Programs

As discussed earlier, vertical resistance governed by major genes generally provides a high level of resistance against insect herbivores. There are several advantages in using vertical resistance in breeding programs; it is easier to transfer from one varietal background to another and provides a high level of resistance. When the insect in question is a vector of a virus, vertical resistance often provides protection against virus infection. The insect does not feed on the resistant plants and thus does not inoculate them with the virus. Its major disadvantage is that it exerts a strong selection pressure on the insect population which may lead to the development of new biotypes that are able to attack the resistant varieties with specific genes.

Several authors, most notably Robinson (1976, 1991), have argued against the use of vertical resistance in the breeding of resistant crop varieties. However, vertical resistance has been employed usefully to control insects in many crop plants. Breeding wheat varieties for resistance to the Hessian fly in the United States (Painter 1968) and rice for resistance to the BPH in Asia (Khush 1992) are excellent examples of the value of vertical resistance in host plant resistance programs. Moreover, not all vertical genes exert selection pressure on the insect populations. Vertical genes for tolerance or moderate resistance may provide stable resistance. Problems of biotypes can also be addressed through a dynamic breeding program. Scientists studying host plant resistance must have procedures to: (1) collect germplasm, (2) evaluate the germplasm for resistance to identify the source of resistance, (3) genetically analyze the resistant germplasm to identify the distinct genes for resistance, and (4) incorporate the diverse genes into elite germplasm. If and when the resistant variety with a specific gene becomes susceptible due to the selection of a new biotype, another variety with a different gene can be immediately released.

The major advantage of horizontal resistance is that it operates against all biotypes of the insect. The level of resistance is generally not very high and it does not exert a selection pressure on the insect and is thus more stable or durable. The main disadvantage of working with horizontal resistance is that it takes a long time to accumulate polygenes from diverse germplasm and to build up the level of resistance. It is comparatively easier to breed for horizontal resistance in outcrossing species where new recombinants are produced each generation and the selection cycle is repeated for several generations in order to increase the frequency of resistance genes. However, the genetic improvement largely depends on the efficiency of the screening techniques and the heritability of the trait. In self-pollinated crops such as rice, special breeding methods are used to generate new favorable recombinants after an initial cycle of crosses between parents with some level of resistance. For example, outcrossing can be induced in self-pollinated crops by introducing male sterility genes in a composite population. By recurrent selection, polygenes from different sources can be accumulated and the level of resistance can be raised.

From the foregoing discussion it is obvious that both types of resistance have a role to play in crop improvement programs. The breeding methods employed and the strategies for deploying the two types of resistance, with examples, are discussed in Chapter 11.

11

BREEDING FOR RESISTANCE TO INSECTS

Varietal resistance is the most economic, least complicated and environmentally soundest approach for protecting crops against insect damage. Insect resistance imparts yield stability and thus ensures food security. Numerous insect-resistant crop varieties have been developed and are planted over millions of hectares of cropland all over the world. The production costs are lowered because farmers save millions of dollars in insecticide costs. Consumers benefit through the availability of less costly and safer products for food, fiber, and shelter. Rice varieties resistant to the brown planthopper and green leafhopper are planted over 20 million hectares of riceland in Asia. Hessian fly-resistant wheats are widely grown in the mid-western United States. These resistant varieties can be grown with a minimum use of insecticides and are an important component of integrated pest management programs.

Good germplasm collections are a prerequisite for the success of breeding programs. Breeding methods and breeding strategies for developing insect-resistant varieties vary from crop to crop. Recent breakthroughs in cellular and molecular biology have provided new tools and approaches for incorporating genes for resistance to insects in improved germplasm.

Genetic Resources

Plant breeding is dependent upon genetic variability within the crop species and related wild relatives. Thus, broad-based germplasm collections are important for the success of crop improvement programs. Access to diverse germplasm is even more important for the host plant resistance programs as pest populations continue to change their virulence patterns, and new genes

324

for resistance must be constantly identified.

Agriculturists have relied upon native genetic sources for selecting new cultivars since times immemorial. However, the Russian geneticist N. I. Vavilov was the first to point out the value of and need for large germplasm collections for crop improvement work. He and his colleagues undertook several expeditions to collect seeds of cultivated plants and their wild relatives from different parts of the world over a 20-year period beginning in 1916. They established the world's first gene bank at St Petersburg, consisting of more than 250,000 accessions collected during their extensive travels. Much research has been done on the germplasm since Vavilov's time and numerous gene banks containing valuable gene pools for on-going crop improvement activities have been established and are now preserved for posterity. Much work remains to be done, however, to complete the germplasm collections of major and minor crop species.

Gene pools for crop improvement

Germplasm collections consist of: (1) wild species, (2) weed races, (3) landraces, (4) unimproved or purified cultivars no longer in cultivation, (5) improved modern cultivars under cultivation, (6) breeding stocks developed by breeders but not released for cultivation, and (7) mutants developed by mutagenic treatments as well as those of spontaneous origin. The breeding value of each category is shown in Table 11.1.

Genetic resources can be classified into primary, secondary, and tertiary gene pools on the basis of difficulty or ease of hybridization and gene introgression (Harlan and De Wet 1971). The primary gene pool consists of improved and unimproved varieties, land and weed races, and wild species that readily hybridize with the cultivated germplasm, produce viable hybrids, and have chromosomes that pair and recombine allowing genetic exchange. Breeders generally work with the primary gene pool as gene transfer from one background to another can readily be made.

The secondary gene pool consists of wild species that are difficult to cross with the cultivated forms because of ploidy differences or other barriers (Stebbins 1958). They have homologous chromosomes that pair poorly with the chromosomes of cultivated species and thus show extremely limited recombination. Thus gene transfer from the secondary gene pools is difficult and time-consuming and breeders shy away from using them. Special techniques for overcoming some of the barriers in gene transfer from secondary gene pools are discussed in a later section.

The tertiary gene pool consists of even more distantly related species belonging to a different subgenus or related genera and are even more difficult to hybridize; fertile progenies are rarely obtained. Thus, gene transfer to cultivated species is almost impossible.

In rice *Oryza sativa* L., for example, the primary gene pool consists of

Table 11.1. Genetic composition, productivity level, and potential value in the breeding of different gene sources (modified from Chang 1976).

Group	Diversity within group	Homogeneity within a strain or population	Agronomic or commercial value	Genetic potential in breeding
Wild species	Moderately low to moderate	Low to moderately low	None	Low
Weed races	Moderately high to high	Low to moderately low	Very low	Low
Landraces	Moderate to high	Moderately low to moderate	Low	Moderately high to high
Unimproved, purified cultivars	Moderately high to high	Moderately low to moderate	Moderate	Moderately high to high
Modern cultivars	Moderately low to moderate	Moderate to high	High	High
Breeding stocks	Moderately low to moderately high	Moderate for lines, low for bulks	Variable	Moderate to high
Mutants	Moderately low to moderate	Moderately high to high	Mostly low, few moderately high	Mostly low

about 150,000 improved and unimproved varieties, land and weed races, and about four wild species: *O. nivara, O. rufipogon, O. barthii,* and *O. longistaminata*. These wild species cross readily with *O. sativa* and chromosomes pair and recombine freely. Several useful genes have been transferred through conventional breeding approaches. The secondary gene pool comprises wild species such as *O. punctata, O. officinalis, O. eichingeri, O. minuta, O. latifolia, O. australiensis,* and *O. brachyantha*. These species are difficult to cross with cultivated rice and have only partially homologous chromosomes. Some gene transfer is possible through special techniques and, in fact, several genes for insect resistance have been transferred (Jena and Khush 1990, Multani *et al* 1994). The tertiary gene pool consists of more distantly related species such as *O. meyeriana* and *O. ridleyi* as well as species of related genera such as *Leersia hexandra* and *Porteresia coarctata*. Although some of them can be crossed with *O. sativa,* to date it has not been possible to transfer any gene from these species to cultivated rice.

Centers of diversity

We owe much to N. I. Vavilov for our knowledge about the global distribution of crop plants and their wild relatives. He and his colleagues undertook several expeditions to collect seeds of cultivated plants and their wild relatives from different parts of the world. On the basis of variation patterns of these collections studied over a 20-year period, Vavilov concluded that varietal wealth was concentrated in great centers of diversity, i.e., China, India, Central Asia, Asia Minor, the Mediterranean, Ethiopia, Central America, and western Central South America (Fig. 11.1). Vavilov suggested that the centers of diversity coincided with the centers of origin of cultivated plants (Vavilov 1926, 1950). The centers of diversity have turned out to be fertile collecting areas for germplasm of various cultivated species and their wild relatives.

The concept of centers of origin has evolved since Vavilov's time. As pointed out by Harlan (1971, 1975), some of the crops did not originate in the Vavilovian centers and some crops do not appear to have defined centers of diversity. For example, plant domestication activities went on almost everywhere south of the Sahara and north of the equator from the Atlantic to the Indian Ocean. Such a large region could hardly be called a center of origin; Harlan (1971) called it a 'noncenter'. Nevertheless, the concept of centers of diversity as proposed by Vavilov has proved extremely useful for the exploration and collection of germplasm resources.

Scientists searching for donors for insect resistance generally look to centers of diversity as sources of germplasm for screening. Donors for resistance are more often found in the areas where the insect is endemic and where the host and the insect have coevolved. For example, much of the brown planthopper-resistant germplasm comes from South India and Sri Lanka where this insect is endemic (Khush 1977). This may not be always

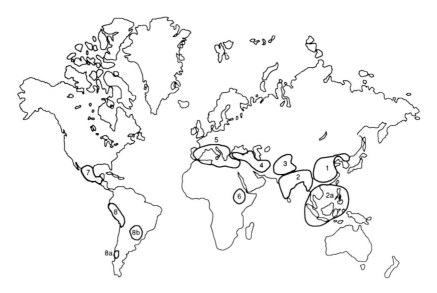

Fig. 11.1. The eight Vavilovian centers of crop plant origin.

true, however. For example, Pathak (1977) reported that 300 accessions of African rice *Oryza glaberrima* were highly resistant to the green leafhopper, which does not occur in Africa.

Sources of genetic diversity

The wild species and weed races are extremely variable as they continue to evolve in response to biotic and abiotic stresses. Even the landraces and unimproved varieties are highly heterogeneous. The introduction of varieties from one region to another and occasional crosses enhance variability. Natural crosses between the domesticated crop and the weed complexes are another source of variability. The third source of variability is the varietal mixtures that farmers plant in primitive agriculture. Occasional intercrosses between component varieties generate more variability. Thus, when seed samples of landraces and unimproved varieties are collected, they contain varying degrees of differences in morphological or physiologic aspects, some of which are obvious to the eye while others are not. Generally, such entities can be differentiated into two or more subpopulations that are distinct in one or more recognizable traits. When tested for resistance to diseases or insects, even the uniform-looking populations may be composed of resistant and susceptible individuals.

Breeding Methods

The choice of a breeding method depends upon the reproductive system of the crop species. Thus, a distinction must be made between self-pollinating species and largely cross-pollinating species. All plants in a population of outcrossing species are highly heterozygous and vigor deteriorates with forced inbreeding. Heterozygosity appears to be an essential feature of commercial varieties of such crops. Plant-breeding programs in such crops are geared towards either maintaining the heterozygosity or restoring it as a final step of the program.

On the other hand, populations of self-pollinated plants have no or very little heterozygosity, but usually consist of many closely related homozygous lines. The genotypes composing the variety of self-pollinating species exist side by side, but remain more or less independent of one another from generation to generation. The individual plants of such species are fully vigorous homozygotes and the goal of most plant-breeding programs is a pure line.

Most of the breeding methods discussed in this chapter are applicable to self-pollinated as well as cross-pollinated crops, with few exceptions. The pedigree method of breeding, for example, is primarily used in breeding self-pollinated crops and is the most commonly used method. Recurrent selection is the method of choice for breeding outcrossing species, but with some modifications it can also be used for self-pollinated crops. Similarly, hybrid breeding is primarily applicable to outcrossing species although hybrids of a few self-pollinated species such as tomato, eggplant, and rice have been developed.

In any breeding program for host plant resistance, the first step is to identify the parents or donors of resistance. These may be commercial cultivars already in use, landraces, weed races, or wild species. If they are commercial varieties, they may be improved by pure line or mass selection or hybridized with elite germplasm. Landraces, weed races, or wild species have to be hybridized with the elite germplasm.

Pure line selection

Sometimes varieties of self-pollinated crops are composed of resistant and susceptible lines that look very similar in morphology. In a pure line breeding program, a large number of plants are harvested individually from a population. A part of the seed of each plant is used in testing for insect resistance in the laboratory or greenhouse (see Chapter 8 for screening techniques).The remnant seeds of the resistant plants identified on the basis of this test are planted in the field in progeny rows. The progeny rows are evaluated for disease resistance, growth duration, and other agronomic and quality characteristics. Lines with obvious defects are discarded. Promising lines are harvested and compared with the parent variety in replicated yield trials. The highest yielding line is released as a variety. Such a purified variety should be equal to or better than the parental variety in yield, but uniformly resistant to the insect. This method of breeding, popular in the early part of this century for purifying heterogeneous varieties, is rarely used in present-day breeding programs.

Mass selection

In this method several hundred, or preferably 2–3000 agronomically desirable plants are selected. Part of the seed of each plant is used in testing for insect resistance. The remnant seeds of the resistant plants are bulked together to form a uniformly resistant improved variety. Thus, the main difference between this method and pure line selection is that a large number of lines rather than just one are selected to make up a new variety. Moreover, yield evaluation of the selected lines or the bulked variety is not required. The varieties developed by this method include fewer genotypes than the parent population but more than the 'single genotype' of varieties developed through pure line selection. Alfalfa germplasm resistant to the spotted alfalfa aphid was developed through mass selection by Hanson *et al* (1972). In highly developed agricultural areas, mass selection is rarely practiced for developing insect-resistant varieties, but it is widely used in pure seed programs for the purification of existing varieties.

Hybridization

As plant breeding progressed and more and more of the natural variability for insect resistance existing within the populations of crop varieties was exploited, it became increasingly important to introduce genes for resistance from various sources into the elite germplasm through hybridization. Generally, a susceptible elite variety or a breeding line is crossed with an insect-resistant variety. In cases where sources of resistance do not exist within the cultivated germplasm, crosses are made with wild or weed species or even species of related genera. Hybrids between resistant and susceptible varieties are allowed to self-pollinate and segregating populations are handled either by the pedigree method, bulk method, single seed descent or backcross method.

Pedigree method

The pedigree method is used most widely with self-pollinated crops in modern-day plant-breeding programs. Its name comes from the records that are kept of the ancestry or pedigree of each of the progenies. The primary selection criterion is resistance to the insect in question, but traits such as vigor, maturity, quality, and resistance to diseases are also considered. In the F_2 generation, selection is limited to individuals. In the F_3 and subsequent generations until homozygosity is reached, selection is made both within and between families. Thereafter, selection is among families until the number of progenies has been reduced to the point where a comprehensive yield evaluation trial can be undertaken.

Parents for the crossing program are carefully selected. Generally, one of the parents is a widely adapted and locally cultivated variety and the other is usually chosen for its resistance to the insect in question. Enough F_1 plants are grown to produce sufficient seeds for an F_2 population of the desired size. The size of the F_2 population depends upon the genetic distance of the parents. If the parents are closely related, the size of the F_2 population may be as small as 1000–1500 plants. If the parents differ widely, it is necessary to grow up to 5000 F_2 plants per cross combination. Selection in the F_2 generation is generally based on the morphological characteristics. Skill and judgment of the breeder acquired through intimate knowledge of many years of experience with the crop are essential for selecting suitable plants from the segregating populations.

Part of the seed of the selected plants is used for growing the F_3 progeny rows. A sufficient number of plants (30–50) of each F_3 family are grown so as to get an idea of the general features of the family and also of its heterozygosity. The remaining seeds of the F_2 plants are used to test for resistance to the insect and for other tests such as quality or resistance to diseases. Selection of the F_3 families is made on the basis of resistance to the insect, agronomic worth, as well as resistance to diseases. Generally, three to six plants are harvested from each of the promising selected F_3 families. The F_4 progenies from the plants of the same F_3 family are grown next to each other for purposes of comparison. The F_4 stage is usually handled in much the same way as the F_3, but with a further shift in emphasis toward selection among superior families. By the F_4 generation, genetic differences among families are much greater than differences within families. Therefore, in this generation, the opportunity exists to reduce the number of families. This is done on the basis of visual observation as well as on the basis of pedigree records pertaining to insect and disease reaction and quality characteristics. By the F_5 generation, the potentialities of individual families usually become fixed and the selection emphasis shifts almost entirely to among families. In the F_6 or F_7 generations, superior progenies are bulk harvested and planted in multirow plots for evaluation of yield and other agronomic characteristics. Selected progenies in each generation are tested for insect resistance. Superior progenies are evaluated in multilocation trials and some are released as varieties (Khush 1977).

Numerous insect-resistant varieties of various crops have been developed through the pedigree method of breeding. Rice varieties for resistance to the green leafhopper *Nephotettix virescens*, three biotypes of the brown plant-hopper *Nilaparvata lugens*, and the stem borer *Scirpophaga incertulas* have been developed at the International Rice Research Institute (IRRI), and some of them are also resistant to the gall midge *Orseolia orzyae* (Table 11.2). These

Table 11.2 Disease and insect reactions of IR varieties of rice.

Variety	Blast	Bacterial blight	Tungro	Grassy stunt	Green leafhopper	Brown plant-hopper	Stem borer	Gall midge
IR5	MR	S	S	S	R	S	MR	S
IR8	S	S	S	S	R	S	S	S
IR20	MR	R	S	S	R	S	MR	S
IR22	S	R	S	S	S	S	S	S
IR24	S	S	S	S	R	S	S	S
IR26	MR	R	MR	S	R	S	MR	S
IR28	R	R	R	R	R	R	MR	R
IR32	MR	R	R	R	R	R	MR	R
IR36	MR	R	R	R	R	R	MR	R
IR38	MR	R	R	R	R	R	MR	R
IR42	MR	R	R	R	R	R	MR	R
IR46	MR	R	R	R	R	R	MR	R
IR50	S	R	R	R	R	R	S	–
IR54	MR	R	R	R	R	R	MR	–
IR58	MR	R	R	R	R	R	S	–
IR60	MR	R	R	R	R	R	MR	–
IR62	MR	R	R	R	R	R	MR	–
IR64	MR	R	R	R	R	R	MR	–
IR66	MR	R	R	R	R	R	MR	–
IR68	MR	R	R	R	R	R	MR	–
IR72	MR	R	R	R	R	R	MR	–

MR, moderately resistant; R, resistant; S, susceptible; – not known.

varieties are planted on millions of hectares of riceland over the world (Khush 1989).

Aids to selection

Following the identification of parents for crossing, selection of individuals in the segregating populations is the most important step in a breeding program. Accurate judgment of the worth of the plants selected in segregating generations requires a detailed knowledge of the crop (Allard 1960). In addition, in a host plant resistance breeding program, the reaction of the plants to the target insect must be known. For effective selection for insect resistance, segregating populations are grown in areas where the natural populations of the insect are high during seasons when the resistant individuals can be easily identified. For example, in endemic areas of the rice gall midge in India and Sri Lanka, segregating populations are planted in August. Peak populations of the insect occur during October when the plants are at maximum tillering and their most vulnerable stage. Resistant plants can be easily identified as they lack the silver shoots that are typical symptoms of gall midge damage.

Most often, however, the breeding materials cannot be screened for insect resistance under field conditions because of sporadic insect outbreaks. The most common approach, therefore, is to evaluate the breeding materials for insect resistance in either a greenhouse or the laboratory. Various screening techniques for insect resistance are discussed in Chapter 8. Screening in the laboratory or greenhouse generally starts in the F_3 generation, although F_2 populations can sometimes be screened. A part of the seed of selected plants is used for growing the next generation and the other part is used to screen for insect resistance under artificial conditions. The screening technique must permit the evaluation of large numbers of progenies and the data on insect resistance must be available before the plant selections are made in the field, usually within 3 months for most crops.

In the rice-breeding program at the IRRI, segregating populations are evaluated for resistance to the green leafhopper and brown planthopper using the seed box techniques (Athwal *et al* 1971). The test is over in two weeks and approximately 60,000 breeding lines from F_3 to F_8 are evaluated every year. Sometimes, the plants must be grown to maturity before they can be evaluated for resistance. Resistance to stem borers is evaluated after the flowering stages of rice. Under such situations, the number of plants that can be evaluated is limited and progress in developing insect-resistant varieties is slower. Selection efficiency may be increased if the linked markers can be employed as aids to selection.

Marker-based selection for insect resistance. Sometimes the gene for insect resistance is tightly linked with a plant morphological trait so that the resistance and the morphological trait segregate together in the breeding populations. Plants selected on the basis of a morphological trait are resistant to the insect. However, such morphological markers are rare and they generally have deleterious effects on plant performance. Moreover, a vast majority of them are dominant. However, the molecular markers (isozymes and restriction fragment length polymorphisms, RFLPs) are very useful for tagging genes of economic importance. They differ from morphological markers in several respects (Tanksley 1983).

1 The genotypes of molecular loci can be determined at the whole plant, tissue, and cellular levels. The phenotypes of most morphological markers can only be distinguished at the whole plant level.

2 A relatively large number of naturally occurring alleles are found at molecular loci. Distinguishable alleles at morphological marker loci occur less frequently.

3 Usually no deleterious effects are associated with alternate alleles of molecular markers. Morphological markers, on the other hand, are often accompanied by undesirable phenotypic effects.

4 Alleles of most molecular markers are codominant, allowing all possible genotypes to be distinguished in any segregating generation. Alleles at morphological marker loci usually interact in a dominant-recessive manner, prohibiting their use in many crosses.

5 With morphological marker loci, strong epistatic effects limit the number of segregating markers that can be unequivocally scored in the same segregating generation. With molecular markers, very few epistatic or pleiotropic effects are observed.

If the genes for insect resistance can be tagged by tight linkage with molecular markers, selection efficiency can be increased and the time and money in moving these genes from one varietal background to another can be greatly reduced. The presence or absence of the associated molecular marker would indicate, at a very early stage, the presence or absence of the desired gene. Codominance of the associated molecular marker allows all possible genotypes to be identified in any breeding scheme even if the economic gene cannot be scored directly. The feasibility of this approach was demonstrated in tomatoes. The gene *Mi* for nematode resistance is closely linked with a rare allele of the *Aps-1* (acid phosphatase) isozyme locus on chromosome 6 (Rick and Fobes 1974). Many tomato-breeding programs screen their segregating progenies for the presence of the *Aps-1* allele instead of the cumbersome evaluation for nematode resistance.

A tight linkage (less than 5 cM) is necessary for tagging a gene with a single molecular marker. For a gene-tagging approach to be successful, molecular markers placed at intervals of less than 5 cM throughout the genome are required. Isozyme markers are not numerous enough in any crop to mark the whole genome. However, RFLP markers are numerous and

Table 11.3. Some examples of tagging insect resistance genes with molecular markers.

Crop	Resistance to	Gene for resistance	Linked molecular marker	Reference
Rice	Whitebacked planthopper	Wbph-1	RG146	McCouch et al (1991)
	Brown planthopper	Bph-10(t)	RG457	Ishii et al (1994)
	Gall midge	Gm-2	RG476, RG776, RG224	Mohan et al (1993)
Mungbean	Bruchid beetle	–	pA 882	Young et al (1992)
Wheat	Hessian fly	H-23 H-24	X Ksu H4, X Ksu G48(A) Xcnl BCD 457, Xcnl CDO 482, X Ksu G48(B)	Ma et al (1993)

–, not known.

saturated maps can be prepared. In maize (Helentjaris 1987), tomatoes (Bernatzky and Tanksley 1986), and rice (Tanksley *et al* 1992), more than 1000 RFLP markers in each have been identified and mapped to cover entire genomes. Genes for insect resistance have been tagged with RFLP markers in several crops (Table 11.3) and marker-based selection is underway.

In a number of cases, resistance to insects is quantitative and is governed by polygenes (Khush and Brar 1991). It is important to establish linkage between the molecular markers and the polygenes (quantitative trait loci or QTL) for insect resistance to increase the selection efficiency for incorporating polygenic resistance to insects. Nienhuis *et al* (1987) have tagged QTL for insect resistance with RFLP markers in the wild tomato species *Lycopersicon hirsutum*.

Bulk method

The bulk method differs from the pedigree method in that the hybrid populations are grown in bulk with no attempt to keep track of the ancestry of individuals. A representative sample of the previous bulk harvest is used for raising the next bulk generation. Through natural selection, the poor competitors are eliminated and shifts in gene frequencies occur (Frey 1975). When artificial selection is practiced – and this is desirable when breeding for insect resistance – it is referred to as a modified bulk method. Artificial selection is carried out during the period of bulk propagation based on individual plant performance (not on the progeny tests of the selected individuals). In breeding for insect resistance, the bulk or modified bulk method may be applied only when the segregating population can be grown in locations where natural populations of the target insect are high, as in the case of the rice gall midge mentioned earlier in this chapter. In most cases, however, the natural populations of insects are not dependable for selection pressure and the bulk method is rarely used in breeding for insect resistance.

Single seed descent

When insect resistance is polygenic, selection for resistance in early generations is complicated by differences in maturity and a small magnitude of differences in resistance. To overcome these difficulties, breeding materials are advanced to the F_6 or F_7 generations without artificial selection. To ensure that natural selection does not eliminate certain genotypes, single seeds from each of the F_2 plants are harvested and bulked together to grow the F_3 bulk population. The process is repeated until sufficient homozygosity is reached (Brim 1966, Knott and Kumar 1975). At this stage, plant selections that give rise to individual families are made. The families are evaluated for insect resistance and the better ones identified.

A modification of single seed descent is the rapid generation advance method where plants are grown under crowded conditions in the greenhouse in small pots and at a high temperature. The plants grow faster, produce only small ears or pods, and thus few seeds. With proper manipulation of photoperiod and temperature, it is possible to grow three to four generations a year instead of one or two (Ikehashi and Fujimaki 1980). Thus, homozygosity is reached more quickly for final evaluation.

Backcross method

This method is particularly suited for transferring specific genes for resistance to a susceptible variety that is otherwise high yielding and well-adapted. The variety to be improved is crossed with the donor for resistance, and a series of backcrosses using the variety as a recurrent parent are made. Each of the backcross F_1 (BCF_1) progenies is evaluated for resistance if the gene for resistance is dominant. Only the resistant BCF_1 plants are used for making the next backcross. If the gene for resistance is recessive, the BCF_1 plants are progeny tested and only those carrying the gene are used for the next backcross. At the end of the backcrossing program, the gene being transferred is in a heterozygous condition. After the last backcross, the backcross progenies are selfed and homozygous individuals are selected. They constitute a variety with exactly the same adaptation, yielding ability, and quality characteristics as the recurrent parent, but are superior to that parent for resistance to the target insect. In contrast to the pedigree and bulk methods of breeding, this method provides the plant breeder with a high degree of genetic control of the populations (Briggs and Allard 1953). Extensive yield-evaluation trials are not required for varieties developed after seven or more backcrosses to a recurrent parent.

The backcross method is profitably employed when the hybrid populations can be readily evaluated for resistance to the insect in question by simple screening tests. The essential feature of this breeding method is the availability of good recurrent parents. If a recurrent parent of superior value is not available, the pedigree method is preferred as improvements in several traits can be affected concurrently using this method.

Recurrent selection

The recurrent method of breeding is more suitable for accumulating poly-genes for resistance to an insect from different varietal backgrounds. It is primarily used to promote recombination and to increase the frequencies of favorable genes for quantitatively inherited traits. It is cyclic, with each cycle encompassing two phases of plant breeding: (1) selection of a group of genotypes that possess polygenes for resistance; and (2) mating among the selected genotypes to obtain genetic recombination. This is the principal breeding method used in cross-pollinated crops. Several parents (up to 50 or even 100) with a better than average level of resistance are allowed to intercross to produce a heterozygous population that is exposed to natural or artificial insect pressure. Susceptible plants are discarded and the resistant plants are allowed to interbreed or are artificially crossed with each other. Progenies from intercrosses are grown and again exposed to insect pressure. This cycle of selection and intermating is repeated four to five times and results in an accumulation of polygenes for resistance from various parents. Widstrom *et al* (1992) developed a maize population with an excellent level of resistance to the fall armyworm *Spodoptera frugiperda* by accumulating polygenes from 50 collections through recurrent selection. Progress in raising the level of resistance to the fall armyworm through five cycles of selection is shown in Fig. 11.2.

The use of recurrent selection in self-pollinated crops has been limited by technical problems of intermating amongst selected individuals. However, the introduction of a male sterility gene can enhance outcrossing amongst the individuals of a composite population. The method is referred to as male-sterile facilitated recurrent selection (MSFRS) (Gilmore 1964, Brim and Stuber 1973). Several donors with low to moderate levels of resistance are crossed with a male-sterile line of desirable genotype. The F_1 plants are grown and roughly equal amounts of the F_2 seeds are mixed to make a composite. The F_2 composite is grown in a place where the target insect population is high. If there is a sufficient insect population, more susceptible plants are damaged and therefore contribute less to the next generation. Seed set on the male-

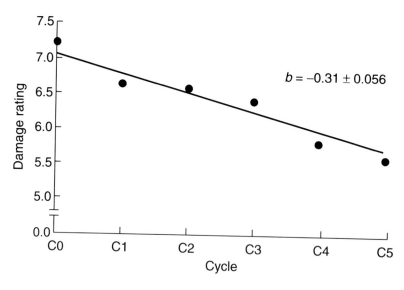

Fig. 11.2. Regression (*b*) of leaf-feeding damage by larvae of the fall armyworm during a recurrent selection cycle, showing progress through five cycles of selection for reduced damage to the maize population. Reproduced from Widstrom *et al* (1992) with permission from the American Society of Agronomy, Madison, USA.

sterile plants is from outcrosses with more resistant plants. Outcrossed seeds from the male-sterile plants are bulked to grow the next composite population. The cycle is repeated four to five times, or longer if necessary. The continuous outcrossing amongst more resistant plants and the elimination of more susceptible ones at each generation leads to an accumulation of polygenes, with additive gene action resulting in a population with a higher frequency of such genes. At the end of the recurrent selection program, individual plant selections are made and the resulting families are evaluated for insect resistance under controlled conditions and for other agronomic and quality characteristics. MSFRS was employed in accumulating polygenes for stem borer resistance in rice from 26 landraces (Chaudhary and Khush 1990).

Wide hybridization

Plant breeders prefer to utilize genetic variability of cultivated germplasm in developing insect-resistant varieties. Sometimes, however, the genetic variability for insect resistance within the cultivated germplasm is either inadequate or even lacking. Wild species generally are more resistant to diseases and insects. However, several barriers are encountered in transferring genes from wild species to cultivated germplasm (Khush and Brar 1992). The most important barrier is the abortion of interspecific hybrid embryos. The embryo rescue technique (Fig. 11.3) is now routinely used to overcome this barrier. Two-week-old embryos are aseptically excised from the developing seed, transferred to a culture medium and incubated in the dark. Embryos start to

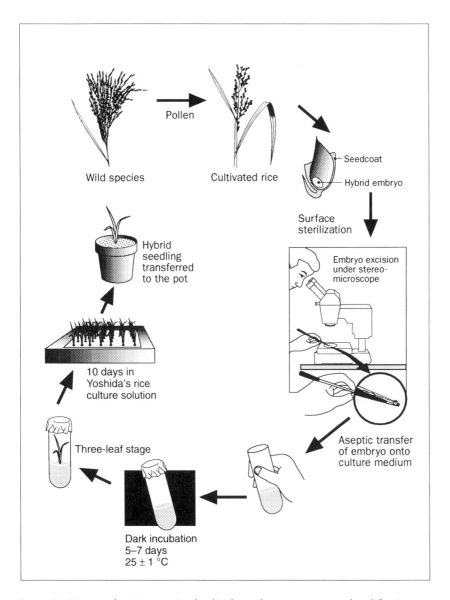

Fig. 11.3. Diagram showing steps involved in the embryo rescue procedure following interspecific hybridization.

Fig. 11.4. Reaction to the brown planthopper by parents and progenies derived from the cross of *O. sativa* (IR31917-45-3-2) and *O. australiensis*. *O. australiensis* and the wide-cross progenies WC1, WC2, WC5, and WC6 are resistant. IR31917-45-3-2, WC3, and WC4 are susceptible. Rathu Heenati and TN1 are resistant and susceptible checks, respectively. Reproduced from Multani *et al* (1994) with permission from Springer-Verlag, New York.

germinate in about 5 days. Plantlets at the three-leaf stage are transferred to a water–culture solution and then to soil in pots (Jena and Khush 1984). The interspecific hybrids are generally sterile and several backcrosses with the cultivated variety or the breeding line as the recurrent parent are made to restore fertility and to eliminate the undesirable features of the wild parents. Genes for resistance to the brown planthopper and whitebacked planthopper were transferred from the wild rice species *Oryza officinalis* (Jena and Khush 1990) to cultivated rice *O. sativa*. The F_1 plants were obtained through the embryo rescue technique and two backcrosses were made to an elite breeding line of cultivated rice to restore fertility. Two brown planthopper-resistant lines originating from the crosses of *O. sativa* and *O. officinalis* were released as varieties in Vietnam (IRRI 1994). Similarly, Multani *et al* (1994) transferred genes for resistance to the brown planthopper from *O. australiensis* to *O. sativa*; the derived lines showed excellent resistance (Fig. 11.4). Four genes (*H-13*, *H-22*, *H-23*, and *H-24*) for resistance to the Hessian fly from *Aegilops squarrosa* (*Triticum tauschii*) have been transferred to bread wheat *Triticum aestivum* (Hatchett *et al* 1981, Gill *et al* 1987, Raupp *et al* 1993). Several other genes for insect resistance have been transferred from wild to cultivated species (Table 11.4).

Table 11.4. Genes for insect resistance transferred from wild species into cultivated crop plants.

Recipient	Alien donor	Resistance to	Reference
Bread wheat (*Triticum aestivum*)	*Secale cereale*	Greenbug	Sebesta and Wood (1978)
	Secale cereale	Hessian fly	Friebe *et al* (1990)
	Aegilops squarrosa	Hessian fly	Hatchett *et al* (1981), Gill *et al* (1987), Raupp *et al* (1993)
Rice (*Oryza sativa*)	*Oryza officinalis*	BPH	Jena and Khush (1990)
	Oryza officinalis	WBPH	Jena and Khush (1990)
	Oryza australiensis	BPH	Multani *et al* (1994)
	Oryza minuta	BPH	D.S. Brar *et al* (unpublished)
Peanut (*Arachis hypogaea*)	*Arachis monticola*	Chewing insects	Hammons (1970)
Lettuce (*Lactuca sativa*)	*Lactuca virosa*	Aphids	Eenink *et al* (1982)
Cotton (*Gossypium hirsutum*)	*Gossypium armourianum*	Boll weevil, leaf worm, bollworm	Meyer (1974)

BPH, brown planthopper; WBPH, whitebacked planthopper.

Mutation breeding

If the donors for insect resistance do not exist within the cultivated or wild germplasm of a crop, mutation breeding may be employed to induce mutants by the use of known mutagenic agents such as X-rays, fast neutrons, gamma-rays, or chemical mutagens. In recent years, MNU (methyl nitrosourea) treatment of fertilized eggs has yielded promising results (Satoh and Omura 1986). Seeds of a high-yielding and adapted but susceptible variety are mutagenized. Sometimes the pollen is irradiated and used to pollinate untreated plants of the same variety. The resulting R_1 or R_2 progenies are evaluated for insect resistance. Selected resistant plants form the starting point of a new variety. This method of breeding is useful in inducing change at a single locus in a highly developed genetic background without disrupting a superior combination of genes. Mutation breeding has often been practiced for improving agronomic traits or disease resistance, but rarely for insect resistance. Mutants for resistance to the brown planthopper were induced by gamma-ray treatment in the background rice variety Pelita 1/1 and released as Atomita 1 and Atomita 2 in Indonesia (Mugiono *et al* 1984).

Hybrid breeding

In hybrid breeding, F_1 hybrids are developed for commercial production. This method exploits the phenomenon of heterosis and involves the use of genetically diverse inbred parental lines. Inbred lines possessing some level of insect resistance are developed by deploying any of the above described breeding methods. These lines are then used as parents for developing the hybrids. If the gene for resistance is dominant, the hybrid variety is resistant. Thus the gene for resistance can be deployed in one step. Several rice hybrids resistant to the brown planthopper and green leafhopper have been developed in China. Although the male-sterile female parents were susceptible, the restorer parents developed at the IRRI (IR26, IR54, and IR9761-19-1) had dominant genes for resistance and the F_1 hybrids were resistant (Lin and Yuan 1980, Yuan *et al* 1985).

Another advantage of hybrids is that two dominant genes for resistance can be deployed simultaneously. One gene is incorporated into the female parent and the other into the male parent. The F_1 hybrid then has a higher level of resistance due to the combined effect of two dominant genes. In this manner, resistance to more than one biotype of the insect can also be built up in the hybrid. Resistance to two insects can also be combined if the female parent is resistant to one insect and the male parent to another. Hybrid breeding also allows a combination of genes for horizontal resistance from one parent and vertical from the other.

Somaclonal variation for insect resistance

Somaclonal variation refers to variation observed in tissue culture-derived progenies. Larkin and Scowcroft (1981) reviewed the occurrence of somaclonal variation in various crop species. Variation was noted for a series of agronomic traits, disease resistance, and biochemical traits. Somaclonal variation may be exploited to select variants for insect resistance. The main steps involved in the isolation of somaclones for insect resistance are: (1) growing calli or cell suspension cultures for several cycles from a high-yielding and well-adapted but susceptible variety, (2) regeneration of plants from such long-term cell lines, and (3) evaluation of a large population of regenerated plants for insect resistance. White and Irvine (1987) produced about 2000 plantlets of sugarcane from calli of a cultivar susceptible to the sugarcane borer *Diatraea saccharalis* (F.) These somaclones were evaluated for resistance to the sugarcane borer under artificial infestation as well as under natural infestation in field plots. Some somaclones showed increased levels of borer resistance. Isenhour *et al* (1991) obtained somaclonal variants of

Fig. 11.5. A somaclonal line of (a) sorghum and (b) the parent variety Hegari. Both entries were grown under natural conditions of fall armyworm infestation in the field. The somaclonal line suffered no damage but the parent variety was badly damaged. Courtesy of M. Nabors, Colorado State University, Fort Collins, USA.

sorghum with a good level of resistance to the fall armyworm compared with the susceptible parent Hegari (Fig. 11.5).

Genetic engineering for insect resistance

One of the most significant breakthroughs in plant science in recent years is the development of techniques to transfer genes from unrelated sources into crop plants. Until recently, plant breeders could manipulate only the primary and secondary gene pools of the cultivated species for crop improvement. However, recent advances in tissue culture and molecular biology have made it possible to introduce genes from diverse sources such as bacteria, viruses, animals, and unrelated plants into crop plants. These novel genes either reinforce the existing functions or add new traits to the 'transformed' or 'transgenic' plants. These developments provide the opportunity to develop transgenic crops with novel genes for insect resistance. In contrast with plant-breeding methods involving sexual hybridization where large segments of the genome are transferred from one parent to the other, genetic engineering techniques permit the introduction of single genes into otherwise highly developed crop varieties. Since the first report of transgenic tobacco plants expressing foreign proteins (Herrera-Estrella *et al* 1984, Horsch *et al* 1985), transgenic plants have been produced in more than 45 species (Uchimiya *et al* 1989, Hiatt 1993). In most of these cases, marker genes of bacterial origin have been introduced. During the last few years, genes of economic

Table 11.5. Transgenic plants with foreign genes for insect resistance.

Transgenic plants	Foreign gene	Resistance to	Reference
Tobacco	*Bt*	Tobacco hornworm	Vaeck *et al* (1987)
		Tobacco hornworm	Barton *et al* (1987)
		Tobacco hornworm	Warren *et al* (1992)
Tomato	*Bt*	Tobacco hornworm	Fischhoff *et al* (1987)
		Tobacco hornworm	Delannay *et al* (1989)
		Tomato fruitworm	Delannay *et al* (1989)
		Tomato pinworm	Delannay *et al* (1989)
Potato	*Bt*	Potato tuber moth	Peferoen *et al* (1990)
Cotton	*Bt*	Cotton bollworm	Perlak *et al* (1990)
		Cabbage looper	Perlak *et al* (1990)
		Beet armyworm	Perlak *et al* (1990)
Populus	*Bt*	Forest tent caterpillar	McCown *et al* (1991)
		Gypsy moth	McCown *et al* (1991)
Maize	*Bt*	European corn borer	Koziel *et al* (1993)
Rice	*Bt*	Striped stem borer	Fujimoto *et al* (1993)
		Leaffolder	Fujimoto *et al* (1993)
Whitespruce	*Bt*	Spruce budworm	Ellis *et al* (1993)
Tobacco	*CpTi*	Tobacco budworm	Hilder *et al* (1987)
		Tobacco budworm	Barfoot and Connett (1989)
		Tobacco hornworm	Gatehouse *et al* (1992)
Tobacco	α-ai	Mealworm	Altabella and Chrispeels (1990)
Tobacco	*P-lec*	Tobacco budworm	Boulter *et al* (1990)
Potato	*P-lec*	Tomato aphid	Gatehouse *et al* (1995)
Pea	α*A1-Pv*	Bruchid beetle	Shade *et al* (1994)

importance such as those for insect resistance have also been introduced
(Table 11.5).

Techniques of genetic transformation

Genetic transformation refers to the introduction of cloned DNA segments or
genes from plants, bacteria, or animals into a new genetic background. The
foreign gene must be integrated stably into the genome of the recipient species
and must express itself in the new genetic background. The transgenic plants
must be fertile and otherwise normal in all respects.

Foreign genes can be transferred through vector-mediated or direct DNA
transfer methods using protoplasts or other tissues. Transformation techniques are dependent upon precise tissue culture methods.

Agrobacterium-mediated DNA transformation

Agrobacterium tumefaciens is a Gram-negative soil bacterium that causes
crown gall disease in many gymnosperms and dicotyledonous angiosperms.
It enters at wound sites and elicits the formation of tumorous calli. These
undifferentiated masses of plant cells can be cultured *in vitro* in the absence
of the bacterium. Unlike calli produced from normal plant cells, crown gall
calli grow indefinitely in the absence of auxins, without losing their tumorous
properties.

Tumor induction is due to the transfer from the bacterium to the plant
of a 23-kilobase (kb) piece of DNA known as t-DNA or transfer DNA (Fig.
11.6). Inside the bacterium, t-DNA forms part of a large plasmid, the Ti or
tumor-inducing plasmid, which is usually greater than 200 kb in size. The
t-DNA segment is located between two 24-kb segments that are almost perfect
copies of one another. They are known as the left and right border repeats.
Although they are not part of the t-DNA, these repeats are important for the
transfer process. Also important are a set of virulence or *vir* genes that are
located on the Ti plasmid but outside the t-DNA segment. The t-DNA carries
genes that are required for the transformed, tumorous, hormone-independent
state of the crown gall callus cells.

The end of the t-DNA closer to the right border repeat enters the plant cell
and integrates into the nuclear genome first. It is believed that the t-DNA is
carried into the nucleus by two proteins that are products of two of the *vir*
genes of the Ti plasmid. t-DNA is integrated essentially at random into the
nuclear genome. *A. tumefaciens* and *A. rhizogenes* (the causal agent of hairy
root syndrome) are the only bacteria known to be able to transfer DNA to
plant cells.

This naturally occurring transformation system has been modified by
scientists to enable almost any gene to be transferred into plant cells. Because
of its large size, the Ti plasmid is unsuitable for genetic manipulation. This
problem has been solved by the creation of a small plasmid that contains the

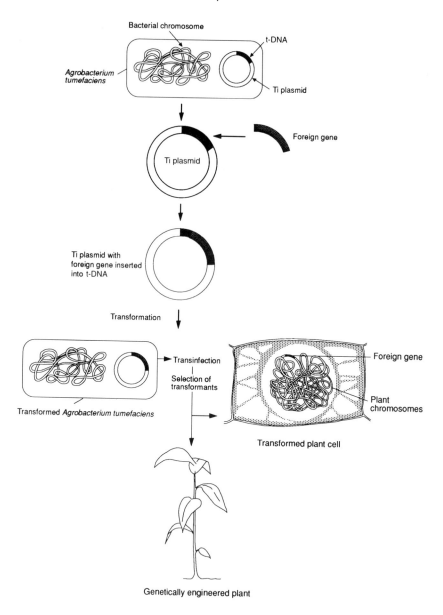

Fig. 11.6. Diagram showing the steps involved in the genetic engineering of plants through the *Agrobacterium* technique.

gene to be transferred flanked by the left and right border repeats of the Ti plasmid. Such a plasmid lacks both the tumor-inducing genes of t-DNA and the *vir* genes of the Ti plasmid. The former are undesirable in the genetic engineering of plants; but the latter are essential. This type of small plasmid is therefore incapable of transformation by itself and must be assisted by a 'helper' plasmid that contains the *vir* genes. The typical helper plasmid consists of a mutated ('disarmed') Ti plasmid that retains its *vir* genes but lacks the tumor-inducing genes of t-DNA.

In addition to the gene of interest flanked by suitable controlling sequences (promoter, terminator) and the left and right border repeats of the Ti plasmid, the small plasmid contains the following features.

1 An *Escherichia coli* origin of replication.
2 An *A. tumefaciens* origin of replication.
3 A selectable marker gene for antibiotic selection in *E. coli* and *A. tumefaciens* containing the plasmid.
4 A selectable marker gene for antibiotic selection in plant cells.

The *E. coli* origin of replication permits construction and manipulation of the plasmid in *E. coli* cells. The *A. tumefaciens* origin of replication permits the plasmid to be maintained in *A. tumefaciens* alongside the disarmed Ti plasmid. The selectable marker gene for use in plants is usually neomycin phosphotransferase, providing resistance to the antibiotic kanamycin.

Horsch *et al* (1985) developed a simple and general procedure for transferring genes into plants with *Agrobacterium* Ti plasmid vectors. Small leaf disks of about 3-mm diameter are punched from a leaf of a suitable plant (petunia, tomato, tobacco, potato), surface sterilized, and inoculated in a medium containing *A. tumefaciens*. The bacterium carries the disarmed Ti plasmid in the form of a cointegrate or binary vector. The discs are cultured for two days and transferred to a medium containing the antibiotic kanamycin. Kanamycin kills all the non-transformed cells, but the cell containing the neomycin phosphotransferase gene continues to multiply. Usually such cells also contain the gene of interest next to the 'neo' gene. Carbenicillin is also present in the medium to kill the *Agrobacterium*. The transformed cells are transferred to the regeneration medium and after some weeks regenerated plants are transferred to pots for growth, analysis, and seed production. In transgenic plants, the foreign gene is integrated into their genome and is stably inherited to the progeny.

Vector-mediated transformation is possible primarily with dicot species. *A. tumefaciens* and *A. rhizogenes* generally do not infect monocots. In the absence of efficient vectors, several direct DNA transfer methods for transformation of monocots, particularly cereals have been developed, and some are described below.

Polyethylene glycol-mediated transformation

Polyethylene glycol (PEG) has been used as a fusion agent in somatic cell hybridization of plant and animal species. It is equally effective in delivering DNA into naked cells or protoplasts. Protoplasts are co-cultured with the foreign gene in the form of plasmid DNA with a selectable marker in the presence of PEG. PEG facilitates the entry of plasmid DNA into the protoplasts where it is eventually incorporated into the nucleus. Transformed cells are selected in the presence of kanamycin and are allowed to produce calli from which transformed plants are regenerated (Paszkowski *et al* 1984).

Electroporation

Electroporation involves the application of high voltage electrical pulses to solutions containing protoplasts and the foreign DNA in the form of plasmids. The DNA enters the protoplasts through reversible pores created in the protoplast membrane by the action of short electrical pulse treatments. Electroporation has been used to produce transgenic rice (Toriyama *et al* 1988), transgenic maize (Rhodes *et al* 1988), and other transgenic cereals.

Biolistic method of transformation

This method consists of delivering DNA into cells of intact plant organs, cultured tissues, or cell suspensions via projectile bombardment. High-density particles (gold or tungsten) coated with plasmid DNA containing the foreign gene are accelerated to high velocity by a particle gun apparatus. These dense particles thus acquire sufficient kinetic energy to penetrate the plant cells through the cell walls and membranes and thus to deliver the DNA into the cells. This method does not require protoplast culture and is thus genotype nonspecific (Christou *et al* 1991).

Microinjection

In this method foreign DNA is delivered into protoplasts using a glass micropipette having an orifice diameter of less than 1 μm. It is a useful technique for the precise delivery of DNA into target cells. In this method, the location of DNA delivery can be controlled with a micromanipulator and the volume of injection can be controlled by the microinjection apparatus. Microinjection of DNA into individual cells using commercially available micromanipulators fixed to the movable stage of an inverted microscope has been used in the transformation of tobacco protoplasts (Crossway *et al* 1986) and microspore-derived embroids of *Brassica* (Neuhaus *et al* 1987).

Novel genes for insect resistance

Several sources of novel genes for insect resistance have been identified and new ones are being discovered. The most well known are the *Bt* genes from *Bacillus thuringiensis* and proteinase inhibitor genes.

Bt genes for insect resistance

Bacillus thuringiensis is a Gram-positive, spore-forming bacterium found commonly in the environment. It produces a number of insect toxins, the most distinctive of which are protein crystals formed during sporulation. The crystal is responsible for the insecticidal activity of the *B. thuringiensis* strains and is composed of one or more insect-control proteins (*Bt* proteins). The *Bt* proteins are active against lepidopteran insects and act by disrupting the midgut cells of the insect. The crystal proteins are dissolved by the alkaline gut juices in the midgut lumen, and are converted by gut proteases into toxic core fragments. Thereafter, the midgut epithelial cells swell and eventually burst, resulting in the death of the insect. Microbial preparations of *Bt* spores and crystals have been used as commercial bioinsecticides for over 20 years.

Genes coding for the crystal proteins have been isolated from the *Bt* strains and introduced into crop plants through transformation techniques (Barton *et al* 1987, Fischhoff *et al* 1987, Vaeck *et al* 1987). However, before introduction in the plants the *Bt* genes are codon corrected or modified for an adequate level of expression in the alien genetic background. Fischhoff *et al* (1987) constructed a chimeric gene consisting of a lepidopteran-specific toxin gene from a *Bt* strain flanked by CaMV35S promoter and the 3'-end of the nopaline synthase gene. The gene was introduced in tomato variety VF36 and the transgenic plants expressing the *Bt* gene were highly resistant to the larvae of the tobacco hornworm *Manduca sexta*. The plants showed little feeding damage and larvae on the transgenic plants died within a few days. Similarly, transgenic potato plants having the *CryIA(b)* gene proved highly resistant to feeding and tunneling damage by the potato tuber moth larvae (Peferoen *et al* 1990).

Perlak *et al* (1990) introduced the *CryIA(b)* and *CryIA(c)* genes into cotton plants. The transformed cotton plants showed a high level of resistance to the cotton bollworm and were thus protected against damage to the squares and bolls. The *Bt* gene was inherited to the progeny and the resistance segregated as a single dominant gene trait. The *Bt* gene has also been introduced into *Populus* and the transformed plants showed a good level of resistance to the forest tent caterpillar as measured by the leaf area consumed by the larvae (McCown *et al* 1991).

Once the nucleotide sequence of a gene is known, it can be synthesized in the laboratory. Koziel *et al* (1993) transformed maize with a synthetic *Bt* gene. The transgenic plants showed a high level of resistance to the European corn borer *Ostrinia nubilalis*, a major pest of maize in North America and Europe.

The first field trial of *Bt*-transformed tobacco plants was carried out in 1986 in North Carolina, USA. Delannay *et al* (1989) evaluated transgenic tomato expressing the *Bt* gene under field conditions. Transgenic plants showed very limited feeding damage after infestation with the tobacco hornworm, whereas the control plants were almost completely defoliated within 2 weeks. Control of both the tomato fruitworm *Helicoverpa zea* and tomato pinworm *Keiferia lycopersicella* was also significant.

About 45 nucleotide sequences of crystal protein genes have been determined. Seventeen different *Bt* crystal protein genes [*CryIA(a)* to *CryIVD*] have been identified (Peferoen 1992). They are effective against different insect species belonging to Lepidoptera, Coleoptera, and Diptera. No *Bt* effective against Homoptera, which represent a major pest problem to crops, have been reported to date.

Proteinase-inhibitor genes for insect resistance

Proteinase inhibitors from plants are of particular interest because they are part of the natural defense system against insect attack. Ryan (1989) reviewed strategies for transformation using proteinase-inhibitor genes to improve plant defenses against herbivores. The storage tissues of most plants contain chemicals that limit predation by insects and other herbivores. Some are protein inhibitors of insect digestive enzymes and the responsible genes confer resistance to insects. The first gene of plant origin to be successfully transferred to another plant species enhancing its resistance to insects was the cowpea trypsin inhibitor (*CpTi*) gene isolated from cowpea (Hilder *et al* 1987). The cowpea trypsin inhibitors provide protection against a number of insects belonging to Lepidoptera, Orthoptera, and Coleoptera (Gatehouse *et al* 1992). Resistance to the bruchid beetle *Callosobruchus maculatus* is partially due to an elevated level of trypsin inhibitors. The trypsin inhibitors have been shown to reduce both the survival and development rates of a range of lepidopteran and coleopteran pests. Hilder *et al* (1987) introduced a serine protease-inhibitor gene of cowpea in tobacco. Transgenic tobacco plants expressing a high level of *CpTi* showed a significant decrease in insect damage (Fig. 11.7).

Johnson *et al* (1989) analyzed the expression of tomato proteinase inhibitor I, tomato proteinase inhibitor II, and potato proteinase inhibitor II in transgenic tobacco. Growth of tobacco hornworms feeding on leaves of transgenic plants containing inhibitor II was significantly retarded compared with the growth of those feeding on leaves of the control plants. The cystein protease inhibitor, oryzacystatin, has been isolated from rice seed. Gut proteases of rice weevil and red flour beetle were strongly inhibited by oryzacystatin. Additional inhibitors from other plants – mungbean trypsin inhibitor, potato proteinase inhibitors I and II, arrowhead protease inhibitors, and towel gourd trypsin inhibitors – are being investigated for their usefulness against digestive enzymes of rice insect pests (Chi 1990). Recently McManus *et al* (1994) showed that the chemotrypsin inhibitor *PPi-II* was effective in protecting transgenic tobacco plants from attack by *Chrysodexis criosoma*.

Other genes for insect resistance

Lectins are carbohydrate-binding proteins found in many plant tissues and are abundant in seeds of some plant species. Some of the lectins have been

Fig. 11.7. Effect of *Heliothis virescens* larvae feeding on tobacco plants transformed with a *CpTi* gene on the right, and control plants on the left. Control plants were severely damaged but transformed plants suffered very little damage. Reproduced from Hilder *et al* (1987) with permission from Macmillan Journals, London.

shown to provide protection against predation by insects (Chrispeels and Raikhel 1991, Gatehouse *et al* 1994). The lectin-like protein, arcelin-1, found in wild accessions of the bean *Phaseolus vulgaris* has toxic effects on the insect pest *Zabrates subfasciatus* (Osborn *et al* 1988, Minney *et al* 1990). It is thought that the deleterious effect of chitin-binding lectins on insect development is mediated by binding to chitin in the peritrophic membrane of the midgut of the insect, thus interfering with the uptake of nutrients. Genes encoding pea lectin (*P-lec*) have been expressed at high levels in transgenic tobacco plants. The larval biomass of *Heliothis virescens* and leaf damage were both reduced on transgenic plants (Boulter *et al* 1990). Transgenic tobacco plants containing both *CpTi* and *P-lec* were obtained through hybridization of two primary transformed lines. Plants expressing two insecticidal proteins showed additive resistance to *Heliothis virescens*.

Breeding Strategies for Insect Resistance

An insect-resistant variety planted on a large scale is likely to become susceptible sooner or later because new insect biotypes can develop. Various strategies can be adopted to prolong the useful life of the resistant varieties or to develop varieties with different genes so that growers can have access to new varieties when the resistance of the current ones break down. Resistance may be governed by polygenes or major genes. Polygenic resistance is also referred to as horizontal resistance, and major gene resistance as vertical resistance (Van der Plank 1963).

Breeding for polygenic resistance

Polygenic resistance is a quantitative trait that is governed by a large number of genes, each with a small contribution to resistance. The level of resistance is generally not very high and it does not exert strong selection pressure on the insect. Thus, a virulent biotype rarely if ever develops and the resistance is more durable. Polygenic resistance is synonymous with tolerance. It does not involve a gene-for-gene relationship (Robinson 1980) and is equally effective against all biotypes. However, there are practical difficulties in incorporating this type of resistance into high-yielding, agronomically acceptable backgrounds. Parents with polygenic resistance are generally landraces with poor agronomic traits. In the process of selecting plants with better agronomic traits in crosses involving such parents, not all the polygenes or QTL are transferred and the level of resistance is diluted. Moreover, most of the screening techniques employed in screening segregating populations usually favor the selection of genotypes with a high level of resistance. Thus, it is important to employ screening techniques that identify segregants with low levels of resistance.

The choice of breeding methods is of crucial importance in breeding for horizontal resistance. Recurrent selection methods are obviously more suited. In self-pollinated crops, MSFRS can be employed in accumulating QTL for resistance from several parents. Alfalfa germplasm resistant to the spotted alfalfa aphid was developed through mass selection for polygenic variation by Hanson *et al* (1972). Marker-aided selection shows great promise in facilitating the breeding for horizontal resistance. Once the QTL for insect resistance are tagged with molecular markers, marker-based selection would facilitate the accumulation of polygenes from diverse parents. Nienhuis *et al* (1987) have tagged QTL for insect resistance with RFLP markers in wild species of tomato.

Breeding for major gene resistance

Major genes are easier to move from one varietal background to the other and generally convey a high level of resistance. Since there is a gene-for-gene relationship between genes for resistance in the host and genes for virulence in the insect, major genes convey resistance to certain biotypes but not to others. Moreover, a high level of resistance exerts selection pressure on the insect populations, leading to the development of biotypes. Several strategies can be employed to maintain varietal resistance and to prolong the useful life of major genes.

Sequential release of varieties with major genes

The sequential release of varieties with major genes involves the incorporation of a single major gene into commercial varieties. When new biotypes capable of damaging these varieties appear, varieties with different major genes are released. This strategy has been successfully followed for controlling the brown planthopper of rice in Asia. Widescale outbreaks of brown planthoppers occurred in 1973–1974 in several rice-growing countries. Rice varieties with the *Bph-1* gene for resistance such as IR26, IR28, and IR30 were released and adopted rapidly by farmers in several countries such as the Philippines, Indonesia, and Vietnam. By 1977–1978, however, a new biotype capable of attacking these varieties appeared. Varieties with *bph-2* such as IR36, IR38, and IR42 were released. These were widely grown for about ten years, but were rendered susceptible by the appearance of yet another biotype. Varieties with *Bph-3* were then released and are now widely grown. Meanwhile, several new genes for resistance to the brown planthopper have been identified and have been incorporated into commercially acceptable germplasm as insurance against future biotypes (Khush 1992). The sequential release strategy has also been followed for resistance to the Hessian fly in wheat (Gallun and Khush 1980).

Pyramiding the major genes

Pyramiding of major genes aims to combine two or more major genes into the same variety. Varieties with two or more major genes are likely to have a longer useful life as the development of new biotypes will be slower. Gould (1986) calculated that a simultaneous release of two genes for resistance to the Hessian fly of wheat would result in better durability than would a sequential release. However, in order to combine two genes one must be able to distinguish plants containing both genes from plants containing either one or the other. This needs more than one biotype to evaluate the materials, which is easily achieved. The most reliable method for producing cultivars possessing more than one resistance gene is to have the genes tagged by molecular markers. Individuals displaying both the markers would have both the genes for resistance. Pyramiding of two *Bt* genes through genetic engineering has also been proposed to prolong the useful life of transgenic resistance (Gould 1988). Combining genes encoding a toxin and a gene encoding a repellant may offer longer lasting resistance than either approach alone (Toenniessen 1991).

Rotation of varieties with major genes

It takes a number of years for the insect population to adapt to a resistant variety and to develop a new biotype. The process of adaptation can be interrupted by growing a resistant variety in one season and another resistant variety with a different gene during the next season. This strategy was followed to protect the rice crop from tungro virus epidemics in South Sulawesi, Indonesia. The tungro virus, transmitted by the green leafhopper, used to be endemic in that area. Varieties resistant to the leafhopper do not get infected with the virus. However, the insect pest adapts to the resistant varieties and can transmit the virus. A varietal rotation strategy was adopted in South Sulawesi in the 1970s. Varieties with one gene for green leafhopper resistance were planted in one season and another variety with a different gene for resistance was planted in the next season. This strategy has been very effective in prolonging the useful life of the vector-resistant varieties (Manwan *et al* 1977).

Developing multiline varieties

The multiline approach originally proposed by Borlaug (1959) envisages the incorporation of several major genes into an isogenic background and the mixing of these lines to form a multiline variety. This strategy was successfully employed for breeding oats with crown rust resistance in the United States (Browning and Frey 1969). The effectiveness of this strategy in developing insect-resistant varieties remains largely unknown. The durability of resistance of multiline varieties would depend on the rate at which biotypes develop, the number of component lines in a mixture, and the extent of the area planted to multiline varieties.

Varietal mixtures

Another strategy that has been proposed to slow down the development of biotypes is varietal mixtures consisting of 80–90% resistant plants and 10–20% susceptible plants of similar varietal background. Such varietal mixtures exert lower selection pressure on the insect as they are able to survive and reproduce on the susceptible plants (Gould 1986).

Genetic engineering for tissue-specific expression

Questions have been raised about the stability of engineered genes such as the *Bt* gene for insect resistance. Gould *et al* (1992) and McGaughey and Whalon (1992) have discussed strategies for deploying insecticidal genes in transgenic crops. Constitutive expression of the transgene may exert great selection pressure on the insect population. The selection pressure may be minimized by expressing the insecticidal genes only in certain plant parts (tissue specific), or at certain growth stages (temporal specific), or only in response to insect feeding (wound specific). Thornburg *et al* (1990) have shown that *pin-2* promoter when attached to the insecticidal gene directs the synthesis of insecticidal proteins in the tissues of transgenic tobacco preferentially consumed by tobacco hornworm larvae. The *pin-2* promoter does not express the insecticidal gene in the leaves, stems, or roots of potato plants until the plants are mechanically wounded (Pena-Cortes *et al* 1988). Wang *et al* (1992) have isolated phloem-specific promoters for use in transgenic rice to control the brown planthopper and other phloem feeders. Transgenic tobacco plants expressing snowdrop lectin coupled to the *RSsI* promoter demonstrate phloem-specific expression of this transgene (Shi *et al* 1994).

REFERENCES

Adkisson P L, Dyck V A (1980) Resistant varieties in pest management systems. Pages 233–251 in Breeding plants resistant to insects. F G Maxwell and P R Jennings, eds. John Wiley and Sons, New York.

Adkisson P L, Gaines J C (1960) Pink bollworm control as related to the total cotton insect control program of Central Texas. Tex. Agric. Exp. Stn. Misc. Publ. 7 pp.

Adkisson P L, Bailey C F, Niles G A (1962) Cotton stock screened for resistance to the pink bollworm. 1960–61. Tex. Agric. Exp. Stn. Misc. Publ. 606.

Agarwal R A, Banerjee S K, Singh M, Katiyar K N (1976) Resistance to insects in cotton. II. To pink bollworm, Pectinophora gossypiella (Saunders). Cotton Fibres Trop. 31:217–221.

Aina O J, Rodriguez J G, Knavel D E (1972) Characterizing resistance to Tetranychus urticae in tomato. J. Econ. Entomol. 65:641–643.

Akeson W R, Haskins F A, Gorj H J (1969) Sweet clover-weevil feeding deterrent. B: Isolation and identification. Science 163:293–294.

All J N, Boerma H R, Todd J W (1989) Screening soybean genotypes in the greenhouse for resistance to insects. Crop Sci. 29:1156–1159.

Allard R W (1960) Principles of plant breeding. John Wiley and Sons, New York.

Altabella T, Chrispeels M J (1990) Tobacco plants transformed with the bean αai gene express an inhibitor of insect α-amylase in their seeds. Plant Physiol. 93:805–810.

Ames B N, Profet M, Gold L S (1990) Dietary pesticides (99.99% all natural). Proc. Nat. Acad. Sci. USA 87:7777–7781.

Anderson E (1960) The evolution of domestication. Pages 67–84 in Evolution after Darwin. S Tax and C Callender, eds. University of Chicago Press, Chicago.

Andow D A (1991) Vegetational diversity and arthropod population response. Ann. Rev. Entomol. 36:561–586.

Argandona V H, Luza J G, Niemeyer H M, Corcuera L J (1980) Role of hydroxamic acids in the resistance of cereals to aphids. Phytochemistry 19:1665–1668.

Argandona V H, Corcuera L J, Niemeyer H M, Campbell B C (1983) Toxicity and feeding deterrency of hydroxamic acids from gramineae in synthetic diets against the greenbug. *Schizaphis graminum.* Entomol. Exp. Appl. 34:134–138.

Arnason J T, Towers G H N, Philogene B J R, Lambertt J D H (1983) The role of natural photosensitizers in plant resistance to insects. Pages 139–151 *in* Plant resistance to insects. P A Hedin, ed., ACS Symp. Series 208. American Chemical Society, Washington, DC.

Arnason J T, Philogene B J R, Neil Towers G H (1992) Phototoxins in plant–insect interactions. Pages 317–341 *in* Herbivores: Their interactions with secondary plant metabolites, Vol. II, 2nd ed. G A Rosenthal and M R Berenbaum, eds. Academic Press, New York.

Asakawa Y, Dawson G W, Griffiths D C, Lallemand J Y, Ley S V, Mori K, Mudd A, Pezechik-Leclaire M, Pickett J A, Watanabe H, Woodcock C M, Zhang Z N (1988) Activity of drimane antifeedants and related compounds against aphids and comparative biological effects and chemical reactivity of (–)- and (+)- polygodial. J. Chem. Ecol. 14:1845–1855.

Athwal D S, Pathak M D, Bacalangco E H, Pura C D (1971) Genetics of resistance to brown planthoppers and green leafhoppers in *Oryza sativa* L. Crop Sci. 11:747–750.

Atsatt P R, O'Dowd D J (1976) Plant defense guilds. Science 193:24–29.

Auclair J L (1963) Aphid feeding and nutrition. Ann. Rev. Entomol. 8:439–490.

Auclair J L (1978) Biotypes of the pea aphid *Acyrthosiphon pisum* in relation to host plants and chemically defined diets. Entomol. Exp. Appl. 24:212–216.

Ave D A, Gregory P, Tingey W M (1987) Aphid repellent sesquiterpenes in glandular trichomes of *Solanum berthaultii* and *S. tuberosum.* Entomol. Exp. Appl. 44:131–138.

Averill A L, Prokopy R J (1987) Residual activity of oviposition-deterring pheromone in *Rhagoletis pomonella* (Diptera: Tephritidae) and female response to infested fruit. J. Chem. Ecol. 13:167–177.

Bailey J A, Mansfield J W (1982) Phytoalexins. Blackie, Glasgow.

Baksh S (1979) Cytogenetic studies on the F_1 hybrid *Solanum incanum* L. × *Solanum melongena* L. variety 'Giant of Banaras.' Euphytica 28:793–800.

Balachowsky A S (1951) La lutte contre les insectes. Payot, Paris. 380 pp.

Baldwin I T (1991) Damage-induced alkaloids in wild tobacco. Pages 47–69 *in* Phytochemical induction by herbivores. D W Tallamy and M J Raupp, eds. John Wiley and Sons, New York.

Baliddawa C W (1985) Plant species diversity and crop pest control. Insect Sci. Applic. 6:479–487.

Barbosa P (1988a) Natural enemies and herbivore–plant interactions: influence of plant allelochemicals and host specificity. Pages 201–229 *in* Novel aspects of insect–plant interactions. P Barbosa and D K Letourneau, eds. John Wiley and Sons, New York.

Barbosa P (1988b) Some thoughts on 'the evolution of host range.' Ecology 69: 912–915.

Barbosa P, Saunders J A, Kemper J, Trumble R, Olechno J, Martinat P (1986) Plant allelochemicals and insect parasitoids: effects of nicotine on *Cotesia congregata* and *Hyposoter annulipes.* J. Chem. Ecol. 12:1319–1328.

Barbosa P, Krischik V A, Jones C G, eds (1991) Microbial mediation of plant–herbivore interactions. John Wiley and Sons, New York.

Barbour J D, Kennedy G G (1991) Role of steroidal glycoalkaloids α-tomatine in host

358   *References*

plant resistance of tomato to Colorado potato beetle. J. Chem. Ecol. 17:989–1005.

Barbour J D, Farrar Jr R R, Kennedy G G (1991) Interaction of fertilizer regime with host-plant resistance in tomato. Entomol. Exp. Appl. 60:289–300.

Barbour J D, Farrar Jr R R, Kennedy G G (1993) Interaction of *Manduca sexta* resistance in tomato with insect predators of *Helicoverpa zea*. Entomol. Exp. Appl. 68:143–155.

Bardner R, Maskell F E, Ross G J S (1970) Measurements of infestations of wheat bulb fly, *Leptohylemyia coarctata* (Fall.), and their relationship with yield. Plant Path. 19:82–87.

Barfoot P D, Connett R J A (1989) AGC's cowpea enzyme inhibitor gene: potential market opportunity. AgBiotech News Inform. 1:177–182.

Barton K A, Whiteley H R, Yang N S (1987) *Bacillus thuringiensis* delta-endotoxin expressed in transgenic *Nicotiana tabacum* provides resistance to lepidopteran insects. Plant Physiol. 85:1103–1109.

Batra L R, ed (1979) Insect–fungus symbiosis, nutrition, mutualism, and commensalism. John Wiley and Sons, New York.

Bazzaz F A (1990) The response of natural ecosystems to the rising global CO_2 levels. Ann. Rev. Ecol. Syst. 21:167–196.

Beck D L, Dunn G M, Routley D G, Bowman J S (1983) Biochemical basis of resistance in corn to the corn leaf aphid. Crop Sci. 23:995–998.

Beck S D (1957) The European corn borer, *Pyrausta nubilalis* (Hubn.) and its principal host plant. VI. Host plant resistance to larval establishment. J. Insect Physiol. 1:158–177.

Beck S D (1965) Resistance of plants to insects. Ann. Rev. Entomol. 10:207–232.

Beck S D (1972) Nutrition, adaptation and environment. Pages 1–6 *in* Insect and mite nutrition–significance and implications in ecology and pest management. J G Rodriguez, ed. North-Holland, Amsterdam.

Beck S D (1980) Insect photoperiodism. Academic Press, New York.

Beck S D, Maxwell F G (1976) Use of plant resistance. Pages 615–636 *in* Theory and practice of biological control. C B Huffaker and P S Messenger, eds. Academic Press, New York.

Beck S D, Reese J C (1976) Insect–plant interactions: nutrition and metabolism. Pages 41–92 *in* Biochemical interaction between plants and insects. J W Wallace and R L Mansell, eds. Recent Adv. Phytochem. Vol. 10. Plenum Press, New York.

Beck, S D, Lilly J H, Stauffer J F (1949) Nutrition of the European corn borer, *Pyrausta nubilalis* (Hbn). Ann. Entomol. Soc. Am. 42:483–496.

Bell A A, Stipanovic R D, Elzen G W, Williams Jr H J (1987) Structural and genetic variation of natural pesticides in pigment glands of cotton (*Gossypium*). Pages 477–490 *in* Allelochemicals–role in agriculture and forestry. G R Waller, ed. ACS Symp. Series 330. American Chemical Society, Washington, DC.

Bell E A (1976) Uncommon amino acids in plants. FEBS Lett. 64:29–35.

Bell W J (1990) Searching behavior patterns in insects. Ann. Rev. Entomol. 35:447–467.

Benedict J H, Hatfield J L (1988) Influence of temperature-induced stress on host–plant suitability to insects. Pages 139–166 *in* Plant stress–insect interaction. E A Heinrichs, ed. John Wiley and Sons, New York.

Bennett S E (1965) Tannic acid as a repellent and toxicant to alfalfa weevil larvae. J. Econ. Entomol. 58:372.

Benson L (1976) Plant classification. Oxford and IBP Publishing Co., New York.

Bentley M D, Leonard D E, Stoddard W F, Zalkow L H (1984a) Pyrrolizidine alkaloids as larval feeding deterrents for spruce budworm, *Choristoneura fumiferana* (Lepidoptera: Tortricidae). Ann. Entomol. Soc. Am. 77:393–397.

Bentley M D, Leonard D E, Reynolds E K, Leach S, Beck A B, Murakashi I (1984b) Lupine alkaloids as larval feeding deterrents for spruce budworm, *Choristoneura fumiferana* (Lepidoptera: Tortricidae). Ann. Entomol. Soc. Am. 77:398–400.

Bentley M D, Leonard D E, Bushway R J (1984c) Solanum alkaloids as larval feeding deterrents for spruce budworm, *Choristoneura fumiferana* (Lepidoptera: Tortricidae). Ann. Entomol. Soc. Am. 77:401–403.

Beregovoy V H, Peters D C (1993) Variation of two kinds of wheat damage characters in seven biotypes of the greenbug, *Schizaphis graminum* (Rondani) (Homoptera: Aphididae). J. Kansas Entomol. Soc. 66:237–244.

Berenbaum M (1980) Adaptive significance of midgut pH in larval lepidoptera. Am. Nat. 115:138–146.

Berenbaum M (1983) Coumarins and caterpillars: a case for coevolution. Evolution 37:163–179.

Berenbaum M (1986) Post ingestive effects of phytochemicals on insects. Pages 121–153 *in* Insect–plant interactions. T A Miller and J Miller, eds. Springer–Verlag, New York.

Berenbaum M (1988) Effects of electromagnetic radiation on insect–plant interactions. Pages 167–185 *in* Plant stress–insect interactions. E A Heinrichs, ed. John Wiley and Sons, New York.

Berenbaum M R (1991) Coumarins. Pages 221–249 *in* Herbivores: Their interactions with secondary plant metabolites, Vol. I, 2nd ed. G A Rosenthal and M R Berenbaum, eds. Academic Press, New York.

Berenbaum M R, Isman M B (1989) Herbivory in holometabolous and hemimetabolous insects: contrasts between Orthoptera and Lepidoptera. Experientia 45:229–236.

Berenbaum M R, Zangerl A (1987) Defense and detente in the coevolutionary arms race: synergisms, syntheses, and sundry other sins. Pages 113–132 *in* Chemical coevolution. K Spencer, ed. Pergamon Press, New York.

Berenbaum M R, Zangerl A R (1992a) Quantification of chemical coevolution. Pages 69–87 *in* Plant resistance to herbivores and pathogens. R S Fritz and E L Simms, eds. The University of Chicago Press, Chicago.

Berenbaum M R, Zangerl A R (1992b) Genetics of secondary metabolism and herbivore resistance in plants. Pages 415–438 *in* Herbivores: Their interactions with secondary plant metabolites. Vol. II. Ecological and evolutionary processes. G A Rosenthal and M R Berenbaum, eds. Academic Press, New York.

Bergamasco R, Horn D H S (1983) Distribution and role of insect hormones in plants. Pages 627–654 *in* Endocrinology of insects. R G H Downer and H Laufer, eds. Alan R Liss, New York.

Bergman J M, Tingey W M (1979) Aspects of interaction between plant genotypes and biological control. Bull. Entomol. Soc. Am. 25:275–279.

Bernatzky R, Tanksley S D (1986) Toward a saturated linkage map in tomato based on isozymes and random cDNA sequences. Genetics 112:887–898.

Bernays E A (1981) Plant tannins and insect herbivores: an appraisal. Ecol. Entomol. 6:353–360.

Bernays E A (1983a) Antifeedants in crop pest managament. Pages 259–271 *in* Natural products for innovative pest management. D L Whitehead and W S Bowers, eds. Pergamon Press, Oxford.

Bernays E A (1983b) Nitrogen in defense against insects. Pages 321–344 *in* Nitrogen as an ecological factor. J A Lee, S McNeill and I H Rorison, eds. Blackwell Scientific Publications, Oxford.

Bernays E A, Chapman R F (1970) Experiments to determine the basis of food selection by *Chorthippus parallelus* (Zetterstedt) (Orthoptera: Acrididae) in the field. J. Anim. Ecol. 39:761–776.

Bernays E A, Chapman R F (1972) The control of changes in the peripheral sensilla associated with feeding in *Locusta migratoria* (L.). J. Exp. Biol. 57:755–763.

Bernays E A, Chapman R F (1975) The importance of chemical inhibition of feeding in host–plant selection by *Chorthippus parallelus* (Zetterstedt). Acrida 4:83–93.

Bernays E A, Chapman R F (1977) Deterrent chemicals as a basis for oligophagy in *Locusta migratoria* (L.). Ecol. Entomol. 2:1–18.

Bernays E A, Chapman R F (1978) Plant chemistry and acridoid feeding behaviour. Pages 99–141 *in* Biochemical aspects of plant and animal coevolution. J B Harborne, ed. Academic Press, New York.

Bernays E A, Chapman R F (1994) Host plant selection by phytophagous insects. Chapman and Hall, London.

Bernays E A, Graham M (1988) On the evolution of host specificity in phytophagous arthropods. Ecology 69:886–892.

Bernays E A, Chamberlain D J, McCarthy P (1980) The differential effects of ingested tannic acid on different species of Acridoidea. Entomol. Exp. Appl. 28:158–166.

Bernays E A, Chapman R F, Woodhead S (1983) Behaviour of newly hatched larvae of *Chilo partellus* (Swinhoe) (Lepidoptera: Pyralidae) associated with their establishment in the host–plant, sorghum. Bull. Entomol. Res. 73:75–83.

Berryman A A (1988) Towards a unified theory of plant defense. Pages 39–55 *in* Mechanisms of woody plant defenses against insects: Search for patterns. W J Mattson, J Levieux and C Bernard–Dagan, eds. Springer-Verlag, New York.

Bicchi C, D'Amato A, Frattini C, Nano G M, Cappelletti E, Caniato R, Filippini R (1990) Chemical diversity of the contents from the secretory structures of *Heracleum sphondylium*. Phytochemistry 29:1883–1887.

Biffen R H (1907) Studies on the inheritance of disease resistance. J. Agric. Sci. Cambridge 2: 109–128.

Birch A N E, Griffiths D W, Hopkins R J, Macfarlane Smith W H, McKinlay R G (1992) Glucosinolate responses of swede, kale, forage and oilseed rape to root damage by turnip root fly (*Delia floralis*) larvae. J. Sci. Food Agric. 60:1–9.

Birch A N E, Staûmlautdler E, Hopkins R J, Simmonds M S J, Baur R, Griffiths D W, Ramp T, Hurter J, McKinlay R G (1993) Mechanisms of resistance to the cabbage and turnip root flies: collaborative field, behavioral and electrophysiological studies. Bull. OILB/SROP 16:1–5.

Bird T G, Hedin P A, Burks M L (1987) Feeding deterrent compounds to the bollweevil, *Anthonomus grandis* Boheman in Rose-of-Sharon, *Hibiscus syriacus* L. J. Chem. Ecol. 13:1087–1097.

Blaney W M, Schoonhoven L M, Simmonds M S J (1986) Sensitivity variations in insect chemoreceptors; a review. Experientia 42:13–19.

Blau P A, Feeny P, Contardo L, Robson D S (1978) Allylglucosinolate and herbivorous caterpillars: a contrast in toxicity and tolerance. Science 200:1296–1298.

Bloem K A, Kelley K C, Duffey S S (1989) Differential effect of tomatine and its alleviation by cholesterol on larval growth and efficiency of food utilization in *Heliothis zea* and *Spodoptera exigua*. J. Chem. Ecol. 15:387–398.

Blom F (1978) Sensory activity and food intake: a study of input–output relationships in two phytophagous insects. Netherl. J. Zool. 28:277–340.

Blum A (1968) Anatomical phenomena in seedlings of sorghum varieties resistant to the sorghum shoot fly (*Atherigona varia soccata*). Crop Sci. 8:388–391.

Blum A (1969) Oviposition preference by the sorghum shootfly (*Atherigona varia soccata*) in progenies of susceptible × resistant sorghum crosses. Crop Sci. 9:695–696.

Blum M S (1983) Detoxification, deactivation, and utilization of plant compounds by insects. Pages 265–275 *in* Plant resistance to insects. P A Hedin, ed. ACS Symp. Series 208. American Chemical Society, Washington, DC.

Blum M S (1987) Biosynthesis of arthropod exocrine compounds. Ann. Rev. Entomol. 32:381–413.

Blum M S (1992) Ingested allelochemicals in insect wonderland: a menu of remarkable functions. Am. Entomol. 38:222–234.

Boeckh J (1977) Aspects of nervous coding of sensory quality in the olfactory pathway of insects. Pages 308–322 *in* Proceedings of the XV International Congress on Entomology, Washington, DC. D White, ed. Entomological Society of America, College Park MD.

Boethel D J, Eikenbary R D, eds (1986) Interactions of plant resistance and parasitoids and predators of insects. Ellis Horwood, Chichester, UK.

Boistel J, Coraboeuf E (1953) L'activité electrique dans l'antenne isolée de Lepidoptere au cours de l'etude de l'olfaction. C. R. Soc. Biol. Paris 147:1172–1175.

Bombosch S (1963) Untersuchungen zur Vermchrung von *Aphis fabae* Scop. in Samenrubenbestanden unter besonderer Berucksichtigung der Schwebfiiegen (Diptera, Syrphidae). Z. Angew. Entomol. 52:105–141.

Boppré M (1990) Lepidoptera and pyrrolizidine alkaloids. Exemplification of complexity in chemical ecology. J. Chem. Ecol. 16:165–185.

Bordasch R P, Berryman A A (1977) Host resistance to the fir engraver beetle, *Scolytus ventralis* (Coleoptera: Scolytidae). 2. Repellency of *Abies grandis* resins and some monoterpenes. Can. Entomol. 109:95–100.

Borer D J, Delong D M, Triplehorn C A (1976) An introduction to the study of insects. Holt, Rinehart and Winston, New York.

Borlaug N E (1959) The use of multi-lineal or composite varieties to control airborne epidemic diseases of self-pollinated crop plants. Pages 12–27 *in* 1st Wheat Genetics Symposium. B C Jenkins, ed. University of Manitoba, Winnipeg, Canada.

Boulter D, Edwards G A, Gatehouse A M R, Gatehouse J A, Hilder V A (1990) Additive protective effects of different plant-derived insect resistance genes in transgenic tobacco plants. Crop Protection 9:351–354.

Bowers M D (1983) The role of iridoid glycosides in host plant specificity of checkerspot butterflies. J. Chem. Ecol. 9:475–493.

Bowers M D (1990) Recycling plant natural products for insect defenses. Pages 353–386 *in* Insect defenses: Adaptive mechanisms and strategies of prey and predators. D L Evans and J O Schmidt, eds. State University New York Press, Albany, NY.

Bowers M D (1991a) Iridoid glycosides. Pages 297–325 *in* Herbivores: Their interactions with secondary plant metabolites, Vol. I, 2nd ed. G A Rosenthal and M R Berenbaum, eds. Academic Press, New York.

Bowers M D, Puttick G M (1988) Response of generalist and specialist insects to qualitative allelochemical variation. J. Chem. Ecol. 14:319–334.

Bowers M D, Puttick G M (1989) Iridoid glycosides and insect feeding preferences: gypsy moths (*Lymantria dispar*, Lymantriidae) and buckeyes (*Junonia coenia*, Nymphalidae). Ecol. Entomol. 14:247–256.

Bowers W S (1991b) Insect hormones and antihormones in plants. Pages 431–456 *in* Herbivores: Their interactions with secondary plant metabolites, Vol. I, 2nd ed. G A Rosenthal and M R Berenbaum, eds. Academic Press, New York.

Bowers W S, Nishida R (1980) Juvocimenes: potent juvenile hormone mimics from sweet basil. Science 209:1030–1032.

Bowers W S, Fales H M, Thompson M J, Uebel E C (1966) Identification of an active compound from balsam fir. Science 154:1020–1021.

Bowers W S, Ohta T, Cleere J S, Marsella P A (1976) Discovery of insect antijuvenile hormones in plants. Science 193:542–547.

Boyd W C, Shapleigh E (1954) Specific precipitating activity of plant agglutinins (lectins). Science 119:419.

Branson T F, Simpson R G (1966) Effects of a nitrogen-deficient host and crowding on the corn leaf aphid. J. Econ. Entomol. 59:290–293.

Brattsten L B (1979) Biochemical defense mechanisms in herbivores against plant allelochemicals. Pages 199–270 *in* Herbivores: Their interaction with secondary plant metabolites. G A Rosenthal and D H Janzen, eds. Academic Press, New York.

Brattsten L B (1988) Potential role of plant allelochemicals in the development of insecticide resistance. Pages 313–355 *in* Novel aspects of insect–plant interactions. P Barbosa and D K Letourneau, eds. John Wiley and Sons, New York.

Brattsten L B, Wilkinson C F, Eisner T (1977) Herbivore–plant interactions: mixed-function oxidases and secondary plant substances. Science 196: 1349–1352.

Bravo H R, Niemeyer H M (1986) A new product from the decomposition of 2,4-dihydroxy-7-methoxy-1,4-benzoxazin-3-one (DIMBOA), a hydroxamic acid from cereals. Heterocycles, 24:335–337.

Brettell J H (1980) Prospects for arthropod pest control utilizing cotton plant resistance. Pages 19–28 *in* Proc. Cotton Pest Control Workshop. Union Carbide South Africa, Agricultural Chemicals, Nelspruit, South Africa.

Brewer G J, Sorensen E L, Hober E, Kreitner G L (1986) Alfalfa stem anatomy and potato leafhopper (Homoptera: Cicadellidae) resistance. J. Econ. Entomol. 79:1249–1253.

Briggs F N, Allard R W (1953) The current status of the backcross method of plant breeding. Agron. J. 45:131–138.

Brim C A (1966) A modified pedigree method of selection in soybeans. Crop Sci. 6:220.

Brim C A, Stuber C W (1973) Application of genetic male sterility to recurrent selection schemes in soybeans. Crop Sci. 13:528–530.

Bristow P R, Doss R P, Campbell R L (1979) A membrane filter bioassay for studying phagostimulatory materials in leaf extracts. Ann. Entomol. Soc. Am. 72:16–18.

Broadway R M, Duffey S S (1986) Plant proteinase inhibitors: mechanism of action and effect on the growth and digestive physiology of larval *Heliothis zea* and

Spodoptera exigua. J. Insect Physiol. 32:827–833.

Broadway R M, Duffey S S, Pearie G, Ryan C A (1986) Plant proteinase inhibitors: a defense against herbivore insects? Entomol. Exp. Appl. 41:33–38.

Brody A K, Karban R (1989) Demographic analysis of induced resistance against spider mites (Acari: Tetranychidae) in cotton. J. Econ. Entomol. 82:462–465.

Brower L P (1969) Ecological chemistry. Sci. Am. 220: 22–29.

Brown Jr K S (1984) Adult-obtained pyrrolizidine alkaloids defend ithomiine butterflies against a spider predator. Nature 309:707–709.

Browning J A, Frey K J (1969) Multiline cultivars as a means of disease control. Ann. Rev. Phytopath. 7:355–382.

Brues C T (1920) Selection of food plants by insects with special reference to Lepidoptera. Am. Nat. 54:313–332.

Brues C T (1946) Insect dietary. Harvard University Press, Cambridge, MA.

Buchner P (1965) Endosymbiosis of animals with plant microorganisms. Interscience, New York.

Burden B J, Norris D M (1992) Role of the isoflavonoid coumestrol in the constitutive antixenosis properties of 'Davis' soybeans against an oligophagous insect, the Mexican bean beetle. J. Chem. Ecol. 18:1069–1081.

Burton R L, Perkins W D (1989) Rearing the corn earworm and fall armyworm for maize resistance studies. Pages 37–45 *in* Toward insect resistant maize for third world. J A Mihm, B R Wiseman and F M Davis, eds. International Wheat and Maize Improvement Center (CIMMYT), El Batan, Mexico.

Burton R L, Schuster D J (1981) Oviposition stimulant for tomato pinworms from surfaces of tomato plants. Ann. Entomol. Soc. Am. 74: 512–515.

Butenandt A, Karlson P (1954) Uber die Isolierung cines metamorphosehormons der Insektan in kristallisierter Form. Z. Naturforsch. 9b:389–391.

Butler L G (1989) Effects of condensed tannin on animal nutrition. Pages 391–402 *in* Chemistry and significance of condensed tannins. R W Hemingway and J J Karchesy, eds. Plenum Press, New York.

Butterworth J H, Morgan E D (1971) Investigation of the locust feeding inhibition of the seeds of the neem tree, *Azadirachta indica.* J. Insect Physiol. 17:969–977.

Buttery R G, Ling L C, Chan B G (1978) Volatiles of corn kernels and husks: possible corn earworm attractants. J. Agric. Food Chem. 28: 866–869.

Buttery R G, Ling L C, Teranishi R (1980) Volatiles of corn tassels: possible corn earworm attractants. J. Agric. Food Chem. 28:771–774.

Caballero P, Shin D H, Khan Z R, Saxena R C, Juliano B O, Zapata F J (1988) Use of tissue culture to evaluate rice resistance to lepidopterous pests. Int. Rice Res. Newsl. 13(5):14–15.

Camm E L, Wat C, Towers G H N (1976) An assessment of the roles of furanocoumarins in *Heracleum lanatum.* Can. J. Bot. 54:2562–2566.

Campbell B C (1989) On the role of microbial symbiotes in herbivorous insects. Pages 1–44 *in* Insect–plant interactions, Vol. I. E A Bernays, ed. CRC Press, Boca Raton, FL.

Campbell B C, Dreyer D L (1985) Host plant resistance of sorghum: differential hydrolysis of sorghum pectic substances by polysaccharides of greenbug biotype (*Schizaphis graminum*: Homoptera: Aphididae) Arch. Insect Biochem. Physiol. 2:203–215.

Campbell B C, Duffey S S (1979) Tomatine and parasitic wasps: potential incompatibility of plant antibiosis with biological control. Science 205:700–702.

Campos F, Atkinson J, Arnason J T, Philogene B J R, Morand P, Werstiuk N H, Timmins G (1988) Toxicity and toxicokinetics of 6-methoxybenzoxazolinone (MBOA) in the European corn borer, *Ostrinia nubilalis* (Hubner) J. Chem. Ecol. 14:989–1002.

Carpenter F M (1977) Geological history and evolution of the insects. Pages 63–70 *in* Proceedings of the XV International Congress on Entomology, Washington, DC. D White ed. Entomological Society of America, College Park MD.

Carroll C R, Hoffman C A (1980) Chemical feeding deterrent mobilized in response to insect herbivory and counteradaptation by *Epilachna tredecimnotata*. Science 209:414–416.

Cartier J J, Isaak A, Painter R H, Sorensen E L (1965) Biotypes of pea aphid *Acyrthosiphon pisum* (Harris) in relation to alfalfa clones. Can. Ent. 97:754–760.

Cartwright W B, Caldwell R M, Compton L E (1946) Relation of temperature to the expression of resistance in wheats to Hessian fly. J. Am. Soc. Agron. 38:259–263.

Casagrande R A, Haynes D L (1976) The impact of pubescent wheat on the population dynamics of the cereal leaf beetle. Environ. Entomol. 5:153–159.

Casida J E, ed (1973) Pyrethrum – The natural insecticide. Academic Press, New York.

Cates R G (1980) Feeding patterns of monophagous, oligophagous, and polyphagous insect herbivores: the effect of resource abundance and plant chemistry. Oecologia 46:22–31.

Centello W W, Jacobsen M (1979) Corn silk volatiles attract many species of moths. J. Environ. Sci. Health A 14:695–709.

Chaboussou F (1972) The role of potassium and of cation equilibrium in the resistance of the plant towards diseases. Potash Rev. 23:29.

Chaloner W G (1970) The rise of the first land plants. Biol. Rev. 45:353–377.

Champagne D E, Arnason J T, Philogene B J R, Morand P, Lam J (1986) Light-mediated allelochemical effects of naturally occurring polyacetylenes and thiophenes from Asteraceae on herbivorous insects. J. Chem. Ecol. 12:835–858.

Chandravadan M V (1987) Identification of triterpenoid feeding deterrent of red pumpkin beetles (*Aulacophora foveicollis*) from *Momordica charantia*. J. Chem. Ecol. 13:1689–1694.

Chang T T (1976) Manual on genetic conservation of rice germplasm for evaluation and utilization. International Rice Research Institute, Manila, Philippines.

Chapman R F (1974) The chemical inhibition of feeding by phytophagous insects: a review. Bull. Entomol. Res. 64:339–363.

Chapman R F (1980) The evolution of insect chemosensory systems and food selection behavior. Pages 131–134 *in* Olfaction and taste. H van der Starre, ed. VII International Retrieval, London.

Chapman R F, Bernays E A (1989) Insect behavior at the leaf surface and learning as aspects of host plant selection. Experientia 45:215–222.

Chapman R F, Bernays E A, Wyatt T (1988) Chemical aspects of host-plant specificity in three Larrea-feeding grasshoppers. J. Chem. Ecol. 14:561–579.

Chaudhary R C, Khush G S (1990) Breeding rice varieties for resistance against *Chilo* spp. of stem borers in Asia and Africa. Insect Sci. Applic. 11:659–669.

Chelliah S, Heinrichs E A (1980) Factors affecting insecticide-induced resurgence of the brown planthopper, *Nilaparvata lugens* on rice. Environ. Entomol. 9:773–777.

Chew F S (1988) Searching for defensive chemistry in the Cruciferae, or, do glucosinolates always control interactions of Cruciferae with their potential herbivores and symbionts? Pages 81–122 *in* Chemical mediation of coevolution. K C Spencer, ed. Academic Press, New York.

Chi C W (1990) Three different kinds of proteinase inhibitors as candidates for insecticidal proteins. Pages 13.1–13.3 *in* Molecular and genetic approaches to plant stress. International Center for Genetic Engineering and Biotechnology, New Delhi, India.

Chiang H C, Holdaway F G (1960) Relative effectiveness of resistance of field corn to the European corn borer, *Pyrausta nubilalis*, in crop protection and in population control. J. Econ. Entomol. 53:918–924.

Chiang H S, Norris D M (1983) Physiological and anatomical stem parameters of soybean resistance to Agromyzid beanflies. Entomol. Exp. Appl. 33:203–212.

Chiang H S, Norris D M, Ciepiela A, Oosterwyk A, Shapiro P, Jackson M (1986) Comparative constitutive resistance in soybean lines to Mexican bean beetle. Entomol. Exp. Appl. 42:19–26.

Chiang H S, Norris D M, Ciepiela A, Shapiro P, Oosterwyk A (1987) Inducible versus constitutive PI 227687 soybean resistance to Mexican bean beetle, *Epilachna varivestis*. J. Chem. Ecol. 13:741–749.

Chiang M S, Hudon M (1973) Inheritance of resistance to the European corn borer in grain corn. Can. J. Plant Sci. 53:779–782.

Chrispeels M J, Raikhel N V (1991) Lectins, lectin genes, and their role in plant defense. Plant Cell 3:1–9.

Christou P, Ford T L, Kofron M (1991) Production of transgenic rice (*Oryza sativa* L.) plants from agronomically important indica and japonica varieties via electric discharge particle acceleration of exogenous DNA into immature zygotic embryos. Bio/Technology 9:957–962.

CIMMYT (International Maize and Wheat Improvement Center) (1989) Toward insect resistant maize for the third world. Proceedings of the International Symposium on Methodologies for Developing Host Plant Resistance to Maize Insects. J A Mihm, B R Wiseman and F A Davis, eds. CIMMYT, El Batan, Mexico. 301 pp.

Cirio U (1971) Reperti sul meccanismo stimoloriposta nell ovideposizione del *Dacus oleae* Gmelin (Diptera: Trypetide). Redia 52:557–600.

Claridge M F, Den Hollander J (1980) The 'biotypes' of the rice brown planthopper, *Nilaparvata lugens*. Entomol. Exp. Appl. 27:23–30.

Claridge M F, Den Hollander J (1983) The biotype concept and its application to insect pests of agriculture. Crop Protection 2:85–95.

Clayton R B (1970) The chemistry of non-hormonal interactions: terpenoid compounds in ecology. Pages 235–280 *in* Chemical ecology. E Sondheimer and J B Simeone, eds. Academic Press, New York.

Cohen R W, Waldbauer G P, Friedman S (1988) Natural diets and self-selection: *Heliothis zea* larvae and maize. Entomol. Exp. Appl. 46:161–171.

Cole M D, Anderson J C, Blaney W M, Fellows L E, Ley S V, Sheppard R N, Simmonds M S J (1990) Neoclerodane insect antifeedant from *Scutellaria galericulata*. Phytochemistry 29:1793–1796.

Cole R A (1987) Intensity of radicle fluorescence as related to the resistance of seedlings of lettuce to the lettuce root aphid and carrot to the carrot fly. Ann. Appl. Biol. 111:629–639.

Cole R A, Riggall W, Morgan A (1993) Electronically monitored feeding behaviour of the lettuce root aphid (*Pemphigus bursarius*) on resistant and susceptible lettuce varieties. Entomol. Exp. Appl. 68:179–185.

Coleman J S, Jones C G (1991) A phytocentric perspective of phytochemical induction by herbivores. Pages 3–45 *in* Phytochemical induction by herbivores. D W Tallamy and M J Raupp, eds. John Wiley and Sons, New York.

Coley P D, Bryant J P, Chapin III F S (1985) Resource availability and plant antiherbivore defense. Science 230:895–899.

Coombe P E (1982) Visual behaviour of the greenhouse whitefly *Trialeurodes vaporariorum*. Physiol. Entomol. 7:243–251.

Cooper-Driver G A (1978) Insect–fern association. Entomol. Exp. Appl. 24:310–316.

Craig R, Mumma R O, Gerhold D L, Winner B L, Snetsinger R (1986) Genetic control of a biochemical mechanism for mite resistance in geraniums. Page 168–176 *in* Natural resistance of plants to pests. M B Green and P A Hedin, eds. ACS Symp. Series 296. American Chemical Society, Washington, DC.

Crossway A, Oakes J V, Irvine J M, Ward B, Knauf V C, Shewmaker C K (1986) Integration of foreign DNA following microinjection of tobacco mesophyll protoplasts. Mol. Gen. Genet. 202:179–185.

Crumpacker D W, Allard R W (1962) A diallel cross analysis of heading date in wheat. Hilgardia 32:275–318.

Cuevas L, Niemeyer H M, Perez F J (1990) Reaction of DIMBOA, a resistant factor from cereals with α-chymotrypsin. Phytochemistry 29:1429–1432.

Curtis C E, Clark J D (1979) Responses on navel orangeworm moths to attractants evaluated as oviposition stimulants in an almond orchard. Environ. Entomol. 8:330–333.

Cuthbert Jr F P, Jones A (1978) Insect resistance as an adjunct or alternative to insecticides for control of sweet potato soil insects. J. Am. Soc. Hort. Sci. 103:443.

Cuthbert Jr F P, Fery R L, Chambliss O L (1974) Breeding for resistance to cowpea curculio in southern peas. HortScience 9:69–70.

Cutler H G, Severson R F, Cole P D, Jackson D M, Johnson A W (1986) Secondary metabolites from higher plants: their possible role as biological control agent. Pages 178–196 *in* Natural resistance of plants to pests. M B Green and P A Hedin, eds. ACS Symp. Series 296. American Chemical Society, Washington, DC.

Czapla T H, Lang B A (1991) Effects of plant lectins on the larval development of European corn borer (Lepidoptera: Pyralidae) and southern corn rootworm (Coleoptera: Chrysomelidae) J. Econ. Entomol. 83:2480–2485.

Da Costa C P, Jones C M (1971) Cucumber beetle resistance and mite susceptibility controlled by the bitter gene in *Cucumis sativus* L. Science 172:1145–1146.

Dadd R H (1973) Insect nutrition: current developments and metabolic implications. Ann. Rev. Entomol. 18:381–419.

Dahms R G, Painter R H (1940) Rate of reproduction of the pea aphid on different alfalfa plants. J. Econ. Entomol. 33:482–485.

Daly H V, Doyen J T, Ehrlich P R (1978) Introduction to insect biology and diversity. McGraw-Hill, New York.

Danielson S D, Manglitz G R, Sorensen E L (1987) Resistance of perennial glandular haired *Medicago* species to oviposition by alfalfa weevils (Coleoptera: Cucurlionidae) Environ. Entomol. 16:195–197.

Danks H V, Rabb R L, Southern P S (1979) Biology of insect parasites of *Heliothis* larvae in North Carolina. J. Georgia Entomol. Soc. 14:36–64.

Darwin C (1882) The variation of animals and plants under domestication, 2nd ed.

John Murray, London.

David W A L, Gardiner B O C (1962) Oviposition and the hatching of the eggs of *Pieris brassicae* (L.) in a laboratory culture. Bull. Entomol. Res. 53:91–109.

David W A L, Gardiner B O C (1966) The effect of sinigrin on the feeding of *Pieris brassicae* L. larvae transferred from various diets. Entomol. Exp. Appl. 9:95–98.

Davis F M (1976) Production and handling of eggs of the south-western corn borer for host plant resistance studies. MISS: Agric. For. Exp. Stn. Tech. Bull. 74.

Davis F M (1985) Entomological techniques and methodologies used in research programmes on plant resistance to insects. Insect Sci. Appl. 6:391–400.

Davis F M (1989) Rearing the southwestern corn borer and fall armyworms at Mississippi State. Pages 27–36 *in* Toward insect resistant maize for third world. J A Mihm, B R Wiseman and F M Davis, eds. International Wheat and Maize Center (CIMMYT), El Batan, Mexico.

Davis F M, Oswalt T G (1979) Hand inoculator for dispensing lepidopterous larvae. United States Department of Agriculture, Agricultural Research Service, Washington DC.

Davis K R, Lyon G D, Darvill A G, Albersheim P (1984) Host–pathogen interactions. XXV. Endopolygalacturonic acid lyase from *Erwinia carotovora* elicits phytoalexin accumulation by releasing plant cell wall fragments. Plant Physiol. 74:52–60.

Day P R (1974) Genetics of host–parasite interaction. W H Freeman, San Francisco. 238 pp.

Day P R (1992) Plant breeding: the next ten years. Pages 515–523 *in* Plant breeding in the 1990s. H T Stalker and J P Murphy, eds. CAB International, Wallingford, UK.

DeBach P (1964) Biological control of insect pests and weeds. Chapman and Hall, London.

De Candolle A P (1804) 'Essai sur les propriétés médicales des plantes, comparées avec leurs formes extérieures et leur classfication naturelle. Essai sur les propriétés médicales.' L'imiprimerie de Didot Jeune, de l'imprimeur de l'ecole de médicine, rue des Macons-Sorbonne à Paris.

De Candolle A P (1886) *Origine des Plantes Cultives* (Origin of cultivated plants). Hafner, New York (reprint from 2nd ed., 1959)

De Ponti O M B (1983) Resistance to insects promotes the stability of integrated pest control. Pages 211–225 *in* Durable resistance in crops. F Lamberti, J M Waller and N A van der Graff, eds. Plenum Press, New York.

Delannay X, Lavallee B J, Proksch R K, Fuchs R L, Sims S R, Greenplate J T, Marrone P G, Dodson R B, Augustine J J, Layton J G, Fischhoff D A (1989) Field performance of transgenic tomato plants expressing the *Bacillus thuringiensis* var. *kurstaki* insect control protein. Bio/Technology 7:1265–1269.

Delobel A G L (1982) Oviposition and larval survival of the sorghum shoot fly, *Atherigona soccata* Rond., as influenced by the size of its host plant (Diptera, Muscidae). Z. Angew. Entomol. 93: 31–38.

Den Hollander J, Pathak P K (1981) The genetics of the 'biotypes' of the rice brown planthopper, *Nilaparvata lugens*. Entomol. Exp. Appl. 29:76–86.

Denno R F, McClure M S, eds (1983) Variable plants and herbivores in natural and managed systems. Academic Press, New York.

Dethier V G (1953) Host plant perception in phytophagous insects. Trans. Int. Cong. Entomol. 9th Amsterdam 2:81–88.

Dethier V G (1954) Evolution of feeding preferences in phytophagous insects. Evolution 8: 33–54.

Dethier V G (1963) The physiology of insect senses. Methuen and Co., London.

Dethier V G (1971) A surfeit of stimuli: a paucity of receptors. Am. Sci. 59:706–715.

Dethier V G (1974) The specificity of the chemoreceptors of the blowfly and the response to natural foods. J. Insect Physiol. 20:1859–1869.

Dethier V G (1980) Evolution of receptor sensitivity to secondary plant substances with special reference to deterrents. Am. Nat. 115:45–66.

Dethier V G (1982a) Mechanisms of host-plant recognition. Entomol. Exp. Appl. 31:49–56.

Dethier V G (1982b) The contribution of insects to the study of motivations. Pages 445–455 *in* Changing concepts of the nervous system. A R Morrison and P L Strick, eds. Academic Press, New York.

Dethier V G (1988) Induction and aversion-learning in polyphagous arctiid larvae (Lepidoptera) in an ecological setting. Can. Entomol. 120:125–131.

Dethier V G, Chadwick L E (1948) Chemoreception in insects. Physiol. Rev. 28:220–254.

Dethier V G, Crnjar R M (1982) Candidate codes in the gustatory system of caterpillars. J. Gen. Physiol. 79:549–569.

Dethier V G, Browne B, Smith C N (1960) The designation of chemicals in terms of the responses they elicit from insects. J. Econ. Entomol. 53:134–136.

Detwiler R P, Hall C A S (1988) Tropical forests and the global carbon cycle. Science 239:42–47.

Dicke F F, Guthrie W D (1988) The most important corn insect. Pages 767–867 *in* Corn and corn improvement, 3rd ed. G F Sprague and J W Dudley, eds. American Society of Agronomy, Madison, WI.

Dicke M, Sabelis M W, Takabayashi J, Bruin J, Posthumus M A (1990) Plant strategies of manipulating predator–prey interactions through allelochemicals: prospects for application in pest control. J. Chem. Ecol. 16:3091–3118.

Dicke M, Van Baarlen P, Wessels R, Dijkman H (1993) Herbivory induces systemic production of plant volatiles that attract predators of the herbivore: extraction of endogenous elicitor. J. Chem. Ecol. 19:581–599.

Diehl S R, Bush G L (1984) An evolutionary and applied perspective of insect biotypes. Ann. Rev. Entomol. 29:471–504.

Dietrick E J (1961) An improved backpack motor fan for suction sampling of insect populations. J. Econ. Entomol. 54:394–395.

Dimock M B (1981) Chemical and physiological studies of antibiosis in the wild tomato *Lycopersicon hirsutum f. glabratum* to the tomato fruitworm, *Heliothis zea*. MS thesis, North Carolina State University, Raleigh, NC. 83 pp.

Dimock M B, Kennedy G G (1983) The role of glandular trichomes in the resistance of *Lycopersicon hirsutum f. glabratum* to *Heliothis zea*. Entomol. Exp. Appl. 33:263–268.

Dodd G D (1973) Integrated control of the Cabbage aphid (*Brevicoryne brassicae* (L.)) PhD thesis, University of Reading, Reading, UK. 91 pp.

Doggett H, Starks K J, Eberhart S A (1970) Breeding for resistance to the sorghum shootfly. Crop Sci. 10:528–531.

Dowd P F (1989) Fusaric acid – a secondary fungal metabolite that synergizes toxicity of co-occurring host allelochemicals to the corn earworm, *Heliothis zea* (Lepidoptera) J. Chem. Ecol. 15:249–254.

Dreyer D L, Jones K C (1981) Feeding deterrency of flavonoids and related phenolics towards *Schizaphis graminum* and *Myzus persicae*: aphid feeding deterrents in wheat. Phytochemistry 20: 2489–2493.

Dreyer D L, Reese J C, Jones K C (1981) Aphid feeding deterrents in sorghum – Bioassay, isolation, and characterization. J. Chem. Ecol. 7:273–284.

Dreyer D L, Jones K C, Jurd L, Campbell B C (1987) Feeding deterrency of some 4-hydroxycoumarins and related compounds: relationship to host plant resistance of alfalfa towards pea aphid (*Acyrthosiphon pisum*) J. Chem. Ecol. 13:925–930.

Dudley N J, Mueller R A E, Wightman J A (1989) Application of dynamic programming for guiding IPM on groundnut leafminer in India. Crop Prot. 8:349–357.

Duffey S S (1980) Sequestration of plant natural products by insects. Ann. Rev. Entomol. 25:447–477.

Duffey S S (1986) Plant glandular trichomes: their partial role in defense against insects. Pages 151–172 *in* Insects and the plant surface. B Juniper and Sir R Southwood, eds. Edward Arnold, London.

Duffey S S, Felton G W (1989) Role of plant enzymes in resistance to insects. Pages 289–313 *in* Biocatalysis in agricultural biotechnology. J R Whittaker and D E Sonnet, eds. American Chemical Society, Washington, DC.

Duke W B, Hagin R D, Hunt J F, Linscott D L (1975) Metal halide lamps for supplemental lighting in greenhouses: crop response and spectral distribution. Agron. J. 67:49–53.

Dyck V A, Thomas B (1979) The brown planthopper problem. Pages 3–17 *in* Brown planthopper: Threat to rice production in Asia. International Rice Research Institute, Manila, Philippines.

Eastop V F (1973a) Deductions from present day host plants of aphids and related insects. Pages 157–178 *in* Insect–plant relationships. H F van Emden, ed. Symp. R. Entomol. Soc. London No. 6. Blackwell Scientific Publications, London.

Eastop V F (1973b) Biotypes of aphids. Bull. Entomol. Soc. NZ 2:40–51.

Edwards P J, Wratten S D (1980) Ecology of insect–plant interactions. Edward Arnold, London.

Edwards P J, Wratten S D, Cox H (1985) Wound-induced changes in the acceptability of tomato to larvae of *Spodoptera littoralis*: a laboratory bioassay. Ecol. Entomol. 10:155–158.

Edwards P J, Wratten S D, Gibberd R M (1991) The impact of inducible phytochemicals on food selection by insect herbivores and its consequences for the distribution of grazing damage. Pages 205–221 *in* Phytochemical induction by herbivores. D W Tallamy and M J Raupp, eds. John Wiley and Sons, New York.

Eenink A H, Groenwold R, Dieleman F L (1982) Resistance of lettuce (*Lactuca*) to the leaf aphid *Nasonovia ribis nigri* I. Transfer of resistance from *L. virosa* to *L. sativa* by interspecific crosses and selection of resistant breeding lines. Euphytica 31:291–300.

Ehrlich P R, Raven P H (1964) Butterflies and plants: a study in coevolution. Evolution 18:586–608.

Eidt D C, Little C H A (1970) Insect control through induced host–insect asynchrony: a progress report. J. Econ. Entomol. 63:1966–1968.

Einhellig F A (1989) Interactive effects of allelochemicals and environmental stress.

Pages 101–118 in Phytochemical ecology: Allelochemicals, mycotoxins and insect pheromones and allomones. C H Chou and G R Waller, eds. Inst. of Botany, Academia Sinica Monog. Series 9. Taipei, Republic of China.

El-Ibrashy M T (1987) Juvenoids and related compounds in tropical pest management: a review. Insect Sci. Applic. 8:743–753.

Elliger C A, Wong Y, Chan B G, Waiss Jr A C (1981) Growth inhibitors in tomato (Lycopersicon) to tomato fruitworm (Heliothis zea) J. Chem. Ecol. 7:753–758.

Ellis D D, McCabe D E, McInnis S, Ramachandran R, Russell D R, Wallace K M, Martinell B J, Roberts D R, Raffa K F, McCown B H (1993) Stable transformation of Picea glauca by particle acceleration. Bio/Technology 11:84–89.

Elzen G W, Williams H J, Vinson S B (1983) Response by the parasitoid Campoletis sonorensis (Hymenoptera: Ichneumonidae) to chemicals (synomones) in plants: implications for host habitat location. Environ. Entomol. 12:1872–1876.

Endress A G, Post S L (1985) Altered feeding preference of Mexican bean beetle, Epilachna varivestris for ozonated soybean foliage. Environ. Pollut. Ser. A 39:9.

Erickson J M, Feeny P (1974) Sinigrin: a chemical barrier to the black swallowtail butterfly. Ecology 55:103–111.

Eriksson C (1975) Aroma compounds derived from oxidized lipids. Some biochemical and analytical aspects. J. Agric. Food Chem. 23:126–128.

Espelie K E, Bernays E A, Brown J J (1991) Plant and insect cuticular lipids serve as behavioral cues for insects. Arch. Insect Biochem. Physiol. 17:223–233.

Etzler M E (1985) Plant lectins: molecular and biological aspects. Ann. Rev. Plant Physiol. 36:209–234.

Evans S V, Fellows L E, Shing T K M, Fleet G W J (1985) Glycosidase inhibition by plant alkaloids which are structural analogues of monosaccharides. Phytochemistry 24:1953–1955.

Everson E H, Gallun R L (1980) Breeding approaches in wheat. Pages 513–533 in Breeding plants resistant to insects. F G Maxwell and P R Jennings, eds. John Wiley and Sons, New York.

Fajer E D, Bowers M D, Bazzaz F A (1989) The effects of enriched carbon dioxide atmospheres on plant-insect herbivore interactions. Science 243:1198–1200.

Farrar Jr R R, Kennedy G G (1987) 2-Undecanone, a constituent of the glandular trichomes to Lycopersicon hirsutum f. glabratum. Effects on Heliothis zea and Manduca sexta growth and survival. Entomol. Exp. Appl. 43:17–23.

Febvay G, Bonnin J, Rahbe Y, Bournoville R, Delrot S, Bonnemain J L (1988) Resistance of different lucerne cultivars to the pea aphid Acyrthosiphon pisum: influence of phloem composition on aphid fecundity. Entomol. Exp. Appl. 48:127–134.

Feeny P (1969) Inhibitory effect of oak leaf tannins on the hydrolysis of proteins by trypsin. Phytochemistry 8:2119–2126.

Feeny P (1970) Seasonal changes in oak leaf tannins and nutrients as a cause of spring feeding by winter moth caterpillars. Ecology 51:565–581.

Feeny P (1982) Ecological aspects of insect–plant relationships – round-table discusssion. Pages 275–283 in Proceedings of the 5th International Symposium on Insect–Plant Relationships. J H Visser and A K Minks, eds. Centre for Agricultural Publications and Documentation, Wageningen.

Feeny P, Lorraine R, Maureen C (1983) Chemical aspects of oviposition behavior in butterflies. Pages 27–75 in Herbivorous insects: Host seeking behavior and mechanisms. S Ahmad, ed. Academic Press, New York.

Feeny P, Sachdev K, Rosenberry L, Carter M (1988) Luteolin 7-*O*-(6'-*O*-malonyl)-β-D-glucoside and *trans*-chlorogenic acid: oviposition stimulants for the black swallowtail butterfly. Phytochemistry 27:3439–3448.

Fehr W R, Caviness C E (1977) Stages of soybean development. Iowa State University Special Report 80. Iowa State University, Ames, IA.

Fellows L E, Evans S V, Nash R J, Bell E A (1986) Polyhydroxy plant alkaloids as glycosidase inhibitors and their possible ecological role. Pages 72–78 *in* Natural resistance of plants to pests. M B Green and P A Hedin, eds. ACS Symp. Series 296. American Chemical Society, Washington, DC.

Fellows L E, Kite G C, Nash R J, Simmonds M S J, Scofield A M (1989) Castanospermine, swainsonine and related polyhydroxy alkaloids: structure, distribution, and biological activity. Pages 395–427 *in* Plant nitrogen metabolism. J E Poulton, J T Romeo and E E Conn, eds. Plenum Publishing, New York.

Felton G W, Duffey S S (1990) Inactivation of baculovirus by quinones formed in insect-damaged plant tissues. J. Chem. Ecol. 16:1221–1236.

Finch S (1980) Chemical attraction of plant-feeding insects to plants. Pages 67–143 *in* Applied biology. T H Coaker, ed. Academic Press, New York.

Finch S, Spinner G (1982) Trapping cabbage root flies in traps baited with plant extracts and with natural and synthetic isothiocyanates. Entomol. Exp. Appl. 31:133–139.

Fischer D C, Kogan M, Greany P (1990) Inducers of plant resistance to insects. Pages 257–280 *in* Safer insecticides: Development and use. E Hodgson and R J Kuhr, eds. Marcel Dekker, New York.

Fischhoff D A, Bowdish K S, Perlak F J, Marrone P G, McCormic S M, Niedermyer J G, Dean D A, Kusano-Kretzmer K, Mayer E J, Rochester D E, Rogers S G, Fraley R T (1987) Insect tolerant transgenic tomato plants. Bio/Technology 5:807–813.

Flath R A, Forrey R R, John J O, Chan B G (1978) Volatile components of corn silk (*Zea mays* L.): possible *Heliothis zea* (Boddie) attractants. J. Agric. Food Chem. 26:1290–1293.

Flemion F, McNear B (1951) Reduction of vegetative growth and seed yield in umbelliferous plants by *Lygus oblineatus*. Contributions of the Boyce Thompson Institute 16:279–283.

Flor H H (1942) Inheritance of pathogenecity in *Melampsora lini*. Phytopathology 32:653–669.

Flor H H (1956) The complementary genetic systems in flax and flax rust. Adv. Genet. 8:29–54.

Fox L R, Macauley B J (1977) Insect grazing on *Eucalyptus* in response to variation in leaf tannins and nitrogen. Oecologia 29:145–162.

Fox L R, Morrow P A (1981) Specialization: species property or local phenomenon? Science 211:887–893.

Fraenkel G S (1959) The *raison d'etre* of secondary plant substances. Science 129:1466–1470.

Fraenkel G S (1969) Evaluation of our thoughts on secondary plant substances. Entomol. Exp. Appl. 12:473–486.

Fraenkel G S, Nayer J K, Nalbandov O, Yamamoto R T (1960) Further investigations into the chemical basis of the insect–host plant relationship. Proc. Int. Congr. Entomol. Vienna 3:122–126.

Frazer B D (1972) Population dynamics and recognition of biotypes in the pea aphid (Homoptera: Aphididae) Can. Entomol. 10:1729–1733.

Frazier J L (1986) The perception of plant allelochemicals that inhibit feeding. Pages 1–42 in Molecular aspects of insect–plant associations. L B Brattsten and S Ahmad, eds. Plenum Press, New York.

Frazier J L (1992) How animals perceive secondary plant compounds. Pages 89–134 in Herbivores: Their interactions with secondary plant metabolites, Vol. II, 2nd ed. G A Rosenthal and M R Berenbaum, eds. Academic Press, New York.

Frey K J (1975) Breeding concepts and techniques for self-pollinated crops. Pages 257–278 in Proceedings of the International Workshop on Grain Legumes. International Crop Research Institute for Semi-arid Tropics, Patancheru, Andhra Pradesh, India.

Friebe B, Hatchett J H, Sears R G, Gill B S (1990) Transfer of Hessian fly resistance from 'Chaupon' rye to hexaploid wheat via a 2 BS/2 RL wheat-rye chromosome translocation. Theor. Appl. Genet. 79:385–389.

Friend J (1976) Lignification of infected tissue. Pages 291–304 in Biochemical aspects of plant–parasite relationships. J Friend and D R Therlfall, eds. Academic Press, New York.

Frost S W (1942) General entomology. McGraw-Hill, New York.

Fujimoto H, Itoh K, Yamamoto M, Kyozuka J, Shimamoto K (1993) Insect resistant rice generated by introduction of a modified delta-endotoxin gene of Bacillus thuringiensis. Bio/Technology 11:1151–1155.

Fukutoku Y, Yamada Y (1982) Accumulation of carbohydrates and proline in water-stressed soybean (Glycine max L.) Soil Sci. Plant Nutri. 28:147–151.

Futuyma D J (1983) Evolutionary interactions among herbivorous insects and plants. Pages 207–231 in Coevolution. D J Futuyma and M Slatkin, eds. Sinauer, New York.

Futuyma D J, Keese M C (1992) Evolution and coevolution of plants and phytophagous arthropods. Pages 439–475 in Herbivores: Their interactions with secondary plant metabolites, Vol. II, 2nd ed. G A Rosenthal and M R Berenbaum, eds. Academic Press, New York.

Futuyma D J, Slatkin M, eds (1983) Coevolution. Sinauer, New York.

Gallagher K D (1992) IPM development in the Indonesian national IPM program. Pages 82–89 in Pest management and environment in 2000. A A S A Kadir and H S Barlow, eds. CAB International, Wallingford, UK.

Gallaher R N, Brown R H (1977) Starch storage in C_4 vs C_3 grass leaf cells as related to nitrogen deficiency. Crop Sci. 17:85–88.

Gallardo F, Boethel D J, Fuxa J R, Richter A (1990) Susceptibility of Heliothis zea (Boddie) larvae to Nomuraea ridleyi (Farlow) Samson. Effects of α-tomatine at the third trophic level. J. Chem. Ecol. 16:1751–1759.

Galliard T, Mercer E I (1975) Recent advances in the chemistry and biochemistry of plant lipids. Academic Press, New York.

Gallun R L (1965) The Hessian fly. USDA Leaflet No. 553. US Government Printing Office, Washington DC. 8 pp.

Gallun R L (1972) Genetic inter-relationship between host plants and insects. J. Environ. Qual. 1:259–265.

Gallun R L (1977) Genetic basis of Hessian fly epidemics. Pages 223–229 in The genetic basis of epidemic in agriculture. P R Day, ed. Ann. NY. Acad. Sci. No. 287. New York.

Gallun R L, Hatchett J H (1968) Inter-relationships between races of Hessian fly, *Mayetiola destructor* (Say) and resistance in wheat. Pages 258–262 *in* Proceedings of the Third International Wheat Genetics Symposium. K W Finlay and K W Shepherd, eds. Australian Academy of Sciences, Canberra.

Gallun R L, Khush G S (1980) Genetic factors affecting expression and stability of resistance. Pages 63–85 *in* Breeding plants resistant to insects. F G Maxwell and P R Jennings, eds. John Wiley and Sons, New York.

Gallun R L, Patterson F L (1977) Monosomic analysis of wheat for resistance to hessian fly. J. Hered. 68:223–226.

Games D E, James D H (1972) The biosynthesis of the coumarins of *Angelica archangelica*. Phytochemistry 11:868–869.

Gardenhire J H, Tuleen N A, Stewart K W (1973) Trisomic analysis of greenbug resistance in barley, *Hordeum vulgare* L. Crop Sci. 13:684–685.

Gardner D R, Stermitz F R (1988) Host plant utilization and iridoid glycoside sequestration by *Euphydryas anicia* (Lepidoptera: Nymphalidae) J. Chem. Ecol. 14:2147–2168.

Gardner W A, Duncan, R R (1982) Influence of soil pH on fall armyworm (Lepidoptera: Noctuidae) damage to whorl-stage sorghum. Environ. Entomol. 11:908–912.

Gatehouse A M R, Dewey F M, Dove J, Fenton K A, Pusztai A (1984) Effect of seed lectins from *Phaseolus vulgaris* on the development of larvae of *Callosobruchus maculatus*; mechanism of toxicity. J. Sci. Food Agric. 35:373–380.

Gatehouse A M R, Boulter D, Hilder V A (1992) Potential of plant-derived genes in the genetic manipulation of crops for insect resistance. Pages 155–181 *in* Plant genetic manipulation for crop protection. A M R Gatehouse, V A Hilder and D Boulter, eds. CAB International, Wallingford, UK.

Gatehouse A M R, Powell K S, Van Damme E, Peumanns W, Gatehouse J A (1995) Insecticidal properties of plant lectins: their potential in plant protection. Pages 39–57 *in* Lectins–Biomedical perspectives. A J Pusztai, ed. Taylor and Francis, London.

Gates D M (1980) Biophysical ecology. Springer-Verlag, New York.

Geier P W and Clark L R (1961) An ecological approach to pest control. Pages 10–18 *in* Proceedings of the Eighth Technical Meeting, 1960. J Kuenen, ed. International Union for Conservation of Nature and Natural Resources, Warsa.

Geissman T A, Crout D H G (1969) Organic chemistry of secondary plant metabolism. Freeman, Cooper and Co., San Francisco.

Gentile A G, Stoner A K (1968) Resistance in *Lycopersicon* and *Solanum* species to the potato aphid. J. Econ. Entomol. 61:1152–1154.

Gentile A G, Webb R E, Stoner A K (1968) Resistance in *Lycopersicon* and *Solanum* to greenhouse whiteflies. J. Econ. Entomol. 61:1355–1357.

Georghiou G P (1972) The evolution of resistance to insecticides. Ann. Rev. Ecol. Syst. 17:133–168.

Georghiou G P (1986) The magnitude of the resistance problems. Pages 14–43 *in* Pesticide resistance, strategies and tactics for management. National Academy Press, Washington, DC.

Georghiou G P, Taylor C E (1977) Genetic and biological influences in the evolution of insecticide resistance. J. Econ. Entomol. 70:319–323.

Gerhold, D L, Craig R, Mumma R O (1984) Analysis of trichome exudate from mite resistant geraniums. J. Chem. Ecol. 10:713–722.

Gershenzon J (1984) Changes in the levels of plant secondary metabolites under water and nutrient stress. Pages 273–320 *in* Phytochemical adaptations to stress. B N Timmermann, C Steelink and F A Loewus, eds. Rec. Adv. Phytochem. Vol. 18. Plenum Press, New York.

Gershenzon J, Croteau R (1991) Terpenoids. Pages 165–219 *in* Herbivores: Their interactions with secondary metabolites, Vol. I, 2nd ed. G A Rosenthal and M R Berenbaum, eds. Academic Press, New York.

Gershenzon J, Rossiter M, Mabry T J, Rogers C E, Blust M H, Hopkins T L (1985) Insect antifeedant terpenoids in wild sunflower: a possible source of resistance to the sunflower moth. Pages 433–445 *in* Bioregulators for pest control. P A Hedin, ed, ACS Symp. Series 276. American Chemical Society, Washington, DC.

Gibson R W (1971) Glandular hairs providing resistance to aphids in certain wild potato species. Ann. Appl. Biol. 68:113–119.

Gibson R W (1978) Resistance in glandular haired wild potatoes to flea-beetles. Am. Potato J. 55:595–599.

Gibson R W, Turner R H (1977) Insect-trapping hairs on potato plants. Pest Articles and News Summaries 23:272–277.

Gibson R W, Valencia L (1978) A survey of potato species for resistance to the mite *Polyphagoparsomemus latus*, with particular reference to the protection of *Solanum berthaultii* and *S. tarijense* by glandular hairs. Potato Res. 21:217–223.

Gilbert B L, Norris D M (1968) A chemical basis for bark beetle (*Scolytus*) distinction between host and non-host trees. J. Insect Physiol. 14:1063–1068.

Gilbert L E (1979) Development of theory in the analysis of insect–plant interactions. Pages 117–154 *in* Analysis of ecological systems. D J Horn, R D Mitchell and G R Stairs, eds. Ohio State University Press, Columbus, OH.

Gildow F E (1980) Increased production of alate by aphids reared on oats infected with barley yellow dwarf virus. Ann. Entomol. Soc. Am. 73:343–347.

Gill B S, Hatchett J H, Raupp W J (1987) Chromosomal mapping of Hessian fly-resistance gene *H13* in D genome of wheat. J. Hered. 78:97–100.

Gill J S, Lewis C T (1971) Systemic action of an insect feeding deterrent. Nature (London) 232:402–403.

Gilmore Jr E C (1964) Suggested method of using reciprocal recurrent selection in some naturally self-pollinated species. Crop Sci. 4:323–325.

Glass E H (1975) Integrated pest management rationale, potential, needs and implementation. Entomological Society of America, College Park, MD.

Goertzen L R, Small E (1993) The defensive role of trichomes in black medick (*Medicago lupulina* Fabaceae) Plant Syst. Evol. 184:101–111.

Gomez K A, Bernardo R C (1974) Estimation of stem borer damage in rice fields. J. Econ. Entomol. 67:509–513.

Gonzalez D, Gordh G, Thompson S N, Adler J (1979) Biotype discrimination and its importance to biological control. Pages 129–136 *in* Genetics in relation to insect management. M A Hoy and J J McKelvey Jr, eds. Rockefeller Foundation, New York.

Gorz H J, Haskins F A, Manglitz G R (1972) Effect of coumarin and related compounds on blister beetle feeding in sweetclover. J. Econ. Entomol. 65:1632–1635.

Gossard T W, Jones R E (1977) The effects of age and weather on egg-laying in *Pieris rapae* L. J. Appl. Ecol. 14:65–71.

Gould F (1984) Role of behavior in the evolution of insect adaptation to insecticides and resistant host plants. Bull. Entomol. Soc. Am. 30:33–41.

Gould F (1986) Simulation models for predicting durability of insect resistant germplasm: a deterministic diploid, two-locus model. Environ. Entomol. 15:1–10.

Gould F (1988) Evolutionary biology and genetically engineered crops. BioScience 38:26–33.

Gould F (1991) The evolutionary potential of crop pests. Am. Sci. 79:496–507.

Gould F, Carroll C R, Futuyma D J (1982) Cross-resistance to pesticides and plant-defenses: a study of the two-spotted spider mite. Entomol. Exp. Appl. 31:175–180.

Gould F, Kennedy G G, Johnson M T (1991) Effects of natural enemies on the rate of herbivore adaptation to resistant host plants. Entomol. Exp. Appl. 58:1–14.

Gould F, Martinez-Ramirez A, Anderson A, Ferre J, Silva F J, Moar W J (1992) Broad spectrum resistance to *Bacillus thuringiensis* toxins in *Heliothis virescens*. Proc. Natl. Acad. Sci. USA 89:7986–7990.

Green T R, Ryan C A (1972) Wound-induced proteinase inhibitor in plant leaves: a possible defense mechanism against insects. Science 175:776–777.

Gregory P (1984) Glycoalkaloid composition of potatoes: diversity and biological implications. Am. Potato J. 61:115–122.

Guerin P M, Visser J H (1980) Electroantennogram responses of the carrot fly, *Psila rosae*, to volatile plant components. Physiol. Entomol. 5:111–119.

Guerra D J (1981) Natural and *Heliothis zea* (Boddie) induced levels of specific phenolic compounds in *Gossypium hirsutum* (L.) MS thesis, University of Arkansas, Fayetteville, AR.

Guha J, Sen S P (1975) The cucurbitacins – a review. Indian Biochem. J. 2:12–28.

Guthrie W D (1989) Advances in rearing the European corn borer on a meridic diet. Pages 46–59 *in* Toward insect resistant maize for third world. J A Mihm, B R Wiseman and F M Davis, eds. International Wheat and Maize Improvement Center (CIMMYT), El Batan, Mexico.

Guthrie W D, Dicke F F, Neiswander C R (1960) Leaf and sheath feeding resistance to the European corn borer in eight inbred lines of dent corn. Ohio Agr. Exp. Stn. Res. Bull. 860. 38 pp.

Guthrie W D, Huggans J L, Chatterji S M (1970) Sheath and collar feeding resistance to the second-brood European corn borer in six inbred lines of dent Corn. Iowa State J. Sci. 44:297–311.

Guthrie W D, Rathore Y S, Cox D F, Reed G L (1974) European corn borer: virulence on crop plants of larvae reared for different generations on a meridic diet. J. Econ. Entomol. 67:605–606.

Guthrie W D, Russell W A, Reed G L, Hallauer A R, Cox D F (1978) Methods of evaluating maize for sheath-collar feeding resistance to the European corn borer. Maydica 23:45–53.

Gutierrez A P (1987) Systems analysis in integrated pest management. Pages 71–84 *in* Integrated pest management protection integree, *quo vadis?* An international perspective V Delucchi, ed. Parasitis, Geneva.

Haglund B M (1980) Proline and valine–cues which stimulate grasshopper herbivory during drought stress? Nature 288:697–698.

Hahlbrock K, Knobloch K H, Kreuzaler F, Potss J R M, Wellmann E (1976) Coordinated induction and subsequent activity changes of two groups of metabolically interrelated enzymes: light-induced synthesis of flavonoid glycosides in cell suspension culture of *Petroselinum hortense*. Eur. J. Biochem. 61:199–206.

Halkier B A, Moller B L (1989) Biosynthesis of the cyanogenic glucoside dhurrin in seedlings of *Sorghum bicolor* (L.) Moench. and partial purification of the enzyme system involved. Plant Physiol. 90:1552–1559.

Hall P K, Parrott W L, Jennings J N, McCarty J C (1980) Use of tobacco budworm eggs and larvae for establishing field infestations on cotton. J. Econ. Entomol. 73:393–395.

Hamamura Y (1959) Food selection by silkworm larvae. Nature (London) 183:1746–1747.

Hamilton R J, Munro J, Rowe J M (1979) The identification of chemicals involved in the interaction of *Oscinella frit* with *Avena sativa*. Entomol. Exp. Appl. 25:328–341.

Hamm J J, Wiseman B R (1986) Plant resistance and nuclear polyhedrosis virus for suppression of the fall armyworm (Lepidoptera: Noctuidae). Florida Entomol. 69:541–549.

Hammond A M, Hardy T N (1988) Quality of diseased plants as hosts for insects. Pages 381–432 *in* Plant stress–insect reactions. E A Heinrichs, ed. John Wiley and Sons, New York.

Hammons R O (1970) Registration of spancross peanuts. Crop Sci. 10:459.

Hanson A D, Hitz W D (1983) Whole-plant response to water deficits: water deficits and the nitrogen economy. Pages 331–343 *in* Limitations to efficient water use in crop production. H M Taylor, W R Jordan and T R Sinclair, eds. American Society of Agronomy, Madison, WI.

Hanson C H, Busbice T H, Hill R R, Hunt O J, Oakes A J (1972) Directed mass selection for developing multiple pest resistance and conserving germplasm in alfalfa. J. Environ. Qual. 1:106–111.

Hanson F E (1970) Sensory response of phytophagous lepidoptera to chemical and tactile stimuli. Pages 81–91 *in* Control of insect behavior by natural products. D L Wood, R M Silverstein and M Nakajima, eds. Academic Press, New York.

Hanson F E, Dethier V G (1973) Role of gustation and olfaction in food plant discrimination in the tobacco hornworm, *Manduca sexta*. J. Insect Physiol. 19:1019–1034.

Harborne J B (1980) Plant phenolics. Pages 329–402 *in* Secondary plant products. Encyclopedia of plant physiology, New Series, Vol. 8. E A Bell and B V Charlwood, eds. Springer-Verlag, Berlin.

Harborne J B (1988a) Introduction to ecological biochemistry, 3rd ed. Academic Press, New York.

Harborne J B (1988b) The flavonoids: Advances in research since 1980. Chapman and Hall, London.

Harborne J B, Turner B L (1984) Plant chemosystematics. Academic Press, London.

Hardy T N, Clay K, Hammond Jr, A M (1985) Fall armyworm (Lepidoptera: Noctuidae): a laboratory bioassay and larval preference study for the fungal endophyte of perennial ryegrass. J. Econ. Entomol. 78:571–575.

Hare J D (1980) Impact of defoliation by the Colorado potato beetle on potato yields. J. Econ. Entomol. 73:369–373.

Hare J D (1983) Manipulation of host suitability for herbivore pest management. Pages 655–679 *in* Variable plants and herbivores in natural and managed systems. R F Denno and M S McClure, eds. Academic Press, New York.

Hare J D (1992) Effects of plant variation on herbivore – natural enemy interactions.

Pages 278–298 *in* Plant resistance to herbivores and pathogens. R S Fritz and E L Simms, eds. The University of Chicago Press, Chicago.

Harlan J R (1969) Evolutionary dynamics of plant domestication. Jap. J. Genet. 44(Suppl.):337–343.

Harlan J R (1971) Agricultural origins: centers and noncenters. Science 174:468–474.

Harlan J R (1975a) Our vanishing genetic resources. Science 188:618–621.

Harlan J R (1975b) Crops and man. American Society of Agronomy, Madison, WI. 295 pp.

Harlan J R, De Wet J M J (1971) Toward a rational classification of cultivated plants. Taxon 20:509–517.

Harris M K (1979) Arthropod–plant interactions related to agriculture, emphasizing host plant resistance. Pages 23–51 *in* Biology and breeding for resistance to arthropods and pathogens in agricultural plants. M K Harris, ed. Publication MP 1451. Texas A & M University, College Station, TX.

Harris M O, Miller J R (1982) Synergism of visual and chemical stimuli in the oviposition behaviour of *Delia antiqua* (Meigen) (Diptera: Anthomyiidae) Pages 117–121 *in* Proceedings of the 5th International Symposium on Insect–Plant Relationships. J H Visser and A K Minks, eds. PUDOC, Wageningen.

Harris M O, Miller J R, de Ponti O M B (1987) Mechanisms of resistance to onion fly egg-laying. Entomol. Exp. Appl. 43:279–286.

Harris P (1974) A possible explanation of plant yield increases following insect damage. Agro-Ecosystems 1:219–225.

Harrison G D (1987) Host plant discrimination and evolution of feeding preference in the Colorado potato beetle, *Leptinotarsa decemlineata* (Say). Physiol. Entomol. 12:407–415.

Hart S V, Kogan M, Paxton J D (1983) Effect of soybean phytoalexins on the herbivorous insects Mexican bean beetle and soybean looper. J. Chem. Ecol. 9:657–672.

Hartley S E, Firn R D (1989) Phenolic biosynthesis, leaf damage and insect herbivory in birch (*Betula pendula*) J. Chem. Ecol. 15:275–283.

Hartley S E, Lawton J H (1991) Biochemical aspects and significance of the rapidly induced accumulation of phenolics in birch foliage. Pages 105–132 *in* Phytochemical induction by herbivores. D W Tallamy and M J Raupp, eds. John Wiley and Sons, New York.

Harvey T L, Hackerott H L (1969) Recognition of a greenbug biotype injurious to sorghum. J. Econ. Entomol. 62:776–779.

Haslam E (1974) The shikimate pathway. Butterworth, London.

Hassell M P, Southwood T R E (1978) Foraging strategies of insects. Ann. Rev. Ecol. Syst. 9:75–98.

Hatchett J H, Gallun R L (1970) Genetics of the ability of the Hessian fly, *Mayetiola destructor*, to survive on wheats having different genes for resistance. Ann. Entomol. Soc. Am. 63:1400–1407.

Hatchett J H, Martin T J, Livers R W (1981) Expression and inheritance of resistance to Hessian fly in synthetic hexaploid wheats derived from *Triticum tauschii* (Coss) Schmal. Crop Sci. 21:731–734.

Haukioja E (1991) Induction of defenses in trees. Ann. Rev. Entomol. 36:25–42.

Haukioja E, Niemela P (1977) Retarded growth of a geometrid larva after mechanical damage to leaves of its host tree. Ann. Zool. Fennici. 14:48–52.

Hayman B I (1954) The theory of analysis of diallel crosses. Genetics 39:789–809.

Heck W W, Taylor O C, Heggestad H (1973) Air-pollution research needs: herbaceous and ornamental plants and agriculturally generated pollutants. J. Air Pollut. Control Assoc. 23:257–266.

Hedin P A, Jenkins J N, Maxwell F G (1977) Behavioral and developmental factors affecting host plant resistance to insects. Pages 231–275 *in* Host plant resistance to insect pests. P A Hedin, ed. ACS Symp. Series 62. American Chemical Society, Washington, DC.

Hedin P A, Jenkins J N, Collum D H, White W H, Parrot W L, MacGown M W (1983) Cyanidin-3-β-glucoside, a newly recognized basis for resistance in cotton to the tobacco budworm *Heliothis virescens* (Fab.) (Lepidoptera: Noctuidae). Experientia 39:799–801.

Hedin P A, Davis F M, Williams W P (1993) 2-Hydroxy-4,7-dimethoxy-1,4-benzoxazin-3-one (N-O-ME-DIMBOA), a possible toxic factor in corn to the southwestern corn borer. J. Chem. Ecol. 19:531–542.

Heftmann E (1973) Steroids. Pages 171–226 *in* Phytochemistry, organic metabolites, Vol. II. L P Miller, ed. Van Nostrand Reinhold Co., New York.

Heinrichs B (1971) The effect of leaf geometry on the feeding behavior of the caterpillar of *Manduca sexta* (Sphingidae) Anim. Behav. 19:119–124.

Heinrichs E A (1986) Perspectives and directions for the continued development of insect-resistant rice varieties. Agric. Ecosyst. Environ. 18:9–36.

Heinrichs E A (1988) Role of insect-resistant varieties in rice IPM systems. Pages 43–54 *in* Pesticide management and integrated pest management in Southeast Asia. P S Teng and K L Heong, eds. International Crop Protection, College Park MA.

Heinrichs E A, Pathak P K (1981) Resistance to the rice gall midge *Orseolia oryzae* in rice. Insect Sci. Appl. 1:123–132.

Heinrichs E A, Rapusas, H R (1985) Cross virulence of *Nephotettix virescens* (Homoptera: Cicadellidae) biotypes among some rice cultivars with the same major resistance gene. Environ. Entomol. 14:696–700.

Heinrichs E A, Aquino G B, Chelliah S, Valencia S L, Reissig W H (1982) Resurgence of *Nilaparvata lugens* (Stal) populations as influenced by method and timing of insecticide applications in lowland rice. Environ. Entomol. 11:78–84.

Heinrichs E A, Fabellar L T, Basilio R P, Cheng-Wen Tu, Medrano F (1984) Susceptibility of rice planthoppers, *Nilaparvata lugens* and *Sogatella furcifera* (Homoptera: Delphacidae) to insecticides as influenced by level of resistance in the host plant. Environ. Entomol. 13:455–458.

Heinrichs E A, Medrano F G, Rapusas H R (1985) Genetic evaluation for insect resistance in rice. International Rice Research Institute, Los Banos, Philippines.

Heinrichs E A, Rapusas H R, Aquino G B, Pallis F (1986) Integration of host plant resistance and insecticides in the control of *Nephotettix virescens* (Homoptera: Cicadellidae), a vector of rice tungro virus. J. Econ. Entomol. 79:437–443.

Helentjaris T (1987) A genetic linkage map for maize based on RFLPs. Trends Genet. 3:217–221.

Helms D (1980) Revision and reversion: changing cultural control practices for the cotton boll weevil. Agric. Hist. 54:108–125.

Henning W (1981) Insect phylogeny. John Wiley and Sons, Chichester, UK.

Herbert R B (1989) The biosynthesis of secondary metabolites, 2nd ed. Chapman and Hall, London.

Herrera-Estrella L, Van Den Broeck G, Maenhaut R, Van Montagu M, Schell J, Timko M, Cashmore A (1984) Light-inducible and chloroplast-associated expression of a chimaeric gene introduced into *Nicotiana tabacum* using a Ti plasmid vector. Nature 310:115–120.

Hiatt A, ed (1993) Transgenic plants: fundamentals and applications. Marcel Dekker Inc., New York.

Hilder V A, Gatehouse A M R, Sheerman S E, Baker R F, Boulter D (1987) A novel mechanism of insect resistance engineered into tobacco. Nature 330:160–163.

Hinton H E (1977a) Enabling mechanisms. Pages 71–83 *in* Proceedings of the XV International Congress on Entomology, Washington, DC. D White, ed. Entomological Society of America, College Park MD.

Hinton H E (1977b) Subsocial behavior and biology of some Mexican membracid bugs. Ecol. Entomol. 2:61–79.

Hodges J D, Elam W W, Watson W F, Nebeker T E (1979) Oleoresin characteristics and susceptibility of four southern pines to southern pine beetle (Coleoptera: Scolytidae) attacks. Can. Entomol. 111:889–896.

Hodgson E S, Lettvin J Y, Roeder K D (1955) Physiology of a primary chemoreceptor unit. Science 122:417–418.

Hogan M E, Schulz W W, Slaytor M, Czolij R T, O'Brien R W (1988) Components of termite and protozoal cellulases from the lower termite, *Coptotermes lacteus* Froggatt. Insect Biochem. 18:45.

Hokkanen H M T (1991) Trap cropping in pest management. Ann. Rev. Entomol. 36:119–138.

Hollingsworth C S, Berry R E (1982) Two spotted spider mite (Acari: Tetranychidae) in peppermint: population dynamics and influence of cultural practices. Environ. Entomol. 11:1280–1284.

Holmes N D (1984) The effect of light on the resistance of hard red spring wheats to the wheat stem sawfly, *Cephus cinctus* (Hymenoptera: Cephidae) Can. Entomol. 116:677–684.

Holtzer T O, Archer T L, Norman J M (1988) Host plant suitability in relation to water stress. Pages 111–137 *in* Plant stress–insect interactions. E A Heinrichs, ed. John Wiley and Sons, New York.

Honda K (1986) Flavonone glycosides as oviposition stimulants in a papilionid butterfly, *Papilio protenor*. J. Chem. Ecol. 12:1999–2010.

Hopkins R M (1991) Feeding behaviour of the brown planthopper on susceptible and resistant cultivars. Int. Rice Res. Newsl. 16:10.

Horber E (1972) Plant resistance to insects. Agric. Sci. Rev. 10:1–10.

Horber E (1980) Types and classification of resistance. Pages 15–21 *in* Breeding plants resistant to insects. F G Maxwell and P R Jennings, eds. John Wiley and Sons, New York.

Hori K (1973) Studies on the feeding habits of *Lygus disponsi* Linnavuori (Hemiptera: Miridae) and the injury to its host plant. III. Phenolic compounds, acid phosphatase and oxidative enzymes in the injured tissue of sugar beet leaf. Appl. Entomol. Zool. 8:103–112.

Hori K, Atalay R (1980) Biochemical changes in the tissue of Chinese cabbage injured by the bug *Lygus disponsi*. Appl. Entomol. Zool. 15:234–241.

Horsch R B, Fry J E, Hoffman N L, Eichholtz D, Rogers S G, Fraley R T (1985) A simple and general method for transferring genes into plants. Science 227:1229–1231.

House H L (1969) Effects of different proportions of nutrients on insects. Entomol. Exp. Appl. 12:651–669.

Hovanitz W (1969) Inherited and/or conditioned changes in host plant preference in *Pieris*. Pages 729–735 *in* Proceedings of the International Symposium on Insect and Host Plant. J de Wilde and L M Schoonhoven, eds. North Holland, Amsterdam.

Hsiao T H (1969) Chemical basis of host selection and plant resistance in oligophagous insects. Entomol. Exp. Appl. 12:777–788.

Hsiao T H, Fraenkel G (1968a) Selection and specificity of the Colorado potato beetle for solanaceous and non-solanaceous plants. Ann. Entomol. Soc. Am. 61:493–503.

Hsiao T H, Fraenkel G (1968b) The influence of nutrient chemicals on the feeding behavior of the Colorado potato beetle, *Leptinotarsa decemlineata* (Coleoptera: Chrysomelidae) Ann. Entomol. Soc. Am. 61:44–54.

Huber L L, Neiswander C R, Salter R M (1928) The European corn borer and its environment. Ohio Agric. Exp. Stn. Bull. 429. 196 pp.

Huesing J E, Murdock L L, Shade R E (1991) Rice and stinging nettle lectins: insecticidal activity similar to wheatgerm agglutinin. Phytochemistry 30:3565–3568.

Huffaker C B, Smith R F (1980) Rationale, organization, and development of a national integrated pest management project. Pages 1–24 *in* New technology of pest control. C B Huffaker, ed. John Wiley and Sons, New York.

Hughes P R (1988) Insect populations on host plants subjected to air pollution. Pages 249–319 *in* Plant stress–insect interactions. E A Heinrichs, ed. John Wiley and Sons, New York.

Hughes P R, Dickie A I, Penton M A (1983) Increased success of Mexican bean beetle on field-grown soybeans exposed to sulfur dioxide. J. Environ. Qual. 12:565–568.

Hulley P E (1988) Caterpillar attacks plant mechanical defense by mowing trichomes before eating. Ecol. Entomol. 13:239–241.

Hunt D W A, Borden J H (1989) Terpene alcohol pheromone production by *Dentroctonus ponderosae* and *Ips paraconfusus* (Coleoptera: Scolytidae) in the absence of readily culturable microorganisms. J. Chem. Ecol. 15:1433–1463.

Hüppi G, Siddall J B (1968) Synthetic studies on insect hormones, Part VI. The synthesis of ponasterone A and its stereochemical identity with crustecdysone. Tetrahedron Lett. 9:1113–1114.

Hurter J, Boller E F, Stadler E, Blattmann B, Buser H R, Bosshard N U, Damm L, Kozlowski M W, Schoni R, Raschdorf F, Dahinden R, Schlumpf E, Fritz H, Richter W J, Schreiber J (1987) Oviposition deterring pheromone in *Rhagoletis cerasi* L. Purification and determination of the chemical constitution. Experientia 43:157–164.

Husain M A, Lal K B (1940) The bionomics of *Empoasca devastans* Distant on some varieties of cotton in the Punjab. Indian J. Entomol. 2:123–136.

ICRISAT (International Crops Research Institute for the Semi-Arid Tropics) (1989) Recommendations. Pages 179–181 *in* Proceedings of the International Workshop on Sorghum Stemborers, 17–20 Nov. 1987. ICRISAT, Patancheru, India.

Ikeda R (1985) Studies on the inheritance of resistance to the rice brown planthopper (*Nilaparvata lugens* Stal) and the breeding of resistant rice cultivars. Bull. Natl. Agric. Res. Cent. 3:1–54.

Ikeda R, Kaneda C (1981) Genetic analysis of resistance to brown planthopper,

Nilaparvata lugens Stal, in rice. Jap. J. Breed. 31:279–285.

Ikeda R, Kaneda C (1983) Trisomic analysis of the gene *Bph-1* for resistance to the brown planthopper, *Nilaparvata lugens* Stal, in rice. Jap. J. Breed. 33:40–44.

Ikeda T, Matsumura F, Benjamin D M (1977) Chemical basis for feeding adaptation of pine sawflies *Neodiprion rugifrons* and *Neodiprion swainei*. Science 197:497–499.

Ikehashi H, Fujimaki H (1980) Modified bulk population method for rice breeding. Pages 163–182 *in* Innovative approaches to rice breeding. International Rice Research Institute, Manila, Philippines.

Ikeshoji T, Ishikawa Y, Matsumoto Y (1980) Attractants against the onion maggots and flies, *Hylemya antiqua* in onions inoculated with bacteria. J. Pesticide Sci. 5:343–350.

Inouye H (1978) Neuere Ergebnisse uber die Biosynthese der Glucoside der Iridoidreiche. Planta Med. 33:193–216.

IRRI (International Rice Research Institute) (1988) Standard evaluation systems for rice. IRRI, Manila, Philippines.

IRRI (International Rice Research Institute) (1993) Program report for 1992. IRRI, Manila, Philippines.

IRRI (International Rice Research Institute) (1994) Program report for 1993. IRRI, Manila, Philippines.

Isaak A, Sorensen E L, Painter R H (1965) Stability of resistance to pea aphid and spotted alfalfa aphid in several alfalfa clones under various temperature regimes. J. Econ. Entomol. 58:140–143.

Isenhour D J, Duncan R R, Miller D R, Waskom R M, Hanning G E, Wiseman B R, Nabors M W (1991) Resistance to leaf-feeding by fall armyworm (Lepidoptera: Noctuidae) in tissue culture derived sorghums. J. Econ. Entomol. 84:680–684.

Ishaaya I, Birk Y, Bondi A, Tencer Y (1969) Soyabean saponins. IX. Studies of their effect on birds, mammals and cold-blooded organisms. J. Sci. Food Agric. 20:433–436.

Ishii T, Brar D S, Multani D S, Khush G S (1994) Molecular tagging of genes for brown planthopper resistance and earliness introgressed from *Oryza australiensis* into cultivated rice, *O. sativa*. Genome 37:217–221.

Ishikawa S (1966) Electrical response and function of a bitter substance receptor associated with the maxillary sensilla of the silkworm, *Bombyx mori* L. J. Cell Physiol. 67:1–11.

Ishimoto M, Kitamura K (1989) Growth inhibitory effects of an α-amylase inhibitor from kidney bean, *Phaseolus vulgaris* (L.) on the three species of bruchids (Coleoptera: Bruchidae). Appl. Entomol. Zool. 24:281–286.

Isman M B, Duffey S S (1982) Toxicity of tomato phenolic compounds to the fruitworm, *Heliothis zea*. Entomol. Exp. Appl. 31:370–376.

Isman M B, Rodriguez E (1983) Larval growth inhibitors from species of *Parthenium* (Asteraceae). Phytochemistry 22:2709–2713.

Jackai L E N (1982) A field screening technique for resistance of cowpea (*Vigna unguiculata* W.) to the pod borer, *Maruca testulalis* (Geyer) (Lepidoptera: Pyralidae). Bull. Entomol. Res. 72:145–156.

Jackai L E N (1991) Laboratory and screenhouse assays for evaluating cowpea resistance to the legume pod borer. Crop Protection 10:48–52.

Jackson D M, Severson R F, Johnson A W, Herzog G A (1986) Effects of cuticular duvane diterpenes from green tobacco leaves on tobacco budworm (Lepidoptera:

Noctuidae) oviposition. J. Chem. Ecol. 12:1349–1359.

Jacobson M (1965) Insect sex attractants. John Wiley and Sons, New York.

Jaenike J (1978) On optimal oviposition behavior in phytophagous insects. Theor. Pop. Biol. 14:350–356.

Janzen D H (1969) Seed-eaters versus seed size, number, toxicity and dispersal. Evolution 23:1–27.

Janzen D H (1975) Ecology of plants in the tropics. Studies in Biology No. 58. Edward Arnold, London.

Janzen D H (1980) When is it coevolution? Evolution 34:611–612.

Jena K K, Khush G S (1984) Embryo rescue of interspecific hybrids and its scope in rice improvement. Rice Genet. Newsl. 1:133–134.

Jena K K, Khush G S (1990) Introgression of genes from *Oryza officinalis* Well × Watt to cultivated rice, *O. sativa* L. Theor. Appl. Genet. 80:737–745.

Jenkins J N, Parrot W L (1971) Effectiveness of frego bract as a boll weevil resistance character in cotton. Crop Sci. 11:739–743.

Jenkins J N, Parrot W L, McCarthy Jr J C (1973) The role of a boll weevil resistant cotton in pest management research. J. Environ. Quality 2: 337–340.

Jenkins J N, McCarthy Jr J C, Callahan F E, Berberich S A, Deaton W R (1993) Growth and survival of *Heliothis virescens* (Lepidoptera: Noctuidae) on transgenic cotton containing a truncated form of the delta endotoxin gene from *Bacillus thuringiensis*. J. Econ. Entomol. 86:181–185.

Jennings C W, Russell W A, Guthrie W D, Grindeland R L (1974) Genetics of resistance in maize to second brood European corn borer. Iowa State J. Res. 48:267–280.

Jermy T (1961) On the nature of oligophagy in *Leptinotarsa decimlineata* Say (Coleoptera: Chrysomelidae). Acta Zool. Acad. Sci. Hung. 7:119–132.

Jermy T (1966) Feeding inhibitors and food preference in chewing phytophagous insects. Entomol. Exp. Appl. 9:1–12.

Jermy T (1984) Evolution of insect/host plant relationships. Am. Nat. 124: 609–630.

Jermy T (1987) The role of experience in the host selection of phytophagous insects. Pages 143–157 *in* Perspectives in chemoreception and behavior. R F Chapman, E A Bernays and J G Stoffolano, eds. Springer-Verlag, New York.

Jermy T (1990) Prospects of antifeedant approach to pest control. A critical review. J. Chem. Ecol. 16:3151–3166.

Jermy T, Szentesi A (1978) The role of inhibitory stimuli in the choice of oviposition site by phytophagous insects. Entomol. Exp. Appl. 24:458–471.

Johnson K J R, Sorenson E L, Horber E K (1980) Resistance in glandular-haired annual *Medicago* species to feeding by adult alfalfa weevils (*Hypera postica*) Environ. Entomol. 9:133–136.

Johnson M W, Welter S C, Toscano N C, Ting I P, Trumble J T (1983) Reduction of tomato leaflet photosynthesis rates by mining activity of *Liriomyza sativae* (Diptera: Agromyzidae). J. Econ. Entomol. 76:1061–1063.

Johnson R, Narvaez J, An G, Ryan C (1989) Expression of proteinase inhibitors I and II in transgenic tobacco plants: effects on natural defense against *Manduca sexta* larvae. Proc. Nat. Acad. Sci. USA 86:9871–9875.

Johnson T (1961) Man-guided evolution in plant rusts. Science 133:357–362.

Jones B F, Sorensen E L, Painter R H (1968) Tolerance of alfalfa clones to the spotted alfalfa aphid. J. Econ. Entomol. 61:1046–1050.

Jones C G (1983) Microorganisms as mediators of plant resource exploitation by insect herbivores. Pages 53–99 *in* A new ecology: Novel approaches to interactive systems. P W Price, C N Slobodchikoff and W S Gaud, eds. John Wiley and Sons, New York.

Jones J E, James D, Sistler F E, Stringer S J (1986) Spray penetration of cotton canopies as affected by leaf and bract isolines. Lousiana Agric. 29:15–17.

Judy K J, Schooley D A, Dunham L L, Hall M S, Bergot B J, Siddall J B (1973) Isolation, structure, and absolute configuration of a new natural insect juvenile hormone from *Manduca sexta*. Proc. Nat. Acad. Sci. USA 70:1509–1513.

Juvik J A, Babka B A, Timmermann E A (1988) Influence of trichome exudates from species of *Lycopersion* on oviposition behavior of *Heliothis zea* (Boddie). J. Chem. Ecol. 14:1261–1278.

Kabir M A, Khush G S (1988) Genetic analysis of resistance of brown planthopper in rice, *Oryza sativa* L. Plant Breed. 100:54–58.

Kafka W A (1974) Physicochemical aspects of odor reception in insects. Ann. NY Acad. Sci. 237:115–128.

Kaissling K E (1974) Sensory transduction in insect olfactory receptors. Pages 243–273 *in* Biochemistry of sensory functions. L Jaenicke, ed. Springer-Verlag, Berlin.

Kaissling K E, Thorson J (1980) Insect olfactory sensilla: structural, chemical and electrical aspects of the functional organization. Pages 261–282 *in* Receptors for neurotransmitters, hormones and pheromones in insects. D B Sattelle, L M Hall and J Hildebrand, eds. Elsevier/North Holland, Amsterdam.

Kalode M B, Karim A N M, Pongprasert S, Heinrichs E A (1986) Varietal improvement and resistance to insect pests. Pages 241–252 *in* Progress in rainfed lowland rice. International Rice Research Institute, Los Banos, Philippines.

Karban R (1991) Inducible resistance in agriculture systems. Pages 403–419 *in* Phytochemical induction by herbivores. D W Tallamy and M J Raupp, eds. John Wiley and Sons, New York.

Karban R (1992) Plant variation: its effects on populations of herbivorous insects. Pages 195–215 *in* Plant resistance to herbivores and pathogens. R S Fritz and E L Simms, eds. The University of Chicago Press, Chicago.

Karban R, English-Loeb G M (1988) Effects of herbivory and plant conditioning on the population dynamics of spider mites. Exp. Appl. Acar. 4:225–246.

Karban R, Myers J (1989) Induced plant resistance to herbivory. Ann. Rev. Ecol. Syst. 20:331–348.

Kartohardjono A, Heinrichs E A (1984) Populations of the brown planthopper, *Nilaparvata lugens* (Stal) (Homoptera: Delphacidae), and its predators on rice varieties with differing levels of resistance. Environ. Entomol. 13:359–365.

Kasting R, McGinnis A J (1962) Quantitative relationship between consumption and excretion of dry matter by larvae of the pale western cutworms, *Agrotis orthogonia* Morr. (Lepidoptera: Noctuidae). Can. Entomol. 94:441–443.

Kawabe S, McLean D L, Tatsuki S, Ouchi T (1981) An improved electronic measurement system for studying ingestion and salivation activities of leafhoppers. Ann. Entomol. Soc. Am. 74:222–225.

Keep E (1989) Breeding red raspberry for resistance to diseases and pests. Pages 245–321 *in* Plant breeding review, Vol. 6. J Janick, ed. Timber Press, Portland, OR.

Keep E, Knight R L, Parker J H (1969) Further data on resistance to the Rubus aphid, *Amphorophora rubi* (Kalt.). Rep. E. Malling Res. Stn. 129–131.

Keller C J, Bush L P, Grunwald C (1969) Changes in content of sterols, alkaloids, and phenols in fluecured tobacco during conditions favoring infestation by molds. J. Agric. Food Chem. 17:331–334.

Kelsey R G, Reynolds G W, Rodriguez E (1984) The chemistry of biologically active constituents secreted and stored in plant glandular trichomes. Pages 187–241 in Biology and chemistry of plant trichomes. E Rodriguez, P L Healey and I Mehta, eds. Plenum Press, New York.

Kennedy G G (1976) Host plant resistance and spread of plant viruses. Environ. Entomol. 5:827–832.

Kennedy G G, Barbour J D (1992) Resistance variation in natural and managed systems. Pages 13–41 in Plant resistance to herbivores and pathogens. R S Fritz and E L Simms, eds. The University of Chicago Press, Chicago.

Kennedy G G, Dimock M B (1983) 2-Tridecanone: a natural toxicant in a wild tomato responsible for insect resistance. Pages 123–128 in IUPAC pesticide chemistry (human welfare and the environment). J Miyamoto, ed. Pergamon Press, New York.

Kennedy G G, Schaefers G A (1975) Role of nutrition in the immunity of red raspberry to Amphorophora agathonica Hottes. Environ. Entomol. 4:115–119.

Kennedy G G, Yamamoto R T (1979) A toxic factor causing resistance in a wild tomato to the tobacco hornworm and some other insects. Entomol. Exp. Appl. 26:121–126.

Kennedy G G, Yamamoto R T, Dimock M B, Williams W G, Bordner J (1981) Effect of daylength and light intensity on 2-tridecanone levels and resistance in Lycopersicon hirsutum f. glabratum to Manduca sexta. J. Chem. Ecol. 7:707–716.

Kennedy G G, Gould F, de Ponti O M B, Stinner R E (1987) Ecological, agricultural, genetic and commercial considerations in the deployment of insect resistant germplasms. Environ. Entomol. 16:327–338.

Kennedy J S (1953) Host selection in Aphididae. Trans. IXth Int. Congr. Ent. Amsterdam 2:106–113.

Kennedy J S (1965) Mechanisms of host plant selection. Ann. Appl. Biol. 56:317–322.

Kennedy J S (1975) Insect dispersal. Pages 103–119 in Insects, science and society. D Pimentel, ed. Academic Press, New York.

Kennedy J S (1977) Olfactory responses to distant plants and other odor sources. Pages 67–91 in Chemical control of insect behavior: Theory and application. H H Shorey and J J McKelvey Jr, eds. John Wiley and Sons, New York.

Kennedy J S, Booth C O (1951) Host alternation in Aphis fabae Scop. I. Feeding preferences and fecundity in relation to age and kind of leaves. Ann. Appl. Biol. 38:25–64.

Kennedy J S, Fosbrooke I H M (1973) The plant in the life of an aphid. Pages 129–140 in Insect–plant relationships. Symposium of the Royal Entomological Society, London. H F van Emden, ed. Blackwell Scientific Publications, Oxford.

Kevan P G, Chaloner W G, Savile D B O (1975) Inter-relationships of early terrestrial arthropods and plants. Paleontology 18:391–417.

Khan Z R, Saxena R C (1985a) Behaviour and biology of Nephotettix virescens (Homoptera: Cicadellidae) on tungro virus-infected rice plants: epidemiology implications. Environ. Entomol. 14:297–304.

Khan Z R, Saxena R C (1985b) Effect of steam distillate extract of a resistant rice variety on feeding behaviour of *Nephotettix virescens* (Homoptera: Cicadellidae) J. Econ. Entomol. 78:562–566.

Khan Z R, Norris D M, Chiang H S, Weiss N, Oosterwyk A S (1986) Light-induced susceptibility in soybean to cabbage looper, *Trichoplusia ni* (Lepidoptera: Noctuidae) Environ. Entomol. 15:803–808.

Khan Z R, Ciepiela A, Norris D M (1987) Behavioral and physiological responses of cabbage looper, *Trichoplusia ni* (Hubner), to steam distillates from resistant versus susceptible soybean plants. J. Chem. Ecol. 13:1903–1915.

Khush G S (1977) Disease and insect resistance in rice. Adv. Agron. 29:265–341.

Khush G S (1989) Multiple disease and insect resistance for increased yield stability in rice. Pages 79–92 *in* Progress in irrigated rice research. International Rice Research Institute, Manila, Philipines.

Khush G S (1992) Selecting rice for simply inherited resistances. Pages 303–346 *in* Plant breeding in the 1990s. H T Stalker and J P Murphy, eds. CAB International, Wallingford, UK.

Khush G S, Brar D S (1991) Genetics of resistance to insects in crop plants. Adv. Agron. 45:223–274.

Khush G S, Brar D S (1992) Overcoming the barriers in hybridization. Pages 47–61 *in* Distant hybridization of crop plants. G Kalloo and J B Chowdhury, eds. Theoretical and Applied Genetics Monograph Series 16. Springer-Verlag, Berlin.

Khush G S, Rezaul Karim A N M, Angeles E R (1985) Genetics of resistance of rice cultivar ARC 10550 to Bangladesh brown planthopper biotype. J. Genet. 64:121–125.

Kimmins F M (1989) Electrical penetration graphs from *Nilaparvata lugens* on resistant and susceptible rice varieties. Entomol. Exp. Appl. 50:69–79.

Kindler S D, Staples R (1970a) Nutrients and the reaction of two alfalfa clones to the spotted alfalfa aphid. J. Econ. Entomol. 63:938–940.

Kindler S D, Staples R (1970b) The influence of fluctuating and constant temperatures, photoperiod, and soil moisture on the resistance of alfalfa to the spotted alfalfa aphid. J. Econ. Entomol. 63:1198–1201.

King E G, Leppla N C, eds (1984) Advances and challenges in insect rearing. USDA-ARS (Southern Region), New Orleans.

Klijnstra J W (1982) Perception of the oviposition deterrent pheromone in *Pieris brassicae*. Pages 145–151 *in* Proceedings of the International Symposium on Insect-Plant Relationships. J H Visser and A K Minks, eds. Centre for Agricultural Publication and Documentation, PUDOC, Wageningen.

Klingauf F (1970) Zur Wirtswahl der grunen Erbsenlaus, *Acyrthosiphon pisum* (Harris) (Homoptera: Aphididae). Z. Angew. Entomol. 65:419–427.

Klingauf F (1971) Die Wirkung des Glucosids Phlorizin auf des Wirtswahlverhalten von *Rhopalosiphum insertum* (Walk.) und *Aphis pomi* de Geer (Homoptera: Aphididae). Z. Angew. Entomol. 68:41–55.

Klingauf F, Nocker-Wenzel K, Klein W (1971) Einflup einiger Wachskomponenten von *Vicia faba* L. auf das Wirtsuahlverhalten von *Acyrthosiphon pisum* (Harris) (Homoptera: Aphididae). Z. Pflanzenkrankh. Pflanzenschutz 78:641–648.

Klingauf F, Nocker-Wenzel K, Rottger U (1978) Die Rolle peripherer Pflanzenwachse fur den Befall durch phytophage Insekten. Z. Pflanzenkrankh. Pflanzenschutz 85:228–237.

Klun J A, Brindley T A (1966) Role of 6-methoxybenzoxazolinone in inbred resistance of host plant (maize) to first brood larvae of European corn borer. J. Econ. Entomol. 59:711–718.

Klun J A, Robinson J F (1969) Concentration of two 1,4-benzoxazinones in dent corn at various stages of development of the plant and its relation to resistance of the host plant to the European corn borer. J. Econ. Entomol. 62:214–220.

Klun J A, Tipton C L, Brindley T A (1967) 2-4-Dihydroxy-7-methoxy-1,4-benzoxazin-3-one (DIMBOA) an active agent in the resistance of maize to the European corn borer. J. Econ. Entomol. 60:1529–1533.

Knapp J L, Hedin P A, Douglas W A (1966) A chemical analysis of corn silk from single crosses of dent corn rated as resistant, intermediate, and susceptible to the corn earworm. J. Econ. Entomol. 59:1062–1064.

Knipling E F (1966) Some basic principles in insect population suppression. Bull. Ent. Soc. Am. 12:7–15.

Knipling E F (1979) The basic principles of insect population suppression and management. USDA Agric. Handbook, No. 512. USDA, Washington DC.

Knoll A H, Rothwell G W (1981) Paleobotany: perspectives in 1980. Paleobiology 7:7–35.

Knott D R, Kumar J (1975) Comparison of early generation yield testing and a single seed descent procedure in wheat breeding. Crop Sci. 15:295–299.

Knyazeva N I (1970) Receptors of the wing apparatus regulating the flight of the migratory locust, *Locusta migratoria* L. Entomol. Rev. 49:311–317.

Kogan M (1972) Fluorescence photography in the quantitative evaluation of feeding by phytophagous insects. Ann. Entomol. Soc. Am. 65:277–278.

Kogan M (1977) The role of chemical factors in insect/plant relationships. Pages 211–227 *in* Proceedings of the XV International Congress on Entomology. Washington, DC. D White, ed. Entomological Society of America, College Park MD.

Kogan M (1982) Plant resistance in pest management. Pages 93–134 *in* Introduction to insect pest management, 2nd ed. R L Metcalf and W H Luckmann, eds. John Wiley and Sons, New York.

Kogan M (1986) Natural chemicals in plant resistance to insects. Iowa State J. Res. 60:501–527.

Kogan M, Fischer D C (1991) Inducible defenses in soybean against herbivorous insects. Pages 347–378 *in* Phytochemical induction by herbivores. D W Tallamy and M J Raupp, eds. John Wiley and Sons, New York.

Kogan M, Goeden R D (1969) A photometric technique for quantitative evaluation of feeding preferences of phytophagous insects. Ann. Entomol. Soc. Am. 62:319–322.

Kogan M, Ortman E E (1978) Antixenosis – a new term proposed to replace Painter's 'nonpreference' modality of resistance. Bull. Entomol. Soc. Am. 24:175–176.

Kogan M, Paxton J (1983) Natural inducers of plant resistance to insects. Pages 153–171 *in* Plant resistance to insects. P A Hedin, ed. ACS Symp. Series 208. American Chemical Society, Washington, DC.

Kojima M, Poulton J E, Thayer S S, Conn E E (1979) Tissue distribution of dhurrin and of enzymes involved in its metabolism in leaves of *Sorghum bicolor*. Plant Physiol. 63:1022–1028.

Kolodny-Hirsch D M, Saunders J A, Harrison F P (1986) Effects of simulated tobacco hornworm (Lepidoptera: Sphingidae) defoliation on growth dynamics and physiology of tobacco as evidence of plant tolerance to leaf consumption. Environ.

Entomol. 15:1137–1144.

Koritsas V M, Lewis J A, Fenwick G R (1991) Glucosinolate responses of oilseed rape, mustard and kale to mechanical wounding and infestation by cabbage stem flea beetle (*Psylliodes chrysocephala*). Ann. Appl. Biol. 118:209–221.

Koziel M G, Beland G L, Bowman C, Carozzi N B, Crenshaw R, Crossland L, Dawson J, Desai N, Hill M, Kadwell S, Launis K, Lewis K, Maddox D, McPherson K, Meghji M R, Merlin E, Rhodes R, Warren G W, Wright M, Evola S V (1993) Field performance of elite transgenic maize plants expressing an insecticidal protein derived from *Bacillus thuringiensis*. Bio/Technology 11:194–200.

Koziol M J, Whatley F R (1984) Gaseous air pollutants and plant metabolism. Butterworth, London.

Kramer P J (1983) Water relations of plants. Academic Press, New York.

Krischik V A, Barbosa P, Reichelderfer C F (1988) Three trophic level interactions: allelochemicals, *Manduca sexta*, and *Bacillus thuringiensis* var. *kurstaki*. Environ. Entomol. 17:476–482.

Kubo I, Matsumoto T (1985) Potent insect antifeedants from the African medicinal plant *Bersama abyssinica*. Pages 183–200 *in* Bioregulator for pest control. P A Hedin, ed. ACS Symp. Series 276. American Chemical Society, Washington, DC.

Kubo I, Nakanishi K (1977) Insect antifeedants and repellents from African plants. Pages 165–178 *in* Host plant resistance to pests. P A Hedin, ed. ACS Symp. Series 62. American Chemical Society, Washington, DC.

Kuc J, Rush J S (1985) Phytoalexins. Arch. Biochem. Biophys. 236:455–472.

Kumar H (1986) Enhancement of oviposition by *Chilo partellus* (Swinhoe) (Lepidoptera: Pyralidae) on maize plants by larval infestation. Appl. Entomol. Zool. 21:539–545.

Kumar H, Asino G O (1993) Resistance of maize to *Chilo partellus* (Lepidoptera: Pyralidae): effect on plant phenology. J. Econ. Entomol. 86:969–973.

Lakshminarayana A, Khush G S (1977) New genes for resistance to brown planthopper in rice. Crop Sci. 17:96–100.

Lamb R J (1980) Hairs protect pods of mustard (*Brassica hirta* 'Gisilba') from flea beetle feeding damage. Can. J. Plant Sci. 60:1439–1440.

Lamberti F, Waller J M, van der Graaff N A (1983) Durable resistance in crops. Plenum Press, New York.

Lambou M G, Shaw R L, Decossas K M, Vix H L E (1966) Cottonseed's role in a hungry world. Econ. Bot. 20:256–267.

Lampert E P, Haynes D L, Sawyer A J, Jokinen D P, Wellso S G, Gallun R L, Roberts J J (1983) Effects of regional releases of resistant wheats on the population dynamics of the cereal leaf beetle (Coleoptera: Chrysomelidae). Ann. Entomol. Soc. Am. 76:972–980.

Langer H, Hamann B, Meinecke C C (1979) Tetrachromatic visual system in the moth *Spodoptera exempta* (Insecta: Noctuidae). J. Comp. Physiol. 129:235–239.

Larkin P J, Scowcroft W R (1981) Somaclonal variation – a novel source of variability from cell cultures for plant improvement. Theor. Appl. Genet. 60:197–214.

Laster M L, Meredith Jr W R (1974) Evaluating the response of cotton cultivars to tarnished plant bug injury. J. Econ. Entomol. 67:686–688.

Latheef M A, Irwin R D (1979) Factors affecting oviposition of *Pieris rapae* on cabbage. Environ. Entomol. 8:606–609.

Lawton J H, Schroeder D (1977) Effects of plant type, size of geographical range and taxonomic isolation on number of insect species associated with British plants. Nature (London) 265:137–140.

Leather S R (1987) Pine monoterpenes stimulate oviposition in the pine beauty moth, *Panolis flammea*. Entomol. Exp. Appl. 43:295–303.

Leather S R, Watt A D, Forrest G L (1987) Insect-induced chemical changes in young lodgepole pine (*Pinus contorta*): the effect of previous defoliation on oviposition, growth and survival of the pine beauty moth, *Panolis flammea*. Ecol. Entomol. 12:275–281.

Lee S C, Matias D M, Mew T W, Heinrichs E A (1984) Interaction between brown planthopper infestation and sheath blight pathogen infection in rice plant. Phil. Phytopath. 20:19.

Lee Y I (1983) The potato leafhopper, *Empoasca fabae*, soybean pubescence, and hopperburn resistance. PhD thesis, University of Illinois.

Lemke C A, Mutschler M A (1984) Inheritance of glandular trichomes in crosses between *Lycopersicon esculentum* and *Lycopersicon pennellii*. J. Am. Soc. Hort. Sci. 109:592–596.

Leszczynski B, Urbanska A, Matok H, Dixon A F G (1993) Detoxifying enzymes of the grain aphid. Bull. OILB/SROP 16:165–172.

Letourneau D K (1986) Associational resistance in squash monocultures and polycultures in tropical Mexico. Environ. Entomol. 15:285–292.

Levin D A (1976) The chemical defenses of plants to pathogens and herbivores. Ann. Rev. Ecol. Syst. 7:121–159.

Levinson H Z (1976) The defensive role of alkaloids in insects and plants. Experientia 32:408–411.

Lewis W J, Jones R L, Sparks A N (1972) A host seeking stimulant for the egg parasite *Trichogramma evanescens*: its source and a demonstration of its laboratory and field activity. Ann. Entomol. Soc. Am. 65:1087–1089.

Lewontin R C (1965) Selection for colonizing ability. Pages 77–94 *in* The genetics of colonizing species. H G Baker and G L Stebbins, eds. Academic Press, New York.

Liener I E (1991) Lectins. Pages 327–352 *in* Herbivores: Their interactions with secondary plant metabolites, Vol. I. G A Rosenthal and M R Berenbaum, eds. Academic Press, New York.

Liener I E, Sharon N, Goldstein I J, eds (1986) The lectins, properties, and applications in biology and medicine. Academic Press, New York.

Lieutier F, Berryman A A (1988) Elicitation of defensive reactions in conifers. Pages 313–319 *in* Mechanisms of woody plant defenses against insects. W J Mattson, J Levieux and C Bernard-Dagan, eds. Springer-Verlag, New York.

Lin H, Kogan M (1990) Influence of induced resistance in soybean on the development and the nutrition of the soybean looper and the Mexican bean beetle. Entomol. Exp. Appl. 55:131–138.

Lin S C, Yuan L P (1980) Hybrid rice breeding in China. Pages 35–51 *in* Innovative approaches to rice breeding. International Rice Research Institute, Manila, Philippines.

Lindley G (1831) A guide to the orchard and kitchen garden. Longman, Rees, Orme, Brown & Green Publishing Co., London.

Liu S, Norris D M, Xu M (1993) Insect resistance and glyceollin concentration in seedling soybeans support resistance ratings of fully developed plants. J. Econ. Entomol. 86:401–406.

Loper G M (1968) Effect of aphid infestation on the coumestrol content of alfalfa varieties differing in aphid resistance. Crop Sci. 8:104–106.

Louda S, Mole S (1991) Glucosinolates: chemistry and ecology. Pages 124–164 *in* Herbivores: Their interactions with secondary plant metabolites, Vol. I, 2nd ed. G A Rosenthal and M R Berenbaum, eds. Academic Press, New York.

Lovett Doust J (1989) Plant reproductive strategies and resource allocation. Trends Ecol. Evol. 8:230–234.

Lowe H J B (1974) Testing sugar beet for aphid-resistance in the glasshouse: a method and some limiting factors. Z. Angew. Entomol. 76:311–321.

Lowe H J B (1975) Infestation of aphid-resistant and susceptible sugar beet by *Myzus persicae* in the field. Z. Angew. Entomol. 79:376–383.

Luginbill P (1969) Developing resistant plants – The ideal method of controlling insects. USDA-ARS Prod. Res. Rep. 111. United States Department of Agriculture Agricultural Research Service, Washington DC.

Ma W C (1972) Dynamics of feeding responses in *Pieris brassicae* as a function of chemosensory input: a behavioral, ultrastructural and electrophysiological study. Meded. Landbouwhogesch. Wageningen 72:162.

Ma W C (1977a) Electrophysiological evidence for chemosensitivity to adenosine, adenine, and sugars in *Spodoptera exempta* and related species. Experientia 33:356–358.

Ma W C (1977b) Alterations of chemoreceptor function in armyworm larvae (*Spodoptera exempta*) by a plant-derived sesquiterpenoid and by sulfhydryl reagents. Physiol. Entomol. 2:199–207.

Ma W C, Schoonhoven L M (1973) Tarsal contact chemosensory hairs of the large white butterfly *Pieris brassicae* and their possible role in oviposition behaviour. Entomol. Exp. Appl. 16:343–357.

Ma Z Q, Gill B S, Sorrells M E, Tanksley S D (1993) RFLP markers linked to two Hessian fly–resistance genes in wheat (*Triticum aestivum* L.) from *Triticum tauschii* (coss.) Schmal. Theor. Appl. Genet. 85:750–754.

Mabry T J, Burnett Jr W C, Jones Jr S B, Gill J E (1977) Antifeedant sesquiterpenes lactones in the Compositae. Pages 179–184 *in* Host plant resistance to pests. P A Hedin, ed. ACS Symp. Series 62. American Chemical Society, Washington, DC.

MacKenzie D R (1980) The problem of variable pests. Pages 183–213 *in* Breeding plants resistant to insects. F G Maxwell and P R Jennings, eds. John Wiley and Sons, New York.

Madden J L (1977) Physiological reactions of *Pinus radiata* to attack by wood wasp, *Sirex noctilio* F. (Hymenoptera: Siricidae). Bull. Entomol. Res. 67:405–426.

Maddox G D, Root R B (1990) Structure of the selective encounter between goldenrod (*Solidago altissima*) and its diverse insect fauna. Ecology 71:2115–2124.

Maloney P J, Albert P J, Tulloch A P (1988) Influence of epicuticular waxes from white spruce and balsam fir on feeding behaviour of the eastern spruce budworm. J. Insect Behavior 1:197–208.

Mamaev B M (1968) Evolution of gall forming insects. Gall midges. British Library Lending Division, Boston Spa, UK. (Translated by A. Crozy 1975.)

Manuwoto S, Scriber J M (1985) Neonate larval survival of European corn borer, *Ostrinia nubilalis*, on high and low DIMBOA genotypes of maize: effects of light intensity and degree of insect inbreeding. Agric. Ecosyst. Environ. 14:221–236.

Manwan I, Sama S, Rizvi S A (1977) Management strategy to control tungro in

Indonesia. Pages 92–97 *in* Proceedings of the Workshop on Rice Tungro virus. Ministry of Agriculture, Jakarta, Indonesia.

Markl H (1966) Schwerkraftdressuren an Honigbienen. I. Die Geomenotaktische Fehlorientierung. Z. Vergl. Physiol. 53:328–352.

Martin M M, Martin J S (1984) Surfactants: their role in preventing the precipitation of proteins by tannins in insect guts. Oecologia 61:342–345.

Martinez C R, Khush G S (1974) Sources and inheritance of resistance to brown planthopper in some breeding lines of rice. Crop Sci. 14:264–267.

Marvel J T (1985) Biotechnology in crop improvement. Pages 477–509 *in* Bioregulators for pest control. P A Hedin, ed. ACS Symp. Series 276. American Chemical Society, Washington, DC.

Mather K, Jinks J L (1982) Biometrical genetics: The study of continuous variation, 3rd ed. Chapman and Hall, London.

Matsumoto Y (1962) A dual effect of coumarin, olfactory attraction and feeding inhibition, on the vegetable weevil adult, in relation to the uneatability of sweet clover leaves. Appl. Entomol. Zool. 6:141–149.

Matsumoto Y (1970) Volatile organic sulfur compounds as insect attractants with special reference to host selection. Pages 133–160 *in* Control of insect behavior by natural products. D L Wood, R M Silverstein and M Nakajiama, eds. Academic Press, New York.

Mattocks A R (1986) Chemistry of toxicology of pyrrolizidine alkaloids. Academic Press, London.

Mattson Jr W J (1980) Herbivory in relation to plant nitrogen content. Ann. Rev. Ecol. Syst. 11:119–161.

Mattson W J, Scriber J M (1987) Feeding ecology of insect folivores of woody plants: water, nitrogen, fiber, and mineral considerations. Pages 105–146 *in* Nutritional ecology of insects, mites, spiders and related invertebrates. F Slansky, Jr and J G Rodgriguez, eds. John Wiley and Sons, New York.

Maxwell F G, Lefever H N, Jenkins J N (1965) Blister beetles on glandless cotton. J. Econ. Entomol. 58:792–793.

McCouch S R, Khush G S, Tanksley S D (1991) Tagging genes for disease and insect resistance via linkage to RFLP markers. Pages 443–449 *in* Rice genetics II. International Rice Research Institute, Manila, Philippines.

McCown B H, McCabe D E, Russell D R, Robison D J, Barton K A, Raffa K F (1991) Stable transformation of *Populus* and incorporation of pest resistance by electric discharge particle acceleration. Plant Cell Rep. 9:590–594.

McGaughey W H, Whalon M E (1992) Managing insect resistance to *Bacillus thuringiensis* toxins. Science 258:1451–1455.

McIndoo N E (1931) Tropism and sense organs of Coleoptera. Smithsonian Misc. Coll. 82:1–70.

McIntyre J L, Dodds J A, Hare J D (1981) Effects of localized infections of *Nicotiana tabaccum* by tobacco mosaic virus on systemic resistance against diverse pathogens and insects. Phytopathology 71:297–301.

McIver S B (1985) Mechanoreception. Pages 71–132 *in* Comprehensive insect physiology, biochemistry and pharmacology, Vol. 6. G A Kerkut and L I Gilbert, eds. Pergamon Press, New York.

McKee H S (1962) Nitrogen metabolism in plants. Clarendon Press, Oxford.

McKey D (1974) Adaptive patterns in alkaloid physiology. Am. Nat. 108:305–320.

McKey D (1979) The distribution of secondary compounds within plants. Pages 55–133 *in* Herbivores: Their interaction with plant metabolites. G A Rosenthal and D H Janzen, eds. Academic Press, New York.

McKibben G H, Thompson M J, Parrot W L, Thompson A C, Lusby W R (1985) Identification of feeding stimulants for boll weevils from cotton buds and anthers. J. Chem. Ecol. 11:1229–1238.

McLean D L, Weigt Jr W A (1968) An electronic measuring system to record aphid salivation and ingestion. Ann. Entomol. Soc. Am. 61:180–185.

McManus M T, White D W R, McGregor P (1994) Accumulation of a chymotrypsin inhibitor in transgenic tobacco affects the growth of closely related insect pests differently. Transgenic Res. 3:50–58.

McMichael S C (1960) Combined effects of glandless genes gl_2 and gl_3 on pigment glands in the cotton plant. Agron. J. 52:385–386.

McMillian W W, Wiseman B R, Widstrom N W, Harrell E A (1972) Resistant sweet corn hybrid plus insecticide to reduce losses from corn earworms. J. Econ. Entomol. 65:229–231.

McMurtry J A (1962) Resistance of alfalfa to spotted alfalfa aphid in relation to environmental factors. Hilgardia 32:501–539.

McNeill S, Aminu-Kano M, Houlden G, Bullock J M, Citrone S and Bell S B (1987) The interaction between air pollution and sucking insects. Page 602 *in* Acid rain: Scientific and technical advances. R Perry, R M Harrington, J N B Bell and J N Lester, eds. Selper, London.

Medrano F G, Heinrichs E A (1985) A simple technique of rearing yellow stem borer *Scirpophaga incertulas* (Walker). Int. Rice Res. Newsl. 10:14–15.

Meisner J, Ascher K R S, Lavie D (1974) Factors influencing the attraction to oviposition of the potato tuber moth, *Gnorimoschema operculella* Zell. Z. Angew. Entomol. 77:179–189.

Meisner J, Navon A, Zur M, Ascher K R S (1977a) The response of *Spodoptera littoralis* larvae to gossypol incorporated in an artificial diet. Environ. Entomol. 6:243–244.

Meisner J, Kehat M, Zur M, Ascher K R S (1977b) The effect of gossypol on the larvae of the spiny bollworm, *Earias insulana*. Entomol. Exp. Appl. 22:301–303.

Meredith Jr W R, Schuster M F (1979) Tolerance of glabrous and pubescent cottons to tarnished plant bug. Crop Sci. 19:484–488.

Metcalf R L (1980) Changing role of insecticides in crop protection. Ann. Rev. Entomol. 25:219–256.

Metcalf R L (1986a) Coevolutionary adaptations of rootworm beetles (Coleoptera: Chrysomelidae) to cucurbitacins. J. Chem. Ecol. 12:1109–1124.

Metcalf R L (1986b) Plant volatiles as insect attractants. Crit. Rev. Plant Sci. 5:251–301.

Metcalf R L (1986c) The ecology of insecticides and the chemical control of insects. Pages 251–297 *in* Ecological theory and integrated insect pest management practice. M Kogan, ed. John Wiley and Sons, New York.

Metcalf R L, Luckmann W H (1982) Introduction to insect pest management, 2nd ed. John Wiley and Sons, New York.

Metcalf R L, Metcalf R A, Rhodes A M (1980) Cucurbitacins as kairomones for diabroticite beetles. Proc. Nat. Acad. Sci. USA 77: 3769–3772.

Metcalf R L, Rhodes A M, Metcalf R A, Ferguson J, Metcalf E R, Lu P (1982) Cucurbitacin contents and diabroticite (Coleoptera: Chrysomelidae) feeding upon

Cucurbita spp. Environ. Entomol. 11:931–937.

Meyer V G (1974) Interspecific cotton breeding. Econ. Bot. 28:56–60.

Mihm J A (1983a) Efficient mass-rearing and infestation techniques to screen for host plant resistance to fall armyworm, *Spodoptera frugiperda*. International Wheat and Maize Improvement Center (CIMMYT), El Batan, Mexico.

Mihm J A (1983b) Efficient mass-rearing and infestation techniques to screen for host plant resistance to maize stem borers, *Diatraea* sp. International Wheat and Maize Improvement Center (CIMMYT), El Batan, Mexico.

Mihm J A (1985) Breeding for host plant resistance to maize stem borers. Insect Sci. Applic. 6:369–377.

Mihm J A (1989) Evaluating maize for resistance to tropical stem borers, armyworms, and earworms. Pages 109–121 *in* Toward insect resistant maize for the third world. J A Mihm, B R Wiseman and F M Davis, eds. International Wheat and Maize Improvement Center (CIMMYT), El Batan, Mexico.

Mihm J A, Peairs F B, Ortega A (1978) New procedures for efficient mass production and artificial infestation with lepidopterous pests of maize. CIMMYT Review. International Wheat and Maize Improvement Center (CIMMYT), El Batan, Mexico.

Mikolajczak K L, Madrigal R V, Smith Jr C R, Reed D K (1984) Insecticidal effects of cyanolipids on three species of stored product insects, European corn borer (Lepidoptera: Pyralidae) larvae, and striped cucumber beetle (Coleoptera: Chrysomelidae). J. Econ. Entomol. 77:1144–1148.

Miles D H, Hankinson B L, Randle S A (1985) Insect antifeedants from the Peruvian plant *Alchornea triplinervia*. Pages 469–476 *in* Bioregulators for pest control. P A Hedin, ed. ACS Symp. Series. 276. American Chemical Society, Washington, DC.

Miles P W, Aspinall D, Rosenberg L (1982) Performance of the cabbage aphid *Brevicoryne brassicae* (L.) on water–stressed rape plants, in relation to changes in their chemical composition. Aust. J. Zool. 30:337–345.

Miller J R, Strickler K. L (1984) Finding and accepting host plants. Pages 127–157 *in* Chemical ecology of insects. W J Bell and R T Carde, eds. Sinauer Associates Inc., Sunderland, MA.

Miller R H, El-Masri S, Al–Jundi K (1993) Plant density and wheat stem sawfly (Hymenoptera: Cephidae) resistance in Syrian wheats. Bull. Entomol. Res. 83:95–102.

Minney B H P, Gatehouse A M R, Dobie P, Dendy J, Cardona C, Gatehouse J A (1990) Biochemical bases of seed resistance to *Zabrotes subfasciatus* (bean weevil) in *Phaseolus vulgaris* (common bean): a mechanism for arcelin toxicity. J. Insect Physiol. 36:757–767.

Mitchell B K (1985) Specificity of an amino acid–sensitive cell in the adult Colorado potato beetle, *Leptinotarsa decemlineata*. Physiol. Entomol. 10:421–429.

Mitchell B K (1987) Interactions of alkaloids with galeal chemosensory cells of Colorado potato beetle. J. Chem. Ecol. 13:2009–2022.

Mitchell B K, Harrison G D (1985) Effects of *Solanum* glycoalkaloids on chemosensilla in the Colorado potato beetle: a mechanism of feeding deterrence. J. Chem. Ecol. 11:73–83.

Mitchell H C, Cross W H, McGovern W L, Dawson E M (1973) Behavior of the boll weevil on frego bract cotton. J. Econ. Entomol. 66:677–680.

Mitchell N D (1977) Differential host selection by *Pieris brassicae* L. (the large white butterfly) on *Brassica oleracea* subsp. *oleracea* (the wild cabbage) Entomol. Exp. Appl. 12:208–219.

Mitchell R (1981) Insect behavior, resource exploitation, and fitness. Ann. Rev. Entomol. 26:373–396.

Mitra R, Bhatia C R (1982) Bioenergetic considerations in breeding for insect and pathogen resistance in plants. Euphytica 31:429–437.

Mochida O (1992) Management of insect pests of rice in 2000. Pages 67–81 *in* Pest management and the environment in 2000. A A S A Kadir and H S Barlow, eds. CAB International, Wallingford, UK.

Moericke V (1969) Host plant specific colour behavior by *Hyalopterus pruni* (Aphididae). Entomol. Exp. Appl. 13:524–534.

Mohan M, Nair S, Bentur J S, Rao U P, Bennett J (1993) RFLP and RAPD mapping of the rice *Gm2* gene that confers resistance to biotype 1 of gall midge (*Orseolia oryzae*). Theor. Appl. Genet. 87:782–788.

Montgomery M E, Arn H (1974) Feeding response of *Aphis pomi*, *Myzus persicae*, and *Amphorophora agathonica* to phlorizin. J. Insect Physiol. 20:413–421.

Mooney H A, Gulmon S L, Johnson N D (1983) Physiological constraints on plant chemical defenses. Pages 21–36 *in* Plant resistance to insects. P A Hedin, ed. ACS Symp. Series 208. American Chemical Society, Washington, DC.

Mori M (1982) *n*-Hexacosanol and *n*-octacosanol: feeding stimulants for larvae of the silkworm, *Bombyx mori*. J. Insect Physiol. 28:969–973.

Mugiono P S, Heinrichs E A, Medrano F G (1984) Resistance of Indonesian mutant lines to the brown planthopper (BPH) *Nilaparvata lugens*. Int. Rice Res. Newsl. 9:8.

Muid B (1977) Host plant modification of insecticide resistance in *Myzus persicae*. PhD thesis, Univeristy of Reading, UK.

Muller K O, Borger H (1940) Experimentelle Untersuchungen uber die *Phytopthora* Resistenz der Kartoffel. Arb. Biol. Anst. Reichsanst. (Berl.) 23:189–231.

Multani D S, Jena K K, Brar D S, delos Reyes B G, Angeles E R, Khush G S (1994) Development of monosomic alien addition lines and introgression of genes from *Oryza australiensis* Domin. to cultivated rice *O. sativa* L. Theor. Appl. Genet. 88:102–109.

Munakata K (1977) Insect antifeedants of *Spodoptera litura* in plants. Pages 185–196 *in* Host plant resistance to pests. P A Hedin, ed. ACS Symp. Series 62. American Chemical Society, Washington, DC.

Munakata K, Okamoto D (1967) Varietal resistance to rice stem borer in Japan. Pages 419–430 *in* The major insect pests of the rice plant. Proceedings of the Symposium of the IRRI. Johns Hopkins Press, Baltimore.

Murray R D H, Mendez J, Brown S A (1982) The natural coumarins. John Wiley and Sons, Chichester, UK.

Musajo L, Rodighiero G, Colombo G, Torlone V, Dall'Acqua F (1965) Photosensitizing furocoumarins: interaction with DNA and photo–inactivation of DNA containing viruses. Experientia 21:22–24.

Mustaparta H (1984) Olfaction. Pages 37–70. *in* Chemical ecology of insects. W J Bell and R T Carde, eds. Chapman and Hall, London.

Myers J H, Bazely D (1991) Thorns, spines, prickles and hairs: are they stimulated by herbivory and do they deter herbivores? Pages 325–344 *in* Phytochemical induction by herbivores. D W Tallamy and M J Raupp, eds. John Wiley and Sons, New York.

Nair K S S, McEwen F L (1976) Host selection by the adult cabbage maggot, *Hylemya brassicae* (Diptera: Anthomyiidae): effect of glucosinolates and common nutrients on oviposition. Can. Entomol. 108:1021–1030.

Nakanishi K, Koreeda M, Sasaki S, Chang M L, Hsu H Y (1966) Insect hormones. The structure of ponasterone A, an insect–moulting hormone from the leaves of *Podocarpus nakaii* Hay. J. Chem. Soc. Chem. Commun. 915–917.

Nakashima M, Enomoto K, Kijima H, Morita H (1982) Discrepancy between the affinities of certain inhibitors for the membrane-bound alpha-glucosidases and those for the sugar receptor of flies. Insect Biochem. 12:579–585.

Nath P, Hall C V (1963) Inheritance of resistance to the striped cucumber beetle in *Cucurbita pepo*. Indian J. Genet. 23:342–345.

National Academy of Science (1969) Plant and animal resistance to insects. Pages 64–99 *in* Principles of plants and animal pest control, Vol. 3. National Academy of Science, Washington, DC.

Naven A, Zur M, Arcan L (1988) Effects of cotton leaf surface alkalinity on feeding of *Spodoptera littoralis* larvae. J. Chem. Ecol. 14:839–844.

Nemoto H, Ikeda R, Kaneda C (1989) New genes for resistance to brown planthopper, *Nilaparvata lugens* Stal, in rice. Jap. J. Breed. 39:23–28.

Neuhaus G, Spangenberg G, Scheid O M, Schweizer H G (1987) Transgenic rapeseed plants obtained by the microinjection of DNA into microspore-derived embryoids. Theor. Appl. Genet. 75:30–36.

Neupane F P, Norris D M (1990) Indo-acetic acid alteration of soybean resistance to the cabbage looper (Lepidoptera: Noctuidae) Environ. Entomol. 19:215–221.

Newsom L D, Kogan M, Miner F D, Rabb R L, Turnipseed S G, Whitcomb W H (1980) General accomplishments toward better pest control in soybean. Pages 51–98 *in* New technology of pest control. C B Huffaker, ed. John Wiley and Sons, New York.

Ng S S, Davis F M, Williams W P (1990) Ovipositional response to southwestern corn borer (Lepidoptera: Pyralidae) and fall armyworm (Lepidoptera: Noctuidae) to selected maize hybrids. J. Econ. Entomol. 83:1575–1577.

Nicol D, Wratten S D, Eaton N, Copaja S V (1993) Effects of DIMBOA levels in wheat on the susceptibility of the grain aphid *Sitobion avenae* to deltamethrin. Ann. Appl. Biol. 122:427–433.

Nielsen J K, Dalgaard L, Larsen L M, Sorensen H (1979a) Host plant selection of the horse-radish flea beetle *Phyllotreta armoraciae* (Coleoptera: Chrysomelidae): feeding responses to glucosinolates from several crucifers. Entomol. Exp. Appl. 25:227–239.

Nielsen J K, Larsen L M, Sorensen H (1979b) Host plant selection of the horse-radish flea beetle *Phyllotreta armoraciae* (Coleoptera: Chrysomelidae): identification of two flavonol glycosides stimulating feeding in combination with glucosinolates. Entomol. Exp. Appl. 26:40–48.

Nielson M W, Lehman W F (1980) Breeding approaches in alfalfa. Pages 277–312 *in* Breeding plants resistant to insects. F G Maxwell and P R Jennings, eds. John Wiley and Sons, New York.

Nielson M W, Schonhorst M H, Don H, Lehman W F, Marble V L (1971) Resistance in alfalfa to four biotypes of the spotted alfalfa aphids. J. Econ. Entomol. 64:506–510.

Nienhuis J, Helentjaris T, Slocum M, Ruggero B, Schaefer A (1987) Restriction fragment length polymorphism analysis of loci associated with insect resistance in tomato. Crop Sci. 27:797–803.

Nishida R, Fukami H (1990) Sequestration of distasteful compounds by some pharmacophagous insects. J. Chem. Ecol. 16:151–164.

Nishio S (1980) The fates and adaptive significance of cardenolides sequestered by larvae of *Danaus plexippus* (L.) and *Cycnia inopinatus* (Hy. Edwards) PhD dissertation, University of Georgia, Athens, GA.

Nordlund D A (1981) Semiochemicals: a review of the terminology. Pages 13–28 *in* Semiochemicals: Their role in pest control. D A Nordlund, R L Jones and W J Lewis, eds. John Wiley and Sons, New York.

Norris D M (1970) Quinol stimulation and quinone deterrency of gustation by *Scolytus multistriatus* (Coleoptera: Scolytidae). Ann. Entomol. Soc. Am. 63:476–478.

Norris D M (1977) The role of repellents and deterrents in the feeding of *Scolytus multistriatus*. Pages 215–230 *in* Host plant resistance to pests. P A Hedin, ed. ACS Symp. Series 62. American Chemical Society, Washington, DC.

Norris D M (1986) Anti-feeding compounds. Pages 97–146 *in* Chemistry of plant protection. G Haug and H Hoffman, eds. Springer-Verlag, Berlin.

Norris D M, Kogan M (1980) Biochemical and morphological bases of resistance. Pages 23–61 *in* Breeding plants resistant to insects. F G Maxwell and P R Jennings, eds. John Wiley and Sons, New York.

Norris D M, Chiang H S, Ciepiela A, Khan Z R, Sharma H, Neupane F, Weiss N, Liu S (1988) Soybean allelochemicals affecting insect orientation, feeding, growth, development and reproductive processes. Pages 27–31 *in* Endocrinological frontiers in physiological insect ecology. F Sehnal, A Zabza and D L Denlinger, eds. Wroclaw Technical University Press, Wroclaw, Poland.

Nottingham S F (1988) Host-plant finding for oviposition by adult cabbage root fly, *Delia radicum*. J. Insect Physiol. 34:227–234.

Nottingham S F, Hardie J, Dawson G W, Hick A J, Pickett A J, Wadhams L J, Woodcock C M (1991) Behavioral and electrophysiological responses of aphids to host and nonhost plant volatiles. J. Chem. Ecol. 17:1231–1242.

Nwanze K F, Reddy Y V R (1991) A rapid method for screening sorghum for resistance to *Chilo partellus* (Swinhoe) (Lepidoptera: Pyralidae). J. Agric. Entomol. 8:41–49.

Obanni M, Patterson F L, Foster J E, Ohm H W (1988) Genetic analysis of resistance of durum wheat PI 428435 to the Hessian fly. Crop Sci. 28:223–226.

Obanni M, Ohm H W, Foster J E, Patterson F L (1989) Reactions of eleven tetraploid and hexaploid wheat introductions to Hessian fly. Crop Sci. 29:267–269.

Obrycki J J, Tauber M J, Tingey W M (1983) Predator and parasitoid interaction with aphid-resistant potatoes to reduce aphid densities: a two-year field study. J. Econ. Entomol. 76:456–462.

Ohsugi T, Nishida R, Fukami H (1985) Oviposition stimulant of *Papilio xuthus*, a citrus-feeding swallow tail butterfly. Agric. Biol. Chem. 49:1897–1900.

O'Keefe L E, Callenbach J A, Lebsock K L (1960) Effect of culm solidness on the survival of the wheat stem sawfly. J. Econ. Entomol. 53:244–246.

Oksanen H, Perttunen V, Kangas E (1970) Studies on the chemical factors involved in the olfactory orientation of *Blastophagus piniperda* (Coleoptera: Scolytidae) Contrib. Boyce Thompson Inst. 24:275–282.

Onukogu F A, Guthrie W D, Russell W A, Reed G L, Robins J C (1978) Location of genes that condition resistance to maize sheath collar feeding by second generation European borer. J. Econ. Entomol. 7:1–4.

Ortman E E, Branson T F, Gerloff E D (1974) Techniques, accomplishments, and future potential of host plant resistance to *Diabrotica*. Pages 344–358 *in* Proceedings of the

Summer Institute on Biological Control of Plant Insects and Diseases. F G Maxwell and F A Harris, eds. University of Mississippi Press, Jackson, MI.

Osborn T C, Alexander D C, Sun S S M, Cardona C, Bliss F A (1988) Insecticidal activity and lectin homology of arcelin seed protein. Science 240:207–210.

Paguia P, Pathak M D, Heinrichs E A (1980) Honeydew excretion measurement techniques for determining differential feeding activity of biotypes of *Nilaparvata lugens* on rice varieties. J. Econ. Entomol. 73:35–40.

Paine T D, Stephen F M (1988) Induced defenses of loblolly pine, *Pinus taeda*: potential impact on *Dendroctonus frontalis* within-tree mortality. Entomol. Exp. Appl. 46:39–46.

Painter R H (1930) Biological strains of Hessian fly. J. Econ. Entomol. 23:322–326.

Painter R H (1951) Insect resistance in crop plants. Macmillan, New York.

Painter R H (1958) Resistance of plants to insects. Ann. Rev. Entomol. 3:267–290.

Painter R H (1966) Plant resistance as a means of controlling insects and reducing their damage. Pages 138–148 *in* Pest control by chemical, biological, genetic and physical means: A symposium. USDA Res. Serv. No. 33, 110, 224.

Painter R H (1968) Crops that resist insects provide a way to increase world food supply. Kans. State Agric. Exp. Stn. Bull. 520, 22 pp.

Painter R H, Salmon J C, Parker J H (1931) Resistance of varieties of winter wheat to Hessian fly, *Phytophaga destructor* (Say). Kans. State Agr. Exp. Stn. Tech. Bull. 27:58.

Pair S D, Laster M L, Martin D F (1982) Parasitoids of *Heliothis* spp (Lepidoptera: Noctuidae) larvae in Mississippi associated with sesame interplanting in cotton, 1971–1974: implications of host–habitat interaction. Environ. Entomol. 11:509–512.

Palmer D F, Windels M B, Chiang H C (1979) Artificial infestation of corn with western corn rootworm eggs in agar-water. J. Econ. Entomol. 70:277–278.

Panda N (1969) Pubescence as a major factor in the resistance of soybean genotype to oviposition and damage by the corn earworm, *Heliothis zea* (Boddie). PhD dissertation, University of Missouri.

Panda N (1979) Principles of host–plant resistance to insect pests. Allanheld Osmun and Co. Montclair and Universe Books, New York.

Panda N, Daugherty D M (1978) Ovipositional preference of *Heliothis zea* (Hubner) on glabrous and dense soybean genotypes. Madras Agric. J. 63:227–230.

Panda N, Heinrichs E A (1983) Levels of tolerance and antibiosis in rice varieties having moderate resistance to the brown planthopper, *Nilaparvata lugens* (Stal) (Hemiptera: Delphacidae) Environ. Entomol. 12:1204–1214.

Panda N, Mahapatra A, Sahoo M (1971) Field evaluation of some brinjal varieties for resistance to shoot and fruit borer (*Leucinodes orbonalis* Guen). Indian J. Agric. Sci. 41:597–601.

Panda N, Sahoo B C, Jena B C (1983) Rice pest management in India. Pages 326–342 *in* Proceedings of the Rice Pest Management Seminar, TNAU, Coimbatore, India. Tamil Nadu Agricultural University, Coimbatore, India.

Papaj D R, Rausher M D (1983) Individual variation in host location by phytophagous insects. Pages 77–124 *in* Herbivorous insects: Host seeking behavior and mechanisms. S Ahmad, ed. Academic Press, New York.

Parnell F R (1935) Origin and development of the U4 cotton. Empire cotton growing. Crop Rev. 12:177–182.

Parnell F R, King H E, Ruston D F (1949) Jassid resistance and hairiness of the cotton plant. Bull. Entomol. Res. 39:539–575.

Parsons L R (1982) Plant responses to water stress. Pages 175–192 *in* Breeding plants for less favorable environments. M N Christiansen and C F Lewis, eds. John Wiley and Sons, New York.

Pasteels J M (1977) Evolutionary aspects in chemical ecology and chemical communication. Pages 281–293 *in* Proceedings of the 15th International Congress on Entomology. Washington, DC. D White, ed. Entomological Society of America, College Park MD.

Pasteels J M, Rowell–Rahier M, Braekman R J, Dupont A (1983) Salicin from host plant as precursor of salicylaldehyde in defensive secretion of Chrysomeline larvae. Physiol. Entomol. 8:307–314.

Paszkowski J, Shillito R D, Saul M, Mandak V, Hohn T, Hohn B, Potrykus I (1984) Direct gene transfer to plants. EMBO J. 3:2712–2722.

Patanakamjorn S, Pathak M D (1967) Varietal resistance of rice to the Asiatic rice borer, *Chilo suppressalis* (Lepidoptera: Crambidae), and its association with various plant characters. Ann. Entomol. Soc. Am. 60:287–292.

Pathak M D (1970) Genetics of plants in pest management. Pages 138–157 *in* Concepts of pest management. R L Rabb and Guthrie F E, eds. North Carolina State University Press, Raleigh, NC.

Pathak M D (1977) Defense of rice crop against insects. Pages 287–295 *in* The genetic basis of epidemics in agriculture. Ann. NY Acad. Sci. Vol. 287. P R Day, ed. Annals of the New York Academy of Sciences, New York.

Pathak M D, Khush G S (1979) Studies of varietal resistance in rice to the brown planthopper at the International Rice Research Institute. Pages 285–301 *in* Brown planthopper: Threat to rice production in Asia. International Rice Research Institute, Manila, Philippines.

Pathak M D, Cheng C H, Fortuno M E (1969) Resistance to *Nephotettix impicticeps* and *Nilaparvata lugens* in varieties of rice. Nature 223:502–504.

Pathak P K, Saxena R C, Heinrichs E A (1982) Parafilm sachet for measuring honeydew excretion by *Nilaparvata lugens* on rice. J. Econ. Entomol. 75:194–195.

Patterson C G, Thurston R, Rodriguez J G (1974) Two-spotted spider mite resistance in *Nicotiana* species. J. Econ. Entomol. 67:341–343.

Pedersen M W, Barnes D K, Sorensen E L, Griffin G D, Nielson M W, Hill Jr R R, Frosheiser F I, Sonoda R M, Hanson C H, Hunt O J, Peaden R N, Elgin Jr J H, Devine T E, Anderson M J, Goplen B P, Elling L J, Howarth R E (1976) Effects of low and high saponin selection in alfalfa on agronomic and pest resistance traits and interrelationship of these traits. Crop Sci. 16:193–199.

Peferoen M (1992) Engineering of insect–resistant plants with *Bacillus thuringiensis* crystal protein genes. Pages 135–153 *in* Plant genetic manipulation for crop protection. A M R Gatehouse, V A Hilder and D Boulter, eds. CAB International, Wallingford, UK.

Peferoen M, Jansens S, Reynaerts A, Leemans J (1990) Potato plants with engineered resistance against insect attack. Pages 193–204 *in* Molecular and cellular biology of the potato. M E Vayda and W C Park, eds. CAB International, Wallingford, UK.

Pena-Cortes H, Sanchez-Serrano J, Rocha-Sosa M, Willmitzer L (1988) Systemic induction of proteinase-inhibitor-II gene expression in potato plants by wounding. Planta 174:84–89.

Penny L H (1981) Vertical-pull resistance of maize inbreds and their test crosses. Crop Sci. 21:237–240.

Penny L H, Scott G E, Guthrie W D (1967) Recurrent selection for European corn borer resistance in maize. Crop Sci. 7:407–409.

Pereyra P C, Bowers M D (1988) Iridoid glycosides as oviposition stimulants for the buckeye butterfly, *Junonia coenia* (Nymphalidae). J. Chem. Ecol. 14:917–928.

Perkins H C (1974) Air pollution. McGraw-Hill, New York.

Perlak F J, Deaton R W, Armstrong T A, Fuchs R L, Sims S R, Greenplate J T, Fischhoff D A (1990) Insect resistant cotton plants. Bio/Technology 8:939–943.

Perlak F J, Fuchs R L, Dean D A, McPherson S L, Fischhoff D A (1991) Modification of the coding sequence enhances plant expression of insect control protein genes. Proc. Nat. Acad. Sci. USA 88:3324–3328.

Perlak F J, Stone T B, Muskopf Y M, Peterson L J, Parker G B, McPherson S A, Wyman J, Love S, Reed G, Biever D, Fischhoff D A (1993) Genetically improved potatoes: projection from damage by Colorado potato beetles. Plant Mol. Biol. 22:313–321.

Perrin D R, Cruickshank I A M (1965) Studies on the phytoalexins. VII. Chemical stimulation of pisatin formation in *Pisum sativum*. Aust. J. Biol. Sci. 18:803–816.

Person C, Samborski D J, Rohringer R (1962) The gene-for-gene concept. Nature 194:561–562.

Pfaffmann C (1941) Gustatory afferent impulses. J. Cell. Comp. Physiol. 17:243–258.

Phillips D R, Matthew J A, Reynolds J, Fenwick G R (1979) Partial purification and properties of a *cis-3:trans-2*-enal isomerase from cucumber fruit. Phytochemistry 18:401–404.

Pillemer E A, Tingey W M (1978) Hooked trichomes and resistance of *Phaseolus vulgaris* to *Empoasca fabae* (Harris) Entomol. Exp. Appl. 24:83–94.

Pimentel D, ed (1981) Handbook of pest management in agriculture, Vol. I. CRC Press, Boca Raton.

Pluthero F G, Singh R S (1984) Insect behavioral responses to toxins: practical and evolutionary considerations. Can. Entomol. 116:57–68.

Porter J W, Spurgeon S L, eds (1981) Biosynthesis of isoprenoid compounds. John Wiley, New York.

Powell J A (1980) Evolution of larval food preferences in microlepidoptera. Ann. Rev. Entomol. 25:133–159.

Prakasa Rao P S (1975) Some methods of increasing field infestation of rice gall midge. Rice Entomol. Newsl. 2:16–17.

Prakasa Rao P S, Rao Y S, Israel P (1971) Factors favouring incidence of rice pests and methods of forecasting outbreaks: gall midge and stem borers. Oryza 8:337–344.

Price P W (1977) General concepts on the evolutionary biology of parasites. Evolution 31:405–420.

Price P W (1986) Ecological aspects of host plant resistance and biological control: interactions among three trophic levels. Pages 11–30 *in* Interactions of plant resistance and parasitoids and predators of insects. D J Boethel and R D Eikenbary, eds. Ellis Horwood, Chichester, UK.

Price P W, Bouton C E, Gross P, McPheron B A, Thompson J N, Weis A E (1980) Interactions among three trophic levels: influence of plants on interactions between insect herbivores and natural enemies. Ann. Rev. Ecol. Syst. 11:41–65.

Pringle J W S (1957) Insect flight, Cambridge Monog. Exp. Biol. Cambridge University Press, Cambridge, UK.

Prokopy R J (1972a) Response of apple maggot flies to rectangles of different colors and shades. Environ. Entomol. 1:720–726.

Prokopy R J (1972b) Evidence for a marking pheromone deterring repeated oviposition in apple maggot flies. Environ. Entomol. 1:326–332.

Prokopy R J, Owens E D (1983) Visual detection of plants by herbivorous insects. Ann. Rev. Entomol. 28:337–364.

Prokopy R J, Aluja M, Green Th A (1987) Dynamics of host odor and visual stimulus interaction in host finding behavior of apple maggot flies. Pages 161–166 *in* Insects–plants. Proceedings of the 6th International Symposium on Insect–Plant Relationships (PAU 1986). V Labeyrie, G Fabres and D Lachaise, eds. Dr W Junk Publishers, Dordrecht, the Netherlands.

Puterka G J, Peters D C (1989) Inheritance of greenbug, *Schizaphis graminum* (Rondani) virulence to *Gb2* and *Gb3* resistance genes in wheat. Genome 32:109–114.

Puttick G M, Bowers M D (1988) Effect of qualitative and quantitative variation in allelochemicals on a generalist insect: iridoid glycosides and southern armyworm. J. Chem. Ecol. 14:335–351.

Pyke G H, Pulliam H R, Charnov E L (1977) Optimal foraging: a selective review of theory and tests. Q. Rev. Biol. 52:137–154.

Rabb R L, Bradley J R (1968) The influence of host plants on parasitism of eggs of the tobacco hornworm. J. Econ. Entomol. 61: 1249–1252.

Rabbinge R, de Wit C T (1989) Systems, models and simulation. Pages 3–15 *in* Simulation and systems management in crop protection. R Rabbinge, S A Ward and van Laar H H, eds. PUDOC, Wageningen.

Raffa K F, Berryman A A (1983) The role of host plant resistance in the colonization behavior and ecology of bark beetles (Coleoptera: Scolytidae) Ecol. Monogr. 53:27–49.

Raikhel N V, Lee H I, Broekaert W F (1993) Structure and function of chitin–binding proteins. Ann. Rev. Plant Physiol. Plant Mol. Biol. 44:591–615.

Raina A K (1981) Deterrence of repeated oviposition in sorghum shootfly, *Atherigona soccata*. J. Chem. Ecol. 7:785–790.

Rains D W (1972) Salt transport by plants in relation to salinity. Ann. Rev. Plant Physiol. 23:367–388.

Ramachandran R, Khan Z R, Caballero P, Juliano B O (1990) Olfactory sensitivity of two sympatric species of rice leaf folders (Lepidoptera: Pyralidae) to plant volatiles. J. Chem. Ecol. 16:2647–2666.

Ramachandran R, Norris D M, Phillips J K, Phillips T W (1991) Volatiles mediating plant–herbivore–natural enemy interactions: soybean looper frass volatiles, 3-octanone and guaiacol, as kairomones for the parasitoid *Microplitis demolitor*. J. Agric. Food Chem. 39:2310–2317.

Ramalho F S, Parrot W L, Jenkins J N, McCarty Jr J C (1984) Effects of cotton leaf trichomes on the mobility of newly hatched tobacco budworms (Lepidoptera: Noctuidae). J. Econ. Entomol. 77: 619–621.

Raman K V, Tingey W M, Gregory P (1979) Potato glycoalkaloids: effect on survival and feeding behavior of the potato leafhopper. J. Econ. Entomol. 72:337–341.

Ramaswamy S B (1988) Host finding by moths: sensory modalities and behaviours. J. Insect Physiol. 34:235–249.

Ramayya N, Rao S R S (1976) Morphology, phylesis and biology of the peltate scale stellate and tufted hairs in some Malvaceae. J. Indian Bot. Soc. 55:75–79.

Rana B S, Singh B U, Rao N G P (1985) Breeding for shootfly and stem borer resistance in sorghum. Pages 347–360 *in* Proceedings of the International Sorghum Entomology Workshop, July 15–21, 1984. Texas A & M University, College Station, TX/ICRISAT, Patancheru, India.

Raupp M J (1985) Effects of leaf toughness on mandibular wear of the leaf beetle, *Plagiodera versicolora*. Ecol. Entomol. 10:73–79.

Raupp M J, Denno R F (1983) Leaf age as a predictor of herbivore distribution and abundance. Pages 91–124 *in* Variable plants and herbivores in natural and managed systems. R F Denno and M S McClure, eds. Academic Press, New York.

Raupp M J, Amri A, Hatchett J H, Gill B S, Wilson D L, Cox T S (1993) Chromosomal location of Hessian fly-resistance genes *H22*, *H23*, and *H24* derived from *Triticum tauschii* in the D genome of wheat. J. Hered. 84:142–145.

Rausher M D (1983) Ecology of host selection behavior in phytophagous insects. Pages 223–257 *in* Variable plants and herbivores in natural and managed systems. R F Denno and M S McClure, eds. Academic Press, New York.

Rausher M D, Papaj D R (1983) Demographic consequences of discrimination among conspecific host plants by *Battus philenor* butterflies. Ecology 64:1402–1410.

Raven J A (1977) Evolution of vascular land plants. Adv. Bot. Res. 154:129.

Read D P, Feeny P P, Root R B (1970) Habitat selection by the aphid parasite *Diaeretiella rapae* (Hymenoptera: Braconidae) and hyperparasite *Charips brassicae* (Hymenoptera: Cynipidae). Can. Entomol. 102:1567–1578.

Reed D K, Webster J A, Jones B G, Burd J D (1991) Tritrophic relationships of Russian wheat aphid (Homoptera: Aphididae), a hymenopterous parasitoid (*Diaeretiella rapae* McIntosh), and resistant and susceptible small grains. Biological Control 1:35–41.

Reed D W, Pivnick K A, Underhill E W (1989) Identification of chemical oviposition stimulants for the diamondback moth, *Plutella xylostella*, present in three species of Brassicaceae. Entomol. Exp. Appl. 53:277–286.

Rees S B, Harborne J B (1985) The role of sesquiterpene lactones and phenolics in the chemical defense of the chicory plant. Phytochemistry 24:2225–2231.

Reese J C (1969) Chemoreceptor specificity associated with choice of feeding site by the beetle, *Chrysolina brunsvicensis* on its food plant, *Hypericum hirsutum*. Entomol. Exp. Appl. 12:565–583.

Reese J C (1979) Interactions of allelochemicals with nutrients in herbivore food. Pages 309–330 *in* Herbivores – Their interaction with secondary plant metabolites. G A Rosenthal and D H Janzen, eds. Academic Press, New York.

Reese J C (1983) Nutrient–allelochemicals interactions in host plant resistance. Pages 231–243 *in* Plant resistance to insects. P A Hedin, ed. ACS Symp. Series 208. American Chemical Society, Washington, DC.

Reese J C, Schmidt D J (1986) Physiological aspects of plant insect interactions. Iowa State J. Res. 60:545–567.

Reinecke J P (1985) Nutrition: artificial diets. Pages 391–419 *in* Comprehensive insect physiology, biochemistry and pharmacology, Vol. 4. G A Kerkut and L I Gilberts, eds. Pergamon Press, New York.

Reitz L P, Hamlin W G (1978) Distribution of the wheat varieties and classes of wheat in the United States in 1974. USDA Dep. Agric. Stat. Bull. 604:1–93.

Renwick J A A (1983) Nonpreference mechanisms: plant characteristics influencing insect behavior. Pages 199–213 *in* Plant resistance to insects. P A Hedin, ed. ACS Symp. Series 208. American Chemical Society, Washington DC.

Renwick J A A (1988) Plant constituents as oviposition deterrents to lepidopterous insects. Pages 378–385 *in* Biologically active natural products for potential use in Agriculture. H G Cutler, eds. ACS Symp. Series 380. American Chemical Society, Washington, DC.

Renwick J A A (1989) Chemical ecology of oviposition in phytophagous insects. Experientia 45:223–228.

Renwick J A A, Radke C D (1980) An oviposition deterrent assocated with frass from feeding larvae of the cabbage looper, *Trichoplusia ni* (Lepidoptera: Noctuidae). Environ. Entomol. 9:318–320.

Renwick J A A, Radke C D (1981) Host plant constituents as oviposition deterrents for the cabage looper, *Trichoplusia ni*. Entomol. Exp. Appl. 30:201–204.

Renwick J A A, Radke C D (1983) Chemical recognition of host plants for oviposition by the cabbage butterfly, *Pieris rapae* (Lepidoptera: Pieridae) Environ. Entomol. 12:446–450.

Renwick J A A, Radke C D (1988) Sensory cues in host selection for oviposition by the cabbage butterfly, *Pieris rapae*. J. Insect Physiol. 34:251–257.

Renwick J A A, Radke C D, Sachdev-Gupta (1989) Chemical constituents of *Erysimum cherianthoides* deterring oviposition by the cabbage butterfly, *Pieris rapae*. J. Chem. Ecol. 15:2161–2168.

Renwick J A A, Radke C D, Sachdev-Gupta K, Stadler E (1992) Leaf surface chemicals stimulating oviposition by *Pieris rapae* on cabbage. Chemoecology 3:33–38.

Rey J R, McCoy E D, Strong Jr D R (1981) Herbivore pests, habitat islands, and the species-area relationship. Am. Nat. 117:611–622.

Reynolds G W, Smith C M (1985) Effects of leaf position, leaf wounding, and plant age of two soybean genotypes on soybean looper (Lepidoptera: Noctuidae) growth. Environ. Entomol. 14:475–478.

Rhoades D F (1979) Evolution of plant chemical defense against herbivores. Pages 4–54 *in* Herbivores: Their interaction with secondary plant metabolites. G A Rosenthal and D H Janzen, eds. Academic Press, New York.

Rhoades D F (1985) Offensive–defensive interactions between herbivores and plants: their relevance in herbivore population dynamics and ecological theory. Am. Nat. 125:205–238.

Rhoades D F, Cates R G (1976) Toward a general theory of plant antiherbivore chemistry. Recent Adv. Phytochem. 10:168–213.

Rhodes C A, Pierce D A, Mettler I J, Mascarenhas D, Detmer J J (1988) Genetically transformed maize plants from protoplasts. Science 240:204–207.

Richard D S, Applebaum S W, Sliter T J, Baker F C, Schooley D A, Reuter C C, Henrich V C, Gilbert L I (1989) Juvenile hormone bisepoxide biosynthesis *in vitro* by the ring gland of *Drosophila melanogaster*: a putative juvenile hormone in the higher Diptera. Proc. Nat. Acad. Sci. USA 86:1421–1425.

Rick C M, Fobes J F (1974) Association of an allozyme with nematode resistance. Rep. Tomato Genet. Coop. 24:25.

Risch S J, Andow D, Altieri M A (1983) Agroecosystem diversity and pest control: data tentative conclusions and new research directions. Environ. Entomol. 12:625–629.

Roberts D W, Tyrrell C (1961) Sawfly resistance in wheat. IV. Some effects of light intensity on resistance. Can. J. Plant. Sci. 41:457–465.

Roberts J J, Foster J E (1983) Effect of leaf pubescence in wheat on the bird cherry oat aphid (Homoptera: Aphidae). J. Econ. Entomol. 76:1320–1322.

Roberts J J, Gallun R L, Patterson F L, Foster J E (1979) Effects of wheat leaf pubescence on the Hessian fly. J. Econ. Entomol. 72:211–214.

Robertson D S, Walter E V (1963) Genetic studies of earworm resistance in maize: utilizing a series of chromosome-nine translocations. J. Hered. 54:267–272.

Robinson J (1993) Conditioning host plant affects antixenosis and antibiosis to Russian wheat aphid (Homoptera: Aphididae). J. Econ. Entomol. 86:602–606.

Robinson R A (1976) Plant pathosystems. Springer-Verlag, New York, 184 pp.

Robinson R A (1980) New concepts in breeding for disease resistance. Ann. Rev. Phytopath. 18:189–210.

Robinson R A (1991) The genetic controversy concerning vertical and horizontal resistance. Revista Mexicana de Fitopatologia 9:57–63.

Robinson T (1980) The organic constituents of higher plants, 4th ed. Cordus Press, North Amherst, MA.

Roessingh P, Stadler E, Schöni R, Feeny P (1991) Tarsal contact chemoreceptors of the black swallowtail butterfly *Papilio polyxenes*: responses to phytochemicals from host- and non-host plants. Physiol. Entomol. 16:485–495.

Roessingh P, Stadler E, Fenwick G R, Lewis J A, Kvist Nielsen J, Hurter J, Ramp T (1992) Oviposition and tarsal chemoreceptors of the cabbage root fly are stimulated by glucosinolates and host plant extracts. Entomol. Exp. Appl. 65:267–282.

Rogers C E, Gershenzon J, Ohno N, Mabry T J, Stipanovic R D, Kreitner G L (1987) Terpenes of wild sunflower (*Helianthus*): an effective mechanism against seed predation by larvae of the sunflower moth, *Homoeosoma electellum* (Lepidoptera: Pyralidae). Environ. Entomol. 16:586–592.

Rohdendorf B B, Raznitsin A P, eds (1980) The historical development of the class Insecta. Trudy Paleont. Inst. (Moscow) 175:1–268.

Rojanaaridpiched C, Gracen V E, Everett H L, Coors J G, Pugh B F, Bouthyette P (1984) Multiple factor resistance in maize to European corn borer. Maydica 29: 305–315.

Romanow L R, de Ponti O M B, Mollema C (1991) Resistance in tomato to the greenhouse whitefly: analysis of population dynamics. Entomol. Exp. Appl. 60:247–259.

Roots R B (1973) Organization of a plant–arthropod association in simple and diverse habitats. The fauna of collards (*Brassica oleracea*). Ecol. Monogr. 43:95–124.

Rosenthal G A (1977) The biological effects and mode of action of L-canavanine a structural analogue of L-arginine. Q. Rev. Biol. 52:155–178.

Rosenthal G A (1982) Secondary plant metabolites–round-table discussion. Pages 331–334 *in* Insect plant relationships. Proceedings of the 5th International Symposium on Insect Plant Relationships. J H Visser and A K Minks, eds. Centre for Agricultural Publishing and Documentation, Wageningen.

Rosenthal G A (1983) L-Canavanine and L-canaline: protective allelochemicals of certain leguminous plants. Pages 279–290 *in* Plant resistance to insects. P A Hedin, ed. ACS Symp. Series 208. American Chemical Society, Washington, DC.

Rosenthal G A (1991) Non protein amino acids as protective allelochemicals. Pages 1–34 *in* Herbivores: Their interactions with secondary plant metabolites, Vol. I, 2nd ed. G A Rosenthal and M R Berenbaum, eds. Academic Press, New York.

Rosenthal G A, Janzen D H, eds (1979) Herbivores: their interaction with secondary plant metabolites. Academic Press, New York.

Rossiter M C, Schultz J C, Baldwin I T (1988) Relationships among defoliation, red oak phenolics, and gypsy moth growth and reproduction. Ecology 69:267–277.

Rothschild M (1973) Secondary plant substances and warning colouration in insects. Pages 59–83 *in* Insect–plant relationships. H F van Emden, ed. Symp. R. Entomol. Soc. London No. 6. Blackwell Scientific Publications, Oxford.

Rowan G B, Boerma H R, All J N, Todd J W (1993) Soybean maturity effect on expression of resistance to lepidopterous insects. Crop Sci. 33:433–436.

Russell E P (1989) Enemies hypothesis: a review of the effect of vegetational diversity on predatory insects and parasitoids. Environ. Entomol. 18:590–599.

Ruzicka L (1953) Isoprene rule and the biogenesis of terpenic compounds. Experientia 9:357–367.

Ryan C A (1983) Insect–induced chemical signals regulating natural plant protection responses. Pages 43–60 *in* Variable plants and herbivores in natural and managed systems. R F Denno and M S McClure, eds. Academic Press, New York.

Ryan C A (1989) Proteinase inhibitor gene families: strategies for transformation to improve plant defenses against herbivores. BioEssays 10:20–24.

Ryan C A, Farmer E E (1991) Oligosaccharide signals in plants: a current assessment. Ann. Rev. Plant Physiol. Mol. Biol. 42:651–674.

Ryan C A, Bishop P D, Graham J S, Broadway R M, Duffey S S (1986) Plant and fungal cell wall fragments activate expression of proteinase inhibitor genes for plant defense. J. Chem. Ecol. 12:1025–1036.

Ryan J D, Gregory P, Tingey W M (1982) Phenolic oxidase activities in glandular trichomes of *Solanum berthaultii*. Phytochemistry 21: 1885–1887.

Ryan M F, Bryne O (1988) Plant–insect coevolution and inhibition of acetyl-cholinesterase. J. Chem. Ecol. 14:1965–1975.

Sachdev-Gupta K, Renwick J A A, Radke C D (1990) Isolation and identification of oviposition deterrents to cabbage butterfly, *Pieris rapae* from *Erysimum cheiranthoides*. J. Chem. Ecol. 16:1059–1067.

Salim M, Heinrichs E A (1987) Insecticide–induced changes in the levels of resistance to rice cultivars to the whitebacked planthopper, *Sogatella furcifera* (Horvath) (Homoptera: Delphacidae). Crop Prot. 6:28–32.

Salim M, Saxena R C, Akbar M (1990) Salinity stress and varietal resistance in rice: effects on whitebacked planthopper. Crop Sci. 30:654–659.

Sarkanen K V, Ludwig C H (1971) Lignins, occurrence, formation, structure and reactions. John Wiley, New York.

Satoh H, Omura T (1986) Mutagenesis in rice by treating fertilized egg cells with nitroso compounds. Pages 707–717 *in* Rice genetics. International Rice Research Institute, Manila, Philippines.

Saunders J A, Conn E E, Chin Ho Lin, Stocking R C (1977) Subcellular localization of the cyanogenic glucoside of sorghum by autoradiography. Plant Physiol. 59:647–652.

Saxena K N, Basit A (1982) Inhibition of oviposition by volatiles of certain plants and chemicals in the leafhopper *Amrasca devastans* (Distant). J. Chem. Ecol. 8:329–338.

Saxena K N, Goyal S (1978) Host-plant relations of the citrus butterfly *Papilio demoleus* L.: orientation and ovipositional responses. Entomol. Exp. Appl. 24:1–10.

Saxena R C (1986) Biochemical bases of insect resistance in rice varieties. Pages 142–159 *in* Natural resistance of plants to pests: Roles of allelochemicals. M B

Green and P A Hedin, eds. ACS Symp. Series 296. American Chemical Society Washington, DC.

Saxena R C (1989) Insecticides from neem. Pages 110–135 in Insecticides of plant origin. J T Arnason, B J R Philogene and P Morand, eds. ACS Symp. Series 387. American Chemical Society, Washington, DC.

Saxena R C, Okech S H (1985) Role of plant volatiles in resistance of selected rice varieties to brown planthopper, Nilaparvata lugens (Stal) (Homoptera: Delphacidae). J. Chem. Ecol. 11:1601–1616.

Schalk J M, Stoner A K (1976) A bioassay differentiates resistance to the Colorado potato beetle on tomatoes. J. Am. Soc. Hortic. Sci. 101:74–76.

Schillinger Jr J A, Gallun R L (1968) Leaf pubescence of wheat as a deterrent to the cereal leaf beetle, Oulema melanopus. Ann. Entomol. Soc. Am. 61:900–903.

Schmialek V P (1961) Die Identifizierung zweier im Tenebriokot und in Hefe vorkommender Substanzen mit Juvenilhormonwirkung. Z. Naturforschg 16b:461–464.

Schmutterer H (1990) Properties and potential of natural pesticides from the neem tree, Azadirachta indica. Ann. Rev. Entomol. 35:271–297.

Schneider D (1969) Insect olfaction: deciphering system for chemical messages. Science 163:1031–1037.

Schneider D (1987) Plant recognition by insects: a challenge for neuro–ethological research. Pages 117–123 in Insects–plants. Proceedings of the 6th International Symposium on Insect Plant Relationships (PAU 1986). V Labeyrie, G Fabres and D Lachaise, eds. Dr W Junk Publishers, Dordrecht, the Netherlands.

Schneider D, Kaissling K E (1957) Der Bau der Antenne des Seidenspinners Bombyx mori L. II Sensillen, cuticulare Bildungen und innerer Bau. Zool. Jahrb. Abt. (Anat. Ontog. Tiere) 76:223–250.

Schneider D, Steinbrecht R A (1968) Checklist of insect olfactory sensilla. Symp. Zool. Soc. London 23:279–297.

Schoener T W (1971) Theory of feeding strategies. Ann. Rev. Ecol. Syst. 11:369–404.

Schoener T W (1974) Resource partitioning in ecological communities. Science 185:27–39.

Schoni R, Stadler E, Renwick J A A, Radke C D (1987) Host and non-host plant chemicals influencing the oviposition behaviour of several herbivorous insects. Pages 31–36 in Insects–plants. Proceedings of the 6th International Symposium on Insect Plant Relationships (PAU 1986). V Labeyrie, G Fabres and D Lachaise, eds. Dr W Junk Publishers, Dordrecht, the Netherlands.

Schooler A B, Anderson M K (1979) Interspecific hybrids between Hordeum brachyantherum L. × H. bogdanii (Wilensky) × H. vulgare L. J. Hered. 70:70–72.

Schoonhoven L M (1967) Chemoreception of mustard oil glucosides in larvae of Pieris brassicae. Koninkl. Ned. Akad. Wet. Proc. Ser. C 70:556–568.

Schoonhoven L M (1973) Plant recognition by lepidopterous larvae. Pages 87–99 in Insect/plant relationships. H F van Emden, ed. Symp. R. Entomol. Soc. London No. 6. Blackwell Scientific Publications, Oxford.

Schoonhoven L M (1981) Chemical mediators between plants and phytophagous insects. Pages 31–50 in Semiochemicals: Their role in pest control. D A Nordlund, R L Jones and W J Lewis, eds. John Wiley and Sons, New York.

Schoonhoven L M (1982) Biological aspects of antifeedants. Entomol. Exp. Appl. 31:57–69.

Schoonhoven L M (1987) What makes a caterpillar eat? The sensory code underlying feeding behavior. Pages 69–97 *in* Perspectives in chemoreception and behavior. R F Chapman, E A Bernays and J G Stoffolano Jr, eds. Springer-Verlag, New York.

Schoonhoven L M (1991) Insect and host plants: 100 years of botanical instinct. Pages 3–14 *in* Proceedings of the International Symposium on Insect–Plant Relationships. T Jermy and A Szentesi, eds. Akademiai Kiado, Budapest, Hungary.

Schoonhoven L M, Derkson-Koppers I (1976) Effects of some allelochemics on food uptake and survival of a polyphagous aphid, *Myzus persicae*. Entomol. Exp. Appl. 19:52–56.

Schreiber K (1958) Uber unige inhaltsstoffe der Solanaceen und ihre Bedeutung fur die kartoffelkaferresistenz. Entomol. Exp. Appl. 1:28–37.

Schultz J C (1983a) Habitat selection and foraging tactics of caterpillars in heterogenous trees. Pages 61–90 *in* Variable plants and herbivores in natural and managed systems. R F Denno and M S McClure, eds. Academic Press, New York.

Schultz J C (1983b) Impact of variable plant defensive chemistry on susceptibility of insects to natural enemies. Pages 37–54 *in* Plant resistance to insects. P A Hedin, ed. ACS Symp. Series 208. American Chemical Society, Washington, DC.

Schultz J C (1988) Many factors influence the evolution of herbivore diets, but plant chemistry is central. Ecology 69:896–897.

Schulze C, Schnepf E, Mothes K (1967) Uber die Lokalisation der Kautschukpartikel in verschiedenen Typen von Milchrohren. Flora (Jena) 158A:458–460.

Schuster D J, Starks K J (1973) Greenbugs: components of host-plant resistance in sorghum. J. Econ. Entomol. 66:1131–1134.

Schwanitz F (1966) The origin of cultivated plants. Harvard University Press, Cambridge, MA.

Schweissing F C, Wilde G (1978) Temperature influence on greenbug resistance of crops in the seedling stage. Environ. Entomol. 7:831–834.

Schweissing F C, Wilde G (1979a) Temperature and plant nutrient effects on resistance of seedling sorghum to the greenbug. J. Econ. Entomol. 72:20–23.

Schweissing F C, Wilde G (1979b) Predisposition and nonpreference of greenbug for certain host cultivars. Environ. Entomol. 8:1070–1072.

Scott G E, Dicke E F, Pesho G R (1966) Location of genes conditioning resistance in corn to leaf feeding of the European corn borer. Crop Sci. 6:444–446.

Scott J A (1986) The butterflies of North America. Stanford University Press, Stanford, CA.

Scriber J M (1984a) Host plant suitability. Pages 154–201 *in* Chemical ecology of insects. W J Bell and R T Carde, eds. Sinauer Associates Inc., Sunderland, MA.

Scriber J M (1984b) Nitrogen nutrition of plants and insect invasion. Pages 441–460 *in* Nitrogen in crop production. R D Hank, ed. American Society of Agronomy, Madison, WI.

Scriber J M, Slansky F (1981) The nutritional ecology of immature insects. Ann. Rev. Entomol. 26:183–211.

Sebesta E E, Wood F A (1978) Transfer of green bug resistance from rye to wheat with x-rays. Agron. Abst. 61–62.

Seigler D S (1991) Cyanide and cyanogenic glycosides. Pages 35–77 *in* Herbivores: Their interactions with secondary plant metabolites. Vol. I, 2nd ed. G A Rosenthal and M R Berenbaum, eds. Academic Press, New York.

Sen Gupta G C, Miles P W (1975) Studies on the susceptibility of varieties of apple to the feeding of two strains of woolly aphids (Homoptera) in relation to the chemical content of the tissues of the host. Aust. J. Agric. Res. 26:157–168.

Seshu D V, Kauffman H E (1980) Differential response of rice varieties to the brown planthoppers in international screening tests. IRRI Res. Paper Ser. 52:1–13.

Shade R E, Kitch L W (1983) Pea aphid (Homoptera: Aphididae) biology on glandular-haired *Medicago* species. Environ. Entomol. 12:237–240.

Shade R E, Doskocil M J, Maxon N P (1979) Potato leafhopper resistance in glandular-haired alfalfa species. Crop Sci. 19:287–289.

Shade R E, Thompson T E, Campbell W R (1975) An alfalfa weevil larval resistance mechanism detected in *Medicago*. J. Econ. Entomol. 68:399–404.

Shade R E, Schroeder H E, Pueyo J J, Tabe L M, Murdock L L, Higgins T J V, Chrispeels M J (1994) Transgenic pea seeds expressing the α-amylase inhibitor of common bean are resistant to bruchid beetles. Bio/Technology 12:293–296.

Shapiro A M (1984) Microorganisms as contaminants and pathogens in insect rearing. Pages 130–142 *in* Advances and challenges in insect rearing. E G King and N C Leppla, eds. USDA–ARS, Washington DC.

Shapiro A M, DeVay J E (1987) Hypersensitivity reaction of *Brassica nigra* L. (Cruciferae) kills eggs of *Pieris* butterflies (Lepidoptera: Pieridae). Oecologia 71:631–632.

Sharma D (1991) India battles to eradicate major crop pest. New Scientist, 10 August, 131:9.

Sharma H C (1993) Host plant resistance to insects in sorghum and its role in integrated pest management. Crop Protection 12:11–34.

Sharma H C, Vidyasagar P, Leuschner K (1988) Field screening sorghum for resistance to sorghum midge (Diptera: Cecidomyiidae). J. Econ. Entomol. 81:327–334.

Shaver T N, Lukefahr M J (1969) Effect of flavonoid pigments and gossypol on growth and development of bollworm, tobacco budworm and pink bollworm. J. Econ. Entomol. 62:643–646.

Shelton A M, Hoy C W, Baker P B (1990) Response of cabbage head weight to simulated lepidopteran defoliation. Entomol. Exp. Appl. 54:181–187.

Shepard M, Carner G R, Turnipseed S G (1977) Colonization and resurgence of insect pests of soybean in response to insecticides and field isolation. Environ. Entomol. 6:501–506.

Shi Y, Wang M B, Hilder V A, Gatehouse A M R, Boulter D, Powell K S, Gatehouse J A (1994) Phloem-specific expression of GUS and GNA directed by RSs1 promoter in transgenic tobacco plants. J. Expt. Bot. 45:623–631.

Shukle R H, Murdock L L (1983) Lipoxygenase, trypsin inhibitor, and lectin from soybeans: effects on larval growth of *Manduca sexta* (Lepidoptera: Sphingidae). Environ. Entomol. 12:787–791.

Sidhu G S, Khush G S (1978) Genetic analysis of brown planthopper resistance in twenty varieties of rice, *Oryza sativa* L. Theor. Appl. Genet. 53:199–203.

Simmonds M S J, Blaney W M, Fellows L E (1990) Behavioral and electrophysiological study of antifeedant mechanisms associated with polyhydroxy alkaloids. J. Chem. Ecol. 16:3167–3196.

Simms E L (1992) Costs of plant resistance to herbivory. Pages 392–425 *in* Plant resistance to herbivores and pathogens: Ecology, evolution and genetics. R S Fritz and E L Simms, eds. The University of Chicago Press, Chicago.

Simms E L, Rausher M D (1992) Uses of quantitative genetics for studying the evolution of plant resistance. Pages 42–68 *in* Plant resistance to herbivores and pathogens: Ecology, evolution and genetics. R S Fritz and E L Simms, eds. The University of Chicago press, Chicago.

Sinden S L, Sanford L L, Cantelo W W, Deahl K L (1986) Leptine glycoalkaloids and resistance to the Colorado potato beetle (Coleoptera: Chrysomelidae) in *Solanum chacoense*. Environ. Entomol. 15:1057–1062.

Singer M C (1982) Quantification of host specificity by manipulation of oviposition behavior in the butterfly *Euphydryas editha*. Oecologia 52:224–229.

Singer M C (1984) Butterfly–host plant relationships: host quality adult choice and larval success. Pages 82–88 *in* The biology of butterflies. R Vane-Wright and P R Ackery, eds. Academic Press, New York.

Singer M C, Ng D, Thomas C D (1988) Heritability of oviposition preference and its relationship to offspring performance within a single insect population. Evolution 42:977–985.

Singh B U, Rana B S (1986) Resistance in sorghum to the shootfly *Atherigona soccata* Rondani. Insect Sci. Applic. 7:577–587.

Singh P, Moore R F, eds (1985) Handbook of insect rearing, Vol. I. Elsevier, New York.

Singh P, Russell G B, Fredericksen S (1982) The dietary effects of some ecdysteroids on the development of housefly. Entomol. Exp. Appl. 32:7–12.

Singh R, Agarwal R A (1983) Fertilizers and pest incidence in India. Potash Rev. 23:1–4.

Singh R, Ellis P R (1993) Sources, mechanisms and bases of resistance in Cruciferae to the cabbage aphid, *Brevicoryne brassicae*. Bull. OILB/SROP 16:21–35.

Singh S R, Painter R H (1964) Effect of temperature and host plants on progeny production of four biotypes of corn leaf aphid *Rhopalosiphum maidis*. J. Econ. Entomol. 57:348–350.

Slama K, Williams C M (1965) Juvenile hormone activity for the bug *Pyrrhocoris apterus*. Proc. Nat. Acad. Sci. USA 54:411–414.

Slansky Jr F (1976) Phagism relationships among butterflies. J. NY Entomol. Soc. 84:91–105.

Slansky Jr F (1992) Allelochemical–nutrient interactions in herbivore nutritional ecology. Pages 135–174 *in* Herbivores: Their interactions with secondary plant metabolites, Vol. II, 2nd ed. G A Rosenthal and M R Berenbaum, eds. Academic Press, New York.

Slansky Jr F, Feeny P (1977) Stabilization of the rate of nitrogen accumulation by larvae of the cabbage butterfly on wild and cultivated food-plants. Ecol. Monogr. 47:209–228.

Slansky Jr F, Rodriguez J G (1987) Nutritional ecology of insects, mites, spiders, and related invertebrates: an overview. Pages 1–69 *in* Nutritional ecology. F Slansky Jr and J G Rodriguez, eds. John Wiley and Sons, New York.

Slansky Jr F, Wheeler G S (1989) Compensatory increases in food consumption and utilization efficiencies by velvetbean caterpillars mitigate impact of diluted diets on growth. Entomol. Exp. Appl. 51:175–187.

Slifer E H (1970) The structure of arthropod chemoreceptors. Ann. Rev. Entomol. 15:121–142.

Slosser J E (1983) Potential of *Heliothis* spp. (Lepidoptera: Noctuidae) resistant cotton in limited-irrigation situations. J. Econ. Entomol. 76:864–868.

Smart J, Hughes N F (1973) The insect and the plant: progressive palaeoecological integration. Pages 143–155 *in* Insect–plant relationships. H F van Emden, ed.

Symp. R. Entomol. Soc. London No. 6. Blackwell Scientific Publications, Oxford.

Smiley J (1978) Plant chemistry and the evolution of host specificity: new evidence for *Heliconius* and *Passiflora*. Science 201:745–747.

Smith B D (1966) Effect of the plant alkaloid sparteine on the distribution of the aphid *Acrythosiphon spartii*. Nature (London) 212:213–214.

Smith C M (1989) Plant resistance to insects: A fundamental approach. John Wiley and Sons, New York.

Smith C M, Brim C A (1979) Field and laboratory evaluations of soybean lines for resistance to corn earworm leaf feeding. J. Econ. Entomol. 72:78–80.

Smith C M, Wilson R F, Brim C A (1979) Feeding behavior of Mexican bean beetle on leaf extracts of resistant and susceptible soybean genotypes. J. Econ. Entomol. 72:374–377.

Smith C M, Khan Z R, Pathak M D (1994) Techniques for evaluating insect resistance in crop plants. Lewis Publishers, Boca Raton, FL.

Smith D H, Webster J A (1973) Resistance to cereal leaf beetle in Hope substitution lines. Pages 761–764 *in* Proceedings of the 4th International Wheat Genetics Symposium. E R Sears and L M S Sears, eds. University of Missouri, Columbia, MO.

Smith R L, Wilson R L, Wilson F D (1975) Resistance of cotton plant hairs to mobility of first instars of the pink bollworm. J. Econ. Entomol. 68:679–683.

Smith S G F, Kreitner G L (1983) Trichomes in *Artemisia ludoviciana* Asteraceae and their ingestion by *Hypochlora alba* (Orthoptera: Acrididae). Am. Midland Natur. 110:118–123.

Snodgrass R E (1935) Principles of insect morphology. McGraw-Hill, New York.

Snyder J C, Carter C D (1984) Leaf trichomes and resistance of *Lycopersicon hirsutum* and *L. esculentum* to spider mites. J. Am. Soc. Hort. Sci. 109:837–843.

Sogawa K, Pathak M D (1970) Mechanisms of brown planthopper resistance in Mudgo variety of rice (Hemiptera: Delphacidae) Appl. Entomol. Zool. 5:145–158.

Solomon M G (1989) The role of natural enemies in top fruit. Pages 70–85 *in* Proceedings of the Symposium on Insect Control Strategies and Environment, Amsterdam. ICI Agrochemicals, Fernhurst, UK.

Sorensen E L, Horber E (1974) Selecting alfalfa seedlings to resist the potato leafhopper. Crop Sci. 14:85–86.

Sosa Jr O (1979) Hessian fly: resistance of wheat as affected by temperature and duration of exposure. Environ. Entomol. 8:280–281.

Sosa Jr O, Foster J E (1976) Temperature and the expression of resistance in wheat to the Hessian fly. Environ. Entomol. 5:333–336.

Southgate B J (1982) The importance of bruchids (Coleoptera: Bruchidae) as pests of pigeon peas (*Cajanus cajan*). Trop. Grain Legume Bull. 24:21–24.

Southon I W, Buckingham J, eds (1989) Dictionary of alkaloids. Chapman and Hall, London.

Southwood S R (1986) Plant surfaces and insects – an overview. Pages 1–22 *in* Insects and the plant surface. B Juniper and Sir R Southwood, eds. Edward Arnold, London.

Southwood T R E (1973) The insect/plant relationship – evolutionary perspective. Pages 3–30 *in* Insect–plant relationships. H F van Emden, ed. Symp. R. Entomol. Soc. London No. 6. Blackwell Scientific Publications, Oxford.

Southwood T R E (1975) The dynamics of insect populations. Pages 151–199 *in* Insects, science, and society. D Pimentel, ed. Academic Press, New York.

Southwood T R E (1978a) The components of diversity. Symp. R. Entomol. Soc. Lond. 9:19–40.

Southwood T R E (1978b) Ecological methods with particular reference to the study of insect populations, 2nd ed. Chapman and Hall, London.

Sparks T C, Lockwood J A, Byford R L, Graves J B, Leonard B R (1989) The role of behavior in insecticide resistance. Pestic. Sci. 26:383–399.

Sprague G F, Dahms R G (1972) Development of crop resistance to insects. J. Environ. Qual. 1:28–34.

Staal G B (1986) Antijuvenile hormone agents. Ann. Rev. Entomol. 31:391–429.

Stadler E (1978) Chemoreception of host plant chemicals by ovipositing females of *Delia* (*Hylemya*) *brassicae*. Entomol. Exp. Appl. 24:711–720.

Stadler E (1984) Contact chemoreception. Pages 3–35 *in* Chemical ecology of insects. W J Bell and R T Carde, eds. Sinauer Associates Inc., Sunderland, MA.

Stadler E (1986) Oviposition and feeding stimuli in leaf surface waxes. Pages 105–121 *in* Insects and the plant surface. B E Juniper and T R E Southwood, eds. Edward Arnold, London.

Stadler E (1992) Behavioral responses of insects to plant secondary compounds. Pages 45–88 *in* Herbivores: Their interactions with secondary plant metabolites, Vol II, 2nd ed. Ecological and evolutionary processes. G A Rosenthal and M R Berenbaum, eds. Academic Press, New York.

Stadler E, Buser H R (1984) Defense chemicals in leaf surface wax synergistically stimulate oviposition by phytophagous insects. Experientia 40:1157–1159.

Stadler E, Hanson F (1978) Food discrimination and induction of preference for artificial diets in the tobacco hornworm, *Manduca sexta*. Physiol. Entomol. 3:121–133.

Stadler E, Schoni R (1990) Oviposition behavior of the cabbage root fly, *Delia radicum* influenced by host plant extracts. J. Insect Behav. 3:195–209.

Stafford H A (1974) The metabolism of aromatic compounds. Ann. Rev. Plant Physiol. 25:459–486.

Stahl E (1888) Pflanzen and Schnecken. Biologische Studie uber die Schutzmittel der Pflanzen gegen Schneckenfrass. Jena Z. Med. Naturwissenchaft 22:557–684.

Stanton M L (1983) Spatial patterns in the plant community and their effects upon insect search. Pages 125–157 *in* Herbivorous insects: Host seeking behavior and mechanisms. S Ahmad, ed. Academic Press, New York.

Starks K J, Burton R L (1977a) Greenbugs: a comparison of mobility on resistant and susceptible varieties of four small grains. Environ. Entomol. 6:331–332.

Starks K J, Burton R L (1977b) Greenbugs: determining biotypes, culturing, and screening for plant resistance. USDA-ARS Tech. Bull. 1556. 12 pp.

Starks K J, McMillan W W, Sekul A A, Cox H C (1965) Corn earworm larval feeding response to corn silk and kernel extracts. Ann. Entomol. Soc. Am. 58:74–76.

Starks K J, Muniappan R, Eikenbary R D (1972) Interaction between plant resistance and parasitism against the greenbug on barley and sorghum. Ann. Entomol. Soc. Am. 65:650–655.

Starks K J, Burton R L, Merkle O G (1983) Greenbugs (Homoptera: Aphididae) plant resistance in small grains and sorghum to biotype E. J. Econ. Entomol. 76:877–880.

Stebbins G L (1958) The inviability, weakness, and sterility of interspecific hybrids. Adv. Genet. 9:147–215.

Steidi R P, Webster J A, Smith Jr D H (1979) Cereal leaf beetle plant resistance: antibiosis in an *Avena sterilis* introduction. Environ. Entomol. 8:448–450.

Steinbrecht R A, Kasang G (1972) Capture and conveyance of odour molecules in an insect olfactory receptor. Pages 193–199 *in* Olfaction and taste IV. D Schneider, ed. Wissenschaftliche Verlagsgesellschaft, Stuttgart.

Stephanou G S, Alahiotis N, Marmaras V J, Christodoulou C (1983) Heat shock response in *Ceratitis capitata*. Comp. Biochem. Physiol. 74:425–432.

Stern V M (1969) Interplanting alfalfa in cotton to control lygus bugs and other insect pests. Tall Timbers Conf. Ecol. Animal Control Habitat Manage. 1:55–69.

Stern V M, Dietrick E J, Mueller A (1965) Improvements on self-propelled equipment for collecting, separating, and tagging mass numbers of insects in the field. J. Econ. Entomol. 58:949–953.

Stewart C R (1981) Proline accumulation: biochemical aspects. Pages 243–259 *in* The physiology and biochemistry of drought resistance in plants. L G Paleg and D Aspinall, eds. Academic Press, Sydney.

Stipanovic R D (1983) Function and chemistry of plant trichomes and glands in insect resistance: protective chemicals in plant epidermal glands and appendages. Pages 69–100 *in* Plant resistance to insects. P A Hedin, ed. ACS Symp. Series 208. American Chemical Society, Washington, DC.

Stipanovic R D, Williams H J, Smith L A (1986) Cotton terpenoid inhibition of *Heliothis virescens* development. Pages 79–94 *in* Natural resistance of plants to pests. M B Green and P A Hedin, eds. ACS Symp. Series 296. American Chemical Society, Washington, DC.

Stoner A K, Frank J A, Gentile A G (1968) The relationship of glandular hairs on tomatoes to spider mite resistance. Proc. Am. Soc. Hort. Sci. 93:532–538.

Stork N E (1980) Role of wax blooms in preventing attachment to brassicas by the mustard beetle, *Phaedon cochleariae*. Entomol. Exp. Appl. 28:100–107.

Strack D, Reznik H (1976) Dynamics of flavonol glycosides during germination of *Cucurbita maxima* Duchense. Z. Pflanzenphysiol. 79:95–108.

Strong Jr D R (1979) Biogeographic dynamics of insect–host plant communities. Ann. Rev. Entomol. 24:89–119.

Strong Jr D R, Lawton J H, Southwood S R (1984) Insects on plants: Community patterns and mechanisms. Harvard University Press, Cambridge, MA.

Sturchkow B, Low I (1961) Die Wirkung einiger Solanum-alkaloids-glycoside auf den Kartofielkafer, *Leptinotarsa decemlineata* Say. Entomol. Exp. Appl. 4:133–142.

Subramanian R, Balasubramanian M (1976) Effect of potash nutrition on the incidence of certain insect pests of rice. Madras Agric. J. 63:561–564.

Sutter G R, Branson T F (1980) A procedure for artificially infesting field plots with corn rootworm eggs. J. Econ. Entomol. 73:135–137.

Suzuki K, Ishimoto M, Kikuchi F, Kitamura K (1993) Growth inhibitory effect of an α-amylase inhibitor from the wild common bean resistant to the Mexican bean weevil (*Zabrotes subfasciatus*). Jap. J. Breed. 43:257–265.

Swain T (1976) Angiosperm–reptile coevolution. Pages 107–122 *in* Morphology and biology of reptiles. D A Bellairs and C B Cox, eds. Linn. Soc. Sym. Series 3. Academic Press, New York.

Swain T (1977) Secondary compounds as protective agents. Ann. Rev. Plant Physiol. 28:479–501.

Swain T (1979) Tannins and lignins. Pages 657–682 *in* Herbivores: Their interaction with secondary plant metabolites. G A Rosenthal and D H Janzen, eds. Academic Press, New York.

Swain T, Cooper-Driver G A (1974) Biochemical systematics in the Filicopsida. Pages 111–134 *in* Phylogeny and classification of ferns. A C Jermy, ed. Linnean Society, London.

Tabashnik B E (1985) Deterrence of diamondback moth (Lepidoptera: Plutellidae) oviposition by plant compound. Environ. Entomol. 14:575–578.

Tabashnik B E (1987) Plant secondary compounds as oviposition deterrents for cabbage butterfly, *Pieris rapae* (Lepidoptera: Pieridae). J. Chem. Ecol. 13:309–316.

Tabashnik B E, Croft B A (1982) Managing pesticide resistance in crop–arthropod complexes: interactions between biological and operational factors. Environ. Entomol. 11:1137–1144.

Tabashnik B E, Wheelock H, Rainbolt J D, Watt W B (1981) Individual variation in oviposition preference in the butterfly, *Colias eurytheme*. Oecologia 50:225–230.

Taksdal G (1993) Resistance in swedes to the turnip root fly and its relation to intergrated pest management. Bull. OILB/SROP 16:13–20.

Tallamy D W (1985) Squash beetle feeding behavior: an adaptation against induced cucurbit defenses. Ecology 66:1574–1579.

Tallamy D W (1986) Behavioral adaptations in insects to plant allelochemicals. Pages 273–300 *in* Molecular aspects of insect–plant associations. L B Brattsten and S Ahmad, eds. Plenum Press, New York.

Tallamy D W, Krischik V A (1989) Variation and function of cucurbitacins in *Cucurbita*: An examination of current hypotheses. Am. Nat. 133:766–786.

Tallamy D W, McCloud E S (1991) Squash beetles, cucumber beetles, and inducible cucurbit responses. Pages 155–181 *in* Phytochemical induction by herbivores. D W Tallamy and M J Raupp, eds. John Wiley and Sons, New York.

Taneja S L, Leuschner K (1985) Methods of rearing, infestation, and evaluation for *Chilo partellus* resistance in sorghum. Pages 175–188 *in* Proceedings, International Sorghum Entomology Workshop, 15–21 July 1984. K Leuschner and G L Teetes, eds. Texas A & M University, College Station, and ICRISAT, Patancheru, India.

Tangley L (1988) Preparing for climate change. Bio Science 38:14–18.

Tanksley S D (1983) Molecular markers in plant breeding. Plant Mol. Biol. Rep. 1:3–8.

Tanksley S D, Causse M, Fulton T, Ahn N, Wang Z, Wu K, Xiao J, Yu Z, Second G, McCouch S (1992) A high density molecular map of rice genome. Rice Genet. Newsl. 9:111–115.

Teetes G L (1980) Breeding sorghums resistant to insects. Pages 457–485 *in* Breeding plants resistant to insects. F G Maxwell and P R Jennings, eds. John Wiley and Sons, New York.

Thielges B A (1968) Altered polyphenol metabolism in the foliage of *Pinus sylvestris* associated with European pine sawfly attack. Can. J. Bot. 46:724–725.

Thompson Jr G A (1980) Plant lipids of taxonomic significance. Pages 535–553 *in* Secondary plant products. E A Bell and B V Charlwood, eds. Springer-Verlag, New York.

Thompson J N (1982) Interaction and coevolution. John Wiley and Sons, New York.

Thompson J N, Burdon J J (1992) Gene-for-gene coevolution between plants and parasites. Nature 360:121–125.

Thompson J N, Pellmyr O (1991) Evolution of oviposition behavior and host preference in Lepidoptera. Ann. Rev. Entomol. 36:65–89.

Thornburg R W, Kernan A, Molin L (1990) Chloramphenicol acetyl transferase (CAT) protein is expressed in transgenic tobacco in field tests following attack by insects. Plant Physiol. 92:500–505.

Thorsteinson A J (1953) The chemotactic basis of host plant selection in an oligophagous insect [*Plutella maculipennis* (Curt.)]. Can. J. Zool. 31:52–72.

Thorsteinson A J (1960) Host selection in phytophagous insects. Ann. Rev. Entomol. 5:193–218.

Thottappilly G, Tsai J H, Bath J E (1972) Differential aphid transmission of two bean yellow mosaic virus strains and comparative transmission by biotypes and stages of the pea aphid. Ann. Entomol. Soc. Am. 65:912–915.

Thurston R (1970) Toxicity of trichome exudates of *Nicotiana* and *Petunia* species to tobacco hornworm larvae. J. Econ. Entomol. 63:272–274.

Thurston R, Smith W T, Cooper B P (1966) Alkaloid secretion by trichomes of *Nicotiana* species and resistance to aphids. Entomol. Exp. Appl. 9:428–432.

Tichenor L H, Seigler D S (1980) Electroantennogram and oviposition responses of *Manduca sexta* to volatile components of tobacco and tomato. J. Insect Physiol. 56:309–314.

Tinbergen N (1951) The study of instinct. Oxford University Press, London.

Ting I P (1982) Plant physiology. Addison-Wesley, Menlo Park, CA.

Tingey W M (1986) Techniques for evaluating plant resistance to insects. Pages 251–284 *in* Insect–plant interactions. J R Miller and T A Miller, eds. Springer-Verlag, New York.

Tingey W M, Gibson R W (1978) Feeding and mobility of the potato leafhopper impaired by glandular trichomes of *Solanum berthaultii* and *S. polyadenium*. J. Econ. Entomol. 71:856–858.

Tingey W M, Sinden S L (1982) Glandular pubescence, glycoalkaloid composition and resistance to the green peach aphid, potato leafhopper, and potato flea beetle in *Solanum berthaultii*. Am. Potato J. 59:95–106.

Tingey W M, Singh S R (1980) Environmental factors influencing the magnitude and expression of resistance. Pages 89–113 *in* Breeding plants resistant to insects. F G Maxwell and P R Jennings, eds. John Wiley and Sons, New York.

Tingey W M, van de Klashorst G (1976) Green peach aphid: magnification of field populations on potatoes. J. Econ. Entomol. 69:363–364.

Tingey W M, Plaisted R L, Laubengayer J E, Mehlenbacher S A (1982) Green peach aphid resistance by glandular trichomes in *Solanum tuberosum* and *S. berthaultii* hybrids. Am. Potato J. 59:241–251.

Tjallingii W F (1978) Electronic recording of penetration behaviour of aphids. Entomol. Exp. Appl. 24:721–730.

Tjallingii W F (1990) Continuous recording of stylet penetration activities by aphids. Pages 89–99 *in* Aphid–plant genotype interactions. R K Campbell and R D Eikenbary, eds. Elsevier Science Publishers, Amsterdam.

Toenniessen G H (1991) Potentially useful genes for rice genetic engineering. Page 253–280 *in* Rice biotechnology. G S Khush and G H Toenniessen, eds. CAB International, Wallingford, UK/International Rice Research Institute, Manila, Philippines.

Toldine T E (1984) Relationship between DIMBOA content and *Helminthosporium turcicum* resistance in maize (in Hungarian with English abstract). Novenytermeles 33:213–218.

Toong Y C, Schooley D A, Baker F C (1988) Isolation of insect juvenile hormone III from a plant. Nature 333:170–171.

Toriyama K, Arimoto Y, Uchimiya H, Hinata K (1988) Transgenic rice plants after direct gene transfer into protoplasts. Bio/Technology 6:1072–1074.

Towers G H N (1984) Interactions of light with phytochemicals in some natural and novel systems. Can. J. Bot. 62:2900–2911.

Traynier R M M (1967) Effect of host plant odour on the behaviour of the adult cabbage root fly, *Erioischia brassicae*. Entomol. Exp. Appl. 10:321–328.

Traynier R M M, Truscott R J W (1991) Potent natural egg-laying stimulant for cabbage butterfly *Pieris rapae*. J. Chem. Ecol. 17:1371–1380.

Treacy M F, Zummo G R, Benedict J H (1985) Interactions of host-plant resistance in cotton with predators and parasites. Agric. Ecosyst. Environ. 13:151–157.

Treshow M (1970) Environment and plant response. McGraw-Hill, New York.

Trouvelot B (1958) L'Immuniate chez les Solanacus a l'egard du Doryphore. Entomol. Exp. Appl. 1:9–13.

Trumble J T, Hare J D (1989) Acidic fog-induced changes in host-plant suitability: interactions of *Trichoplusia ni* and *Phaseolus lunatus*. J. Chem. Ecol. 15:2379–2390.

Trumble J T, Hare J D, Musselman R C, McCool P M (1987) Ozone-induced changes in host plant suitability: interaction of *Keiferia lycopersicella* and *Lycopersicon esculentum*. J. Chem. Ecol. 13:203–218.

Trumble J T, Kolodny-Hirsch D M, Ting I P (1993) Plant compensation for arthropod herbivory. Ann. Rev. Entomol. 38:93–119.

Turlings T C J, Tumlinson J H, Lewis W J (1990) Exploitation of herbivore-induced plant odors by host-seeking parasitic wasps. Science 250:1251–1253.

Tuttle A F, Ferro D N, Idoine K (1988) Role of visual and olfactory stimuli in host finding of adult cabbage root flies, *Delia radicum*. Entomol. Exp. Appl. 47:37–44.

Uchimiya H, Handa T, Brar D S (1989) Transgenic plants. J. Biotechnology 12:1–20.

Unnithan G C, Nair K K, Kooman C J (1978) Effects of precocene II and juvenile hormone III on the activity of neurosecretory A-cells in *Oncopeltus fasciatus*. Experientia 34:411–412.

Uritani I, Saito T, Honda H, Kim W K (1975) Induction of furano-terpenoids in sweet potato roots by the larval components of the sweet potato weevils. Agric. Biol. Chem. 39:1857–1862.

Vaeck M, Reynaerts A, Hofte H, Jansens S, De Beuckeleer M, Dean C, Zabeau M, Van Montagu M, Leemans J (1987) Transgenic plants protected from insect attack. Nature 328:33–37.

Vaishampayan S M, Waldbauer G P, Kogan M (1975) Visual and olfactory responses in orientation to plants by the greenhouse whitefly, *Trialeurodes vaporariorum* (Homoptera: Aleyrodidae). Entomol. Exp. Appl. 18:412–422.

van der Klashorst G, Tingey W M (1979) Effect of seedling age, environmental temperature, and foliar total glycoalkaloids on resistance of five *Solanum* genotypes to the potato leafhopper. Environ. Entomol. 8:690–693.

Van der Plank J E (1963) Plant diseases: Epidemics and control. Academic Press, New York. 349 pp.

Van der Plank J E (1968) Disease resistance in plants. Academic Press, New York. 206 pp.

van Drongelen W (1978) The significance of contact chemoreceptor sensitivity in the larval stage of different *Yponomeuta* species. Entomol. Exp. Appl. 24:343–347.

van Drongelen W, van Loon J J (1980) Inheritance of gustatory sensitivity in F1 progeny of crosses between *Yponomeuta cagnagellus* and *Y. malinellus*. Entomol. Exp. Appl. 28:199–203.

van Emden H F (1965) The role of uncultivated land in the biology of crop pests and beneficial insects. Sci. Hort. 17:121–136.

van Emden H F (1972) Aphids as phytochemists. Pages 25–43 *in* Phytochemical ecology. J B Harborne, ed. Academic Press, New York.

van Emden H F (1990) The interaction of host plant resistance with other control measures. Proc. Brighton Crop Prot. Conference 3:939–949.

van Emden H F, Wearing C H (1965) The role of the aphid host plant in delaying economic damage levels in crops. Ann. Appl. Biol. 56:323–324.

van Emden H F, Eastop V F, Hughes R D, Way M J (1969) The ecology of *Myzus persicae*. Ann. Rev. Entomol. 14:197–270.

van Lenteren J C (1991) Biological control in a tritrophic system approach. Pages 3–28 *in* Aphid–plant interactions: Populations to molecules. D C Peters and J A Webster, eds. Oklahoma State University Press, Stillwater, OK.

van Lenteren J C, Ramakers, P M J, Woets J (1980) Integrated control of vegetable pests in greenhouses. Pages 109–118 *in* Integrated control of insect pests in the Netherlands. A K Minks and P Gruys, eds. Centre for Agricultural Publishing and Documentation, Wageningen, The Netherlands.

Van Soest P J (1982) Nutritional ecology of ruminants. O and B Books, Corvallis, OR.

Vavilov N I (1926) Studies on the origins of cultivated plants [in Russian]. Bull. Appl. Bot. Plant Breed. 16:1–245.

Vavilov N I (1950) Phytogeographic basis of plant breeding. Pages 14–54 *in* The origin, variation, immunity and breeding of cultivated plants (translated by K. Starr Chester). Ronald Press, New York.

Velusamy R, Heinrichs E A (1986) Electronic monitoring of feeding behavior of *Nilaparvata lugens* (Homoptera: Delphacidae) on resistant and susceptible rice cultivars. Environ. Entomol. 15:678–682.

Velusamy R, Thayumanavan B, Sadasivam S, Jayaraj S (1990a) Effect of steam distillate extract of resistant wild rice *Oryza officinalis* on behavior of brown planthopper *Nilaparvata lugens* (Stal) (Homoptera: Delphacidae). J. Chem. Ecol. 16:809–817.

Velusamy R, Thayumanavan B, Sadasivam S (1990b) Effect of steam distillate extracts of selected resistant cultivated and wild rices on behavior of leaffolder (*Cnaphalocrocis medinalis* (Guenee) (Lepidoptera: Pyralidae). J. Chem. Ecol. 16:2291–2296.

Verschaffelt E (1911) The cause determining the selection of food in some herbivorous insects. Proc. Acad. Sci. Amsterdam 13:536–542.

Vidyabhushanam R V (1972) Breeding for shoot fly resistance in India. Pages 218–232 *in* Control of sorghum shoot fly. M G Jotwani and W R Young, eds. Oxford and IBH Publishing Co., New Delhi, India.

Vinson S B, Barbosa P (1987) Interrelationships of nutritional ecology of parasitoids. Pages 673–695 *in* Nutritional ecology of insects, mites, spiders and related invertebrates. F Slansky Jr and J G Rodriguez, eds. John Wiley and Sons, New York.

Visser J H (1979) Electroantennogram responses of the Colorado beetle, *Leptinotarsa decemlineata* to plant volatiles. Entomol. Exp. Appl. 25:86–97.

Visser J H (1983) Differential sensory perceptions of plant compounds by insects. Pages 215–230 *in* Plant resistance to insects, P A Hedin, ed. ACS Symp. Series 208.

American Chemical Society, Washington, DC.

Visser J H (1986) Host odor perception in phytophagous insects. Ann. Rev. Entomol. 31:121–144.

Visser J H, Ave D A (1978) General green leaf volatiles in the olfactory orientation of the Colorado potato bettle, *Leptinotarsa decemlineata*. Entomol. Exp. Appl. 24:738–749.

Visser J H, Thiery D (1985) Behavioral responses of Colorado potato beetle to stimulation by wind and plant odors. Bull. Agric. Exp. Stn. Univ. Mass. 704:117–125.

Vogt E A, Nechols J R (1993) Responses of the squash bug (Hemiptera: Coreidae) and its egg parasitoid, *Gryon pennsylvanicum* (Hymenoptera: Scelionidae) to three *Cucurbita* cultivars. Environ. Entomol. 22:238–245.

Wada K, Munakata K (1968) Naturally occurring insect control chemicals, Isoboldine, a feeding inhibitor, and Cocculolidine, an insecticide in the leaves of *Cocculus trilobus* DC. J. Agric. Food Chem. 16:471–474.

Wagner M R, Benjamin D M, Clancy K M, Schuh B A (1983) Influence of diterpene resin acids on feeding and growth of larch sawfly *Pristiphora erichsonii* (Hartig) J. Chem. Ecol. 9:119–127.

Waiss Jr A C, Chan B G, Elliger C A, Wisemann B R, McMillian W W, Widstrom N W, Zuber M S, Keaster A J (1979) Maysin, a flavone glycoside from corn silk with antibiotic activity toward corn earworm. J. Econ. Entomol. 72:256–258.

Waldbauer G P, Friedman S (1991) Self-selection of optimal diets by insects. Ann. Rev. Entomol. 36:43–63.

Walker Jr J K, Niles G A (1971) Population dynamics of the boll weevil and modified cotton types: implication for pest management. Tex. Agric. Exp. Stn. Bull. 1109. 14 pp.

Wallace L E, McNeal F H, Berg M A (1974) Resistance to both *Oulema melanopus* and *Cephus cinctus* in pubescent-leaved and solid stemmed wheat selections. J. Econ. Entomol. 67:105–107.

Wallner W E, Walton G S (1979) Host defoliation: a possible determinant of gypsy moth population quality. Ann. Entomol. Soc. Am. 72:62–67.

Walter E V (1957) Corn earworm lethal factor in silks of sweet corn. J. Econ. Entomol. 50:105–106.

Wang M, Boulter D, Gatehouse J A (1992) A complete sequence of the rice sucrose synthase I (RS_{s1}) gene. Plant Mol. Biol. 19:881–885.

Warren G W, Carozzi N B, Desai N, Koziel M G (1992) Field evaluation of transgenic tobacco containing a *Bacillus thuringiensis* insecticidal protein gene. J. Econ. Entomol. 85:1651–1659.

Way M J, Cammell M (1970) Aggregation behavior in relation to food utilization by aphids. Pages 229–247 *in* Animal population in relation to their food resources. A Watson, ed. Blackwell Scientific Publications, Oxford.

Wearing C H (1972) Selection of Brussels sprouts of different water status by apterous and alate *Myzus persicae* and *Brevicoryne brassicae* in relation to the age of leaves. Entomol. Exp. Appl. 15:139–154.

Wearing C H, Hutchins R F N (1973) α-Farnesene, a naturally occurring oviposition stimulant for the codling moth, *Laspeyresia pomonella*. J. Insect Physiol. 19:1251–1256.

Webster J A (1975) Association of plant hairs and insect resistance. An annotated bibliography. USDA-ARS Misc. Publ. 1297:1–18.

Webster J A, Smith Jr D H (1983) Developing small grains resistant to the cereal leaf beetle. USDA Tech. Bull. 1673.

Wegorek W, Krzymanska J (1970) Investigation on the biochemical causes of some lupine varieties resistant to pea aphid (*Acyrthosiphon pisum* Harris) Biol. Abstr. 51:6216.

Weinstein L H (1977) Fluoride and plant life. J. Occupational Med. 19:49–78.

Weiss M J, Morrill W L (1992) Wheat stem sawfly (Hymenoptera: Cephidae) revisited. Am. Entomol. 38:241–245.

Wellso S G, Hoxie R P (1982) The influence of environment on the expression of trichomes in wheat. Crop Sci. 22:879–886.

Welter A, Jadot J, Dardenne G, Marlier M, Casimir J (1976) 2,5-Dihydroxymethyl 3,4-dihydroxypyrrolidine dans les feuilles de *Derris elliptica*. Phytochemistry 15:747–749.

Wensler R J D (1962) Mode of host selection by an aphid. Nature (London) 195:830–831.

Wensler R J D, Filshie B K (1969) Gustatory sense organs in the food canal of aphids. J. Morphol. 129:473–491.

Whalley P (1977) Lower Cretaceous Lepidoptera. Nature (London) 266:526.

Whalley P, Jarzembowski E A (1981) A new assessment of *Rhyniella*, the earliest known insect from the Devonian of Rhynie, Scotland. Nature (London) 291:317.

White W H, Irvine J E (1987) Evaluation of variation in resistance to sugarcane borer (Lepidoptera: Pyralidae) in a population of sugarcane derived from tissue culture. J. Econ. Entomol. 80:182–184.

Whitham T G (1981) Individual trees as heterogenous environments: adaptation to herbivory or noise? Pages 9–27 *in* Insect life history patterns: Habitat and geographic variation. R F Denno and H Dingle, eds. Springer-Verlag, New York.

Whitman D W (1988) Allelochemical interactions among plants, herbivores, and their predators. Pages 11–64 *in* Novel aspects of insect–plant interactions. P Barbosa and D K Letourneae, eds. John Wiley and Sons, New York.

Whittaker R H, Feeny P (1971) Allelochemicals: chemical interactions between species. Science 171:757–770.

Whitten M J (1992) Pest management in 2000: what we might learn from the twentieth century. Pages 9–44 *in* Pest management and the environment in 2000. A A S A Kadir and H S Barlow, eds. CAB International, Wallingford, UK.

Widstrom N W, Wiseman B R, McMillian W W (1973) Evaluation of selection potential for earworm resistance in two corn populations and their cross. Crop Sci. 15:183–184.

Widstrom N W, Williams W P, Wiseman B R, Davis F M (1992) Recurrent selection for resistance to leaf feeding fall armyworm on maize. Crop Sci. 32:1171–1174.

Wieczorek H (1976) The glycoside receptor of the larvae of *Mamestra brassicae* L. J. Comp. Physiol. 106:153–176.

Wiermann R (1981) Secondary plant products and cell and tissue differentiation. Pages 85–116 *in* The biochemistry of plants: A comprehensive treatise, Vol. 7. Secondary plant products. E E Conn, ed. Academic Press, New York.

Wigglesworth V B (1965) Water and temperature. Pages 594–628 *in* The principles of insect physiology. Methuen, London.

Wilbert H (1980) The effect of resistant plants on pest population dynamics. Z. Angew.

Entomol. 89:298–314.

Wilde G, Feese H (1973) A new corn leaf aphid biotype and its effect on some cereal and small grains. J. Econ. Entomol. 66:570–571.

Wilhoit L R (1991) Modelling the population dynamic of different aphid genotypes in plant variety mixtures. Ecol. Modelling 55:257–283.

Wilkund C (1981) Generalist vs. specialist oviposition behavior in *Papilio machaon* and functional aspects on the hierarchy of oviposition preference. Oikos 36:163–170.

Williams A L, Mitchell E R, Heath R R, Barfield C S (1986) Oviposition deterrents for fall armyworm (Lepidoptera: Noctuidae) from larval frass, corn leaves, and artificial diet. Environ. Entomol. 15:327–330.

Williams C M (1970) Hormonal interactions between plants and insects. Pages 103–132 *in* Chemical ecology. E Sondheimer and J B Simeone, eds. Academic Press, New York.

Williams H J, Elzen G W, Vinson S B (1988) Parasitoid–host-plant interactions, emphasizing cotton (*Gossypium*). Pages 171–200 *in* Novel aspects of insect–plant interactions. P Barbosa and D K Letourneau, eds. John Wiley and Sons, New York.

Williams W G, Kennedy G G, Yamamoto R T, Thacker J D, Bordner J (1980) 2-Tridecanone: a naturally occurring insecticide from the wild tomato, *Lycopersicon hirsutum f. glabratum*. Science 207:888–889.

Williams W P, Buckley P M, Taylor V N (1983) Southwestern corn borer growth on callus initiated from corn genotypes with different levels of resistance to plant damage. Crop Sci. 23:1210–1212.

Williams W P, Buckley P M, Davis F M (1987) Feeding response of corn earworm (Lepidoptera: Noctuidae) to callus and extracts of corn in the laboratory. Environ. Entomol. 16:532–534.

Willmer P G (1980) The effects of a fluctuating environment on the water relations of larval lepidoptera. Ecol. Entomol. 5:271–292.

Willmer P G (1982) Microclimate and the environmental physiology of insects. Pages 1–57 *in* Advances in insect physiology, Vol. 16. M J Berridge, J E Treherne, V B Wigglesworth, eds. Academic Press, New York.

Wilson F D, Shaver T N (1973) Glands, gossypol content, and tobacco budworm development in seedlings and floral parts of cotton. Crop Sci. 13:107–110.

Wilson R L, Wissink K M (1986) Laboratory method for screening corn for European corn borer (Lepidoptera: Pyralidae) resistance. J. Econ. Entomol. 79:274–276.

Winstanley C, Blaney W M (1978) Chemosensory mechanisms of locusts in relation to feeding. Entomol. Exp. Appl. 24:750–758.

Wiseman B R (1989) Technological advances for determining resistance in maize to *Heliothis zea*. Pages 94–100 *in* Toward insect resistant maize for the third world. J A Mihm, B R Wiseman and F M Davis, eds. International Wheat and Maize Improvement Center (CIMMYT), El Batan, Mexico.

Wiseman B R (1990) Plant resistance: a logical component of sustainable agriculture. Ann. Plant Resist. Insects Newsl. 16:40.

Wiseman B R, Harrell E A, McMillian W W (1973) Continuation of tests of resistant sweet corn hybrid plus insecticides to reduce losses from corn earworm. Environ. Entomol. 2:919–920.

Wiseman B R, Widstrom N W, McMillian W W (1974) Methods of application and numbers of eggs of the corn earworm required to infest ears of corn artificially. J. Econ. Entomol. 67:74–76.

Wiseman B R, Widstrom N W, McMillian W W, Waiss Jr A C (1985) Relationship between maysin concentration in corn silk and corn earworm (Lepidoptera: Noctuidae) growth. J. Econ. Entomol. 78:423–427.

Wiseman B R, Lynch R E, Mikolajczak K L, Gueldner R C (1986) Advancements in the use of a laboratory bioassay for basic host plant resistance studies. Florida Entomol. 69:559–565.

Wolcott G W (1958) New termite repellent wood extractives. Proc. 10th Int. Congr. Entomol. 4:417–421.

Wong B L, Berryman A A (1977) Host resistance to the fir engraver beetle. 3. Lesion development and containment of infection by resistance *Abies grandis* inoculated with *Trichosporium symbioticum*. Can. J. Bot. 55:2358–2365.

Wong E (1968) The role of chalcones and flavanones in flavonoid biosynthesis. Phytochemistry 7:1751–1758.

Wood Jr E A, Starks K J (1972) Effect of temperature and host plant interaction on the biology of three biotypes of the greenbug. Environ. Entomol. 1:230–234.

Woodhead S (1981) Environmental and biotic factors affecting the phenolic content of different cultivars of *Sorghum bicolor*. J. Chem. Ecol. 7:1035–1047.

Woodhead S (1982) Leaf surface chemicals of seedling sorghum and resistance to *Locusta migratoria*. Pages 375–376 *in* Proceedings of the 5th International Symposium on Insect–Plant Relationships. J H Visser and A K Minks, eds. Centre for Agricultural Publishing and Documentation, Wageningen, the Netherlands.

Woodhead S (1983) Surface chemistry of *Sorghum bicolor* and its importance in feeding by *Locusta migratoria*. Physiol. Entomol. 8:345–352.

Woodhead S, Bernays E (1977) Changes in release rates of cyanide in relation to palatability of *Sorghum* to insects. Nature 270:235–236.

Woodhead S, Bernays E A (1978) The chemical basis of resistance of *Sorghum bicolor* to attack by *Locusta migratoria*. Entomol. Exp. Appl. 24:123–144.

Woodhead S, Padgham D (1988) The effect of plant surface characteristics on resistance of rice to the brown planthopper, *Nilaparvata lugens*. Entomol. Exp. Appl. 47:15–22.

Woodruff H B (1980) Natural products from microorganisms. Science 208:1225–1229.

Wootton R J (1981) Palaeozoic insects. Ann. Rev. Entomol. 26:319–344.

Wratten S D, Edwards P J, Dunn I (1984) Wound-induced changes in the palatability of *Betula pubescens* and *B. pendula*. Oecologia (Berlin) 61:372–375.

Wu G, Wu Z F, Zhao S X, Xu C J (1993) The effects of resistance of rice varieties on carboxyl-esterase and phosphatase activity of the white backed rice planthopper. Acta Phyto. Sinica 20:139–142.

Yamamoto R T, Fraenkel G (1959) Common attractants for the tobacco hornworm *Protoparce sexta* (Johan) and the Colorado potato beetle *Leptinotarsa decemlineata* (Say). Nature 184:206–207.

Young N D, Kumar L, Menancio-Hautea D, Danesh D, Talekar N S, Shanmuga-sundarum S, Kim D H (1992) RFLP mapping of a major bruchid resistance gene in mungbean (*Vigna radiata* L. Wilczek) Theor. Appl. Genet. 84:839–844.

Yu S J, Berry R E, Terriere L C (1979) Host plant stimulation of detoxifying enzymes in a phytophagous insect. Pestic. Biochem. Physiol. 12:280–284.

Yuan L P, Virmani S S, Khush G S (1985) Wei You 64 – an early duration hybrid for China. Int. Rice Res. Newsl. 10(5):11–12.

Zacharuk R Y (1980) Ultrastructure and function of insect chemosensilla. Ann. Rev. Entomol. 25:27–47.

Zangerl A R (1990) Furanocoumarin induction in wild parsnip: evidence for an induced defense against herbivores. Ecology 71:1926–1932.

Zangerl A R, Berenbaum M R, Levine E (1989) Genetic control of seed chemistry and morphology in wild parsnip (*Pastinaca sativa*). J. Hered. 80:404–407.

Zawarzin A (1912) Histologische studien uber insekten III Uber das sensible Nervensystem der Larven von *Melolontha vulgaris*. Z. Wiss. Zool. 100:447–458.

Zhody N (1976) On the effect of the food of *Myzus persicae* Sulz on the hymenopterous parasite *Aphelinus asychis* Walker. Oecologia 26:185–191.

Zohren E (1968) Laboruntersuchungen zu Massenanzucht, Lebenweise, Eiablage und Elablageverhalten de Kohlfliege, *Chortophila brassicae* Bouche (Diptera: Anthonoyidae). Z. Angew. Entomol. 62:139–188.

Zucker W V (1983) Tannins: does structure determine function? An ecological perspective. Am. Nat. 121:335–365.

Zuniga G E, Salgado M S, Corcuera L J (1985) Role of an indole alkaloid in the resistance of barley seedlings to aphids. Phytochemistry 24:945–947.

Zwolfer H (1978) Mechanismen und Ergebnisse der Co-Evolution von phytophagen und entomophagen insekten und hoheren Pflanzen. 20. Phylogenetisches Symposium (Hamburg). Sonderland des Naturwissenschaftlichen Vereins Hamburg 2:7–50.

INDEX

FINDING THE PATH

THEMES AND METHODS FOR THE TEACHING OF MATHEMATICS IN A WALDORF SCHOOL

Bengt Ulin

Published by The Association of Waldorf Schools of North America

Swedish Text: ATT FINNA ETT SPÅR

Motiv och metoder i matematikundervisningen

Copyright © 1991 by The Association of Waldorf Schools of North America,

c/o Pine Hill Waldorf School, Wilton, NH 03086

Edited by David Mitchell

Library of Congress Cataloging-in-Publication Data

Ulin, Bengt

FINDING THE PATH

THEMES AND METHODS FOR THE TEACHING OF MATHEMATICS

IN A WALDORF SCHOOL

Translation of: *Att Finna Ett Spår.*

ISBN: 0-9623978-1-4 ✓

Designed by Robin Brooks

ACKNOWLEDGEMENT

The editor wishes to express gratitude to Bengt Ulin, a mathematics teacher at the Kristoffer School in Stockholm, Sweden, for making it possible for this edition of his book, *Att Finna ett Spår*, to be available to English speaking audiences. I would also like to thank members of the A.W.S.N.A. Publications Committee for their assistance, most notably to Susan Demanett for proofreading and to Jan Baudendistel for distribution.

February, 1991
Wilton, N.H.

David Mitchell

CONTENTS

PREFACE TO THE
THIRD SWEDISH EDITION

For the individual who wishes to find out what mathematics is all about there are no simple shortcuts. One must work with the mathematics itself. The job can be made easier if one succeeds in finding a cicerone with feeling for both the subject and the innately human, a guide who can stimulate one's thoughts and joy of discovery.

Bengt Ulin is such a cicerone. He has understood that mathematical discovery need not at all be seen as something for only a small exclusive group of great mathematicians, whose results in polished and suitably humble form are presented to students as "facts." He understands that within mathematics there is something for each of us to discover. There is then, of course, nothing new for humanity. There is, in a sense, something more important than that, something which gives the student true knowledge, which shows how knowledge takes form within the student, how thinking develops, and what creativity means.

Just as in the case of the person who wishes to know what travel is all about, being presented a few facts is not enough. One must tread the road himself. A good travel guide is usually a help. *Finding the Path* is an excellent guide.

It ought to find a very broad readership. Within the school system it will be useful for both active teachers and new teachers-to-be, at different levels, as well as for many high school students. Here I am thinking not primarily of those who go the finals in mathematical contests but rather of a large group who might become interested in mathematics if they only got the right stimulus. Apart from the school system I believe there are many of the general public who through this book could find a path, perhaps even a whole new hobby, in mathematics.

Bengt Ulin follows a thinking tradition which leads one to associations with George Pólya. He encourages experimentation while not

11

forgetting precision when it is time for it. Whether this is then called Waldorf pedagogy or something else is of no matter. Of importance is that we in this book have examples showing that mathematics does not need to be a mechanical manipulation of symbols according to certain rules which must be memorized, that mathematics can lead to other than a fixation with answers, a phenomenon which hinders the student from seeing the actual road. *Finding the Path* helps the reader to discover patterns and structure and leads the reader to much of the beauty which mathematics has to offer.

It is pleasing to note that through this new edition *Finding the Path* will become available to a larger public, and I can only wish all readers good luck on their fascinating journey under the enthusiastic leadership of the very knowledgeable Bengt Ulin.

Andrejs Dunkels
Professor of Mathematics
University of Luleå

1

INTRODUCTION

The presentations in this book are built on experiences from mathematics teaching in the grades 7-12 at the Kristoffer School, a Rudolf Steiner School, and from teaching at the Rudolf Steiner Seminar at Jaerna (Sweden).

Visitors of the school, of the teacher seminar, and of Waldorf school exhibitions have asked many questions about the teaching in the Waldorf school and about mathematics. Some visitors have desired printed material.

During the evaluation of the Kristoffer School by the Swedish state school authorities 1976/77, the question about publication of the methods in some subjects was actualized, especially since the evaluation could not comprise mathematics and science to an extent which had been proper.

Largely as a result of the stimulating interest of Karl-Georg Ahlströem, professor at the Department of Pedagogy at the University of Uppsala and leader of the evaluation group, I began to collect glimpses and experiences from the lessons.

Dr. Georg Unger, leader of the Institute for Mathematics and Physics at Dornach, Switzerland, gave me important impulses. Ingemar Wik (University of Umeå), Hans Brolin (The High school for teachers, Uppsala), and Sven-Erik Gode (the publishing firm Natur och Kultur, Stockholm) took the trouble to read the first version, and I am thankful for their advice concerning the work which was to follow.

The book is primarily written for younger mathematics teachers or teachers-to-be. It does not offer a pedagogical collection of recipes, but should simply give examples of how one might engage the pupils in heterogeneous (undifferentiated) classes.

Most of the book is taken up by Chapter 3, which shows a number of themes from the teaching. In the Waldorf school, mathematics is taught in all grades, during periods, every morning for some weeks during each period, and during fixed, weekly exercise hours.

What is presented in the twelve sections of Chapter 3 is not reports from the whole period or a whole lesson but examples of themes able to engage the pupils.

With the purpose of also enabling generally interested readers with varying prerequisites to follow the expositions, I have simplified the text as much as possible and avoided unnecessary mathematical terms. Hoping to reach different categories of readers, I have given Chapter 3 a lot of space. Its sections also offer pedagogical material for the remarks in Chapters 4–8.

The presentation is as little a textbook in mathematics as a pedagogical handbook. There are several domains of school mathematics which do not appear. The purpose has not been to give a total view of the school curriculum but to contribute to the pedagogical development. I am aware that experienced teachers will recognize much of the content, but there may be variations of some themes which might be interesting even for them, perhaps aspects on teaching in rather heterogeneous classes.

Besides the persons already mentioned, I want to thank my colleagues Arne Nicolaisen (Oslo), Lars Hallqvist and Rüdiger Neuschütz (Bromma) for advice and control work.

At last I wish to thank the Swedish state school authorities for financing the first edition as a project work, entitled "Methods in Mathematics Teaching from the View of Waldorf Pedagogy."

2

MATHEMATICS AS A PATH FOR DEVELOPMENT OF THINKING — PAST AND PRESENT

2.1 Foreign Cultures

The papyrus scrolls with mathematics texts which were found in Egypt bear witness to well developed abilities in problem-solving among an elite in ancient Egypt. The texts, particularly the Rhind Papyrus and the Moscow Papyrus, date back to originals from the Middle Kingdom (2000-1800 B.C.) and thus are almost 4000 years old. The Rhind Papyrus begins by explaining that its contents concern the art of "penetrating into things" and that it will provide "knowledge of everything existing, of all secrets." In actual fact the Rhind Papyrus is a collection of recipes for calculating grain requirements, distributing wages, farm field areas and storehouse volumes, conversion of units, etc. In short, the text is a collection of methods for the solution of various practical problems.

B.L. van der Waerden believes these mathematical texts were intended for teaching in a school for scribes, the royal public servants who were the master calculators and "undersecretaries" to the Pharaoh. A limestone sculpture from the 5th dynasty of the Ancient Kingdom (2500 B.C.), now in the Louvre, shows the intensity of concentration of a scribe as he sits ready to take notes. The scribe could lose his life if he made an error.

The ancient Egyptians were particularly successful in geometry. They used the very good approximation

$$4 \cdot \left(\frac{8}{9}\right)^2 = 3.16049...$$

for π and they performed the masterly feat of calculating the volume of a truncated square pyramid using the correct formula

$$V = (a^2 + ab + b^2) \cdot \frac{h}{3}$$

where a and b are the sides of the square top and bottom sections and h is the height between them.

The Babylonians developed algebra surprisingly far. Cuneiform texts reveal that they could solve quadratic equations and knew of methods for treating certain third degree equations. They solved systems of equations, even of the second degree, and succeeded in developing a number of arithmetic formulas, including among others the formula for sums of squares (see Section 3.5.6).

In geometry the Babylonians were well acquainted with the concept of similarity and with triplets of numbers which form right triangles (Pythagorean numbers). They even calculated the volume of the prism and the cylinder.

According to research in the history of mathematics the Greeks had probably acquainted themselves with the mathematics of both Egyptians and Babylonians. Much points to Greeks having spent considerable time in Egypt and in the Tigris-Euphrates valley and there having the opportunity of thoroughly studying that which had been developed. Various accounts report of Thales, Pythagoras, Democritus and Eudoxus — all prominent mathematicians in ancient Greece — undertaking travels to Egypt and Babylonia.

This in no way, however, entitles the conclusion that Greek mathematics was simply a product of what they had found in other cultures. Van der Waerden quotes Plato in the posthumously published dialogue "Epinomis", words which van der Waerden finds particularly apt:

...whatever Greeks acquire from foreigners is finally turned by them into something nobler.

What the Greeks could make use of in the area of mathematics were methods for problem-solving, formulas, and reference tables — in short, instructions and data. But how had the Egyptians and Babylonians come to these results? Were they reliable? Among the ancient works were incorrect formulae and methods. For example, the Babylonians calculated the volume of a truncated cone with the formula:

$$V = \frac{3h}{2} (R^2 + r^2) \cdot$$

It was the Greeks' great service that they took the step from calculation to mathematics. They sought proofs for all of the results which they had come upon and developed the art of problem-solving to an eminent degree. They showed an impressive ability at finding constructive methods and at developing different forms of proof when needed to give a rigorous foundation for constructions or other methods of calculation. Definitions, assumptions, and axioms were formulated — all with an admirable precision.

It was Thales (ca: 600 B.C.) who brought proof into geometry. That which he received from the East and the South were results which had once been borne up by a living culture but which in his time existed only as document. From these collections of formulae, Thales structures a logical geometry. Consciousness makes its entry into mathematics: "one knows that one knows."

From studies of the pyramids we know that the Egyptians were good at applying geometry, as early as 2000 B.C.; they could build geometrical constructions on even the enormous scale required for the pyramids. Nor were th Greeks one-sided theoreticians. The 1 kilometer long Eupalinos tunnel through the Kastro mountain on Samos, approximately 530 B.C., containing a water channel arranged with ventilation shafts, speaks for itself. It was dug from both directions simultaneously! The two work crews met in the middle of the mountain with an error of less than 10 meters sideways and 3 meters vertically.

For the Greeks such practical tasks were, however, a by-product. The important thing for them was to develop thinking. The Pythagorean School, like Thales, attached great importance to the development of logical structure in geometry and proved with the aid of parallel lines, among other things, that the sum of angles in a triangle is 180°. They were well acquainted with the right triangle and developed a system of representing problems in arithmetic as geometrical problems — the reverse of Descartes' idea using a co-ordinate system to transform geometrical problems into algebraic ones.

Of major importance to the Greeks were clarity and lucidity. They were fond of summarizing a geometrical proof with a figure and the simple text: "Observe!" But for the Pythagoreans it was also a logical necessity to transform arithmetical problems into geometrical ones: they had not mastered the irrational numbers. The Pythagoreans' whole system of thought was long based on the belief that all numbers are either whole or made up of ratios of whole numbers (we call such numbers "rational"). When the study of the square's diagonal and the proportions in the regular pentagram led to the discovery of other "non-expressible" numbers, Pythagorean mathematics came to a crisis. It turned out that in geometry, with the aid of compass and straight edge, one could very easily construct lengths corresponding to certain irrational numbers; for example, the diagonal of a square in the case of $\sqrt{2}$.

With this, Greek mathematics had come into deep water, where neither inherited knowledge nor studies of the physical world were of any help. Mathematics came to be the science, where thought's own powers were developed, independent of sensory knowledge. Plato (427-348 B.C.) writes in his famous work "Republic" :

> Through mathematics is the instrument of the soul cleansed as though awakened to a new life force in a tempering fire; while other occupations consume it and remove it from its power of sight, a power which would, however, be far more deserving of retention than a thousand bodily eyes, since only through such a mental instrument may the truth be seen.

Plato had been initiated into Pythagorean mathematics and other exact sciences by Archytas from Taras, the mathematician who solved the famous doubling of the cube problem from Delos through an ingenious spatial construction.

According to a number of Greek sources, the inhabitants of Delos had been advised by an oracle to double the size of an altar in order to rid themselves of a pestilence which prevailed upon their island. The architects were said to have come to despair to Plato, who told them that the oracle meant to criticize the Greeks for neglect of mathematics and geometry. For the mathematicians the problem became finding the side of a cube which has twice the volume of a given cube. Two other classical problems come from Greece: to construct a square with the same area as a given circle ("squaring the circle") and to divide a given angle into three equal parts ("trisecting an angle").

These three problems came to play an exceptional role not only in Greek culture but also in the continuing development of mathematics. Not until the 1800's was it proven that neither the cube's doubling, nor the circle's squaring, can be carried out using *only* compass and straight edge, and that trisection of an angle with these instruments is only possible for certain special angles but not in general. The problem of squaring the circle was solved first in 1882 when Ferdinand Lindemann was able to show that $\sqrt{\pi}$ is a transcendental number.

Greek mathematicians succeeded in solving the three classical problems with *other* tools than only ruler and compass. In the previously mentioned spatial construction for doubling the cube, Archytas implicitly makes use of the principle of continuity. This and other solutions to these problems are recounted by van der Waerden; for example, the use of ready-made curves. Let us here examine a few examples of solutions of a "mechanical" nature. Figure 2.1.1 shows two lines at right angles to each other with intersection at O. A and B are two given points located such that $OB = 2 \cdot OA$.

If we set OA = 1 unit, then OB = 2 units. We wish to find the side X of a cube which has a volume of 2 units (twice as large as a given cube with volume 1 unit) and seek, therefore, to construct the length $X = \sqrt[3]{2}$.

If we can construct points X and Y in Figure 2.1.2 so that AX and BY are at right angles to XY, then it follows from similar triangles that x (= OX) and y (=OY) satisfy the relations

$$\frac{1}{x} = \frac{x}{y} = \frac{y}{2}$$

from which it follows that

$$y = x^2 \text{ and } xy = 2.$$

Consequently we have $x^3 = 2$, i.e. $x = \sqrt[3]{2}$.

OX thus gives the side of the sought-after cube. A Greek showed how one can "mechanically" construct X: one makes use of two U-shaped right angles as in Figure 2.1.3, of which the one slides inside the other. The larger right angle is placed so that the inner edge passes through B and the corner H falls on the y-axis OY. Apart from this it may be freely rotated. At the same time the smaller U-angle is moved so that its outer edge k goes through A and the corner K falls on the x-axis. When these two

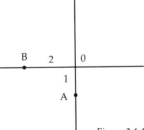

Figure 2.1.1

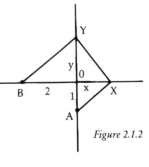

Figure 2.1.2

conditions are satisfied simultaneously, k's intersection with the x-axis will give the point sought, X.

In a dramatically written dialogue, "Platonicus," Eratosthenes has Plato utter the following criticism of Archytas, Eudoxus, and Menaichmos:

> Have you found mechanical solutions? There's no trick in that; even I who am no mathematician can do that. — With such methods the good in geometry will fall to ruin, and the eye will be directed from the purely geometrical to merely observable things... ·

For Plato it was obviously important that solutions be purely logical — constructive and preferably carried out within constructions which may be done exactly with straight edge and compass. These tools were merely aids with which the solution might be made observable to the senses. The important thing was the train of thought, which showed that the

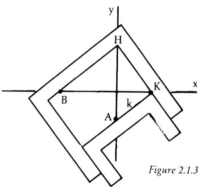

Figure 2.1.3

solution had been found and was correct. Let us now have a look at a problem which admits a solution with compass and straight edge and which in a typical manner satisfies Plato's demand for freedom from worldly tools:

A circle C and a point P (not the center) are given. How can one draw a chord through P such that the chord has a given length k? (Figure 2.1.4)

What do we need to construct? We need to find an end-point of the desired chord. Since the chord must go through P, we have then only to lay the straight edge through P and the end-point and draw the line.

How shall we find an end-point? We have it, if we can determine the length of one of the two subdivisions of the chord lying on either side of P. How can we get such a length?

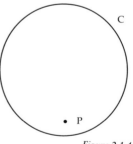

Figure 2.1.4

We utilize now the circle's symmetry of rotation and draw a chord k′ with length k anywhere on the circle: using the compass we mark off the distance k starting from an arbitrary point X and thereby get the other endpoint Y on the circle (Figure 2.1.5). Can we rotate this chord, so that it passes through P? Or can we rotate P to a corresponding point P′?

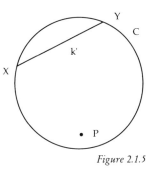

Figure 2.1.5

The latter is easily done with the compass placed at circle center and P, drawing the concentric circular arc around until it intersects the chord k′. We call the intersection P′(Figure 2.1.6).

We may now measure one of the two chord sublengths, say the shortest one, on k′. The length we get there can then be marked off from P out onto the circle and we have found an end-point of the desired chord. (The length may be marked off in two ways on the circle, the one solution is a mirror image of the other.)

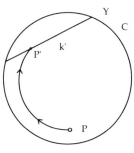

Figure 2.1.6

The discussion we have just carried out utilized the figures only to make the train of thought more easily understandable for another person. And the straight edge and compass have only served in the practical realization of the construction. The whole line of reasoning is conveyed in the world of thought.

It was such pure mathematics which Plato wanted to encourage. Mathematics would be a training ground for pure thinking, which need not rely on sensory perception (as it does in the case of "mechanical" solutions). This strictness in Greek geometry has been one of the major foundations for that logical clarity, that exactitude, and that consciousness which has been built up in western science, especially in the last 400 years.

Let us look at another example, taken from Archimedes (287?-212 B.C.), who lived considerably later than Plato and who must be seen as one of the greatest mathematicians of all time. Archimedes shows how a given angle can be trisected using a mechanical method.

In Figure 2.1.7 the given angle v is placed at the centre of a circle C of arbitrary radius. Archimedes extends the one angle leg MA (the line a) outside the circle and seeks a point X on that line so that the portion of the line segment BX lying outside the circle is of the same length as the radius. We can mark off the radius' length on the ruler and then mechanically adjust the ruler's position through B and to the line a so that the condition is satisfied (Figure 2.1.8). Then the angle u between BX and the line a will be one-third of this angle v.

It is not difficult to prove this with the theorem on external triangles: using the nomenclature in the figure we have

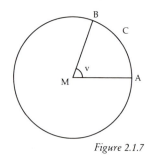

Figure 2.1.7

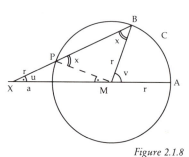

Figure 2.1.8

$$x = 2u \text{ and } v = x + u$$

from which $v = 3u$.

Archimedes was not the one to take lightly the logical, proof-finding side of mathematics, but he had a surprisingly ingenious talent for getting ideas and heuristic methods of problem solution. It is interesting to see how often he is led to pictures from mechanic, when he works with a problem. Archimedes describes, in his book *The Method*, how he approached certain difficult problems such as the area of a parabolic segment, the volume of a sphere, and the center of gravity of a hemisphere.

"Certain things," writes Archimedes, "first became clear to me by a mechanical method, although they had to be demonstrated by geometry afterwards, because their investigation by the said method did not furnish an actual demonstration. But it is, of course, easier when we have previously acquired, by the method, some knowledge of the question, to supply the proof than it is to find it without any previous knowledge."

Archimedes gave here an approach which would show itself to be very fertile. Most of the major results in mathematics have been achieved with an approach which is similar, in principle, to that given by Archimedes, even if the intuitive means of finding solutions have taken many expressions other than analogies in mechanics.

Archimedes' description also points out something important about mathematical activity: proof is the concluding phase. We will return to this in various contexts later on but for the moment simply ask: in what ways can we today make use of mathematics to develop thinking during school years?

2.2 Our Times

Technical development during recent decades places important questions before the "common man." Ignoring the complex of problems around war, the arms race, weapons, etc., we might name questions concerning the world's food production, its energy requirements and the necessity of environmental policy, particularly to safeguard the earth's own resources. Other important problems might be brought forth. Not least with energy and environmental questions have the difficulties in relations between experts and laymen become highly current. Politicians must here often be regarded as laymen. For him who is neither expert nor politician, but has given some thought to the importance of a particular problem, there are two extremes to take: one relies entirely on the experts and politicians, or one troubles oneself to penetrate into the available facts, studies, and evaluations. Depending on abilities, some laymen may go even further and more or less become experts.

Should we leave the thinking to the experts? Would not that be a form of resignation, authority worship, or the easy way out — all rather antiquated in our day?

The need in each of us to be able to think critically has certainly never been so current as it is today. Many scandals such as the neurosedyne catastrophe in Sweden might have been avoided if awareness had been greater of the role of one's own thinking. The individuals entrusted with evaluating the drug chose to execute their responsibility by relying on studies done abroad and assurances given there, instead of considering the question as their own task.

Everywhere in society we find the need to be able to delve into accounts, reports, papers, studies, etc. The task often concerns not only learning the report's contents but perhaps even more importantly reading between the lines and trying to evaluate the author's motives, values, possible subjectivity, etc., to the extent that he hasn't accounted for them openly and honestly in the text. In popular presentations and in school books, facts are often mixed together with interpretations, the point of departure and the goals are only hinted at or not mentioned at all.

What is required here is the ability to analyze and listen. The Swedish school board's recommendation that education shall contribute to development of alert and logical thinking is more appropriate now than ever before. Development of thinking embraces, however, much more than the ability to criticize; for that mater, even much more than being able to think logically. The stronger and more important demand for creativity in the schools must naturally include even creativity in thinking. There is a need for creative fantasy everywhere, and when criticism is expressed, it ought always to suggest something constructive which can replace or improve the object being criticized.

Mathematics with its strongly contoured concepts, clearly delimited problem areas, and meager demands for materials. has the potential more than any other school subject for giving the pupils valuable development in creative and logical thinking.

The 2000 year old inheritance from th logical school in Greece has been invaluable, but during our century it has also been the cause of rigidity in forms of thinking. Liberation from the demonstrative method of proof, which goes back to Euclid and other Greeks, must continue within education.

In recent years there has developed a strong interest in the broad contributions to the field of mathematical heuristics made by the mathematician Georg Pólya. Two volumes were published in the United States by Pólya in 1954 on "Mathematics and Plausible Reasoning" (vol. I "Induction and Analogy in Mathematics", vol. II "Patterns of Plausible Inference"). Further work on problem-solving were published in 1961 and 1965.

Pólya has long been recognized as an active mathematician, but also as the author of pedagogically selected collections of problems, "Aufgaben und Lehrsätze aus der Analysis," which he wrote together with G. Szegö. Pólya trained teachers in mathematics for more than 20 years. The books mentioned here have been translated to German as well as other languages.

A predecessor to these books was published in 1945 as a pocket book with the title *How to Solve It*. In a thoughtful review of the Swedish version in 1975, Dag Prawitz, professor of Philosophy at Oslo University, calls heuristics "a woefully ignored field."

> All that work which leads to the idea upon which a proof is based has no official status in the mathematical literature and is usually not mentioned at all. Modern school mathematics would look very different if consideration had been given to this field instead of so singularly concentrating on the logical part within the philosophy of mathematics.

Let us consider a few of Pólya's words:

> Mathematics is regarded as a demonstrative science. Yet this is only one of its aspects. Finished mathematics presented in a finished form appears as purely demonstrative, consisting of proofs only. Yet mathematics in the making resembles any other human knowledge in the making. You have to guess a mathematical theorem before you prove it; you have to guess the idea of the proof before you carry through the details. You have to combine observations and follow analogies; you have to try and try again. The result of the mathematician's creative work is demonstrative reasoning, a proof; but the proof is discovered by plausible reasoning, by guessing. If the learning of mathematics reflects to any degree the invention of mathematics, it must have a place for guessing, for plausible inference.
>
> The general or amateur student should also get a taste of demonstrative reasoning: he may have little opportunity to use it directly, but he should acquire a standard with which he can compare alleged evidence of all sorts aimed at him in modern life. But in all his endeavors he will need plausible reasoning. At any rate, an ambitious student of mathematics, whatever his further interests may be, should try to learn both kinds of reasoning, demonstrative and plausible.
>
> (From the preface of
> "Mathematics and Plausible Reasoning", vol.I, 1954)

Pólya does not mean to imply that there exists an infallible method for learning the art of guessing. But he addresses "teachers of mathematics at all levels" when he says, "Let us teach guessing."

Pólya's works contain a large number of examples from his experience as university teacher and mathematician, with which he illustrates how fantasy, trial, and investigation can proceed forward to reach the solution. The books are directed primarily to students of mathematics, and the examples are often taken from first year university courses but correspond to high school level in other cases. They have exceptionally much to give to the subject teacher of mathematics, especially to those who have a few years experience teaching at different levels.

The following questions should have priority when revising the mathematics teaching plan:

1. How to develop fantasy, the ability to get ideas, to guess?
2. How can the students learn to use experience they have gained?
3. How can they train self-discipline concerning logical thinking?

More valuable than any particular mathematical fact or trick, theorem, or technique, is for the student to learn two things:

First, to distinguish a valid demonstration from an invalid attempt, a proof from a guess.

Second, to distinguish a more reasonable guess from a less reasonable guess.

(From "Mathematics and Plausible Reasoning," vol. II, Chap. 16, Section 9.)

For years the high school curriculum has been designed basically with continued education at the university, teachers colleges, and other institutions in mind. Those recipients played a large role in the development of the 1970 Swedish School Curriculum Plan, even if in principle the school was to be organized such that the lower level would be independent of the higher levels.

The question now is if a change in mathematics curriculum in the direction of Pólya's guidelines would make high school students less capable of carrying on advanced studies. It is quite possible that what is lost in special knowledge and technique is more than compensated for by an ability in constructive thinking, thinking which would be more independent than what appears to be the case today.

Before we go on to examples and experience from our teaching, I would like to give here a few suggestions as to the direction which answers to the above three questions take:

1. The student must be given sufficient time to understand the problem itself, as presented. The first and immediate understanding directly following presentation of a problem is usually rather superficial. It ought to be deepened through questions, examples, and answers. The second understanding is much more likely to motivate the students. After this there must be time allowed for the students to dig into and explore the mathematical material offered – integers, proportions, triangles, or whatever it may be. They ought to seek personal experience, or best of all, their own discoveries through individual or group work. It is always exciting to go exploring.

2. Students should be given opportunity to vary the problems, simplify, choose special cases, generalize, or in other ways study related variants. By all means let them formulate and investigate problems of their own, choosing a problem of a similar nature. It is especially interesting trying to reach goals one has chosen for oneself.

(Once a girl in the 9th grade was indefatigable in carrying out divisions to investigate the length of the period in decimal fractions. The interesting thing is that she got going on this without my or the other pupils' being aware of it until she happily presented her results to us. The class was working on other things at the time.)

Results and partial results should be noted down in some kind of notebook so that comparisons can be easily made. It is well known that Michael Faraday carefully kept accurate and complete diary records of all his experiments. Even before he began his career as a researcher, he had taken copious notes and made detailed drawings at lectures he attended.

3. Development of self-control in logical thinking should be integrated with other subjects, particularly physics, chemistry, and English. It is a question here of observing one's own thinking and formulating it in words, activities which also are a part of the natural sciences and English.

A part of self-control includes checking, both while solving the problem and after reaching the solution, if all of the facts and conditions given in the text have been used. Problems with unnecessary, excess data occur. In applied mathematics it is often a part of problem formulation first to sift out those facts which are not relevant to the problem at hand. Much practice should be done describing what one has seen or thought. This gives good training in the objectivity and concentration which mathematics requires.

Terminology should be reduced to a minimum but consistently followed.

The pedagogical problem lies to a great extent in the question of how to awaken the pupils' interest. The methods which teachers in the lower classes use play a large role later on in the upper classes when the mathematics teacher takes over. If the lower class teacher feels enthusiasm when confronting problems and for different ways of approaching them, then this will rub off on the pupils. The more the teacher follows old routines, the less interest will be generated in the pupils. Cheap tricks to entertain the students are not necessary at all. "Bingo" games need not be used. When one of the textbooks for the "new math" came out here during the 1970's, it lacked a pedagogical method for the introduction of geometry. The children (in the 4th grade) were expected to entertain themselves by drawing people figures using circles and triangles, i.e. to do exercises which were entirely lacking in both artistic and mathematical content.

Many years' experience refutes the belief that motivation must be awakened with practical problems. Pupils can, of course, often advantageously be enthused through everyday occurrence, but they can and should be enthused even with purely mathematical questioning, as the next chapter will illustrate.

We shall return in Chapter 4 to the question of mathematics as a practice-arena for the thinking of young pupils after we have looked at teaching experiences and examples in the following chapter. The first section of Chapter 6, 6.1., and Chapter 8 discuss the introductory, orientation aspect of teaching mathematics.

3

THEMES FROM THE CLASSROOM

3.1 "How Many Are There?"
— Numbers and Number Systems

3.1.1 How Many Are There?

Most children get well acquainted with our numbers in a natural fashion before they come to school. We need not and should not exercise them in counting. Children meet numbers in many games and otherwise in various daily situations, and eventually learn to experience them. They know that they have one nose, two eyes; they can see that three loaves are in the oven; they learn to identify the numbers 1 to 6 with the dot figures on dice, and so on. They may be able to count to 7, or perhaps to 17, when they start school. Some children find it exciting to count how many there are of various things — a joy which can grow during the first years of school. It can be a "sport" for a child to count the number of cars in a train passing by. They are not difficult to count when it is a passenger train; the cars are long enough to allow counting at a comfortable speed. But when we see an ore train pass by at a distance, it can be really hard. We must concentrate, strain ourselves to be as attentive as we possibly can in order not to lose count. There are so many cars and exactly alike. The difficulty seems to lie in keeping the cars separate from each other while counting. We need to "stop" them and retain individuality of each car as it passes by. Our glance must be strictly controlled by a determined will. When we count a pile of oranges, we often lay each one aside as we count it, but the ore cars cannot be put off to the side. Our whole awareness must be concentrated on our seeing.

This ore-train example shows that counting is always an act of will, and in actual fact when we count, we are counting our will impulses.

29

These impulses are so weak in most cases that the counting seems to go completely automatically.

In school we learn how to write the different numbers. We get more and more acquainted with the decimal system, learning its notation and arithmetic procedures. We become so used to our number system that we do not get any perspective on its elegant effectiveness until we are given the chance to build up another number system. Should we lose our memory and suddenly only be able to count to ten, then we would be troubled to construct a new system — very likely a 10-system — and we would appreciate it since we had constructed it through our own efforts.

During the 1st through 8th grades children have practiced the 10-system properly, so that arithmetic goes mechanically. One should not need to stop and think, for example, when given 25 · 35 to multiply. One works entirely within the system, although in fact quite without thinking, according to what one has learned earlier.

It is then of value to put oneself in another situation; for example, why not the Stone Age? We imagine that we can only count five stones or five arrows, etc. The concepts "six", "seven", etc. we do not have. How can we then describe the number of animals or arrows when that number is more than 5? Assume there are 38 arrows. How can we state that number when we only can count to 5?

The class will certainly have suggestions, for example, that we can draw a symbol for number 5, perhaps a hand (stylized) and state "units" up to four with slash marks:

How would, for example, 23 be written? Of course, as four hands and 3 lines:

And 38? As seven hands and 3 lines? No! No one knows how to count to 7, we are simply not able to grasp a group of 7 hands; we can only count to 5. What do we then do?

Our answer to this question will determine whether or not we eventually construct a more or less effective system. If we stop with only the two symbols – the hand and the slash — we could indicate 38 with this figure:

But what a lot of space this figure takes! And how much space would be needed for even larger numbers, assuming that we could learn to interpret them at all? What might help us toward a better system? We invent a new symbol, a symbol for 5 hands. In one ninth grade class they agreed upon using a shoe for symbol: in another they chose a five-pointed star. The picture for 38 would then be:

How do we go on? Will any problems arise? It soon becomes apparent that a new symbol is needed for five shoes or five stars. Otherwise the same dilemma as earlier will occur. The above classes chose, respectively, a boot and a star with extended rays.

In the latter class there was another interesting suggestion for the symbols for 25 and 125 (compare Fig. 3.1.1):

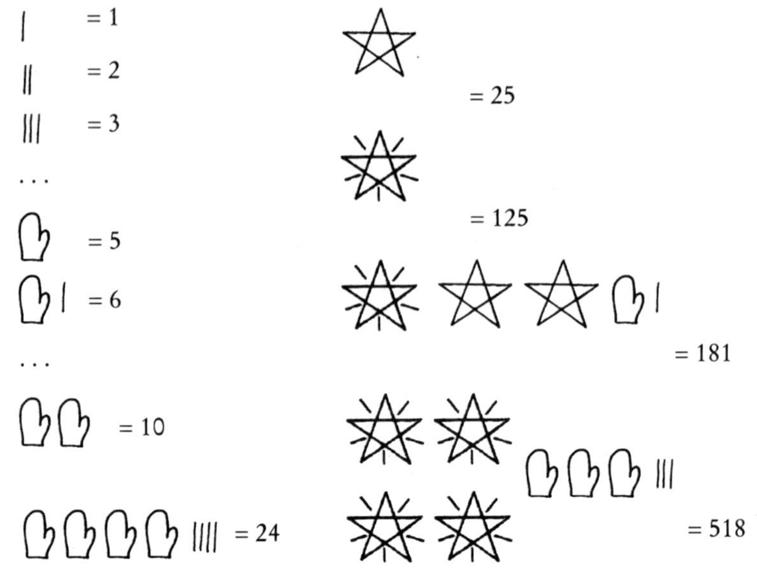

| | = 1

‖ = 2

‖| = 3

. . .

= 5

| = 6

. . .

= 10

‖‖ = 24

= 25

= 125

= 181

‖| = 518

Figure 3.1.1

Extending is now easy. We have understood that the numbers 1, 5, 25, 5 · 25 = 125, 5 · 125 = 625, etc., build the base in our five-system. If we now restrict ourselves to the world of picture-symbols, then a separate symbol will be needed for each one of these numbers. Assume we have

for 1, 5, 25, 125, and up to 625. What is the largest number we will be able to write pictorially? Obviously

$$4 \cdot 625 + 4 \cdot 125 + 4 \cdot 25 + 4 \cdot 5 + 4 = 3124.$$

We get the number just below he next symbol number, 3125!

Now we should ask: can every number from 1 to 3124 be written unambiguously with our five symbols? Are there possibly one or more numbers which cannot be handled at all in our newly constructed five-system – or which can be written in more than one way?

It is not difficult to convince ourselves: our 5-system, extended indefinitely, gives unique representations of all the integers. (Exercise 3)

3.1.2 Number Systems in Early Cultures

How did the ancient Egyptians construct their number systems?

Numbers and arithmetic operations were expressed, as was their normal writing, with picture-symbols. When numbers were inscribed on a relatively hard material, they used special hieroglyphs for the numbers. But when the arithmetic was done with Indian ink or a similar liquid on papyrus leaves, they wrote a relatively flowing style, so-called hieratic (holy) writing. Here we limit ourselves to the hieroglyphic writing. They used a 10-system where the base was composed of the following hieroglyphs:

= 1	⌐ = 10 000 (finger)
∩ = 10	⌐⌐ = 100 000 (tadpoles?)
ℓ = 100 (measuring rope)	and
⚘ = 1000 (lotus flower)	= 1 000 000 (the god of infinity)

The Egyptians used each symbol to represent as many units as the base number specified. To describe a number of cows one drew each individual symbol as many times as was required. For example, 3508 was written

|||| ℓ ℓ ℓ   ⚘ ⚘ ⚘
|||| ℓ ℓ   ⚘ ⚘ ⚘

Since each symbol represented a certain fixed number, the symbols could be placed any which way. This too:

must mean 3508.

What corresponds to our zero? Is it needed? No, in this pictorial writing there is no need for a zero symbol.

The best documents showing us ancient Egyptian writing are the Rhind Papyrus and the Moscow Papyrus, kept at the British Museaum in London and in Moscow, respectively. Rhind is the name of the purchaser of the first-mentioned document. Both rolls have their roots in the epoch around 1800 B.C. in ancient Egypt. They contain a series of solved problems and could be called collections of recipes of the type "do this," "do that," "you will find it to be right." In a later period the Egyptians took the first step toward a more abstract way of writing numbers. One wrote out a sort of table where the columns correspond to different base numbers. For example, the number 705,000 would be written:

Strange as it may seem, the people of the Fertile Crescent around the Tigris-Euphrates valleys (the Akkadians, Babylonians, Sumerians, and Assyrians) used a completely different system for representing numbers. Wedge-shaped prism marks were stamped in soft clay (see Figure 3.1.2).

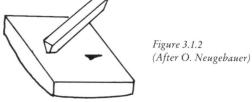

Figure 3.1.2
(After O. Neugebauer)

The writing is called cuneiform.

The base in the Babylonians' system was 60; the basic numbers were 1, 60, 60 · 60 = 3600, 60 · 3600 = 216,000, etc.

All of.these numbers were indicated by the wedge-shaped stamp:

It was the context which showed the particular value the stamp had in particular place. In this system the number 59 would require 59 stamp marks, which would be difficult to read and understand. One would have to count the number of wedge marks and one could count wrongly, arriving at 58 instead of 59. They therefore had a helping symbol with the value of ten, a wing-shaped mark:

This figure corresponded to 10 wedges and thus could represent 10, 10 · 60 = 600, 10 · 3600 = 36 000, 10 · 216 000 = 2 160 000, etc.

The following are some examples of number representations:

From these examples we see that the wedge's placement, its position, plays a determining role. For example, the wedge on the left could indicate the value 60 while the wedge on the right could mean 1. Here we have the beginning of a so-called positional system, a system in which the symbol's position determines which value that symbol adds to the number being represented. Our own system is a positional system.

In the number 5358 both of the "5 s" represent 5, of course, but the left five contributes the value 5000 to the number while the right-hand five contributes 50. In this way we are able to use the same symbol to account for different values.

Was it always possible to determine from the context what the wedge marks mean? E.g. in

Could one decide whether the value was 2, 61, 120, or 3601?

From the fact that various clarifying symbols were introduced it seems clear that mistakes were made. One was to write a word beside the symbol. Another method was to use a sort of double-wing as a symbol for an "empty-space." For example:

$$= 3600 + 0.60 + 4 = 3604.$$

Symbols for the empty space are predecessors to the number zero. In ancient India the empty space was marked with a dot. Opinions are divided on when zero (naught) began to be used as a number, that is, on when zero was given the same status as the numbers 1, 2, 3, etc. According to van der Waerden it was the Greek astronomers who introduced the number zero, roughly around the time of Christ.

Where do our numbers come from?

We often say that we use arabic numerals, but the fact is that our number system came from ancient India. The individual forms of the numerals have in some cases undergone radical metamorphoses (Figure 3.1.3).

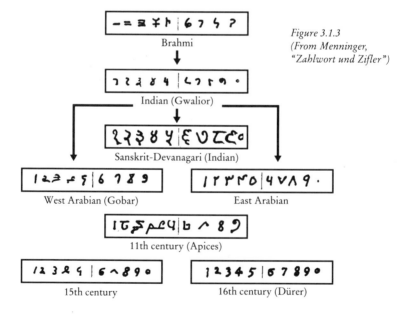

Brahmi

Indian (Gwalior)

Sanskrit-Devanagari (Indian)

West Arabian (Gobar)

East Arabian

11th century (Apices)

15th century

16th century (Dürer)

*Figure 3.1.3
(From Menninger,
"Zahlwort und Zifler")*

The Indian numerals were brought to the West by the Arabs, primarily during Islam's expansion westward in the first centuries after Mohammed's death. Albrecht Dürer (1471-1528) gave the numbers the forms which they by and large have retained to this day.

We have thus received our numerals from India and the number zero from Greece. The 10-system has its roots in old Egypt (and certainly further back in time) and the position-principle in its early form comes from the people of the Tigris-Euphrates valley.

Does the 10-system have connections to our 10 fingers (and toes)? Most certainly, and therefore ought its roots to be as old as man himself? Language research tells us:

- in German 10 is "zehn" and "toes" are "Zehen"
- in English "digit" means both finger and numeral
  (computers which give results as numbers are called digital
  computers – and in recent years we have digital clocks and watches)
- according to Tobias Dantzig, 10 is the base for the numbers in all
  the Indo-European languages and in some other languages as well

It may appear from the words "eleven" and "twelve" that we have had a twelve-system. Language research shows, however, that the German words for eleven and twelve, from which our English words come, "elf" and "zwölf" respectively, come from the compounds *ein-lif* and *zwo-lif*, where *lif* is an old Germanic word for ten.

In the following table we see some examples of number names from systems based on 20 and 5.

1. The Mayans used the system which you get if you choose 20 as your base, i.e. a 20-system. The day was divided into 20 hours. An army division was comprised of $20 \cdot 20 \cdot 20 = 8000$ soldiers and so on. Basic numbers:

1	hun	=	1
20	kal	=	20
20^2	bak	=	400
20^3	pic	=	8 000
20^4	calab	=	160 000
20^5	kinchel	=	3 200 000
20^6	alce	=	64 000 000

2. Typical primitive base-5-systems have been found with the Api people in the New Hebrides (the island group east of Australia) as well as with the Wanyassa tribe in Central Africa. (The information on the Wanyassas stems from Stanley.) We can note that the systems are structured in the same way, despite the great distance between these two cultures!

Api culture		*Wanyassa*
1	tai	kimodzi
2	lua	vioviri
3	tolu	vitatu
4	vari	vinyé
5	luna	visiano
6	otai	visiano na kimodzi
7	olua	visiano na vioviri
8	otolu	visiano na vitatu
9	ovari	visiano na vinyé
10	lua luna	visiano na visiano

Luna means both five and hand in the Api language. The prefix "o" means "more", e.g. otai = "more one" or "one more", i.e. 1 + 5.

3.1.3 The Five-System with Numbers

We constructed earlier a five-system with the help of symbols for the basic numbers, i.e. with pictures for the numbers 1, 5, 25, 125, etc.

What do base-5-numbers look like if we refrain from our pictures and use our numerals instead? Can we make a table from right to left with columns in the order for 1, 5, 25, 125, ... just as in our 10-system with its columns for 1, 10, 100, 1000, etc.?

The numbers 1, 2, 3, and 4 can be written unchanged. But 5 must be written as the sum of 1 "five" and 0 "ones":

125	25	5	1
		1	0

5 is therefore written 10 in the five-system. To avoid confusion we attach little number tags to specify which number system we mean:

(Five in the ten-system is written "one-zero" in the five-system. We shouldn't read "one-zero" as "ten" because "ten" already has its meaning for us within the decimal system.)

In the same way we get:

$$6_{10} = 11_5$$
$$7_{10} = 12_5$$
$$. . . .$$
$$10_{10} = 20_5$$
$$11_{10} = 21_5$$

etc.

$$25_{10} = 100_5$$
$$26_{10} = 101_5 \text{ etc.}$$

Those who wish to give a try on their own may do some of the exercises following. From the system's structure it is clear that no digit higher than 4 should appear – nor is there any need. The system has the five digits 0, 1, 2, 3, and 4. Adding the numbers 6 and 8 with 5-system addition gives:

11
13
24 (with the value fourteen)

Adding instead 78 to 113 we get with the aid of a table:

	125	25	5	1
78		3	0	3
113		4	2	3
	1	2	3	1

Here we begin with 3 + 3 getting the sum 11_5, in which the left-most one specifies a "five" and is carried over to the fives-column. This is why the fives-column adds up to 3 (1+0+2). Finally, the sum 7 in the 25-column gives a carry-over 1 in the 125s-column.

Ninth graders usually don't mind trying out the other three arithmetic operations in the fives-system. Those who are interested in seeing how these work out can take one one or another of the exercises at the end of the section.

3.1.4 The Binary System

By this time the pupils have asked which system computers use. There are computers built for the 10-system, but the majority of them as well as pocket calculators are based on the 2-system, or as it is called, the binary system ("bi" = two).[1]

Which basic numbers do we get if we choose 2 as our base?

We get 1, 2, 4 = 2 · 2, 8 = 2 · 4, 16 = 2 · 8, etc., in other words: a sequence starting with 1 in which each number doubles the previous number on and on.

The table here shows the numbers from 1 to 17 "translated" into binary language. Here too avenues open up for discovering on one's own how the operations addition, subtraction, multiplication and division work out.

Why are computers based on the binary system? There might be a student in the class who wishes to prepare a sufficiently detailed yet easily understood answer to this question to present later to the class.

	16	8	4	2	1
1 =					1
2 =				1	0
3 =				1	1
4 =			1	0	0
5 =			1	0	1
6 =			1	1	0
7 =			1	1	1
8 =		1	0	0	0
9 =		1	0	0	1
10 =		1	0	1	0
11 =		1	0	1	1
12 =		1	1	0	0
13 =		1	1	0	1
14 =		1	1	1	0
15 =		1	1	1	1
16 =	1	0	0	0	0
17 =	1	0	0	0	1

[1] Space does not allow here going into *how* the 2-system is used but examples in the classroom teaching may be to advantage.

Pupils should understand at the least that it is the basic 2-way polarity of electricity and magnetism and their phenomena which makes the 2-system so natural for technical calculation. The 2-system has, of course, only the digits 0 and 1. Corresponding to these digits could be: high and low voltage respectively, current flow versus no-current, or clockwise magnetization versus counter-clockwise magnetization.

We come naturally into a discussion of the advantages and disadvantages of binary numbers compared to decimal (the 10-system). Look how easy the multiplication table is in the two-system:

$$0 \cdot 0 = 0$$
$$0 \cdot 1 = 0$$
$$1 \cdot 0 = 0$$
$$1 \cdot 1 = 1$$

But, on the other hand, — oh, such long numbers! The number one hundred must be written 1100100 in the 2-system. Doing arithmetic in the 2-system with pen and paper would be, in fact, a trouble, but with the incredibly fast computers it doesn't particularly matter. An addition of 2 digits can be done in approximately 60 billionths of a second.

Why work with this material in grade 9? We take up this question in Chapter 7 (Section 7.2). See also section 3.2.7.

To conclude this orientation on number systems we will take a look at two primitive binary systems.

Similar primitive base-2-systems have been found as far apart as in a tribe at Torres Straits, Australia, and in the Bakairi tribe in central Brazil:

	At Torees Straits	*In the Bakairi tribe*
1	urapun	tokale
2	okosa	ahage
3	okosa-urapun	ahage-tokale
4	okosa-okosa	agahe-ahage
5	okosa-okosa-urapun	ahage-ahage-tokale
6	okosa-okosa-okosa	ahage-ahage-ahage

For numbers larger than 6 the Bakairis have only the word "more," which means "many."

Again, we have an example of two systems from completely differ-ent parts of the world which despite the distance are the same with respect to the structure of the words for numbers. In what way are these primitive? The answer comes directly from the words themselves; the system is structured additively: one, one-plus-one, two-one, two-two, etc. No wonder that they didn't get very far along this route. The num-ber sixteen would sound (in English):

two - two - two - two - two - two - two - two !

The binary system which our culture uses is based upon the *multi-plicative* principle, in a similar fashion to our decimal system and also to the 5-system we constructed earlier. After 1 and 2 we do not go to 1 plus 2 but to 2 "times" 2 = 4 and after that to $2 \cdot 4 = 8$ etc. Multiplication by 2 repeats, and the basic numbers are what we call the powers of 2.

2 is the first power of 2 and is written 2^1
4 is the second power of 2 and is written 2^2 (meaning $2 \cdot 2$)
8 is the third power of of 2 and is written 2^3 ($2 \cdot 2 \cdot 2$) etc.

It is natural to complete this system with the definition $1 = 2^0$.

Among the Thimshian people in British Columbia have been found 7 different sets of number words. One set for flat objects and ani-mals, one for round objects and time, one for counting people, a set for long objects and trees, a set for canoes, a set for measurements, and a set for counting when no particular objects are specified. The first six sets point to times long ago when this people's counting was very much sen-sory-bound. As T. Dantzig points out in his book, *Numbers — the Language of Science*: "The concrete preceded the abstract."

We have not touched here upon fractions in the different systems. At school we have to some extent gone into such questions as what kind of binary language would correspond to tenths, hundredths, etc., what sym-bols could we use for binary fractions, etc.? Here I will content myself with referring the interested reader to exercise 7. Quick pupils can be given the task of building up systems with bases larger than 10, for exam-ple a 12-system or a 16-system. The 12-system requires two new numerals for ten and eleven, while the 16-system requires characters for the six num-bers 10-15. One can use letters, for example A = 10, B = 11, etc.

3.1.5 Figurated Numbers

During the later phases of the ancient Greek civilization, especially among the Pythagoreans (Pythagoras was active in Greece during a long period around 500 B.C.), the whole numbers were presented as figures, so-called figurated numbers. Triangle numbers, square numbers, rectangles, etc. were formed (Figure 3.1.4).

The square numbers are determined by the number of dots in the squares and are thus 1, 4, 9, 16, 25, etc., the result of squaring $1 \cdot 1$, $2 \cdot 2$, $3 \cdot 3$, $4 \cdot 4$, $5 \cdot 5$, etc. The rectangle numbers are 2, 6, 12,20, 32, ... (consider: the area of a rectangle is "length times width"). The eleventh rectangle number is apparently $11 \cdot 12 = 132$, the 26th would be $26 \cdot 27 = 702$, and the n-th would be $n \cdot (n + 1)$.

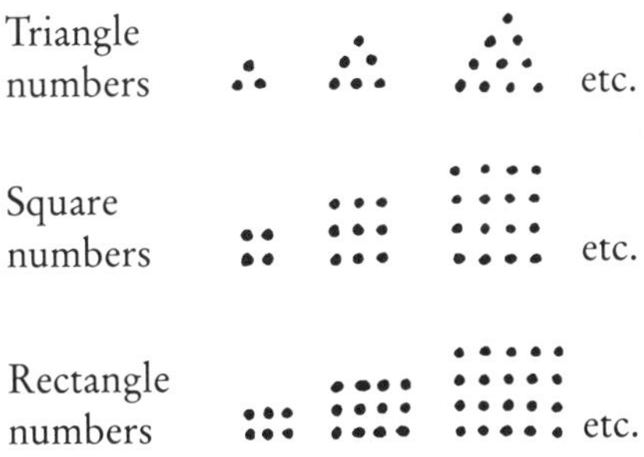

Now what values do we get for the triangle numbers?

In the beginning we have 1, 3, 6, 10, 15, ...

How does the series continue? Can we find a better method than drawing triangle after triangle and counting the dots?

In a 7th or 8th year class

Figure 3.1.4

— or wherever else the question is raised — there will be no shortage of proposals. Some pupils look at the structure of the triangles and find that each triangle is made of rows: the top row has 1 dot, below it 2 dots, and so on with each lower row having one more dot until the last row which has as many dots as the number of the triangle (its order in the series of triangles). Thus we should write, for example,

$$1 + 2 + 3 + \dots + 11 \text{ for the eleventh triangle number,}$$

call it t_{11}, and calculate the sum.

Other pupils have examined only the numbers themselves, from the first few triangles:

$$t_1 = 1$$
$$t_2 = 3$$
$$t_3 = 6$$
$$t_4 = 10$$
$$t_5 = 15$$

and found that the increases from triangle to triangle are 1, 2, 3, 4, 5. They say "the next increase will be 6, the following 7, etc." Can we be sure of this? "Yes," say those who have studied the triangles. "It's right, because for each step we add a new bottom row which has one dot more than the bottom row of the previous triangle." So in principle we can work out any triangle number we want to, for example

$$t_{100} = 1 + 2 + 3 + \ldots + 100.$$

If it hasn't been discussed earlier one may here take up the question of how sums of this type, of so-called arithmetic series, are calculated. And with that, tell the classical story of how Gauss as a 10-year-old schoolboy solved a similar, although probably more difficult, problem. In such a way we can obtain a formula which directly calculates the triangle number without adding up a sum.

Is there another way, a way to directly calculate the triangle numbers? Can we find a geometrical solution by putting together two dot-triangles? Perhaps a little hint is needed. Can we put two identical triangles together? Of course, making a parallelogram (Figure 3.1.5a).

Figure 3.1.5a    Figure 3.1.5b

And the parallelograms can be judiciously shifted into rectangles – without changing the number of dots (Fig. 3.1.5b).

Now we can see the triangle numbers as half of the corresponding rectangle numbers. We thus get, for example,

$$t_5 = \frac{5 \cdot 6}{2} = 15 \qquad \text{and} \qquad t_{100} = \frac{100 \cdot 101}{2} = 5050.$$

And the general formula for the triangular number is

$$t_n = \frac{n\,(n + 1)}{2}.$$

Let us look back upon the different levels in our calculations of the triangular numbers:

1st level: we draw the triangle and count its dots

2nd level: we note the triangle structure and form a sum to be calculated. We no longer need draw the triangle. The crucial point is that we identify the base row's length with the number of the triangle (its order in the series). From this we draw the conclusion that the sum includes all the integers from 1 up to and including the number of the triangle. We have found a general method.

3rd level: We wish to improve the method and find a geometrical solution which directly gives us the triangle number as half of the rectangular number of the same order. We can state a formula:

$$t_n = \frac{n\,(n + 1)}{2}.$$

All pupils can work at the "physical" level, and many can contribute with ideas and comments which lead to the other, truly mathematical, levels of solution.

This problem, in all its simplicity, gives much insight into how we can let geometrical clarity and experience with numbers lead us to a crucial idea.

3.1.6   Two Problems in Subdividing a Circle

In connection with the figurated numbers, the following two closely related problems are of interest:

1. A circle is subdivided into as many subdivisions as possible by 1 chord, 2 chords, 3 chords, etc. How many subdivisions are possible when the number of chords is

a) 4?   b) 11?   (Figure 3.1.6)

Start by making a table up to n = 6.

2. On the perimeter of a circle we place a point. With this point only we have just one subdivision, the whole circle. We place another point somewhere on the perimeter and draw the line connecting it to the

first point. Now the circle is divided by this chord into 2 areas. Place out a third point, a fourth, etc. in such a way that the number of subdivisions is maximal when all the points are connected by lines (all three points with each other, etc.). Make a table as in Figure 3.1.7 with the maximum number of subdivisions.

The first question is: what do we do, how can we satisfy the condition for maximum number of subdivisions?

A further task: find a formula, in Problem 1, for calculating the number of subdivisions (Exercise 8).

After we have acquainted ourselves with Problem 1 by drawing figures, a little thought brings us to the realization that we must place each new chord so that it intersects all of the old chords but at different points on the different chords.

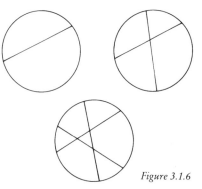

Figure 3.1.6

n Number of chords	A (n) Max. number of areas
1	2
2	4
3	7
4	?
5	?
6	?

p Number of chords	A (p) Max. number of areas
1	1
2	2
3	4
4	?
5	?
6	?

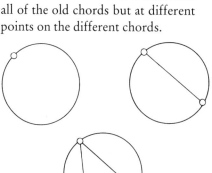

Figure 3.1.7

We must not draw a new chord through an intersection of two old chords.

In Problem 2 we are not concerned with the maximum requirement until we come to the 6th point.

The sixth point must avoid positions (there are 5 possible) such that a connecting line to one of the old points goes through an intersection of two earlier chords (Figure 3.1.8).

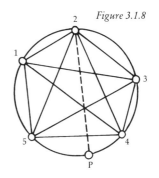

Figure 3.1.8

Students generally come quickly to agreement on the results in Problem 1:

Number of chords	Max. number of areas
1	2
2	4
3	7
4	11
5	16
6	22

During work on Problem 2 opinions have to date always been divided. In one class the quickest pupil jumped up and shouted hotly:

"It has to be 32! It's obvious; it must be 32!" On other occasions pupils have worked quietly for themselves and found 31 subdivisions for 6 points, but believed it to be wrong ("It ought to be 32."). They made new figures. "Perhaps I didn't check off *all* the areas" is a common reflection. But it turns out, after carefully checking off the subdivisions, that the table looks like this:

Number of points	Max. number of areas
1	1
2	2
3	4
4	8
5	16
6	31

Many are surprised that the number of areas does not again double when we come to 6 points — the sequence previously was, to be sure 1, 2, 4, 8 and 16.

Before we had time to discuss the question further, some students have set out 7 points and counted 57 subdivisions. With 8 points the number rises to 99. The sequence thus diverges more and more from the sequence with successive doubling: 1, 2, 4, 8, 16, 32, 64, 128, ...

We will give a name to our new sequence – 1, 2, 4, 8, 16, 31, 57, 99, ... and return to it in a later section (3.2.6). We call it Moser's sequence, because the problem was formulated by Leo Moser (*Scientific American*, August, 1969).

Moser's example gives us useful experience: many times we are deceived by a preconceived idea about the results. The reason might be, as in our example, that we generalize too early, something for all analysts and decision-makers to give thought to.

3.1.7 Prime Number Generators

Often cited examples with surprising results are the following:

1. What values does the expression $x^2 + x + 41$ take on when we successively set x to 0, 1, 2, 3, ...?

Can we see anything in common among the values we obtain?
Putting in x = 0   gives $0^2 + 0 + 41 = 41$
    x = 1 : $1^2 + 1 + 41 = 43$
    x = 2 : $2^2 + 2 + 41 = 47$
    x = 3 : $3^2 + 3 + 41 = 53$
    x = 4 : $4^2 + 4 + 41 = 61$

Do 41, 43, 47, 53, and 61 have anything in common?
Yes, they are so-called *prime numbers* (a number n which is not divisible by an integer between 1 and n).

Is $x^2 + x + 41$ an expression which only generates prime numbers? We test this idea by putting in x = 5 : $5^2 + 5 + 41 = 71$.

A prime number!

Further testing can be easily divided up in the classroom, so that each pupil has his or her number to put into the expression. This will show that new prime numbers continually turn up. The next five are

for   x = 6:   83

       x = 7:   97

       x = 8:   113

       x = 9:   131

       x = 10: 151

We see easily that some prime numbers are skipped over. Between 53 and 61 lies 59, between 83 and 97 lies 89, etc. But – does the expression always give prime numbers, even if there are many missing? Or are there values of x, which do not give prime numbers when substituted in the expression?

There are such x-numbers. Let's put in 41. What happens?

We get $41^2 + 41 + 41$ — a sum which is very obviously divisible by 41: the value can be re-written

          $41 \cdot 41 + 2 \cdot 41$ or more simply $43 \cdot 41$. We see that the number is composed of the two prime number factors 41 and 43.

Leonard Euler (1701-1783), from whom the example comes, found that all the 40 values of x from x = 0 to x = 39 give prime numbers. (It is not difficult to show that x = 40 gives a value which can be written $41 \cdot 41$, so that even x = 40 gives a composed value.)

Let us now form a spiral of squares starting with the number 41 (Figure 3.1.9).

Where do the prime numbers which we have obtained from the x-substitutions appear? (See Exercise 10.)

Euler also gave the expression $x^2 + x + 17$ as another surprising "prime number generator." Since the constant 41 is exchanged here for the lower number 17, the work with the quadratic spiral will be less demanding.

The expression $x^2 + x + 17$ is a so-called polynomial with one variable (x) and of the second degree (the highest

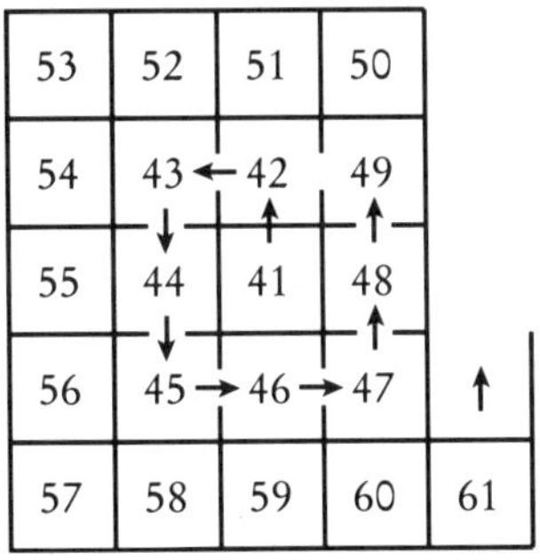

Figure 3.1.9

exponent is 2, in the x^2-term). It has been proved that no polynomial can generate *only* prime numbers, not even if the polynomial contains more variables.

But there are polynomials of 12 variables which manage to generate all the prime numbers, along with some non-primes. A relatively simple formula, (beyond, however, the scope of the school curriculum) was published in 1975 which gives the prime numbers p ($p_1 = 2$, $p_2 = 3$, $p_3 = 5$, etc.) successively when putting in the numbers n = 1, 2, 3, ... Interesting results have been published in the *American Mathematics Monthly*, vol. 83, 1976, and the *Canadian Mathematics Bulletin*, vol. 18 (3), 1975.

3.1.8 Exercises

1. Write the numbers 734 and 12059 with

   a) old Egyptian hieroglyphs
   b) cuneiform (Babylonian wedge-marks)

2. Write the numbers a) 39   b) 150   c) 795   in the fives-system. (Use the numerals 0, 1, 2, 3, 4.)

3. Can every natural number be represented in a fives-system? Uniquely?

4. Convert the following numbers to ordinary decimal form (to 10-system numbers):

   a) 34_5 b) 230_5 c) 304_5 d) 10110_2 e) 11011_2

5. Write out the multiplication table in the 5-system for numbers up to 4 · 4. In other words, finish the table below where a number in the left-most column is multiplied by a number in the top row and the result put in the appropriate place in the table. Some products are already filled in to get you started.

	0	1	2	3	4
0	0	0	0	0	0
1	0	1	2	3	4
2	0	2	4	11	13
3					
4					

6. Carry out the following arithmetic in the given systems without using the 10-system:

a) $31_5 + 33_5$ b) $10_2 + 101_2$ c) $10101_2 - 110_2$ d) $101_2 \cdot 11_2$

Check, if you wish, by converting all numbers (both in the problem and in the answer) to decimal form.

7. Analogous to the decimal comma (decimal point) in the 10-system, we introduce a binary comma (;) in the 2-system and utilize the basic fractions

$$0;1 = \frac{1}{2} \qquad 0;01 = \frac{1}{4} \qquad 0;001 = \frac{1}{8} \text{ etc.}$$

Do the following sum in the 2-system and convert the result to a decimal number: $1;01 + 0;011 + 0;101$

8. Derive a formula for the maximum number of subdivisions in the first circle subdivision problem, where n = the number of chords (3.1.6).

9. Determine the values corresponding to the so-called pentagonal numbers as illustrated below in Fig. 3.1.10. (Try to work out a formula or at least an expression which gives the right values.)

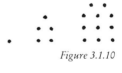

Figure 3.1.10

10. Referring to the Prime Number Generators earlier, can any of the numbers generated by the expression

$$x^2 + x + 41$$

land anywhere but in the corner boxes of the quadratic spiral?

3.2 Pascal's Triangle

3.2.1 Street Network

We continue on our voyage of discovery through the world of numbers and once again examine counting with a concrete example.

Figure 3.2.1 illustrates a street system in a modern city where avenues and boulevards form a network of squares. The avenues in the figure run downward to the right; the boulevards downward to the left. We can label the streets and the intersections. Corner H in the figure is the intersection of Avenue 2 and Boulevard 3 and may thus be called A2, B3. From the top corner AO, BO one may proceed to H by different paths. Some

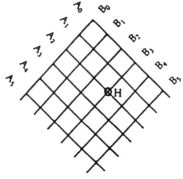

Figure 3.2.1

paths give a route of minimum length, namely 5 blocks long. Ignoring all longer paths we ask: how many paths of shortest possible length go from AO, BO to A2, B3?

Many students choose to be concrete and draw the paths which they find. All pupils have the ability to approach the problem in this way. But soon the question arises: have we found *all* of the paths or is there yet another route left?

The pupils compare their figures, of course. Some have 8 routes, others 9, others 10 and someone has 11 routes. How can we *know* that we have found the last path? We need to *systematize*!

We need to order all the paths according to a system – when we can determine which route will be the last and in this way know the number of paths. How can we come upon an organizing principle such that we can number the paths and let the next path develop logically from the previous one?

Figure 3.2.2 shows a sequence of routes which some of the pupils have found. According to what principle are the routes ordered in the figure? The answer is given in Exercise 1.

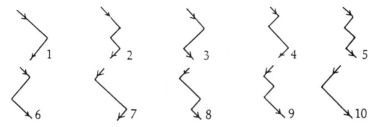

Figure 3.2.2 A systematic sequence of paths in Section 3.2.1 See also Exercise 1.

A problem such as this speaks naturally to the intellect but appeals also to the imagination. We can draw our street system and count successively how many shortest paths go from the top corners to different intersections. We start with the nearest intersections and write in the results on our drawing. To begin with we get, schematically:

$$\begin{array}{ccc} & 1 & 1 \\ 1 & 2 & 1 \end{array}$$

How does the number diagram continue to unfold?

Let us continue on until the shortest paths are of length 6 blocks.

Let us observe the numbers we get. Perhaps we may soon make a discovery and thereby come to a theory of how the numbers increase.

Perhaps we may even see directly during our work exactly how the numbers grow. That is, of course, our real goal.

We can in any case all help out in developing the number scheme. The correct scheme looks like this

$$\begin{array}{ccccccc} & & & 1 & 1 \\ & & 1 & 2 & 1 \\ & & 1 & 3 & 3 & 1 \\ & 1 & 4 & 6 & 4 & 1 \\ 1 & 5 & 10 & 10 & 5 & 1 \\ 1 & 6 & 15 & 20 & 15 & 6 & 1 \end{array} \qquad \text{etc.}$$

Do these numbers show us anything interesting? Is there any regularity to be discovered? Doesn't it appear as if each number of paths is the sum of the nearest numbers in the row above? (The numbers on the edges are, of course, always 1.) We can check the sums in the entire diagram:

$$3 = 1 + 2, \quad 4 = 1 + 3 \text{ or } 3 + 1, \quad 10 = 6 + 4 \text{ or } 4 + 6 \quad \text{etc.}$$

Does this rule apply generally? Could we, for example, prove that the number of paths to A2, B3 has to be 6 + 4, i.e. has to be the sum of the number of paths to A2, B2 and to A1, B3?

And could we thereafter realize that for any arbitrary choice of intersection:

$$z = x + y ?\ (\text{see Figure 3.2.3})$$

The chain of thought in a group – or individually – might go roughly like this: we keep in mind that only the shortest paths are to be considered. How must these look in the network? Well, all the shortest paths to the Z-intersection must pass through either intersection X or intersection Y. How many paths pass through these intersections? Obviously numbers x through X and y through Y respectively. The paths to corner X are

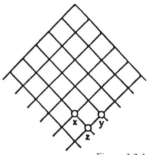

Figure 3.2.3

simply extended by one block, and this does not change their number. "We just go one step further with each path," might be the pupils' expression. In the same way the number of paths to corner Z through corner Y is y. Thus

$$z = x + y$$

and we have proven an additive rule which allows us to further expand the number diagram.

We can leave behind the rather primitive and time-consuming method of looking for each new intersection, for all possible shortest paths and then counting them on our fingers, so to speak. Let us expand the table another 6 or 7 rows down.

```
                    1       1
                1       2       1
            1       3       3       1
        1       4       6       4       1
      1     5     10      10      5     1
    1     6     15      20      15     6     1
  1     7     21     35      35     21     7     1
1     8     28     56     70      56     28     8     1
1   9    36    84    126    126     84    36    9    1
1  10   45   120   210    252    210   120   45   10   1
1 11   55  165  330   462    462   330  165   55  11  1
1 12  66  220  495  792   924   792  495  220  66  12  1
```

"Gosh, this is getting to be work. Isn't there any easier way to do this, Mr. Ulin?" "Isn't there some formula to directly calculate the numbers so that we don't have to add up the numbers in the table, row by row?"

We shall take up this question later but leave it to rest for a while. Now we turn our attention to...

3.2.2 A Completely Different Problem

Of five people, whom we shall call A, B, C, D, and E, we wish to choose two to do a task. How many pairs can be formed from the five persons? How many combinations of two objects are there from five different given objects?

One can approach the problem in many ways. Some pupils write down all combinations systematically in the alphabetical order:

$$AB \quad BC \quad CD \quad DE$$
$$AC \quad BD \quad CE$$
$$AD \quad BE$$
$$AE$$

and arrive at the sum $4 + 3 + 2 + 1$, just as in the study of triangle numbers (Section 3.1).

Someone says: if we write down all 4 combinations for each person, then the table will look like this:

AB	BA	CA	DA	EA
AC	BC	CB	DB	EB
AD	BD	CD	DC	EC
AE	BE	CE	DE	ED

Here every combination appears accounted for twice, for example, BA in addition to AB. The number of combinations is therefore

$$\frac{5 \cdot 4}{2}$$

This method gives directly the sum to which the previous method led.

We can also let the five people be represented by 5 points, regularly arranged as the corners of a pentagon, and seek the number of possible connecting lines between them. Each connecting line then corresponds geometrically to one combination of two people.

If the number of people is n, then the number of choices of two persons — according to the reasoning which just gave us $\frac{5.4}{2}$ as the number in the case of 5 people — will be

$$\binom{n}{2} = \frac{n\,(n-1)}{2}$$

We introduce here the nomenclature $\binom{n}{k}$ for the number of combinations of k people from among n people, or more generally, k objects from n different objects.

$\binom{n}{2}$ gives us the Greek triangle numbers:

$$\binom{2}{2} = 1 \quad \binom{3}{2} = 3 \quad \binom{4}{2} = 6 \quad \binom{5}{2} = 10 \quad \binom{6}{2} = 15 \quad \text{etc.}$$

A look at the number diagram for the street network earlier shows that the numbers there form two symmetrical lines in the triangular table. For example, the number of paths to those intersections lying along Boulevard 2 give us the triangle numbers 1, 3, 6, 10, 15, ... This leads us to ask:

Can we find an equivalence between the street network problem and the combinatorial problem? Are these problems only different in *appearance;* are we here basically concerned with one and the same mathematical problem?

To form a combination of two persons from five means that I must choose two people. Do I encounter any choices when I go from AO, BO to the A3, B2, which, as we remember, has 10 shortest paths to it?

Of course! Right at the start I choose to go left or right (defined, let us say, from the reader's point of view). I will pass another 4 intersections before I arrive at A3, B2. In order to get there I must choose left at two of the intersections. *Where* I choose to turn left does not matter, just that I turn left exactly two times. The situation is the same when choosing two people. I can go to each one of them and say "yes" (selecting them) or "no." Exactly twice must I say "yes" if I am to have two people in my combination.

In this way it is clear that the number of paths from the starting point to the intersection A7, B5, for example, is the same as the number of combinations of 7 (or 5) people from among 12 people. The number of paths to the intersection will thereby be:

$$\binom{12}{5} = \binom{12}{7}.$$

(This equality is understood at once when we consider that "yes" to 5 people means "no" to the other 7 people.)

3.2.3 Pascal's Triangle

We can now take up the question whether there exists a direct formula for the numbers $\binom{n}{k}$ which constitute the number scheme, or Pascal's triangle, as the scheme came to be called after the French mathematician and philosopher Blaise Pascal (1623-1662). Pascal wrote a very stimulating dissertation on the scheme in 1653: "Traité du triangle arithmetique."

This arithmetic triangle was known before Pascal's time, however. Documents have shown that the scheme existed in China early in the 1300's. The triangle begins with a one at the top and thereafter is identical with the scheme for the number paths in the street network (Section 3.2.1).

```
                1
             1     1
          1     2     1
       1     3     3     1
    1     4     6     4     1
 1     5    10    10     5     1
 . . . . . . . . . . . . . . . . . .
```

If we wish, we may re-write the triangle as below, where we let have the value 1, just as we do for all the natural numbers n:

$$
\begin{array}{ccccccc}
 & & & \binom{0}{0} & & & \\
 & & \binom{1}{0} & & \binom{1}{1} & & \\
 & \binom{2}{0} & & \binom{2}{1} & & \binom{2}{2} & \\
\binom{3}{0} & & \binom{3}{1} & & \binom{3}{2} & & \binom{3}{3}
\end{array}
$$
. . . .

Can the $\binom{n}{k}$ -values be calculated directly? When pupils construct Pascal's triangle, they find the additions rather laborious as they come down to rows with 2-digit numbers. Tables may exist for $\binom{15}{k}$ or perhaps even a lower row, but how would you determine $\binom{52}{13}$, for example?

Doing it by addition is far too time-consuming and unsure, even with the help of a pocket calculator.

Direct calculation of $\binom{n}{k}$

Let's take $\binom{9}{4}$ as a concrete example the number of combinations of 4 persons from 9, whom we could call A, B, C, ... H and I.

If we start by trying to follow the method we used for $\binom{5}{2}$, we would begin by asking: How many groups of letters are there with A as the first letter? When A is the first letter, how many choices have we for the second letter? There are 8 choices. And after that, when the second letter is chosen, in how many ways can we select letter number 3? There are 7 choices. And so on. The number of choices possible for letter number 1, 2, 3, etc. is 9, 8, 7, ... respectively down to 2 and 1. "One choice" for the last letter must here be interpreted as meaning that we have to choose the letter remaining. (Some students argue here that we have no choice at all....)

We may illustrate the number of choices of the successive letters with a kind of tree or street network which subdivides and spreads out more and more in all directions (Figure 3.2.4).

The figure illustrates the choice of the second and third letters and we can easily imagine for ourselves this network expanded to include the choice of the 4th letter. The number of groups with A as the first letter must be 8 · 7 · 6.

Since the first letter may be chosen in 9 ways, the total number of 4-letter groupings must be 9 · 8 · 7 · 6 (= 3024)

But wait a minute! Do not the four letters A, B, C, and D occur as a group several times among these 3024 groupings? For example,

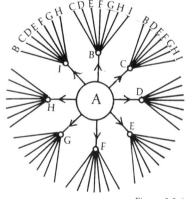

Figure 3.2.4

ABCD, BCAD, CDBA, DBCA, etc., which differ only in their ordering, must be considered as one and the same group of people, as only *one* combination of four elements. How many times do persons A, B, C and D occur together among the 3024 groupings which were formed when we considered the placement of the letters, when we ranked, so to speak, the people into first, second, third, and fourth places?

A, B, C, and D occur as many times together as there are possible placements of these four letters.

How many different placings are there?

For the first position in a 4-letter group we have 4 choices, for the second position 3 choices, for the third position 2 choices and for the fourth position "1 choice." This leads us to the number of different groupings, or permutations as they are usually called: the number is

$$4 \cdot 3 \cdot 2 \cdot 1 = 24.$$

This is how the list of 24 groupings looks, in alphabetical order:

ABCD	BACD	CABD	DABC
ABDC	BADC	CADB	DACB
ACBD	BADC	CADB	DACB
ACDB	BCDA	CBDA	DBCA
ADBC	BDAC	CDAB	DCAB
ADCB	BDCA	CDBA	DCBA

To summarize: of the 9 letters A through I we form $9 \cdot 8 \cdot 7 \cdot 6$ four-letter groups, where each letter elected occurs only once (AABC, for example, is not allowed). But each group of four letters occurs 24 times. The number of combinations must therefore be

$$\frac{9 \cdot 8 \cdot 7 \cdot 6}{24}$$

or more clearly written,

$$\frac{9 \cdot 8 \cdot 7 \cdot 6}{4 \cdot 3 \cdot 2 \cdot 1}$$

which, when evaluated, gives the number 126.

The logic we have followed has shown us that

$$\binom{9}{4} = \frac{9 \cdot 8 \cdot 7 \cdot 6}{4 \cdot 3 \cdot 2 \cdot 1}$$

and could be applied to any other particular numerical example without change. Thanks to the explicit appearance of each factor in expression (1), we can see through the particular case and grasp the general. How would we write, for example, the corresponding expression for $\binom{17}{6}$?

The answer is clear to us in a moment as soon as we become aware of how many factors there should be in the number (above the line); in the new example there are 6 objects to be chosen. We thus get

$$\binom{17}{6} = \frac{17 \cdot 16 \cdot 15 \cdot 14 \cdot 13 \cdot 12}{6 \cdot 5 \cdot 4 \cdot 3 \cdot 2 \cdot 1}$$

The students take pleasure in reducing factors top and bottom and finally arrive at the value 12376. If the example were $\binom{17}{11}$, we would get in the same way

$$\binom{17}{11} = \frac{17 \cdot 16 \cdot 15 \cdot 14 \cdot 13 \cdot 12 \cdot 11 \cdot 10 \cdot 9 \cdot 8 \cdot 7}{11 \cdot 10 \cdot 9 \cdot 8 \cdot 7 \cdot 6 \cdot 5 \cdot 4 \cdot 3 \cdot 2 \cdot 1}$$

which can be simplified at once to expression (2) for $\binom{17}{6}$.

Considering the fact that selecting 11 objects from among 17 different objects is equivalent to choosing 6 to throw out, we might directly have made use of the equality

$$\binom{17}{11} = \binom{17}{6}$$

The general formula becomes

$$\binom{n}{k} = \frac{n(n-1)(n-2) \ldots \{n - (k-1)\}}{k(k-1)(k-2) \ldots 2 \cdot 1}$$

or more simply

$$\binom{n}{k} = \frac{n(n-1)(n-2) \ldots (n-k+1)}{k(k-1)(k-2) \ldots 2 \cdot 1}$$

We note that the number of permutations of n different objects is

$$p_n = n(n-1)(n-2) \ldots 2 \cdot 1$$

The expression for   is usually written n! and called "n factorial." For example:

$$5! = 5 \cdot 4 \cdot 3 \cdot 2 \cdot 1$$

(the 1 may of course be left out, if one wishes.)

The formulas for p_n and $\binom{n}{k}$ can then be written

$$p_n = n!$$

See also Exercises 3-5 below.

3.2.4 Binomial Coefficients

In numerical tables and in the mathematical literature in general, the numbers $\binom{n}{k}$ are called *binomial coefficients*. Where does this name come from?

A binomial is an expression which is formed as the sum of two terms, e.g. a + b, a + x, 1 + x, etc. Oftentimes we need to raise such binomials to powers, for example:

$$\text{second powers} \quad (a + b)^2, \; (1 + x)^2 \quad \text{etc.}$$
$$\text{third powers} \quad (a + b)^3, \; (1 + x)^3 \quad \text{etc.}$$

For example, if the side of a cube having length 1 is increased by x units, the new volume will be the third power of the new side, i.e.

$$\text{volume} = (1 + x)^3$$

Which terms appear if we expand these power expressions?

We perhaps remember the squaring rule which says that

$$(a + b)^2 = a^2 + 2ab + b^2.$$

The coefficients, as they are called, for a^2, ab and b^2 are here 1, 2 and 1 respectively.

(The expression may, of course, be written $1a^2 + 2ab + 1b^2$.)

Those who don't remember the squaring rule could always write (a + b) (a + b) for $(a + b)^2$ and carry out the multiplications step by step:

$$a (a + b) + b(a + b) = a^2 + ab + ab + b^2 = a^2 + 2ab + b^2.$$

Analogously we may write

$$(a + b)^3 = (a + b) (a + b) (a + b).$$

Here we need to form and add up all products which have exactly one factor from each of the parentheses. What kinds of products can appear?

Three a-factors gives $\quad$ a · a · a = a^3
Two a-factors and one b-factor gives $\quad$ a · a · b = a^2b
One a-factor and two b-factors gives $\quad$ a · b · b = ab^2
and finally three b-factors gives $\quad$ b · b · b = b^3

But how many products do we get of each kind?
Obviously one a^3 and one b^3.

In order to get an a^2b product, we need to choose b once out of three choices (for each parentheses we have the choice of a or b). The number of such choices (combinations) is

$$\binom{3}{1} = 3$$

The term ab^2 arises when we choose b from two of the tree parentheses, therefore giving us

$$\binom{3}{2} = 3 \quad \text{of the } ab^2\text{-products.}$$

Finally, if we note that a^3 and b^3 arise with 0 or 3 choices of b, respectively, from the parentheses, then we can write

$$(a + b)^3 = a^3 + 3a^2b + 3ab^2 + b^3$$
$$= \binom{3}{0}a^3 + \binom{3}{1}a^2b + \binom{3}{2}ab^2 + \binom{3}{3}b^3$$

Expression (1) is of course the one we would use in practice but expression (2) helps us to see how the formula would look for higher powers. We can follow analogous choice logic for $(a + b)^5$, to take another example. It may, of course, also be written as

$(a + b) (a + b) (a + b) (a + b) (a + b)$ and gives therefore

which calculates out to

$$(a + b)^5 = \binom{5}{0}a^5 + \binom{5}{1}a^4b + \binom{5}{2}a^3b^2 + \binom{5}{3}a^2b^3 + \binom{5}{4}ab^4 + \binom{5}{5}b^5$$

which calculates out to

$$(a + b)^5 = a^5 + 5a^4b + 10a^3b^2 + 10a^2b^3 + 5ab^4 + b^5.$$

Power expansions of such binomials thus give formulas containing the numbers we encountered in Pascal's triangle. It is in connection with the expansion of powers of binomials that these numbers have received the name "binomial coefficients."

3.2.5 An Application in Physics (Theory of Heat)

A solid cube of steel has sides of dimension 5 cm at 0°C. It is heated to 20°C, expanding so that each side has the length $5 + h$ cm, where h according to heat theory would be

$$h = 5 \cdot 20 \cdot 0,000012.$$

The decimal fraction is the so-called coefficient of expansion (for steel, in this case).

We get $\qquad h = 0,0012.$

What is this new volume?

We must expand $(5 + h)^3$ and get

$$(5 + h)^3 = 5^3 + 3 \cdot 5^2 h + 3 \cdot 5 \cdot h^2 + h^3 = 125 + 75h + 15h^2 + h^3.$$

Of these terms, 125 is the volume before expansion (5^3). 75h has the value $75 \cdot 0,0012 = 0,09$ and is thus quite small. We can ignore the still smaller terms $15h^2$ and h^3. The new volume has, with good accuracy, the value $125 + 75h = 125,1$ cm^3.

3.2.6 The Surprising Triangle

We have seen that Pascal's triangle is made up of the binomial coefficients (3.2.3 and 3.2.4). We have further seen (in 3.2.2) that the Greek triangle numbers form a line in the triangle. Are there other interesting number sequences which appear in Pascal's triangle? There are *many*, and it is often surprising that the triangle in one way or another contains a number sequence which one has come upon in some particular problem.

Let me give two examples to start with.

Example 1: If we add the numbers in Pascal's triangle row by row we obtain the sums:

$$1$$
$$1 + 1 = 2$$
$$1 + 2 + 1 = 4$$
$$1 + 3 + 3 + 1 = 8$$
etc.

The sums thus form the doubling sequence 1, 2, 4, 8, ... Will this continue with each succeeding row? (Exercise 2).

Example 2: In Moser's circle subdivision problem (Section 3.1.6) we obtained the following number of subdivisions:

$$1, 2, 4, 8, 16, 31, 57, 99, \ldots$$

We recall there that the series could fool us when it deviates from the doubling series 1, 2, 4, ... starting with the value 31. Can Moser's sequence, too, be found in Pascal's triangle? In fact, yes. If we add along the rows up to the line drawn in Figure 3.2.5 we get the correct sums. One can show that this applies as far as one wishes to go in the circle subdivision problem, or in Pascal's triangle.

	Maximum number of areas
1	1
1 1	2
1 2 1	4
1 3 3 1	8
1 4 6 4 1	16
1 5 10 10 5 1	31
1 6 15 20 15 6 1	57
1 7 21 35 35 21 7 1	99
1 8 28 56 70 56 28 8 1	163

Figure 3.2.5

3.2.7 A Glimpse of Probability, Chance and Risk

In the 9th grade the students are at an age when they particularly want to test their powers of intelligence, especially in discussions with teachers or with parents at home. It is an important stage in their freeing themselves from dependence on adults. In mathematics class we get into the concept of probability; we pose problems concerning chance and risk. Some students have already met up with questions of the type: What is the chance of guessing all right on the football pools? Is it equally difficult to guess all of them wrong as all of them right? What is the chance of throwing three sixes in a row with a die? And so on. These are problems which they gladly investigate, so that they can feel they have clear-cut answers and that they have a grasp of the basic "foundation." I emphasize "foundation" here because the concept of probability is difficult, a difficulty which it is not easy to become conscious of. On the other hand, students readily note that the actual problems are themselves quite difficult, in fact "sneaky." One can easily be led astray without knowing it.

The students will once again meet Blaise Pascal, the man who together with his countryman Pierre Fermat laid the foundations of classical probability theory, and they will have the chance to tackle basic problems of the same type as Pascal faced.

Let us examine three problems which are related to earlier sections in this chapter.

1. Suppose we guess the answer to each of 5 questions which are to be answered with a "yes" or a "no." The questions are such that we do not have the slightest idea of the right answers. The chance of answering correctly is thus ½.

a) What is the probability of getting 3 correct answers?
b) What are the chances of getting at least 3 correct answers?

We start with a): it does not matter *which* 3 questions we succeed in guessing correctly. We could answer correctly on the first three, on the last three, on questions 1, 3, and 4, etc. How many such combinations of 3 right answers are there?

The number must obviously be equal to the number of ways of choosing 3 questions from out of 5, that is $\binom{5}{3} = 10$. (See 3.2.1-3.2.2).

There are thus 10 so-called successful cases, 10 different 5-row betting pool guesses with exactly 3 correct answers. But how many different 5-row bets are *possible at all*? On every row (question) we have two alternative answers. How many alternative answers does that give for 5 questions? Many pupils have a tendency here to answer 5 · 2 = 10 alternatives. This answer cannot be right, of course, since the number of alternatives with only three correct answers is 10.

In actual fact the number of alternative answers increases *multiplicatively* from row to row: for *each* of the two answers to question 1 we have two possible ways of answering question 2, which gives 2 · 2 alternative ways of answering both questions, etc. Figure 3.2.6 ought to show clearly enough why the growth is multiplicative. The number of *possible* alternative answers is therefore

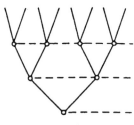

Figure 3.2.6

$$2 \cdot 2 \cdot 2 \cdot 2 \cdot 2 = 2^5 = 32$$

According to classical probability theory, the probability of exactly 3 correct answers is the following fraction:

$$\frac{\text{number of successful cases (with 3 correct answers)}}{\text{number of possible cases (with 0, 1, 2, 3, 4 or 5 correct answers)}}$$

The probability in question is thus

$$\frac{10}{32} \approx 0.31 \quad \text{or } 31\%$$

b) Getting at least 3 right means getting 3, 4, or 5 right. The probability sought after is therefore (according to the investigation above)

$$\frac{\binom{5}{3} + \binom{5}{4} + \binom{5}{5}}{32} = \frac{10 + 5 + 1}{32} = \frac{1}{2} = 50\%$$

(Can examples of this type of problem be solved yet more easily? See example 3 below!)

2. We have a combination lock, outfitted with 10 buttons labeled with the numbers 1, 2, 3, ... 9 and 0, plus a release button marked k. The lock opens only if one pushes down the right combination of number buttons and thereafter pushes k.

It does not matter in which order one pushes the number buttons. How big is the risk that an unauthorized person might succeed in opening the lock on his first try?

What do we mean by "risk"?

We take the same approach as Fermat and Pascal and seek the number of ways of setting the right combination in relation to the number of ways of pushing down *any* combination of buttons.

Let us even include the possibility of pushing k directly, i.e. the possibility of not setting any of the number buttons at all, as one of the possible ways.

The number of button combinations which opens the lock is obviously only one.

How many are then the total number of settings? A strategy which lies close at hand would be to calculate

$$
\begin{aligned}
\text{the number of ways to choose no button} &= 1 \\
\text{plus the number of ways to choose 1 button} &= 10 \\
\text{plus the number of ways to choose 2 buttons} &= \binom{10}{2} \\
\text{plus......} & \\
\text{plus the number of ways to choose 10 buttons} &= \binom{10}{10}
\end{aligned}
$$

This sum is equal to the total number of button combinations possible. But there is a considerably easier way to go about it.

We can look at the choice of buttons to push from another angle: for every button we suppose we face a choice shall we push the button or not? We have these two alternatives for each and every one of the 10 buttons.

Analogous to the study in Example 1, this gives $2^{10} = 1024$ possible combinations. The risk that the lock is opened after only one random try is thus

$$\frac{1}{1024}$$
    or approximately 0,1 %.

3. What are the chances of getting at least 1 six in 4 throws of a die? "At least 1 six" means 1, 2, 3, or 4 sixes. We might choose here the method of calculating

the number of cases with 1 six
the number of cases with 2 sixes
the number of cases with 3 sixes
and the number of cases with 4 sixes (which is 1)

and thereafter total up these numbers. If the total is x, then the chance we are looking for would be

$$\frac{x}{6 \cdot 6 \cdot 6 \cdot 6} = \frac{x}{1296}$$

since the number of possibilities is 6 at every throw.

There is, however, a time-saving trick, namely that of calculating the number of cases which are *opposite* of getting at least 1 six. This means simply determining the number of cases where no six appears at all during the four throws.

There are 5 such possibilities at *each and every* throw, which means the number of possible series of four throws without sixes is

$$5^4 = 625.$$

The number of cases with at least one six must then be

$$1296 - 625 = 671$$

and the possibility of throwing such a case would thus be $\frac{671}{1296}$

which gives a 51.8 % chance.

3.2.8 Exercises

1. Figure 3.2.2 showed the 10 paths between corners AO, BO and A2, B3. According to what principle are these paths ordered?

2. Prove that the sums in Example 1, Section 3.2.6, always give powers of 2, that is, that they give the sequence 1, 2, 4, 8, 16, ... (otherwise written 2^0, 2^1, 2^2, 2^3, 2^4, . . .) by setting x = 1 in the expansion of $(1 + x)^n$ for n = 0, 1, 2, 3, ...

3. In how many ways can 7 people be placed in a row, for example, along one side of a long table?

4. How many committees of four members can one select from seven people?

5. In Sections 3.2.2 and 3.2.3 we had examples of equalities such as

$$\binom{12}{5} = \binom{12}{7} \quad \text{and} \quad \binom{17}{11} = \binom{17}{6}.$$

Our motivation for this implies quite generally that

$$\binom{n}{k} = \binom{n}{n-k}$$

This relation can also be proven with the aid of the formula

$$\binom{n}{k} = \frac{n!}{k! \ (n-k)!}$$

How can this be shown?

6. On a circle lie 10 points. How many chords (connecting lines) can be drawn between these points?

7. 12 points are given. They lie such that no straight line can be drawn through any set of 3 points. How many triangles can be formed with three of the 12 points as corners?

8. How many letter groupings of 5 letters can be formed of the five letters A, B, C, D and E

      a) when each letter may only appear once in any grouping?
      b) when the condition above is relaxed, i.e. repetition of letters is
         allowed?

9. We return to Problem 2, section 3.2.7. As we remember, a lock has 10 buttons, labeled 0, 1, 2, ... and 9 respectively. If certain buttons are pressed, perhaps none at all, then the lock may be opened by pushing the release button. It does not matter in which order the proper buttons are pushed down.

How big is the risk that a thief opens the lock on one of 200 random tries (among which he may repeat some earlier trial combinations)? Example 3 in Section 3.2.7 gives a clue.

3.3 Fibonacci Numbers

3.3.1 Fibonacci

Fibonacci ("son of Bonacci"), or Leonardo of Pisa as he was called, was one of those who most strongly contributed to the penetration into Western Europe of Indian numerals, thereby pushing out the Roman numerals then in use. Fibonacci grew up in Bugia on the North African coast along the Mediterranean. (The town is called Bejaija today.) His father, Guglielmo Bonaccio, had a job which most closely could be called manager of a department store, which he owned. He was influential and wealthy. His son received his basic education from a Moorish teacher. He became acquainted with the Indian numbers, which had been assimilated into the Arabian culture. As a young man the son had the great advantage of visiting different countries in the Mediterranean area. He travelled in Egypt, Syria, Greece, Sicily, and Southern France and concerned himself with the arithmetical methods in use, including usage in commerce. Fibonacci found the Indian-Arabian system far superior to other systems and decided to write a book on the art of arithmetic.

Fibonacci's book *Liber Abaci* (abacus book; abacus = counting board) came out in 1202. It includes 15 chapters covering different areas of arithmetic. The first chapter is devoted to Indian numerals. Fibonacci thereafter takes up arithmetic with whole numbers, with fractions, business arithmetic, roots and square roots, equations, algebra and solution of selected problems. *Liber Abaci* became a great success and came out in a revised new edition in 1228. Beyond this volume, Fibonacci wrote three others, including one on geometry and disputed two of them in the presence of the benign Emperor, Fredrik II. Mathematics did not stand high at the universities of that time, however, and preference for the Roman numerals lived on. Bankers were terrified with the thought that Indian numerals would invite easily-made falsification, for example, with extra zeroes or by erasing zeroes at the end of a number. It was not until the 1400's that bankers succumbed, but today we still have the custom of writing the amount with letters as well as numerals when we write a check.

Yet Fibonacci has become a name in the mathematical world not so much for having promulgated the use of Indian numerals and having shown their advantages, as for having posed an interesting problem. It is called Fibonacci's Rabbit Problem.

3.3.2 *"Fibonacci's Rabbit Problem"*

This problem was introduced in the second edition of *Liber Abaci.* We will now concern ourselves with the rabbit problem which, although it must be said to be rather artificial, nonetheless was shown great interest.

Our starting point is a pair of rabbits, one of each sex. Let us say this pair is newly born in month 0. In month 1 the couple is sexually mature and in month 2 they beget a pair of bunnies, also one of each sex. The original pair goes on giving birth to a new pair each month (and by "pair" we always mean one of each sex). Each newborn pair increases the family in the same manner as the original pair, i.e. beginning in month 2 of their lives they give birth to a new pair each month thereafter. We suppose further that none of the rabbits die.

We now ask: how many rabbit pairs are there after a year, i.e. in month 12? We might perhaps formulate the problem a little more specifically: how does the number of pairs grow month after month up to and including month 12? The task is then to construct a table of the number of rabbit pairs as a function of the month number. If we call the number of pairs A and add a subscript with the number of the month, we can then read the following directly from the Fibonacci text:

$$A_0 = 1 \quad A_1 = 1 \quad A_2 = 1 + 1, \quad \ldots, \quad A_{12} = ?$$

I have posed this problem many times in the tenth grade: "Try to keep track of the rabbits by making an illustrative figure, perhaps like a table, perhaps with different symbols for newborn, etc. Perhaps someone wishes to draw a sort of tree of the rabbit population. Try now to find a good systematic method for organizing and accounting for the rabbit population."

Since the problem's solution does not require any real mathematical knowledge, it is not surprising that everyone in the class considers the problem, starts with some idea or other, and begins to count rabbits. After a while the students begin comparing some of their preliminary results with each other, and groups begin to form. Some students begin perhaps with one method but find it unpromising (usually the work lacks clarity) and look for a new idea. In some classes there would be one student — usually one who had difficulty with the subject — who succeeded in coming up with a clear and correct solution while highly talented students got entangled in their diagrams.

Apart from the degree of success, it was evident that Fibonacci could engage the students' interest, and most of them gained experience of value for problem-solving in general.

We will now consider a few variants of rabbit counting which students presented to each other.

1. Trying to draw a sort of population tree did not work out well. With time, as the months progress, the tree had so many branches and shoots in different directions that the picture was unclear.

2. A method drawing a separate family tree for each pair of rabbits used incredible amounts of paper. The student marked off the rabbit couples with X-es and carried on as in Figure 3.3.1 (but continued farther on). Time seemed too short for this method, but the idea was undeniably direct.

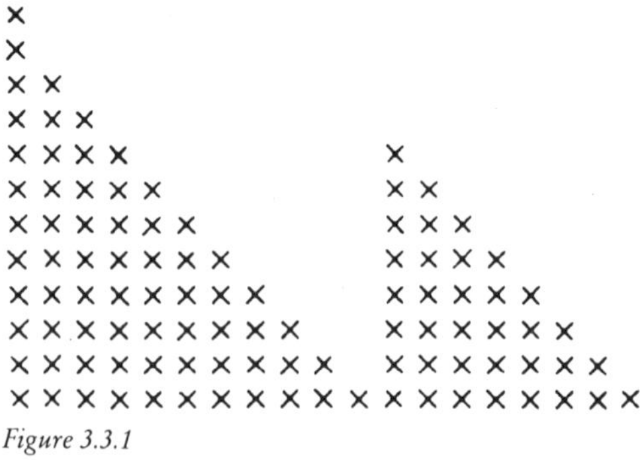

```
×
×
× ×
× × ×
× × × ×                    ×
× × × × ×                  × ×
× × × × × ×                × × ×
× × × × × × ×              × × × ×
× × × × × × × ×            × × × × ×
× × × × × × × × ×          × × × × × ×
× × × × × × × × × ×        × × × × × × ×
× × × × × × × × × × ×  × × × × × × × × ×
```
Figure 3.3.1

3a. Each pair was marked with a one, the birth of a new pair with a line connecting parents to children (Figure 3.3.2). The number of ones in each column is added up and in this way we get the desired A-values.

3b. As in solution 3a. each pair is marked with a 1 in the beginning but here each pair is only represented once with a 1, namely in the month they are born. When later several pairs are born in the same month, for example 5 pairs, then 5 ones are noted in the table as the increase for that month. The numbers which are written into the table are thus always newly born pairs. To get the total number of pairs at a given time, e.g. month 7, we must add all the ones up to and including month 7. See Figure 3.3.3.

4a. Circles represent newborn pairs; circles with dots are mature pairs; and darkened circles are pairs which are at least 2 months old ("aging pairs"). Month by month the aging, mature and newborn pairs are filled in (Figure 3.3.4).

4b. The same method as in 4a. but the letters A, M and N are used instead of the circle symbols for Aging, Mature and Newborn respectively (Figure 3.3.5).

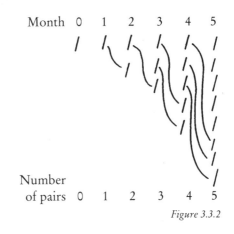

Figure 3.3.2

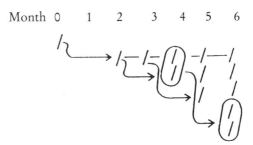

Figure 3.3.3

Month		Number of pairs
0	○	1
1	⊙	1
2	● ○	2
3	● ⊙ ○	3
4	● ● ⊙ ○ ○	5

Figure 3.3.4

The question now lies close to hand: how many surviving (aging and mature) and newborn shall we note down in the step going from one month to the next? Empirically we may have found some values, for example

Month		Number of pairs
0	1N	1
1	1M	1
2	1A + 1N	2
3	1A + 1M + 1N	3
4	2A + 1M + 2N	5

Figure 3.3.5

$$A_3 = 3 \qquad A_4 = 5 \qquad A_5 = 8$$

By this time many pupils, alone or in groups, have already observed a regularity in the numbers obtained. Let us look at these:

$$A_0 = 1 \qquad A_4 = 5$$
$$A_1 = 1 \qquad A_5 = 8$$
$$A_2 = 2 \qquad A_6 = 13$$
$$A_3 = 3 \qquad A_7 = 21$$

These are enough values. Do we see any relation between the numbers? Some pupils, who have discovered a relation, choose to use it "mechanically" in continuing on to month 12, convinced that the regularity always holds.

We discover that each new number is the sum of the two nearest preceding numbers: $1 + 1$ gives A_2, $1 + 2$ gives A_3, $2 + 3$ gives A_4, etc. Finally we have

$$A_{12} = A_{10} + A_{11} = 89 + 144 = 233.$$

But does this addition rule apply unconditionally? Might the answer to his question perhaps lead to a better method of doing the actual rabbit accounting?

5. If not earlier, then now some of the pupils will come upon the idea of asking: how many rabbit pairs live on, and how many are newborn when we go over from one month to the next?

If we consider, for example, the change from month 4 to month 5, then we have:

(1) All pairs of rabbits from month 4 live on, i.e. $A_4 = 5$ pairs

(2) In month 5 as many new are born as there were mature and aging pairs in month 4. This is as many as the total number of pairs in month 3, i.e. $A_3 = 3$ pairs. (See the table in Figure 3.3.4 or Figure 3.3.5.) From this we get $A_5 = A_4 + A_3 = 5 + 3 = 8$.

The same reasoning gives $A_6 = A_5 + A_4$ and generally

$$A_n = A_{n-1} + A_{n-2} \quad (n = 2, 3, 4, \ldots) \tag{1}$$

With this newly proven formula we can successively obtain the number of rabbit pairs in any month we wish, but we must use the formula repeatedly step by step.

Is there not perhaps some formula for *direct* calculation of the number of pairs "so that we can get out of adding up all the lower numbers"? This question lies near at hand to the pupils. Since this problem is quite tricky and would require considerable time to develop, I usually content myself with a formula which the French mathematician J. P.M. Binet published in 1843. In order to keep the formula as simple as possible we renumber the Fibonacci numbers (1,1, 2, 3, 5, 8, ...) with f_1 as the first number of A: $f_1 = 1$, $f_2 = 1$, $f_3 = 2$, $f_4 = 3$, etc.

The formula then is

$$f_n = \frac{1}{\sqrt{5}} \left\{ \left(\frac{1 + \sqrt{5}}{2} \right)^n - \left(\frac{1 - \sqrt{5}}{2} \right)^n \right\}, \ n = 1, 2, 3, \ldots$$

This formula was actually known in the 1730's to the Swiss mathematicians Leonard Euler and Daniel Bernoulli. There is, however, from the standpoint of the pupils, a simpler formula for direct calculation of lower Fibonacci numbers, proven by É. Lucas. Like so many other number relationships it may be found hidden in Pascal's triangle. In Figure 3.3.6 parallel

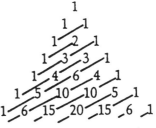

Fibonacci numbers in Pascal's triangle

diagonals with a specific slope are drawn. If we sum the numbers which lie on such a diagonal, beginning from the top, we will obtain precisely the Fibonacci numbers. An appropriate and stimulating task is to prove, with the help of formula (1), that all of the diagonals — not only those in the beginning — give Fibonacci numbers.

We have here a convenient addition formula for the Fibonacci numbers:

$$f_1 = 1 = \binom{1}{0}$$

$$f_2 = 1 = \binom{1}{1}$$

$$f_3 = 2 = 1 + 1 = \binom{2}{0} + \binom{2}{1}$$

$$f_4 = 3 = 1 + 2 = \binom{3}{0} + \binom{2}{1} \text{ etc.}$$

and generally
$$f_n = \binom{n-1}{0} + \binom{n-2}{1} + \binom{n-3}{2} + \ldots \qquad \text{for } n \geq 1,$$

where the summation extends until the difference between the upper and the lower numbers is zero in the case of n being odd, or one, in the case of even n.

It may now be of value to review the different rabbit-counting methods developed by pupils which were presented as variants 1 - 5 above. We can differentiate various degrees of abstraction:

In 3a, 3b and 4a we count up ones or figures which represent rabbit pairs, more or less as we would count several litters of animals. The degree of concreteness is high here; the achievement is almost entirely in the construction of a systematic arrangement of the units.

In 4b, letters are introduced in a first step toward abstraction, as symbols for the different kinds of rabbit pairs.

In method 5 we leave both the symbols and the tangible counting of rabbit pairs and find the additive structure of the Fibonacci numbers. We have here reached the level of thought which gives us a general method, even if it is a step-by-step method. (Formula (1) is called a recursive formula.)

The more levels of concreteness and abstraction with which a particular problem may be solved, the more suitable the problem is for differentiation within a class.

3.3.3 More on Fibonacci Numbers

Why have Fibonacci numbers become so well known? What interesting characteristics do they show? It became apparent that Fibonacci numbers more-or-less unexpectedly appeared in many different contexts and that such numbers have a whole variety of relationships between themselves.

Apart from certain aspects which are taken up in the exercises below, we will now primarily concentrate on the occurrence of Fibonacci numbers in the plant kingdom and show their relation to the so-called "golden section."

Let us first return to our rabbit population and ask: how large are the fractions of newborn pairs and surviving pairs, respectively, month by month? Is it a population which grows younger and younger in the sense that the fraction of newborn increases, or is it an aging population?

We select the following notation:

A_n = number of rabbit pairs in month n (n = 0, 1, 2, ...)
B_n = number of newborn pairs in month n.

The fraction of newborn is then $u_n = \dfrac{B_n}{A_n}$. The fraction of others (surviving) is $v_n = 1 - u_n$.

Taking n up through 12 we get the following table:

n	u_n	v_n	n	u_n	v_n
0	$\frac{1}{1}=1$	$\frac{0}{1}=0$	7	$\frac{8}{21}=0{,}38695\ldots$	$\frac{13}{21}=0{,}61904\ldots$
1	$\frac{0}{1}=0$	$\frac{1}{1}=1$	8	$\frac{13}{34}=0{,}38235\ldots$	$\frac{21}{34}=0{,}61764\ldots$
2	$\frac{1}{2}=0{,}5$	$\frac{1}{2}=0{,}5$	9	$\frac{21}{55}=0{,}38181\ldots$	$\frac{34}{55}=0{,}61818\ldots$
3	$\frac{1}{3}=0{,}33\ldots$	$\frac{2}{3}=0{,}66\ldots$	10	$\frac{34}{89}=0{,}38202\ldots$	$\frac{55}{89}=0{,}61797\ldots$
4	$\frac{2}{5}=0{,}4$	$\frac{3}{5}=0{,}6$	11	$\frac{55}{144}=0{,}38194\ldots$	$\frac{89}{144}=0{,}61805\ldots$
5	$\frac{3}{8}=0{,}375$	$\frac{5}{8}=0{,}625$	12	$\frac{89}{233}=0{,}38197\ldots$	$\frac{144}{233}=0{,}61802\ldots$
6	$\frac{5}{13}=0{,}38461\ldots$	$\frac{8}{13}=0{,}61538\ldots$			

Can we see anything interesting in this table?

Some students notice that the fractions seem to converge toward the values 0.3820 and 0.6180 respectively. Others notice that u_n and v_n swing up and down in value — every other time increasing, every other time decreasing — but the swings become smaller and smaller as n increases. With the aid of a calculator one can easily extend the table for a few more values of n. One finds, for example,

$$A_{18} = 4181, A_{19} = 6765, A_{20} = 10946, \text{ of which}$$

$$u_{20} = \frac{4181}{10946} = 0.38196601$$

$$v_{20} = \frac{6765}{10946} = 0.61803398$$

In fact it does seem as if, decimal by decimal, the numbers are converging to certain values. Judging by everything, the fractions u_n and v_n are approaching limiting values

$$g = 0.38196601 \quad \text{and} \quad G = 0.61803398...$$

The fractions u_n and v_n seem to come as close as we wish to two numbers g and G respectively and their approaches are oscillatory. We let rest the questions of the existence and of the exact values of g and G until Exercises 4 and 5, and ask here instead: *if* limits exist, do they then eventually appear with the same values when one begins with entirely different starting values for the number of pairs in months 0 and 1? For example, if we have 6 newborn pairs and 1 mature pair in month 0? In this case we get a Fibonacci series which begins with

$$A_0 = 7, \quad A_1 = 8 \quad (7 \text{ pairs} + 1 \text{ newly born pair})$$

and which continues with $A_2 = 15$, $A_3 = 23$, $A_4 = 38$, etc.

What happens to the functions u_n and v_n respectively? This problem is taken up in Exercises 3-6.

3.3.4 Leaf Rotation (Phyllotaxis)

Fibonacci numbers lay more-or-less in a "Sleeping Beauty" state for 600 years, until the 1830's. We have seen that Euler, D. Bernoulli, and Binet had studied them the century before, but they were given at least as much attention by the botanists of the 19th century.

Alexander Braun published in 1831 a dissertation with the title "Vergleichende Untersuchung über die Ordnung der Schuppen an den Tannenzapfen als Einleitung zur Untersuchung der Blattstellungen über-haupt" (Comparative study of scale arrangement on pine cones as introduction to a general study of leaf position).

As the title suggests Braun studied the geometry of pinecones. Anyone can understand from the appearance of a pinecone (Figures 3.3.7 and 3.3.8) that the cone has an architecture which one ought to describe mathematically. Braun's research lead to the concept of leaf fraction or phyllotaxis.

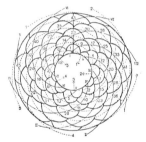

Figure 3.3.7
Pinecone

Pinecone with phyllotaxis 8/21
(From Strasburger, "Lehrbuch der Botanik")

Goethe had emphasized that ordinary upright, growing flowering plants with leaves follow two basic principles during growth: the stalk grows straight upward (the vertical principle) while the leaves (usually) form an upward spiral. As the leaves grow outward from higher and higher levels there occurs a "twisting" around the stalk from one leaf to the next highest neighbor. Besides the vertical principle there is thus also a "rotation" with the stalk as axle (Figure 3.3.9). When the plant flowers, the leaf spiral is completed with a crown of leaves (the petals). Braun showed that leaf rotation is by different amounts in different categories of plants. We shall here take a short look at some phyllotaxis results which are well known to botanists.

1. On tulips and gladiolas the leaves grow out alternately to the left and to the right. The "twisting" from one leaf to the next is thus 1/2 revolution. The phyllotaxis (leaf fraction) is said to be 1/2. (Figure 3.3.10).

2. In 3-bladed grasses and meadow saffron the three leaves distribute themselves around 1 revolution, so that the rotation from one leaf to the next is 1/3 revolution, i.e., the phyllotaxis is 1/3 (Figure 3.3.11).

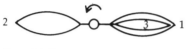

Figure 3.3.10

Figure 3.3.9

3. If we count the leaves in plants of the Rosacae family (roses, raspberries, plums, violets, etc.), we find that 5 successive leaves upwards on the stem twist 2 revolutions. The twist is thus 2/5 revolution per leaf and the phyllotaxis is 2/5 (Figure 3.3.12).

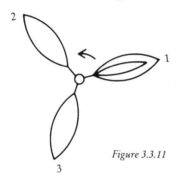

Figure 3.3.11

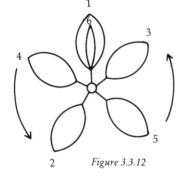

Figure 3.3.12

4. In most cabbages (the Cruciferae family), snap dragons, plantain and monk's hood there are 8 leaves to every 3 revolutions and the phyllotaxis is 3/8.

5. Dandelions, mullein (figwort family), potatoes and other plants have a phyllotaxis 5/13.

6. Now we come to the starting point of Braun's investigations: the fir. It has a phyllotaxis of 8/21, both in the scales on the cone and in the rotation of the needles. The same values apply to the pines and larches.

In summary: phyllotaxes of 1/2, 1/3, 2/5, 3/8, 5/13 and 8/21 have been found.

What numbers do we have in this sequence? Fibonacci's numbers appear, as we see, in the phyllotaxes, and the fractions themselves agree with the population fractions which we computed in the previous section 3.3.3.

If phyllotaxes were to exist with even higher Fibonacci numbers, such fractions should approach the value g = 0.3820... Are there such fractions? Can it be, that the limit g has a very real meaning in the plant kingdom? These are questions which it is very natural to ask. The botanist G. van Iterson counted leaves up the stalk and used a microscope when the naked eye could no longer determine the point of origin of the very tiny stems on leaves which were just coming out. These studies, presented by Iterson in 1907 in the dissertation "Mathematische und mikroskopisch-anatomische Studien über Blattstellungen" and later furthered by M. Hirmer (1922), concentrated on the vegetative cone, that part of the plant which produces the very youngest leaf buds. The results showed that these start, independent of botanical family, with the leaf fraction g ≈ 0.38 and that the phyllotaxis then deviates from this more and more as we go down the stalk to older leaves. There the leaf rotation finally becomes what is typical for that family.

All leaf rotations thus begin with the value g ≈ 0.38 at the top leaf buds. The phyllotaxis thereafter gradually changes toward the family's characteristic fraction, for example 2/5 = 0.40 for the roses.

Further examples of Fibonacci numbers in the plant world are given in Section 3.7.2.

3.3.5 The Golden Section

The results from research on phyllotaxis becomes even more interesting if we relate them to a proportion which was much appreciated during the Middle Ages and known even before that by the Greeks, the proportion which arises from the so-called golden section. Let us take a straight line AB of length 10 cm, for example. A point S on AB divides the length into the golden section if the ration of the smaller interval to the larger is the same as the ration of the larger to the whole interval.

In other words, if AS is the shorter length, the golden section would imply

$$\frac{AS}{SB} = \frac{SB}{AB} \quad \text{or letting } x = SB: \quad \frac{1 - x}{x} = \frac{x}{1}$$

In order to illustrate this sectioning we let the rectangle in Figure 3.3.13 have length 1 unit and the shorter side width x. At one end of the rectangle we mark off a square with side x. According to (1) the rectangle is now divided into golden section proportions if the smaller rectangle is similar (in the geometrical meaning) to the larger, original rectangle. If one solves equation (1), leading to the second degree equation

$$x^2 + x = 1,$$

one gets as the useful (positive) root the number $x = \dfrac{\sqrt{5} - 1}{2} \approx 0.6180\ldots$

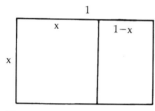

Figure 3.3.13

A   S   B

Figure 3.3.14

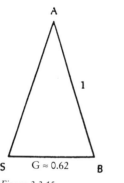

Figure 3.3.15a

This number determines the golden sectioning of the interval AB and agrees exactly with the limit G which arose in the study of leaf rotation (Section 3.3.4). (This agreement is proven in the solution to exercises 4 and 5.)

When S is placed at the golden section's dividing point, the decimeter long interval is divided into two parts g and G, approximately 0.38 and 0.63 respectively. Figure 3.3.14 shows this sectioning.

Let us now construct a triangle in which the base is to the side as G is to 1. We take the side AB = 1 dm (decimeter = 10 centimeters), for example, and the base BS = G dm ≈ 0.62 dm (Figure 3.3.15a). We measure the top angle with a protractor and find that it is 36°. Is this value exact? If that were the case the triangle would comprise one-tenth of a regular ten-sided polygon, since the 10-side polygon's inner triangles have a center angle of

$$\frac{360°}{10} = 36°.$$

Does the golden section produce such triangles?

Let us instead begin with a regular ten-sided polygon, remove one of its triangles, and set its side equal to 1 unit. We ask, how long is the triangle's base? Those who know the basis of trigonometry can directly write the relation

$$\frac{\text{half the base}}{\text{the side}} = \sin 18°$$

from which

$$\begin{aligned} \text{base} &= 2 \cdot \text{side} \cdot \sin 18° \\ &= 2 \sin 18° \\ &\approx 0.618 \end{aligned}$$

We are now well convinced that the length of the base is exactly G, i.e. that our golden-section triangle fits into the ten-sided polygon. But how could one show this exactly?

How large are the angles in the 10-sided polygon's triangles? Obviously the base angles are 72°, since $36° + 2 \cdot 72°$ gives the correct sum of angles, 180°. If we now bisect one of the base angles with a line, we get two similar triangles as in Figure 3.3.15b.

The equation for equal rations of base to side in the small triangle BCD and in the large triangle ABC becomes

$$\frac{1 - x}{x} = \frac{x}{1}$$

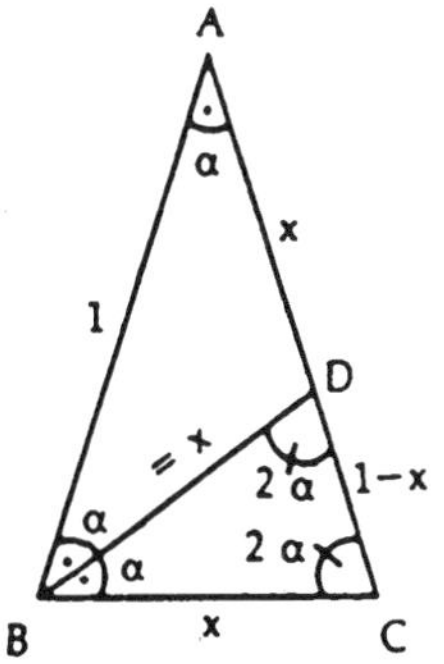

Figure 3.3.15b

This equation is the same as equation (1)! The useful root has already been given,

$$x = G = \frac{\sqrt{5} - 1}{2}$$

Ten golden-section triangles thus form a regular ten-sided polygon.

3.3.6 The Golden Section in Art and Nature

The golden section, as mentioned, was much appreciated during the Middle Ages and was used as the starting point in a variety of applications. A few examples:

The artist Piero della Francesca, who was both painter and mathematician, in all likelihood used geometrical structures as the starting point in a number of his paintings, and made use, in particular, of the golden section. Giotto in his famous painting of Franciscus with the birds placed the painting's center of interest , Franciscus' right eye, at the golden-section proportion along the painting's diagonal. A number of Gothic cathedrals show golden-section proportions, as do the shape of Stradivarius-violins.

In Section 3.3.4 we studied leaf rotations. We can formulate Iterson's and Hirmer's research results in the following way: in the budding leaves of plants with spiraling leaves there is a common leaf rotation equal to the golden section's smaller number, $g \approx 0.38$. From this value the various plants or families of plants then diverge toward the rotation fraction which is characteristic for them.

An interesting research report on proportions came out in 1854 in the dissertation by A. Zeising, "Neue Lehre von den Proportionen des menschlichen Körpers" (New Findings on the Proportions in the human body). As a result of various series of measurements, Zeising shows that on average the navel divides the human body height into the proportions of the golden section. In one class where the pupils became interested by this they measured and calculated their own values and found 0.617 as the mean value of the proportion foot-to-navel divided by body height (FN/BH).

Of course, there are those who dismiss Zeising's results as coincidence, but the golden section recurs repeatedly in the proportions of the arms and of the hands.

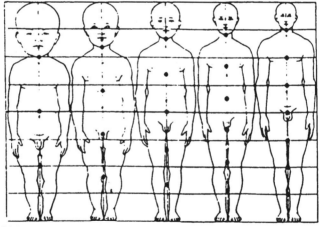

Figure 3.3.16

In the newborn child the proportion FN/BH is roughly 1:2. Figure 3.3.16 from Mörike-Mergenthaler's "Biologie des Menschen" ("Human Biology") illustrates how this proportion changes during growth. It is worth noting that the human being, in contrast to the leaves of the plants, *grows toward* the golden-section proportion.

3.3.7 Exercises

1. Drone bees develop from unfertilized eggs, in contrast to worker bees who grow from fertilized eggs. In other words, drones have only one parent, the mother, while workers and queen bees have two parents.

a) Draw a family tree for a drone bee, going 5 or 6 generations back. The figure here (Figure 3.3.17) shows the start of the tree.

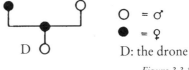

D: the drone

Figure 3.3.17

b) Do the numbers of bees in the tree, generation by generation, make up a Fibonacci series? If so, how can this be proven?

2. Fibonacci Puzzle

In Figure 3.3.18 is a square with 21 units as side (21r) divided into four parts – two right triangles A and B plus two trapezoids C and D. A and B together form a rectangle 8r x 21r. The shortest side in each of the four parts is 8r. In Figure 3.3.18b it appears as if the four pieces of the square form a rectangle with sides 13r and 34r. If this is the case, then the areas of the square and the rectangle should agree. But

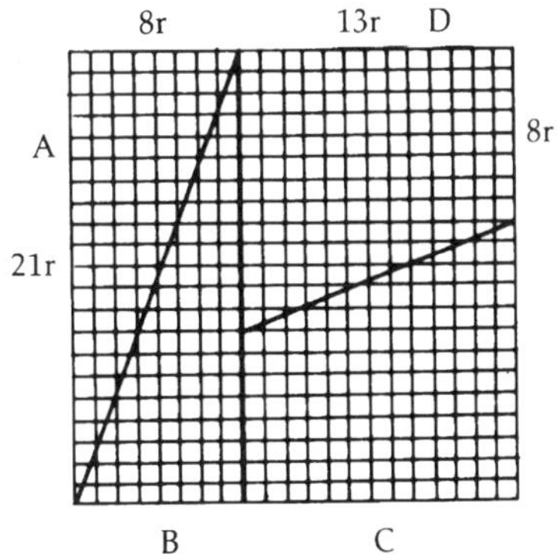

$$(21r)^2 = 441r^2 \text{ and } 13r \cdot 34r = 442r^2.$$

Where is "the catch" here?

Figure 3.3.18a

3. Calculate to 3 decimal places, correctly rounded, the value of the quotient between the 9th and 10th numbers in the Fibonacci series which begins with the numbers 7 and 8. In other words, calculate f_9/f_{10} when $f_1 = 7$ and $f_2 = 8$. ($f_3 = f_2 + f_1$, $f_4 = f_3 + f_2$, etc.)

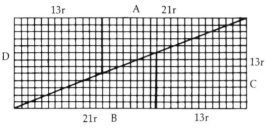

Figure 3.3.18b

4. Show that if the quotient of two successive numbers in the Fibonacci series 1, 1, 2, 3, 5, ... is

$$v_n = \frac{f_n}{f_{n+1}},$$

then the next quotient will be

$$v_{n+1} = \frac{f_{n+1}}{f_{n+2}} = \frac{1}{1+v_n}.$$

Show also, using this result, that if the quotients go toward a limit as n goes to infinity, then the limit is

$$G = \frac{\sqrt{5} - 1}{2}$$

5. (For readers with experience in the convergence series)
How may it be proven that the number G (introduced in Sec. 3.3.4 and above in Exercise 4) truly is the limit of the sequence $\{v_n\}$ in Sec. 3.3.3.?

6. a) Form the 10th Fibonacci number if the first two are $F_1 = a$ and $F_2 = b$.
b) Does the value of the ratio between two successive Fibonacci numbers, F_n/F_{n+1}, approach a limit no matter what the two starting points are? Is this limit, if it exists, independent of the starting numbers, i.e. is there a common limit for all possible Fibonacci series? (Let a and b be positive.)

7. Show that the Fibonacci series which begins with G and 1 (that is the sequence G, 1, G + 1, 2G + 3, 3G + 5, ...) is identical with the geo-

metrical sequence $\qquad G, 1, \dfrac{1}{G}, \dfrac{1}{G^2}, \ldots$

8. For numbers in the Fibonacci sequence 1, 1, 2, 3, 5, ... it may be shown that

$$f_3 = 2 = 1 + 1 = 1^2 + 1^2$$
$$f_4 = 3 = 4 - 1 = 2^2 - 1^2$$
$$f_5 = 5 = 4 + 1 = 2^2 + 1^2$$
$$f_6 = 8 = 9 - 1 = 3^2 - 1^2$$
$$f_7 = 13 = 9 + 4 = 3^2 + 2^2$$

Does this regularity continue – and how can it be written as a general expression?

3.4 1, 2, 3, 4, ... and 1, 2, 4, 8, 16, ...
Two Important Growth Principles

The sequences 1, 2, 3, 4, ... and 1, 2, 4, 8, 16, ... have become a part of everyday folklore through the classical story of the inventor of chess. According to the story, the inventor, who was promised whatever he wanted as a reward by an Indian prince, asked for one kernel of wheat on the first square of the chessboard, two on the second square and each time double the amount on each of the other 62 squares. The total number of kernels would be

$$1 + 2 + 4 + 8 + \ldots + 2^{63}$$

(where the last term is the product of 63 two's multiplied together).

Calculated out this became some 18 billion billion kernels of wheat, or more than the harvest in the whole province!

If the inventor had asked for 1 kernel on the first square, 2 on the second, 3 on the third, etc., then the reward would have remained at the modest level of

$$1 + 2 + 3 + 4 + \ldots 64 = \frac{64 \cdot 65}{2} = 2080 \text{ kernels}$$

We shall soon see that both sequences of numbers in the title play important roles in various topical applications.

3.4.1 The Piano Keyboard

Figure 3.4.1 shows the keyboard of a piano, which most often spans 7 octaves plus 2 extra white keys and 1 extra black on the left.

Each octave has 7 white and 5 black keys which together give 12 tones. The so-called octave "just to the left of the piano's keyhole" goes from middle C to C in the octave above (c^1). If we count the white keys with C as the first, then we come to c^1 as the eighth key (and from this the name octave). c^1 is the first note in the next higher octave. Piano tuners tension the a^1 string so that it vibrates at 440 vibrations per second. The tone then has the frequency, as we call it, of 440 cycles per second, or 440 Hertz (Hz). Going from note a^1 to a^2 in the next octave, that

Figure 3.4.1

is taking a 1-octave step upward from a^1, we get a tone which, when played at the same time as a^1, is in perfect harmony with a^1. Playing the white keys from a^1 up to a^2, it is for our tonal experience as if returning to the starting tone, but on a higher level. We have taken a step upward on the tonal scale which feels completely natural.

The tonal ear is very sensitive to the exact agreement of a^2 relative to a^1. When tuned clearly, a^2 has exactly double the frequency of a^1, that is $2 \cdot 440 = 880$ Hertz.

In this way the frequencies increase octave by octave, and we have (in Hertz)

$$a^1 = 440$$
$$a^2 = 2 \cdot 440 = 880$$
$$a^3 = 4 \cdot 440 = 1760$$
$$a^4 = 8 \cdot 440 = 3520 \text{ Hz}$$

Taking a 1-octave step downward instead, i.e. to the left on the keyboard, we get (using the common nomenclature)

$$a = \frac{1}{2} \cdot 440 = 220 \text{ Hz}$$
$$A = \frac{1}{4} \cdot 440 = 110$$
$$A_1 = \frac{1}{8} \cdot 440 = 55$$
$$A_2 = \frac{1}{16} \cdot 440 = 27.5$$

The a-notes thus have the following frequencies:

$$A_2 = \frac{1}{16} \cdot 440 \qquad a^1 = 1 \cdot 440$$
$$A_1 = \frac{1}{8} \cdot 440 \qquad a^2 - 2 \cdot 440$$
$$A = \frac{1}{4} \cdot 440 \qquad a^3 = 4 \cdot 440$$
$$a = \frac{1}{2} \cdot 440 \quad \text{and} \quad a^4 = 8 \cdot 440$$

If we start with c instead of with a, the sequence will be similar, but 440 is replaced by another frequency. Let us call it f:

$$C_1 = \frac{1}{8} \cdot f \qquad c^1 = 2 \cdot f$$
$$C = \frac{1}{4} \cdot f \qquad c^3 = 4 \cdot f$$
$$c = \frac{1}{2} \cdot f \qquad c^4 = 8 \cdot f$$
$$c^1 = 1 \cdot f \quad \text{and} \quad c^5 = 16 \cdot f \quad \text{(which is the highest tone on our piano)}$$

The coefficients 1, 2, 4, … always appear when we go upward by octaves and the coefficients ½, ¼, ⅛, … when we go downward by octaves. In both cases 1 corresponds to the frequency we start with.

We now leave the world of tones temporarily and take a look at bacteria and radioactive substances.

3.4.2 Bacteria and Radioactive Substances

If a bacterial culture or a culture of yeast cells is allowed to grow under constant and favorable conditions, including sufficient nourishment available the entire time, then the number of bacteria or cells doubles with certain intervals of time. If the number of bacteria at time 0 is A, then this number will grow in the given time interval, let us say 3 hours here, according to the following table:

Time, hours	0	3	6	9	12	15
Number of bacteria	A	2A	4A	8A	16A	32A

But as soon as the nourishing culture begins to dry up, this regularity stops; growth is slowed. Growth can stop; the number of bacteria can even decrease. Figure 3.4.2 illustrates the table in diagram form.

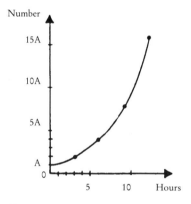

Figure 3.4.2

In yet another case do the doubling numbers 1, 2, 4, 8, ... play an important role.

Not long after the first radioactive elements had been discovered by the husband and wife Curie team (polonium in 1898 and radium in 1902), it was found that radioactive substances disintegrate and that during a certain specific time, for example 5 days, a substance would be reduced to only half of its original amount.

Whenever we begin our measurement, after the 5 days one half of the original amount of the substance will be left. Each radioactive substance has its own so-called half-life.

Uranium (with atomic weight 238) has a half-life of 4.5 billion years, radium (atomic weight 226) has 1600 years, polonium (atomic weight 210) 139 days, and a polonium isotope (atomic weight 214) just 0.00015 seconds.

If the quantity of material weighs m grams to start with and the half-life is T days, we have the following table illustrating the disintegrating principle:

Time (number of days)	0	T	2T	3T	4T	5T
Weight of substance in grams	m	½ m	¼ m	⅛ m	¹⁄₁₆ m	¹⁄₃₂ m

Here we see the halving numbers ½ , ¼ , ⅛ , etc.

A curve of the decreasing substance weight is shown in Figure 3.4.3.

Earlier in Section 3.1 we wrote the numbers 1, 2, 4, 8, ... as powers of 2:

$$1 = 2^0$$
$$2 = 2^1$$
$$4 = 2^2$$
$$8 = 2^3$$
$$16 = 2^4 \quad \text{etc.}$$

If we go from the bottom up in this column, we find that the numbers (on the left) are halved each step upward while the exponents (the powers on the right) decrease by 1 with every step. For this reason 0 has been defined as the exponent of 2 for the number 1, i.e. $2^0 = 1$.

If we continue to halve on the left, we will get values

$$\frac{1}{2}, \frac{1}{4}, \frac{1}{8}, \frac{1}{16}, \text{ etc.}$$

Can we also write these as powers of 2? The natural thing would then be to write

$$\frac{1}{2} = 2^{-1} \quad \frac{1}{4} = 2^{-2} \quad \frac{1}{8} = 2^{-3} \quad \frac{1}{16} = 2^{-4} \quad \text{etc.} \qquad (^*)$$

We now draw up a table which shows how the inverse parts 2 and 1/2, 4 and 1/4, etc. also correspond to each other with regard to their exponents.

Why is the power form (*) so logical and natural?

The answer is quite simply that the rules which are used for arithmetic with positive exponents will also work with 0 and negative integers as exponents.

We content ourselves here with a few examples which show the applicability of the three main rules for exponent arithmetic.

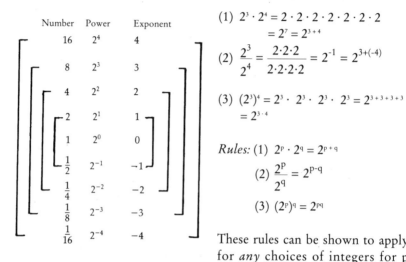

Number	Power	Exponent
16	2^4	4
8	2^3	3
4	2^2	2
2	2^1	1
1	2^0	0
$\frac{1}{2}$	2^{-1}	−1
$\frac{1}{4}$	2^{-2}	−2
$\frac{1}{8}$	2^{-3}	−3
$\frac{1}{16}$	2^{-4}	−4

(1) $2^3 \cdot 2^4 = 2 \cdot 2 \cdot 2 \cdot 2 \cdot 2 \cdot 2 \cdot 2$
$$= 2^7 = 2^{3+4}$$

(2) $\dfrac{2^3}{2^4} = \dfrac{2 \cdot 2 \cdot 2}{2 \cdot 2 \cdot 2 \cdot 2} = 2^{-1} = 2^{3+(-4)}$

(3) $(2^3)^4 = 2^3 \cdot 2^3 \cdot 2^3 \cdot 2^3 = 2^{3+3+3+3}$
$$= 2^{3 \cdot 4}$$

Rules: (1) $2^p \cdot 2^q = 2^{p+q}$

(2) $\dfrac{2^p}{2^q} = 2^{p-q}$

(3) $(2^p)^q = 2^{pq}$

These rules can be shown to apply for *any* choices of integers for p and q. In a number system such as this the exponents are called logarithms, or more precisely, base-2 logarithms, when the base is 2, as here. One writes

$$\log_2 8 = 3 \quad \log_2 4 = 2 \quad \log_2 1 = 0 \quad \log_2 \frac{1}{2} = -1 \quad \text{etc.}$$

Analogously we may construct a table of the basic numbers in the 10-system and introduce base-10 logarithms:

$$\log_{10}1000 = 3, \quad \log_{10}100 = 2, \quad \log_{10}10 = 1, \quad \log_{10}1 = 0, \quad \log_{10}0, 1 = -1, \text{ etc.}$$

Can we also write numbers such as 17, 145, 5 and 2, 8 in power form? Can we determine (define) numbers x and y such that, for example,

$$13 = 10^x \text{ and } 13 = 2^y$$

in the 10- and 2-systems respectively?

Yes, we can and it has been done. The x and y values in question are called the base-10 and base-2 logarithms, respectively, of the number 13.

For every positive number whatsoever one can determine a base-10 logarithm of a base-2 logarithm or a logarithm in any other base so long as it is positive. We cannot go into how one does this computation, but we can take a look at the graphical equivalent (more or less accurate depending on how well we draw the graphs) in the curves $y = 10^x$ and $y = 2^x$ respectively and the figures show how one graphically finds the logarithms $\log_{10}13$ and $\log_2 13$.

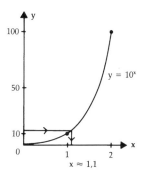

Figure 3.4.4
Graphical determination of $\log_{10}13$

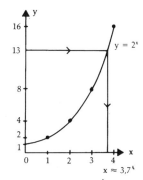

Figure 3.4.5
Graphical determination of $\log_2 13$

If we construct a scale for the numbers in the 10-system alongside a linear scale (equal spacing) for their base-10 logarithms, as in Figure 3.4.6, we call the resulting scale for the numbers logarithmic.

On the slide rule's main scales (usually called C and D and sometimes marked with an x) the numbers go from 1 to 10 and are fitted to a linear scale of logarithms going from 0 to 1 (Figure 3.4.7). [1]

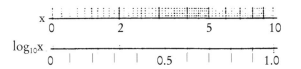

Figure 3.4.6
Logarithmic x-scale

We now return to the world of music and shall get to know something about the difficult task a piano tuner faces.

_____

[1] Nowadays the slide rule is of interest only in connection with the conception of the logarithm and the pupil's understanding of it.

Figure 3.4.7
Slide rule scales

3.4.3 The Piano Tuner's Difficult Task

We know that the musical ear is very sensitive to octaves. The piano tuner must therefore tune pure octave steps; this means physically that the frequency exactly doubles from one tone to the same tone an octave higher.

A violinist, by contrast, tunes his of her four strings in fifths. The "fifth" interval corresponds to 7 so-called half-tones. (See Figure 3.4.8 for an example.)

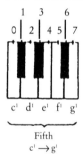

Figure 3.4.8

Counting from c^1, the fifth white key to the right is g^1 which is the next higher fifth.

The violin strings are tuned so that they give g, d^1, a^1, and e^2, forming pure fifth steps upward. A pure fifth step is, second to pure octave steps, that interval which the ear most easily distinguishes. In terms of frequencies the fifth step upward means that the frequency is multiplied by

$$\frac{3}{2} = 1.5.$$

Looking at a piano keyboard we can ask ourselves: how many fifths cover the 7 octaves from A_2 to a^4 or — which is the same thing — from C_1 to c^5?

We can count 7 half-tone steps at a time (equalling a fifth) and find as in Figure 3.4.9 that 7 octaves correspond to 12 fifths.

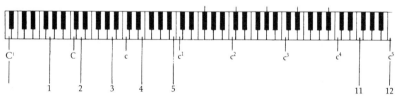

Figure 3.4.9

On the piano, 12 fifth steps (at 7 half-tones each) are equal to 7 octave jumps (at 12 half-tones each).

With A_2 as a starting point we have the frequency 27.5 Hz (cps). After 7 octave-sized steps, that is, after 7 doublings in frequency, we come to

$$a^4 = 128 \cdot 27.5 \text{ Hz.}$$

Let us instead take twelve fifth-sized steps from A_2. Here we get the frequency of a^4 to be

$$a^4 = 1,5 \cdot 1,5 \cdot 1,5 \cdot \ldots \cdot 1,5 \cdot 27.5 \text{ Hz or approximately}$$
$$a^4 \approx 129.75 \cdot 27.5 \text{ Hz}$$

But 129.75 $\cdot$ 27.5 must surely be bigger than 128 $\cdot$ 27.5! The twelve pure fifths give

$$a^4 \approx 129.75 \cdot 27.5 \approx 3568.1 \text{ Hz}$$

while the seven pure octaves give

$$a^4 = 128.27 \cdot 5 \approx 3520 \text{ Hz.}$$

How can this be reconciled? In no other way than that the piano tuner, among other things, makes all the fifths a little lower. His difficult task is in fact to tune the fifths (and other intervals) just impurely enough. When the degree of impurity is just right, the piano sounds good!

We have just seen that the power term $1,5^{12}$ gives the value 129.75 instead of 128. How large should the fifth step be, if not 1.5?

If we call this unknown, slightly lower fifth step "x", then we have the equation

$$x^{12} = 128$$

that is $\quad x \cdot x \cdot x \cdot x \cdot x \cdot x \cdot x \cdot x \cdot x \cdot x \cdot x \cdot x = 128$

The solution to this equation gives the slightly lowered fifth step to be

$$x \approx 1.498.$$

This fifth is called the equal-tempered fifth.

One fifth up from a^1 is e^2. If this is a pure fifth, the frequency of e^2 will be 1.5 $\cdot$ 440 = 660 Hz. If it is tempered, the frequency is 1.498 $\cdot$ 440 Hz or 659.12 Hz.

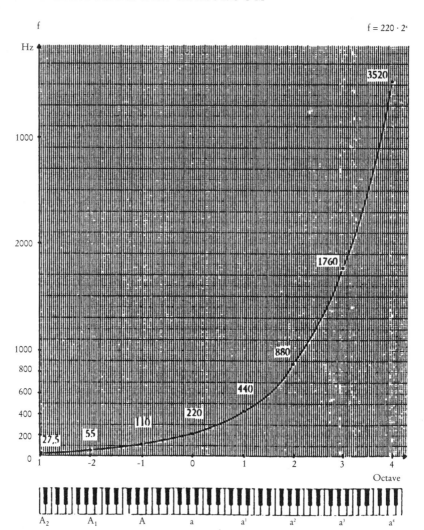

Figure 3.4.10

However, it is not only the fifths which need to be tempered but other intervals as well. And finally it is the half-tones, all of which must be the same size and adjusted so that 12 half-tone steps give an octave. The piano is then "tuned to equal temperature." The equal-tempered piano was introduced during Bach's time.

Thanks to tempering it became possible to go relatively freely between one key and another. With time a new musical form developed beside the older forms (classical, romantic and others): twelve-tone music. Before tempering, each key was, so to speak, walled-in, and modulating between keys caused problems. As a tribute to tempering Bach wrote the well-known collection of preludes and fugues which carry the title "Das wohltemperierte Klavier" (The equal-tempered piano). Many of us probably recognize the relatively easy-to-play Prelude in C major with which the collection begins. Diagram 3.4.10 illustrates graphically how frequency "grows" on the equal-tempered piano.

3.4.4 Some Psychological Observations

If we hit the keys for A_2, A_1, A, a, a^1, a^3 and a^4 on the piano, we hear "the same" a-tone "repeated" but at steadily higher octaves. The experience is truly such that we may rightly speak of octave *steps.* We identify the octaves with each other and feel we are taking equal steps upward as far as frequency goes.

If we go to the external, physical counterpart, we have the vibrating string, for example, as reality. It is that which puts the air into motion, and the air works in its turn on the ear drum. This physical "teasing" is transmitted thereafter to the inner ear (the hammer, the anvil, the stirrup, etc.) and contributes in a puzzling way to the perception of the tone which was sounded through space.

Let us once again look at the frequency of the vibrating string as we take the octave steps on the A-keys. How does the frequency increase step by step?

	Frequency (Hz)	Frequency Increase (Hz)
A_2	27.5	
A_1	55	27.5
A	110	55
a	220	110
a^1	440	220
a^2	880	440
a^3	1760	880
a^4	3520	1760

We see that the increases get bigger and bigger. From A_2 to A_1 the increase is 27.5, from a^3 to a^4 it is 1760 Hz. How does it come to be that we identify these increases with one another?

In the table above, the increases are calculated as absolute, additive increments. If we instead ask about the *ratios* of the frequencies, the answer is that the ratio for the octave step is always determined by the proportion 2:1. (We could, of course, also say that the *relative* increase is always 100%.)

Our musical ear thus relates to and senses the ratios of the frequencies. Our perception does not follow passively the frequencies which the physical stimulation presents to the musical ear.

How do we react to stimulation levels in other kinds of perception? This is a question which fascinated E.H. Weber (1795-1878), professor of anatomy and physiology in Leipzig.

Weber carried out measurements in what we today call experimental psychology. We will consider a few simple examples here in order to illustrate what Weber discovered:

1. We experience not only tone highness and lowness in music but also tone loudness. We have a special knob or button on the radio receiver for "volume," with which we make the sound louder or softer. More and more has noise become a subject of research. We hear now and then of noise damage to the ear, read of sound levels given in decibels, etc. The question now is: how do we *experience* physical sound level increases? Suppose that we hear a jack-hammer from street work in the distance. After a while there are two jack-hammers in operation simultaneously. If one more jack-hammer, a third, is now put into operation we will experience the sound level increase as less than the previous increase, despite the fact that the increase in sound energy must, of course, be the same as the earlier increase.

But if *two* new drills had started up, so that the number of machines had doubled again, then the new increase would be experienced as equal to the earlier, first increase.

In the case of sound level, too, it is the relative increase in the external stimulation (within certain limits) which is decisive for our comparisons.

2. During the time of classical Greece the stars were divided into so-called orders of magnitude. The name is a misnomer, since it is not a

question of the size of the stars but of their apparent brightness, i.e. of the light intensity which we experience.

The brightest were classified to have first order of magnitude, brighter than those of the second order of magnitude, etc. The Greeks used six orders of magnitude for the stars they could see. Stars of the sixth order are quite dim.

In cameras usually a light intensity meter is built in. Earlier it was common to use a separated light meter held in the hand. When astronomers began to measure the intensity of the starlight reaching earth, they soon found that the step from one order of magnitude to the next lower order of magnitude (i.e. to the next *brighter* group) quite accurately corresponded to a factor 2.5-fold increase in brightness of the light.

In other words, if the sixth order of magnitude has light intensity L, then the intensity of

order 5 is	$2.5L$	$=$	$2.5L$
order 4	$2.5 \cdot 2.5L$	$=$	2.5^2L
order 3	$2.5 \cdot 2.5^2 \cdot$	$=$	2.5^3L
order 2	$2.5 \cdot 2.5^3 \cdot$	$=$	2.5^4L
order 1	$2.5 \cdot 2.5^4 \cdot$	$=$	2.5^5L

We once again meet a physical multiplicative factor.

In more recent times orders of magnitude have been defined relative to each other by setting 100 as the ratio of first order light intensity to sixth order intensity. Instead of the five steps from the 6th order up to 1st order, having a factor of

$$2.5^5 \text{ (see table above)} \approx 97.66,$$

they are now given the brightness ration of exactly 100. In order to agree the factor 2.5 changes slightly, becoming about 2.51. (The exact value is the solution to the equation

$$x \cdot x \cdot x \cdot x \cdot x = 100$$

and is written $\sqrt[5]{100}$.

3.4.5 Weber-Fechner's Law or the Psycho-Physical Law

Is it possible to measure inner soul experience, in particular, our perceptions? This and similar questions have undoubtedly been asked and probed into by more than a few minds through the ages.

Galileo strongly advocated the view that science must be based on measurement. His expression, "That which is not measurable must be made measurable" may seem paradoxical if one does not understand it in the sense of: "that which today is not yet measurable must be made measurable in order to be amenable to research." Galileo was one of those who first divided the senses into primary sensory qualities (perception of number, length, shape, etc.) and secondary sensory qualities (experience of color, taste, smell, etc.).

Is it possible to give our experience of tone level or light intensity a certain quantitative or perceptive value? For example, can one set up a perception scale for loudness, as has been done for temperature, length, weight, etc.? Where, in that case, should the zero-point be set? Or must we limit ourselves to comparing different degrees of perception with one another – if that is at all possible?

Such questions engaged Gustav T. Fechner about the year 1850. Fechner had had a dazzling academic career, becoming Professor and Chairman of the Physics Department at the famous University of Leipzig at the age of 33 (in 1834). But he apparently overworked himself in a variety of ways and left his teaching chair in 1839.

Fechner attacked his new research with great energy. After publishing a few short papers in 1858 and 1859, he gave out his book "Elemente der Psychophysik" in 1860, a famous work on "the exact science of the functional relation between body and soul."

Did Fechner succeed in capturing perceptions of the soul as quantities, on scales? The answer is surprisingly more-or-less "yes," if we limit ourselves to a few areas, primarily tone level, loudness and brightness – those areas we have already touched upon. We have actually already learned that which Fechner stated in his psycho-physical "law", that perceptive level grows with equal steps when the physical stimulus grows multiplicatively in equal steps; or, expressed differently, when the stimulus grows relatively in equal steps.

Such a "law" cannot, of course, be expected to apply over an unlimited range of perception. For one thing, the stimulus (tone frequency, loudness or light level) must be strong enough that we clearly perceive something above the so-called stimulus threshold. Secondly, the stimulus must not be so strong that we begin to approach the level of pain or other reactions which disturb our senses.

We cannot here go into a thorough study of Weber-Fechner's law and the criticism of it. Instead we will use a graphical presentation to give a clearer picture of that which Fechner asserted.

In Figure 3.4.10 earlier, where a curve is drawn of tone frequency, we used common linear scales on both axes. How does the diagram look if we replace the vertical frequency scale with a *logarithmic* scale but let the horizontal x-axis remain linear? (For reference to the concept of logarithmic scale, see Section 3.4.2.)

For this we mark off the frequency values for the octaves with *equal* spacing along the vertical axis; the length of the interval can be chosen arbitrarily. How do the points now lie?

We set out the points for -3 and 27.5, for -2 and 55, etc. and finally for 4 and 3520. We note that these points now lie in a straight line. For every step sideways the point moves vertically by the same constant amount, equal to the distance chosen for the vertical

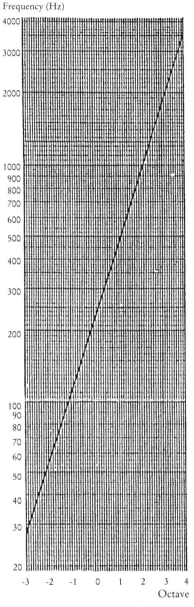

Figure 3.4.11
Graph on logarithmic paper

axis octave markings (see Figure 3.4.11). The octaves correspond to the equal distances between points on the line.

Noise research has grown in importance in recent years, especially that aimed at making homes and places of work more humane, so that people at least avoid damage to their hearing. Measurement of loudness was adapted to the fact that our perception of sound level basically follows the Weber-Fechner law: the loudness was set according to our perceptions and made logarithmic. The scale's unit is called decibel (db). In order to understand how the decibel scale is constructed we shall here make up a new scale for frequencies. We use a^1 as a base and call its frequency 50 "Euler" (honoring the mathematician Leonard Euler, who presented the psycho-physical material concerning tone intervals and frequency). Further, we let 10 Euler correspond to one octave step.

Then a^2 will have the frequency 60 Euler, a^3's frequency will be 70 Euler, etc. Going in the other direction we get

$$a = 40 \text{ E} \quad A = 30 \text{ E} \quad A_1 = 20 \text{ E} \quad A_2 = 10 \text{ E}$$

and arrive at the following table of comparisons:

Tone	Frequency in Hz	Frequency in E
a^2	880	60
a^1	440	50
a	220	40
A	110	30
etc.		

What frequency in Hz would correspond to the Euler frequency 0? Apparently that frequency which is one octave under A_2, so we get

$$0 \text{ E} = 0.5 \cdot 27.5 \text{ Hz} = 13.75 \text{ Hz},$$

and further

$$-10 \text{ E} = 0.5 \cdot 13.75 \text{ Hz} = 6.875 \text{ Hz, etc.}$$

The Euler scale number goes toward negative infinity, as the number of vibrations per second decreases toward zero (see Figure 3.4.12). We have by then left the world of tone, since our perception of tone stops below about 20 Hz. The decibel scale for loudness is constructed in a manner completely analogous to our construction of the Euler scale here. 10 decibels means a doubling of loudness.

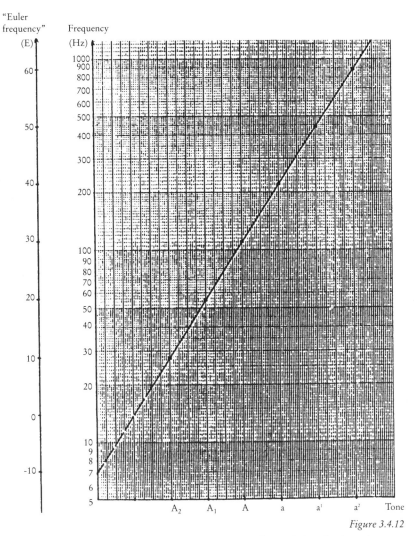

Figure 3.4.12

In summary of this look into our soul's activity during the perception of tones, sound level, and light intensity, we may say, graphically speaking, that we unconsciously transform a linear scale to a
logarithmic one. Those somewhat familiar with the concept of logarithms will understand another formulation of the Weber-Fechner law:
in the process of perception we take the logarithm of the intensity of
the stimulus.

3.4.6 A Little Population Mathematics

When youth are called into the army, their intelligence, among other things, is tested. Some of the test problems have been arithmetic of the following type:

Example 1 A numeric sequence begins with the numbers
3, 6, 9, 12 ...
What comes after 12?

Example 2 What number comes after 3, 6, 12, 24, ... ?

I have chosen two simple examples. We answer with 15 and 48 respectively. More correct would be to say: the *simplest* continuation gives 15 and 48. In the first example we observe a constant increase (3), in the second, a doubling of the numbers, step by step – the sequence has the multiplicative factor of 2. We might remind ourselves here of Moser's circle subdivision problem in Section 3.1 where we successively obtained the numbers 1, 2, 4, 8, 16 and were inclined, perhaps even took it as obvious, to predict 32 as the next following number. In actual fact the next number after 16 was 31.

If in advance we state that we consider only sequences with constant increases (or decreases), or with constant multiplication factors, we may unambiguously state the continuations in the following examples:

Example   3a)   7, 11, 15, ...
          3b)   23, 16, 9 ...
          3c)   80, 20, 5 ...
          3d)   2, -6, -14, ...

(It would have been sufficient to give only the first two numbers.)
The numbers are 19, 2, 1.25, and -22 respectively.

Sequences with constant multiplication factor (Examples 2 and 3c) are called geometric sequences. If the sequence on the other hand has a constant increase (Examples 1 and 3a) or a constant decrease (i.e. a negative increase, as in Examples 3b and 3d) then it is called an *arithmetic* sequence.

Figures 3.4.13a and b illustrate these two types of sequences geometrically (see also Section 3.7.2).

How does the population of a city, a country, a continent, or of the world grow?

This is always a current question. More than anyone else Thomas Robert Malthus (1766-1834) has become known for his investigation of this question.

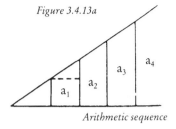

Figure 3.4.13a

Arithmetic sequence

Malthus, who studied at Cambridge, was ordained as a priest in 1798 and that same year gave out a book called "An Essay on the Principle of Population As It Affects the Future Improvement of Society," which brought sharp criticism and started a lively public debate in England.

The book was published anonymously and directed toward an optimism for the future which had come to expression through the anarchist Godwin. The Liberals had introduced poor-laws in England in 1796 to create a better society. Such a measure would, according to Malthus, son lead to worsened conditions for the poor. The supposed improvements

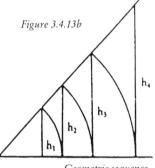

Figure 3.4.13b

Geometric sequence

would stimulate increased nativity and after a time the growth in the numbers of poor would lead to new and even more difficult times of need.

Malthus noted a number of factors which, according to him, acted to limit the population increase: poverty, sickness, and war. If such inhibitors do not exist the population continues to increase geometrically, i.e. the number of people forms a geometric sequence, when equal points in time are taken.

Malthus' starting point was that "a population of a thousand million doubles just as easily in 25 years as a population of a thousand."

> In the Northern States of America..., the population has been found to double itself, for above a century and a half successively, in less than twenty-five years. — In the back settlements, where the sole employment is agriculture, and vicious customs and unwholesome occupation are little known, the population has been found to double itself in fifteen years.
>
> But the food, which shall support the increase of the greater population, is by no means obtained easily.

Here we come to the main thought in Malthus' presentation: while population increases geometrically, food production can only increase arithmetically.

> ... supposing the present population equal to a thousand millions, the human species would increase as the numbers 1, 2, 4, 8, 16, 32, 64, 128, 256, and subsistence as 1, 2, 3, 4, 5, 6, 7, 8, 9. In two centuries the population would be to the means of subsistence as 256 to 9; in three centuries as 4096 to 13; and in two thousand years the difference would be almost incalculable.[2]

In other words: if a population of 1 billion people today divides up 1 unit of food stuff among themselves, then 256 billion people 200 years later would have 9 units of food to live upon. Poverty would increase enormously – if none of the inhibiting factors of hunger, sickness, and war were to slow the increase down.

JAPAN	
Year	Mill. inhab.
1850	27
1870	32
1890	40
1900	44
1910	50
1920	56
1930	64
1940	73
1950	84
1960	94
1970	104
1980	117

Criticism led Malthus to collect empirical evidence for a new edition of the book. It was published in 1803 after Malthus had undertaken study trips to Germany, Sweden, Norway, Finland, and Russia. He stood firm by his basic hypothesis in the new edition but pointed out that other factors than hunger, sickness, and war could hold back the population increase and improve the supply of the means of subsistence. As examples of favorable positive factors to raise the minimum level of existence Malthus suggested late marriage, voluntary restraint, and the formation of new habits.

Does the hypothesis that a population grows geometrically in equal periods (as long as living condi-

---

[2] The meaning here must be that the fraction $2^n/(n-1)$ would be incredibly large.

tions are good) hold true for any country or region of the world? Let us look at Japan:

Has population increased in geometric sequence in Japan over the span 1850-1970 or during any part of that period? (See table). Diagram 3.4.14 shows graphically how Japan's population has increased. In order to study the growth of population we will acquaint ourselves with a graphical test method which reveals very simply whether a growth is geometric or not.

Semi-Log Diagram

In Section 3.4.2 we introduced the concept of a logarithmic scale. On such a scale a geometric series is compressed so that the numbers in the geometric series are put down with equal spacing. What does the curve of a geometric sequence look like on a diagram where the horizontal scale for the order in the sequence is linear, while the vertical scale for the numbers themselves is logarithmic? Such a diagram is said to be semi-logarithmic. Graph paper is available with pre-printed scales to use. This paper is called semi-log graph paper.

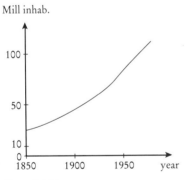

Figure 3.4.14

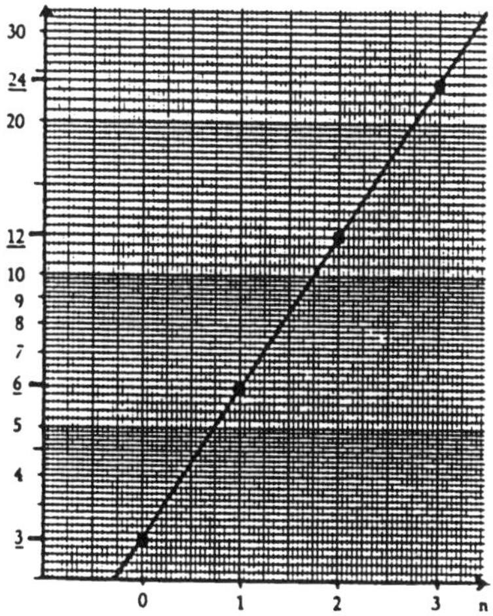

The result may be seen in Figure 3.4.15. The curve is, quite simply, a straight line. Geometric

Figure 3.4.15

growth gives a straight line in a semi-log diagram. This is a practical method of graphically testing whether or not a given growth is geometric.

In Figure 3.4.16 we see the population of Japan on a semi-log plot. The diagram in large measure lends support to the hypothesis that

Mill inhab.

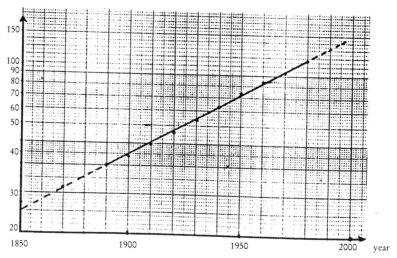

Figure 3.4.16

growth has been geometric. The results are not equally convincing if one studies, in the same way, the population of the United States for the same period...

Malthus is still very timely. In a newspaper account from the International World Population Conference in Bucharest 1974, could be read, among other things, the following:

> The ghost of the English doomsday prophet Malthus hung over the congregation after the opening speech.... Malthus became a weapon at the conference for those who wanted to get at the developmental pessimists and defenders of the privileged. China and every one of the attending African and Latin American nations quickly shot at what they thought were "neo-Malthusian" tones in the plan. This group considered that the plan, developed by the United Nations Commission on Population, placed far too much emphasis on family planning.

It goes without saying that Malthus can spark strong interest in a tenth-grade class.

3.4.7 Exercises

1. What are the fourth and fifth terms in the arithmetic sequence

$$1 \quad 3,4 \quad 5,8 \dots ?$$

2. What are the fourth and fifth numbers in the geometric sequence

$$5 \qquad 8 \qquad 12.8 \dots ?$$

3. The numbers 12, 6, and 3 are the beginning of a decreasing geometric series. Which numbers follow?

4. The angles of successive swings of a pendulum form a decreasing geometric sequence. Determine the angle of the sixth swing if the first three swings are 15°, 12°, and 9,6°.

5. Figure 3.4.17 shows a graphical method of constructing a geometric series using only compass and straight edge. The n-lines are drawn vertically from the base line g up to the line a. Circular arcs are rotated from a down to g, using O as pivot center. The distances from O to the points along g now form a geometric series. The first two numbers in the sequence in the figure

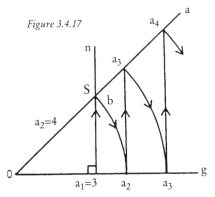

Figure 3.4.17

are 3 and 4. First the distance 3 cm is marked off from O to the line n with the aid of a compass. This intersection along n determines a's slope. Line a can now be drawn, and the method of constructing a geometric series with 3 and 4 as starting numbers may proceed.

Letting 5 and 8 be starting values, determine graphically the fifth number in the sequence. Compare with the results in Exercise 2 above.

6. A mirror is found to reflect 95% of the incoming light. What percent of the light is lost if light is reflected successively in 4 such mirrors?

7. Russia's population grew as follows over the period 1850-1970:

Year	1850	-70	-90	1900	-10	-20	-30	-40	-50	-60	-70
Million people	61	75	99	111	140	134	156	174	181	214	245

Illustrate this growth with a plot. What can be observed between 1910 and 1920?

Now illustrate the population growth on a semi-log plot. If the curve obtained is a straight line for any period of time, it indicates that population growth was exponential (geometric) during that time period.

3.5   The Step from Arithmetic to Algebra

3.5.1 Why "Algebra"?

We use the word "algebra" here in the limited sense of arithmetic with letters and "arithmetic" in the sense of numerical calculations.

Many pupils find calculation with letters a, b, c, x, y, etc. to be abstract. For the majority numerical arithmetic is considerably more concrete, and it is truly an important pedagogic problem of how to introduce arithmetic with letters. Often — for many pupils — algebra seems unnecessary. The teacher will certainly hear the question, "What's this good for? Does it have any use?"

Such a reaction is quite natural if the pupils have not at a rather early stage — as soon as the prerequisites exist — had an experience of the "power of algebra." ·

Let us proceed directly to a problem concerning numbers:

We choose four arbitrary numbers — for simplicity's sake one- or two-digit numbers — and write them beside one another with a little space between, on the blackboard. Suppose we have chosen the numbers 3, 11, 4, and 17. We now add adjacent numbers two at a time and write their sums one row down in the spaces in between:

$$
\begin{array}{cccc}
3 & \quad 11 & \quad 4 & \quad 17 \\
& 14 & \quad 15 & \quad 21
\end{array}
$$

We repeat this procedure and get

	3		33		4		17
		14		15		21	
			29		36		

Finally we do the last sum, which we call the *bottom number*:

	3		11		4		17
		14		15		21	
			29		36		
				65			

Our choice of initial numbers led to an *odd* bottom number.

The question now is: can we find some simple rule with the help of which we can predict whether the bottom number will be odd or even, as soon as we know the four starting numbers?

For example, if they are 1, 9, 16, and 8, will the bottom number be odd or even?

Without exception the students add up the sums to see what bottom number appears. They choose their own examples to try out, and I ask them to be alert concerning the results so that we can come up with a theory, at least a guess, from our collection of examples.

Someone soon points out – perhaps immediately – that if all the given numbers are even, then the bottom number will also be even. The same applies if all the numbers are odd, someone adds. And now groups of pupils go exploring in different directions.

Some groups investigate the effect of the number of even numbers in the beginning set of four. Other groups investigate if the sum of the given numbers is of any use. Someone says that they have discovered that the bottom number is even, if the sum of the initial numbers is even. Will the bottom number be odd, in this case, if the initial numbers' sum is odd? The search becomes more intensive. The conviction that the sum of the initial numbers determines the even-oddness of the bottom number grows stronger. "All our examples thus far agree with that," point out some pupils and they begin to consider the result as certain.

We gather together again as a class and listen to this theory about the sum of the initial numbers. "How many examples would you deem it necessary to exhibit, in order for the theory to be proved?" I once asked in a ninth-grade class which had not seen the problem before and which was not especially advanced in algebra. The class took a long "think" and first after a pause came several cautious opinions. Those who answered believed that four or five examples would be sufficient. This became a lovely prelude to a discussion of what is meant by a proof and of the value of a few examples versus a great number of examples. It soon became clear to the pupils that not even a million examples which all agree are sufficient as a proof – despite the strong conviction we might then have. We even got into a discussion of the value of experiment in science versus "experiment" in mathematics.

3.5.2 We "Experiment" Further

There also came up examples of experiences with bottom numbers for the case when some of the given numbers were even (or odd). Before we seriously took on the task of proving one or the other of our theories we were tempted to "experiment" using 5 starting numbers. If we had 5 given numbers would examples also indicate that the sum of the numbers determines the bottom number's quality? Several examples at first seemed to confirm this, for example 1, 5, 8, 9, 11 with the row sum 34:

$$
\begin{array}{ccccc}
1 & 5 & 8 & 9 & 11 \\
& 6 & 13 & 17 & 20 \\
& & 19 & 30 & 37 \\
& & & 49 & 67 \\
& & & & 116
\end{array}
$$

If the bottom number is determined solely by the top row sum, then the "even" quality should be maintained if we exchange places in the top row, e.g. the 1 and the 8:

$$
\begin{array}{ccccc}
8 & 5 & 1 & 9 & 11 \\
& 13 & 6 & 10 & 20 \\
& & 19 & 16 & 30 \\
& & & 35 & 46 \\
& & & & 81
\end{array}
$$

But here the bottom number is odd!

We have found a counter-example to the theory that the bottom number's even-oddness is determined by the sum of the five given numbers. One single example is enough to overthrow a theory!

But what might the rule then be, if there is a simple rule?

There were still hopeful students who sought for some special rule. A few came with the logical and promising idea that we could try to relate the case with 5 numbers back to the earlier case with four numbers, since the first row of additions gives 4 new numbers. But then studying 6, 7, 8, or more given numbers with successive "backtracking" downwards would be laborious to carry out and difficult to summarize.

Why not try putting letters in place of the numbers?

We decided to go back to the case with 4 given numbers and to call these a, b, c, and d. We formed the sums row by row and obtained the following triangle:

$$a \qquad b \qquad c \qquad d$$
$$a+b \qquad b+c \qquad c+d$$
$$a+2b+c \qquad b+2c+d$$
$$a+3b+3c+d$$

We have thus obtained the bottom sum: B = a + 3b + 3c + d.

Does the even-odd quality of the sum s = a+b+c+d determine the quality of this expression, B? We compare B and s and find that

$$B = s + 2b + 2c$$
$$\text{or} \qquad B - s = 2(b + c) \qquad\qquad (1)$$

Here we see that 2(b + c) is always an even number — it is divisible by 2. Equality (1) tells us that B and s are *simultaneously* even or odd, i.e. either both are even or both are odd. We have hereby come to a clear and proven conclusion for the case with 4 initial numbers:

the bottom sum is even or odd when the sum of the initial numbers is even or odd, respectively.

The class wants now to try out this effective method for the case when we have 5 numbers to start with. What bottom number will we then get, and what conclusion will we be able to draw?

We must now start with 5 letters, each representing a number, any letters we like, say a, b, c, d, and e. Without great difficulty we come up with the bottom sum

$$B = a + 4b + 6c + 4d + e$$

Now what kind of rule can we get out of this expression? It turns out to be a difficult question for the majority. At this point many simply copy, without a closer thought, the comparison between B and s which we did earlier and which led to equality (1). They arrive at the relation

$$B = s + 3b + 5c + 3d$$

but come no further, since the sum $3b + 5c + 3d$ is even for even numbers and odd for odd b, c, d.

But a glance shows us that the rule is hidden within B itself :

$$B = a + 4b + 6c + 4d + e$$

which says that $B = a + e +$ an even number, namely $2(2b + 3c + 2d)$.

With this it is clear that $a + e$, i.e. the sum of the outer numbers determines the bottom number's quality when we have five numbers initially.

The rule is simpler when we have 5 numbers than when we have 4!

The class usually gets so involved that they want to go on: does the rule get even simpler when we have 6 numbers to start with? Or do we get a rule similar to the case with 4 numbers?

The results are once again surprising! And give further opportunity for wrestling with questions towards a more general investigation. Several examples of different students' investigations and partial results are found in the exercises following.

But *one* observation shall be noted here, before we leave the bottom numbers:

2 starting numbers give			$B = a + b$
3 "	"	"	$B = a + 2b + c$
4 "	"	"	$B = a + 3b + 3c + d$
5 "	"	"	$B = a + 4b + 6c + 4d + e$
6 "	"	"	$B = a + 5b + 10c + 10d + 5e + f$
etc.			

Do we perhaps recognize the numbers which appear in the respective rows?

```
              1        1
          1        2        1
      1        3        3        1
   1      4        6      4        1
1       5      10       10       5       1
```

Why, yes, they are actually the same as the numbers in Pascal's triangle! Can we obtain B so simply, by going to Pascal's triangle? (Exercise 2).

3.5.3 The Rule of 9's

Pupils usually learn a rule in grade 4 or 5, of which many of us have heard. It is the answer to the question: when is a whole number evenly divisible by 9? The rule says: a number is divisible by 9 when the sum of the digits is divisible by 9 – and only then. (The corresponding rule applies for 3).

A couple of examples: Is 2169 divisible by 9?
Sum of the digits: $2 + 1 + 6 + 9 = 18$,
divisible by 9.
the number 2169 is therefore divisible by 9.

And if we take 31,478?
Digit sum: $3 + 1 + 4 + 7 + 8 = 23$, *not* divisible by 9, so neither is the number.

How can we prove such a rule?

Let us first limit ourselves to 3-digit numbers. We suppose that the 3 digits in the number are a, b, and c. How would the number be written? abc? No, since abc in algebra means $a \cdot b \cdot c$ – we don't write out the multiplication sign between letters. 538 certainly does not mean $5 \cdot 3 \cdot 8$ but rather 5 *hundred thirty* (three ten) eight, i.e.

$$500 + 30 + 8 \text{ or more clearly } 5 \cdot 100 + 3 \cdot 10 + 8$$

How shall we then write our number?

Correct: $n = a \cdot 100 + b \cdot 10 + c$
or more nicely $n = 100a + 10b + c$

And the sum of the digits? Correct answer: $s = a + b + c$

How can we compare n with s?

We can perhaps write them underneath each other: $n = 100a + 10b + c$

$$s = \quad a + \quad b + c$$

and see that the *difference* between them is $n - s = 99a + 9b$

$$\text{or} \quad n - s = 9\,(11a + b), \quad\quad (2)$$

a number which is *always* divisible by 9.

We "solve" for n and get from (2)

$$n = s + 9\,(11a + b) \quad\quad (3)$$

What does equality (3) tell us?

It says: if s is divisible by 9, then n will also be divisible by 9. But is n divisible by 9 *only* when s is divisible by 9?

Formulated differently: when n is divisible by 9, must s also be so?

We return to (3) and solve for s, getting

$$s = n - 9\,(11a + b) \quad\quad (4)$$

Equality (4) tells us that s is divisible whenever n is – and only then. With this we have completely proven the rule for 3-digit numbers. Since the argument may be carried through analogously with an arbitrary number of digits we understand that any number is divisible by 9 whenever its digit sum is divisible by 9.

3.5.4 *The Example as Teacher*

One day on the Uppsala-Stockholm commuter train I was witness to a dispute between two young men, university students — let us call them A and B — who had differing opinions as to the speed of the train. "Why don't you just calculate the velocity; you've got a calculator?" said A. "The distance is 66 km, and the travel time is 40 minutes." "You'll see I'm right," said B and took out his calculator. He began to punch keys. But no result seemed to be forthcoming. We passed Marsta, then Upplands Vasby and came into the Solna tunnel near Stockholm. I began to guess that B had searched his memory in vain looking for a formula containing distance, time, and velocity (s, t, and v) and that now he was simply experimenting with his calculator.

In such a situation one may come to one's own aid with the help of a simple example, where the three qualities can be seen in relation to each other.

For example: how far does a train travel in 3 hours if it goes 80 km/hour? The train obviously covers a distance of $80 \cdot 3 = 240$ km.

From this simple example we see directly the relation

distance = velocity times time

(assuming that velocity is constant, of course).

In a short version: $\qquad s = v \cdot t \qquad$ or $\qquad s = vt \qquad$ (1)

We need not store this formula in memory. We can have faith in our imagination to find a concrete example and in our thinking ability to extract the formula from the example, once again.

By the very act of writing formula (1) we take the step from numeric example to algebra, limited however, in the sense that we use letters as abbreviations for quantities which occur in the formula. What do we do when velocity is unknown as in the dialogue above?

One might then take the example that a train covers 90 km in 2 hours. How fast did it go?

Naturally $90/2 = 45$ km / hour (most likely a freight train).

It matters not at all if the results of our home-made examples are a little far-fetched. The main thing is that we set our thinking in motion through the choice of concrete, numerical values.

This time we find that

$$\frac{distance}{time} \qquad \text{or} \qquad v = \frac{s}{t} \qquad\qquad (2)$$

If one brings to mind that speeds for trains and cars are usually given in km / hour, one sees that it must be a question of dividing a distance by a time. Those who are used to working with letters need not formulate more than the first example: from (1) we can easily derive formula (2). If in fact we divide both sides of (1) by t we will get

$$\frac{s}{t} = \frac{v \cdot t}{t} \qquad \text{and after reducing} \qquad \frac{s}{t} = v.$$

If we instead divide (1) by v on both sides, we get a third relation,

$$\frac{s}{v} = t.$$

Formulas (2) and (3) are just reformulations of the basic relation in (1).

There are many examples of relations of this kind. A few are:

Area of a rectangle = length x width $\qquad$ A = l · w
Mass = density x volume $\qquad$ m = d · V
Voltage = resistance x current $\qquad$ E = RI (Ohm's law)
Cylinder volume = base area x height $\qquad$ V = B · h

An example can be of value not only when we are looking for a formula. It can sometimes also give us the solution to a problem which is broader than the original task.

Let us look at the following problem, which we imagine using in a sixth or seventh grade class:

What kind of sums do we get when we add three consecutive integers? For example, 23 + 24 + 25, 6 + 7 + 8, etc. We ask each pupil to give at least one example, and take the time to write down all the totals which the pupils obtain. It may happen now and then that an incorrect sum gets in, but this only makes the exercise more interesting. The list might begin like this:

45, 95, 12, 156, 51, 15, 78, 222, ...

Do these numbers have anything in common?

The class soon discovers that the sums are divisible by 3 and are entirely convinced that they *always* can be divided by 3. Might there not be exceptions? How could we prove the divisibility?

At this point it is usually a great help to examine *one* example thoroughly. Someone has taken 13 + 14 + 15 and obtained 42. Might we from the very arrangement itself, 13 + 14 + 15, be able to predict that the total is divisible by 3?

The class begins thinking. We wait with our answers so that everyone has time to think. Soon several students quietly come up and show on paper an idea that they have. For example, 15 lies just as much above 14 as 13 lies under. The sum is therefore 3 · 14, that is, it is divisible by 3. The moment we see the correctness of that idea we understand that it applies *generally:* the same argument can be applied to any sum of 3 consecutive integers. The sum must therefore always be divisible by 3.

We see here that thinking about a particular example can be so fruitful and enlightening that we arrive at a general result.

It is no great task afterwards to clothe the reasoning in algebraic dress: letting the middle number be N, the sum is

$$N - 1 + N + N + 1 = 3 N$$

The class might now be given the task of investigating if similar results can be obtained for sums of four, five, or six integers.

3.5.5 Magic Hexagons

In the lower grades the pupils have eagerly added rows, columns, and diagonals in "magic squares" and perhaps themselves constructed a simple magic square, for example a square with 9 boxes containing the numbers 1 through 9, in which the rows and the columns and even the diagonals all add up to the same sum.

8	1	6
3	5	7
4	9	2

Albrecht Dürer constructed a square with 4 x 4 boxes, containing the numbers 1 through 16, which, by the way, was done in such a way that two adjacent boxes in the bottom row gave the year of construction, 1514.

16	3	2	13
5	10	11	8
9	6	7	12
4	15	14	1

There are many books available which can acquaint us with magic squares of even larger size. Can one construct a square with only 2 x 2 boxes containing four whole numbers? That the numbers 1, 2, 3, and 4 are not a solution requires no great effort to discover. But can we succeed with four other integers? We return to this problem in Exercise 3.

An American office worker, Clifford W. Adams, posed himself the problem of constructing magic hexagons (6-sided figures). We call a single hexagon a hexagon of the first order, a group of 7 hexagons as in Figure 3.5.1 is called a hexagon of the second order and a group of 19 hexagons as in Figure 3.5.2 a third order hexagon. Is it possible in Figure 3.5.1 to put the first seven numbers 1, 2, 3, … 7, and in Figure 3.5.2 to put the 19 numbers 1 to 19, so that one and the same sum is obtained when one adds the numbers in the boxes horizontally, and along each of the two directions for diagonals? (Figure 3.5.3)

Figure 3.5.1
Hexagon of the second order

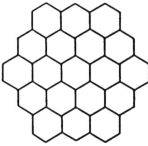

Figure 3.5.2
Hexagon of the third order

If we let a,b, and c be three numbers placed in adjacent boxes in a second order hexagon, as in Figure 3.5.4, then we are to have the row sum a + b be equal to the diagonal sum b + c. Is this possible? What conclusion can we draw from the equality

$$a + b = b + c?$$

The equality applies only if a = c, which means that two boxes contain the same number. Since no number may appear more than once, we see that a magic hexagon of the second order does not exist. But perhaps one exists of third order, with 19 boxes?

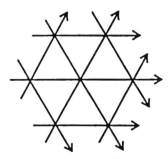

Figure 3.5.3

Adams began his investigations, according to an article in *Scientific American* (No. 8, 1963), the year 1910. He proceeded by trying to place the numbers 1, 2, 3, … 19 in the boxes in different ways. For a long time he did not succeed, but in 1957, after 47 years, the retired Adams found a solution. It must have been a shock for him when some time later he discovered that he had misplaced the paper with the

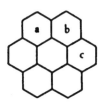

Figure 3.5.4

magic hexagon he had constructed and that he would not remember the placement of the numbers. Should he begin again trying out different combinations, or should he try to find the paper? He looked long and in vain for the paper, but finally found it five years later! The solution then came to Martin Gardner's attention. (Gardner was the editor of the mathematics corner in *Scientific American*.) He turned to W. Trigg with a request for a general mathematical investigation concerning the existence of magic hexagons of arbitrary order. Trigg proved in 1963 that no magic hexagon of higher order than 3 exists. Perhaps it was good intuition which saved Adams from the Sisyphean task of looking for a magic hexagon of order 4.

3.5.6 Mathematical Induction

Example 1: When Galileo studied how far a ball rolls down an inclined plane during a specified time, or how far a marble falls in the air during a given time, he came to the conclusion which he formulated in the following way:

The distances which the ball rolls or falls during successive, equally long time intervals are proportional to each other as the odd integers:

$$1 : 3 : 5 : 7 \dots$$

If a ball rolls 1 "bit" during the first second, then
it will roll 3 "bits" during the second second, and
        5 "bits" during the third second, and
        7 "bits" during the fourth second
        etc. (Fig. 3.5.5)

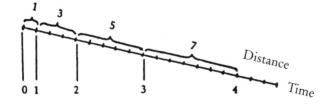

Figure 3.5.5. Galileo's rolling ball

If we now add up the distances from the starting point, we find that the total distance traveled is

1 "bit" in 1 second

$1 + 3 = 4$ "bits" in 2 seconds

$1 + 3 + 5 = 9$ "bits" in 3 seconds

$1 + 3 + 5 + 7 = 16$ "bits" in 4 seconds

etc.

The numbers which determine the total distance are thus 1, 4, 9, 16, ... These numbers are *squares* of 1, 2, 3, 4, etc., i.e. of the numbers which specify the times. Does this pattern continue? And if so, out to infinity? If we leave mechanics and formulate the question purely mathematically, our problem is to find out:

Does the sum of the series

$$1 + 3 + 5 + \ldots + \text{a last odd number}$$

make a perfect square?

$1 + 3$ gives 2^2, $1 + 3 + 5$ gives 3^2, etc. It appears as if the sum is the square of the number of terms.

For example, we would expect

$$1 + 3 + 5 + 7 + 9 + 11 = 6^2 \quad \text{(since there are 6 terms)}$$

We check it: the sum is actually 36.

How can we prove our supposition?

The Greeks found a beautiful geometric proof: Figure 3.5.6 shows a square which is made up of a number of L-shaped right angles ("gnomons"). The number of points in the right angles corresponds to the successive odd numbers. Their sum is the number of points in the whole square. The figure shows directly that $1 + 3 + 5 + 7 + 9 = 5^2$.

Figure 3.5.6

What happens now if we add 11 more? A new, larger right angle piece gets added on, and we see with the help of Figure 3.5.7 that this piece together with the old square makes a *new square* which corresponds to the sum 6^2. If we now continue to add on larger and larger angle pieces, do we always get the next largest square?

The new right angle has two points more than the previous. These two extra points are just

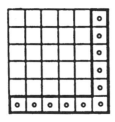

Figure 3.5.7

what are needed so that the larger right angle can go round the corner of the previous square (Figure 3.5.8).

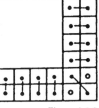

Figure 3.5.8

That brings us to the insight that a square plus a right angle makes the next larger square, step by step.

We can now draw the conclusion that the sum of the old integers always gives a perfect square, taking as many integers as we like.

In summary we may say:

(1) 1 + 3 gives the square of the number 2.

(2) Adding on the next following odd number gives the next larger perfect square.

(3) This can be repeated endlessly.

(4) The sum of the successive odd integers always gives the square of the number of terms.

Concentrating the proof to *one step* which can be repeated an unlimited number of times, starting from a proven beginning point, is the kernel of the elegant means of proof called mathematical induction.

This method of proof was popularized by Blaise Pascal, who even in his early years gave evidence of exceptional mathematical talent (see also 3.2.3). The point in the induction method of proof is that one need not look for a direct proof. In our example we avoid the task of directly summing 1 + 3 + 5 + 7 +...

If we wish to carry out inductive proof purely arithmetically for our odd number series, we can begin by *supposing* that the sum of n terms

$$1 + 3 + 5 + ... + (2n - 1) \quad \text{is } n^2. \tag{1}$$

We know that this is correct for n = 1. (2)

If we add now the next following odd number, 2n + 1, to series (1), the sum will be, according to our supposition

$$n^2 + 2n + 1.$$

Is this a square? Yes, since $n^2 + 2n + 1 = (n + 1)^2$.

If (1) is true, then it is also true that

$$1 + 3 + 5 + \ldots + (2n + 1) = (n + 1)^2 \qquad (3)$$

The *step* of adding the next higher odd integer thus gives once again a sum which is the square of the number of terms. We can now go back to the beginning (2) and carry out the step from (1) to (3) as many times as we like. The induction is complete.

Example 2: Finding a formula for the sums of squares is much more difficult:

$$1^2 + 2^2 + 3^2 + \ldots = ?$$

The first four sums are

$$1^2 = 1$$
$$1^2 + 2^2 = 5$$
$$1^2 + 2^2 + 3^2 = 14$$
$$1^2 + 2^2 + 3^2 + 4^2 = 30$$

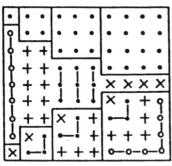

Figure 3.5.9

Compare with the figurated numbers in Figure 3.1.4.

It is interesting that the Babylonians succeeded in finding a formula for the sums of squares and also gave a geometrical proof by induction for the formula. The method is in principle the same as the Greeks' gnomon method, but the level of difficulty is considerably higher.

Figure 3.5.9 shows how the Babylonians set out the dots (points) of the first four squares in a dot rectangle and by this means came to the formula:

$$1^2 + 2^2 + 3^2 + \ldots + n^2 = \frac{1}{3}(1 + 2n)(1 + 2 + 3 + \ldots + n) \qquad (4)$$

The figure shows that

$$1^2 + 2^2 + 3^2 + 4^2 = \frac{1}{3}(1 + 2 \cdot 4)(1 + 2 + 3 + 4) = 30.$$

3.5.7 From Number Riddle to Algebra

I ask the pupils in a seventh grade class to think of a number, double it, add 24 and then divide by 2. From the resulting number they subtract the original number. I then impress them by saying that I can "see" the final result ... It was 12, wasn't it? Did anyone get anything else? Oh well then, whose thinking has gone wrong, yours or mine? We repeat the exercise in similar variations. *Must* the answer in the game above always be 12? Let us try a little algebra: We let the original number be a. It is, of course, different for different people and so we cannot do more than give it a letter a, which could represent any number at all.

And now we write, step by step:

double the number:	we get	$2a$
add 24:	we get	$2a + 24$
divide by 2:	we get	$\dfrac{2a + 24}{2}$

Subtract the original number: the result is

$$\frac{2a + 24}{2} - a.$$

Can this expression be simplified? Yes!

$$\frac{2a + 24}{2} \quad \text{is quite simply} \quad a + 12.$$

From this it follows $a + 12 - a = 12$
and the reading of minds is revealed for what it is!

In this manner one can exercise a class in algebraic simplification in an enjoyable way.

The step up to equations is now not especially large:

I am thinking of a number. I increase it by 8 and divide by 5. I get 7 as my result. What number was I thinking of?

One can solve this problem in the head, or preferably with pen and paper, and notice how the solution came about. It is both interesting and important that we now and then observe how we think.

It usually turns out that not all students (of those who actually solved the problem) can clearly account for *how* they solved the problem. But some students succeed in observing their thought process. They say, "First I took 7 times 5," since division gave 7. 35 must then have been the number before I divided by 5. Then I took away 8 and got 27, which is my answer.

As we see, the solution goes backwards relative to the statement of the problem:

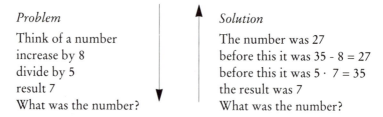

Problem		Solution
Think of a number		The number was 27
increase by 8		before this it was 35 - 8 = 27
divide by 5		before this it was 5 · 7 = 35
result 7		the result was 7
What was the number?		What was the number?

Let us now look at another problem taken from everyday mathematics:

By adding salt to 5 kg of 4 percent salt solution, we wish to increase the concentration to 5 percent by weight. How much salt need be added?

In this problem we do not know in advance how much the salt will weigh, and therefore we cannot do a kind of "backward calculation" to find out the desired amount of salt. The problem is not so amenable to working in the head. For problems of this and similar types, equations are an effective tool. How do we go about setting up an equation?

We return to the riddle above and let x denote the unknown number we wish to calculate. Now we "translate" the problem wording step by step to an algebraic expression:

think of a number:	we call it	x
increase by 8:	we write	$x + 8$
divide by 5:	we write	$\dfrac{x + 8}{5}$
the result is 7:	we can write	$\dfrac{x + 8}{5} = 7$

We have thus come to two different expressions for the result:

$\dfrac{x + 8}{5}$ on the left and simply the number 7 on the right.

We connect the two sides with an equality sign and have then an equation. Now the problem is to *solve* the equation

$$\frac{x + 8}{5} = 7$$

We are to get it to the point x = some number.

In other words, we want to try to "get x by itself."

What does the equation tell us? x + 8 divided by 5 gives 7. We can multiply *both* sides by 5 and in that way maintain equality:

we get
$$5 \cdot \frac{x + 8}{5} = 5 \cdot 7$$

that is,
$$x + 8 = 35.$$

This equation applies just as truly as the equation

$$x + 8 - 8 = 35 - 8$$

which is obtained by subtracting 8 from *both* sides.

We obtain the solution hereby: $\qquad x = 27.$

If we compare solving the equation with our solving of the riddle earlier, we find that we have taken the same steps. The advantage with the equation is that one can (with practice) write the equation down in the same order as the problem's wording and then "technically" go about solving for the desired value. We avoid the necessity of "thinking backwards."

From a *pedagogical* standpoint it is often better to activate students with more "figure-it-out-in-your-head" solution methods than to have them carry through mechanical solutions. But with a problem such as the salt problem the practical advantages of using an equation are considerable:

We let x kg be the amount of salt which the equation asks for.

Now we "translate" according to the text:

We have to begin with:	5 kg 4-percent solution
We add x kg salt:	The solution then weighs (5 + x) kg
This solution has 5% salt by weight	5 % of (5 + x) is 0.05 (5 + x) kg salt (5)

Do we have any value to set equal to 0.05 (5 + x)?

This expression represents the amount of *salt.* Can this be written in another way? Yes. We look back to the original quantity of solution and note that it contained 4% salt, i.e. 0.04 · 5 kg or 0.2 kg salt.

We add x kg salt:

the amount of salt becomes $\qquad$ $(x + 0.2)$ kg $\qquad$ (6)

Both (5) and (6) give the amount of salt, so the equation is:

$$x + 0.2 = 0.05\,(5 + x)$$

Step by step in proper order we now get the following equivalent equations:

$$
\begin{aligned}
x + 0.2 &= 0.25 + 0.05x \\
-0.2 & \quad -0.20 \\
\hline
x &= 0.05 + 0.05x \text{ (note that x means 1.00x)} \\
-0.05x &= \quad -0.05x \\
\hline
0.95x &= 0.05
\end{aligned}
$$

Finally we "get x all by itself" by dividing both sides by 0.95.

$$\frac{0.95x}{0.95} = \frac{0.05}{0.95}$$

$$x = \frac{1}{19} \approx 0.053$$

The answer is therefore: 0.053 kg or 53 g of salt needs to be added.

I have consciously chosen to present a problem leading to a relatively difficult equation. In the sixth or seventh grades in school one must begin with easier equation problems. The risk is then, however, that the problem is so easy to solve in the head that the students protest what they feel is making an easy thing difficult by doing it with equations. The thing here is to find a good middle-of-the-road problem. It is desirable for the class at an early stage to gain an appreciation that equations are a working tool in the solution of problems.

3.5.8 Exercises

1. A cylindrical pipe has a wall thickness of t, inner diameter d and outer diameter D. State some relations between these three quantities.

2. We look back to the end of Section 3.5.2: is the last triangle obtained there identical with Pascal's triangle?

3. Can four numbers a, b, c, and d form a magic square of 2 x 2 boxes, such that the rows, columns, and diagonals (with two numbers each) all give the same sum?

4. What general results can one come upon if one continues the study in Section 3.5.4?

5. Show that every prime number greater than 3 can be written in the form 6n + 1 or 6n - 1, where n is a natural number (i.e. a positive integer).

6. Show with the help of Exercise 5, that if p is a prime number greater than 3, then

$$p^2 + 2$$

cannot possibly be a prime number.

7. Try to prove by induction the beautiful formula

$$1^3 + 2^3 + 3^3 + \ldots + n^3 = (1 + 2 + 3 + \ldots + n)^2$$

$$n = 1, 2, 3 \ldots$$

(Refer to Section 3.5.6)

8. We have 1 kg brass, containing 60 % copper.
How much more copper must we melt down and add if we wish to bring the copper content up to 70%?

3.6 Judgment and Misjudgment

3.6.1 Big Balls and Little Balls

Time after time everyday life asks us to make judgments. It can concern the most widely differing things, from simple comparisons of length to difficult evaluations of quality.

We are very familiar with comparison of lengths and distance. It is not hard to judge , in Figure 3.6.1, how many times longer line AB is than line CD.

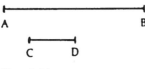

A   B

C   D

Figure 3.6.1

Much more difficult are judgments of distance "in depth," for example, across a bay or in the mountains. As far as time intervals are concerned, we all know how difficult it is to free ourselves from the subjectivity of our own experience. With two- and three-dimensional things we usually also have difficulty making reliable comparisons.

I don't know how many times I have begun a geometry lesson in the eighth grade with the following little story:

For a birthday party the hostess had made marzipan balls, solid, in two sizes. Some of the balls were twice as large in diameter as the others. The hostess offered them to her guests. The first guests each took a small ball; then someone took two small balls. The next guest thought, "I can just as well take a large ball instead of two small; there isn't much difference." After that another guest reasoned, "I'll take three small balls; I'm sure it isn't more than one large."

"What do you say, kids? How many small balls could one politely take, if one didn't wish to take more than the one who took a large ball? Can you estimate? You may very well — if you wish — use decimal fractions. Does a large ball contain just as much marzipan as 3 small, or 5 or 2.9 — or how many?"

Most students have guessed that 1 large equals 4 small. A few answered 3; a very few have answered with a decimal or with 5. Only seldom has someone answered 8, and then he was met with surprise and disparaging glances from his classmates.

Even in those classes which have been generous in their estimates, or where someone has answered 8, it has been advantageous not immediately to say if the answer is correct or incorrect. It is more fruitful to let the class itself judge the reasonableness and correctness of the answers.

I have therefore spun out the tale and told how the hostess also served jellied candies in the shape of solid cubes, some of them with double the edge width of the others. How many small cubes together would make as much candy as one large cube?

The answer "four" usually persists, but it comes a little doubtfully. And it doesn't take long before some students eagerly and with conviction in their voices say: "8 cubes." They are happy to go up to the blackboard. There they quickly sketch a cube, which, like a Christmas package, is divided into quarters on each side (Figure 3.6.2). There is no doubt about it: the large cube corresponds to 8 small cubes.

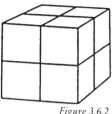

Figure 3.6.2

Everyone now suspects that the same holds true for the marzipan balls. According to the famous and classical "method of exhaustion" by the Greek Eudoxus (408-355B.C.), one can see that doubling the ball diameter gives an 8-fold enlargement in volume: one inscribes an endless series of smaller and smaller cubes within the ball, so that the total volume of all the cubes begins to approach the volume of the sphere as a limit.

In what context does a doubling of a length dimension result in a quadrupling? The class thinks this over, and there may come widely differing answers, which yet are basically correct. For example: if one doubles the radius of a cake but keeps the height unchanged, then the cake will be "4 times as big." Or: if one makes the side of a square twice as long, then the area will be four times as big.

In the continuation of our work, we eventually gain important knowledge about scaling. We set up a table:

When the length is	... then the area is	... and the volume
doubled	4 times as big	8 times as big
tripled	9 times as large	27 times as big
4 times as long	16 times as large	64 times as large
10 times as long	100 times as large	1000 times as big
half as long	¼	⅛
one-third	⅑	1/27
one-fourth	1/16	1/64
1/10	1/100	1/1000

The numbers in the area column are the squares of the length numbers, the numbers in the volume column are the cubes of the length factors.

In summary: Area scale is the square of the length scaling.

Volume scale is the cube of the length scaling.

The most important application is without doubt converting between different units of measure for area and volume.

We have come up with an important rule, but we must time and again practice with what appear to be different examples:

1. I sketch a cylindrical glass on the board: "Here is water up to this height. Now we fill water to double the height. How much water do we have in the glass now?" No one is fooled.

2. "Here is a conical glass (Fig. 3.6.3). It is filled up to half the height. If we now fill it up to the brim, how many times more water have we?"

A pupil remarks: "One can see that you have experience with this stuff!" After that encouragement the hands start going up. The first answer: four times as much. The second, third and fourth answers: four times as much. Are we all agreed on this? Doubt, until someone comes up with the right answer: 8 times as much. This time the pupils are surprised not so much over the answer as over the fact that they let themselves be fooled.

Figure 3.6.3

Repetition is the midwife to understanding, but it ought to be interesting and not just routine.

During a repetition one can readily tack on something new. Figure 3.6.4 is taken from Huff. It shows two sacks of money. The one has double the dimensions of the other. We suppose that the figure is meant to illustrate the doubling of profits in a business and thereby show how "successful" the company is. We might ask the class if they have any comments on the figure. Does it well illustrate the company's doubling of profits?

Figure 3.6.4 (From Huff, "How to Lie with Statistics.")

The class need not think long before they see through the trick which lies behind the figure. What is it?

If the sack stands before us, how much more would you think the big one holds than the small one? When we see the sacks drawn three-dimensionally as in the figure, most people get an *unconscious* feeling that the big sack holds much more than twice the contents of the little one. Herein lies the trick. The volume ratio is, in reality, as we have seen, 8 to 1. More correct would be, for example, to illustrate the doubling of profits with a bar diagram (Fig. 3.6.5).

The three-dimensional drawing occurs quite often but sometimes the illustrator does not seem to have thought of the trick which it contains.

A problem concerning area and volume scaling, which occupied Galileo and can generate good interest in a ninth grade class, is taken up in Exercise 6.

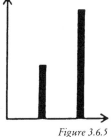

Figure 3.6.5

3.6.2 Beware of Averages

Among types of judgment and comparisons we make not least are averages. Oil consumption, food costs, speeds, etc. are all generally stated as averages. Suppose we are travelling by car, have covered a distance of 600 km. and been driving 8 hours including breaks. For fun we want to know what our average speed has been, including the breaks. This might even be of value with thought of planning for future trips by car. We find that our average sped was 600 km/8 hours, i.e. 75 km/hr.

Let us now consider the following problem: an airplane flies a certain distance at a speed of 800 km/hr. and immediately returns with a speed of 1000 km/hr. What average velocity has the plane kept? It is tempting to answer 900 km/hr. This is incorrect, however. If the distance were 10,000 km, for example, then the flying time would be

$$\frac{10,000}{800} + \frac{10,000}{1,000} = 12.5 + 10 = 22.5 \text{ hours.}$$

An airplane flying at 900 km/hr. the whole time would need the time

$$\frac{20,000}{900} \text{ hours, i.e. } 22.222 \ldots \text{ hours } \quad \text{to make the trip,}$$

which gives too short a flight time. The correct average speed must therefore be less than 900 km/hr. We see here how easy it is to get lost when calculating averages.

The airplane in our example returns at greater speed than during the outbound flight. The return thus goes more quickly, which means that the plane is travelling with speed 1000 km/hr for a shorter time than with speed 800 km/hr. If the plane had flown at these two speeds during equally long *times,* then the average speed would have been 900 km/hr, i.e. the average of 800 km/hr, since the distance covered in a given time is proportional to the speed.

3.6.3 Per Cent Calculations

Imagine how difficult it would be to get anything out of data and statistics, if we could not state the results as per cent!

Let us suppose that a person P owns three sevenths of the stocks in a company X and five elevenths in another company Y. In which company does P have the greater share?

It is not easy to compare directly $\frac{3}{7}$ and $\frac{5}{11}$. We often use per cent values and find that $\frac{3}{7}$ corresponds to 42.9% and $\frac{5}{11}$ to 45.5%. With that the comparison is completely clear.

In the old Swedish grade school the pupils were drilled in calculating how many per cent higher or lower a certain price was, compared to another. We need not long for a return to the old school methods, but considering how often percentage calculation is used in daily life, not the least in commercial advertising, it must be a part of general education to know a bit about such calculation.

If a firm wishes to sell out a lot of goods for half price, it may quite simply advertise: "These goods now selling at half price." Or: "50% discount on these goods."

But a more common way of advertising nowadays is:

"Get double value for your money." Or: "Get 100% more goods for your money."

All these expressions say the same thing, but doesn't it sound more tempting to the customer to get 100% more for his money than a 50% discount?

Per cent calculation is perhaps that area of everyday mathematics which offers us the most varied collection of problems, and on top of that, problems with the fresh taste of real life. Pupils become more aware of what it is one compares with, and they see how the results take on different expressions and have different values depending upon what one uses as a reference.

Per cent calculations also give problems which invite estimation with head calculation. Let us examine the declaration of contents of Falukorv (a favorite Swedish sausage): It states that the meat content is 68 % (by weight). The meat ingredients are then declared separately

beef 45%

pork 35%

and   lard 20%

How much beef and how much lard, respectively, are contained in Falukorv?

Since the meat content is 68%, it represents about ⅔ of the weight. The fraction of beef would then be ⅔ of 45% or about 30%, while the fraction of lard would be ⅔ of 20% or approximately 14%. One can find good examples in books like D. Huff's *How to Lie with Statistics*, but perhaps training in per cent becomes most alive when one picks out interesting articles from the daily newspaper. I reproduce here a few clippings to show what breadth such examples can have.

1. "Enormous mark-up! What does a farmer get for a kilogram of carrots at the wholesaler — and what does the housewife pay for the same carrots at the grocer's or market square? Express Patrol found that there is an enormous mark-up on certain vegetables and flowers. Carrots go up by 350% from wholesaler to retailer and broad beans by 600% during the same journey." (From an article in the Swedish newspaper *Expressen*.)

2. "1900-1977: Divorce up by 800%. Divorces in Sweden have increased markedly, looked at from a historical perspective. The increase is 800-900% since the beginning of the 20th century. Of all marriages entered, 25-30% are dissolved as a result of divorce." (Dagens Nyheter, Sweden)

3. In a 1973 article the ownership of the Swedish domestic airline Linjeflyg was described as complicated (Figure 3.6.6):

Owners of the Domestic Airline

DK = the Danish government
N = Norwegian government
S = Swedish government
Ö = Miscellaneous interests in DK, N, and S.
S INT LT = Swedish Intercontinental Air Traffic
S STORF = Swedish big industry
DANSKT LS = the Danish Airline Co.
NORSKT LS = the Norwegian Airline Co.
AEROTR = Swedish Aero-transport Inc.

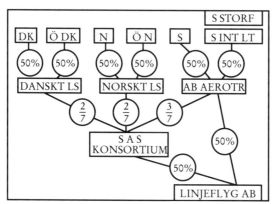

Figure 3.6.6

"Linjefyg's ownership is complicated. Danish and Norwegian interests, through SAS, have influence over the Swedish domestic airline, just as Swedish interests have in neighboring countries."

4."New taxation Shocks" — newspaper headline following release of new property valuations in 1970.

"Nämdö community reports an increase of over 500% in taxation value for an area containing summer cottages. At last taxation the lots were valued at $1000. Now they are raised to $5000."

(Is the per cent figure correct?)

The numbers in these examples are a few years old, but we can find examples daily — equally varied and original — and use them in our work at school.

Opportunities for co-operation with colleagues in the natural sciences and particularly social studies abound. Inflation, wage increases, real increase in earnings, city planning, large scale production, etc., all give rich materials without further ado, materials which motivate the pupils in the higher levels in their studies.

3.6.4 Exercises

1. Refer to Example 3 above on Linjeflyg. What per cent of Linjeflyg is owned by Danes?

2. Refer to Example 4 above. Is the figure 500% correct?

3. The Boliden Company's Aitik mine is one of the "poorest" copper mines in the world: the ore is only 0.5% copper. How many tons of ore must be mined to get one ton of copper?

4. The price of a product is raised 25%. How big a discount can the store now give without the price going below the original level?

5. A community one year reduces its financial contribution to a school by 50%. The next year it raises it by 100%. (No other changes in the school monies are made). At what level, compared with the original year, is the school contribution now?

6. Galileo writes in his essays on "things which float on water":

> ... If one wishes to maintain the same proportions in the bones of the skeleton of a great giant, as we find in the ordinary human, then one must either use a stronger and harder material for the legs or accept a reduction in strength compared to common man, since if our height were to increase normally we would fall and be crushed by our own weight. Therefore, if a body's size is decreased, the strength of the body is not diminished in the same proportion; the smaller the body, the greater the relative strength. A small dog ought to be capable of bearing two or three dogs of the same size upon its back; but I believe that a horse could carry not even one of its own size.

Let us now suppose that adult mammal is 8 times heavier than a young animal of the same kind. How many times larger must the bones that bear weight in the adult then be, if we assume that the mechanical strength in both cases is in proportion to the cross-section of the bones?

7. A boat went 20 knots for 10 minutes and then 15 knots for 30 minutes. What is the boat's average speed during this 40 minute-period?

8. Calculate the average speed for the airplane of Section 3.6.2. It flies a given distance at 800 km/hr., then turns around and flies immediately home at a speed of 1000 km/hr.

3.7   Nature's Geometry and Language of Form

3.7.1 Polyhedra

There is a rich literature on the many wonders of nature. Lovely color photographs show us examples of how nature has solved her problems in form and how she has given minerals, plants and animals color — not only beautiful but also functionally "correct." We have, perhaps, read some of these books or seen films, but how many observations have we ourselves made directly out in nature?

Quite certainly we have looked at plants and let their language of form and color fascinate us. What about minerals and particularly crystals? Possibly very few of us have sought out and scrutinized crystals in nature. Most often our interest is limited to precious jewels or indirect meetings through illustrations in books.

Experience in school has shown us that direct meetings in nature give pupils a much greater return and waken their enthusiasm considerably more, when they have had the opportunity to acquire knowledge of the subject in advance. (The same observation applies for subjects such as architecture, history, painting and other areas which we can meet up with during study trips and camps, for example..)

Let us begin with:

1. Crystal Shapes

Children get great joy from the first snow, sometime before Christmas, when King Bore sends his myriads of snowflakes. The earth is covered in a white shroud, cleaner than most other materials which we can find. When the sun shines on the new snow, it often glimmers in different colors, and if we look carefully we can discover a little of the architecture which forms snowflakes.

The person who studied the snowflake's form probably more patiently than anyone else was the photographer W. A. Bentley, an American who developed a camera technique and photographed a vast number of snowflakes.

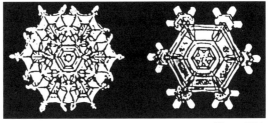

Figure 3.7.1

Together with W. J. Humphreys he gave out the book *Snow Crystals* (Dover Publishing, 1931) containing 2453 pictures of snow flakes — each picture different! One meets here an incredible multiplicity of variations on one and the same underlying theme: the architecture of the number six. Figure 3.7.1 shows some of Bentley's pictures.

Interest in crystals can be traced as far back as the historian's eye reaches. In Hellenic culture, Plato stands in the foreground among mathematicians' study of geometrical form. The huge mathematical work *Elements*, put together over 2000 years ago, comprised 13 volumes whose presentation reached its peak in the thirteenth and last volume, which treated regular three-dimensional figures, so-called regular polyhedra, or, as they also later came to be known, Platonic solids.

2. Regular Polyhedra

Polyhedra are solid bodies, bounded by plane surfaces: triangles, rectangles, and trapezoids, other polygons or a combination of such surfaces. (Poly = many, hedron = side, face, surface).

How many surfaces are necessary as a minimum, in order to make a polyhedron? Two is too few — we just get a "plough" shape. Is three enough? No, we only get a corner. But if we take four surfaces? Yes, we can construct, for example, a so-called tetrahedron with four triangles (tetra = four). We all have seen the "tetrapack" which milk, cream or juice comes packaged in. When the four triangles are equilateral and of the same size, we get a regular tetrahedron, and this we note down as one of the Platonic solids.

Do we know any of the other regular polyhedra? "The cube" is heard immediately from several directions in the room. Yes, it has six identical sides, six squares in a regular arrangement.

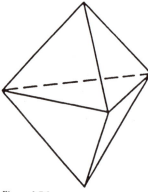

Figure 3.7.2

More examples? Someone says: "If we put two tetrahedra together so that two faces fit together over another, then we get a body which is formed by six equilateral triangles of the same size." This is correct, but … How about the corners? "Are all the corners 'the same'?" — "What do you mean by that?" — "Well, if we look at the corners in the regular tetrahedron we notice that all the corners are the same, and this is also true for the cube. Does this apply to the double tetrahedron which we are now considering?"

We look at a sketch on the blackboard. This time no one wishes to speak first. I must give a clue: look and see how many surfaces meet at each corner. Some now answer "three surfaces," others say "four triangles." Who is right? Proponents of these answers go up and point. Both are right in a way….

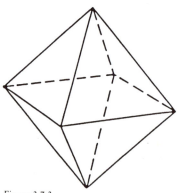

Figure 3.7.3

Let us look at Figure 3.7.2. Where are the corners with three triangles, where with four? At three of the corners, four triangles meet; at two corners (the top and bottom in the figure), only three meet. The corners are thus different. And we may therefore not call the double tetrahedron a regular polyhedron or a Platonic solid. In such a body it is not enough that the surfaces are all the same regular triangle or polygon. Even the corners must have the same form. We will therefore have to throw out our double tetrahedron.

What can we suggest now?

Someone says: "A double pyramid. We take two pyramids which have a square base and put the bases together. The walls are supposed to be equilateral triangles."

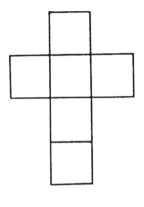

We get a solid bounded by 8 equilateral triangles (Fig. 3.7.3). How many meet in the corners? Around the "base square's corners" 4 triangles meet and at the top and bottom also 4. It looks promising. Can we make such a solid out of cardboard? In Figure 3.7.4 we see networks for cube and a tetrahedron. How would we draw a network for our double pyramid?

In Figure 3.7.5 we see one pupil's proposal, which ought to work well. We cut out the network and fold creases along certain edges. Finally we get the cellophane and the polyhedron stands finished. We twist it and turn it. Yes, all the corners are the same. No one doubts that this solid is entirely regular, and it becomes our third Platonic solid. It is called an octahedron, since it is bounded by eight surfaces.

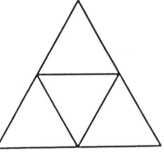

Figure 3.7.4

The class cannot come upon any further regular polyhedra — and it isn't easy. We therefore go on to comparing the three we have: How many corners, edges, and surfaces does each have?

We make up a table of the results:

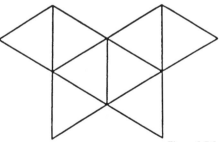

Figure 3.7.5

	Corners	Edges	Surfaces
Cube	8	12	6
Octahedron	6	12	8
Tetrahedron	4	6	4

Someone discovers that the numbers 8 and 6 are just reversed in the cube and the octahedron. The number of edges is 12 in both. It seems as if these two might be related in one way or another.

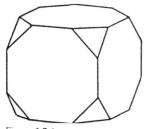

Figure 3.7.6

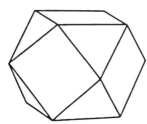

Figure 3.7.7

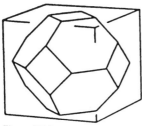

Figure 3.7.8

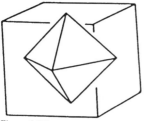

Figure 3.7.9

I now lead over into what appears to be a completely different question: What kind of solid arises if we gradually grind down the corners of a cube making the growing surfaces bigger and bigger?

We grind symmetrically so that equilateral triangles are created at the corners and so that all the corners are ground equally much. For example, what is the solid we get when the ground surfaces meet halfway along each edge of the cube?

Figure 3.7.7 shows this stage and 3.7.6 a previous stage along the way, where each square of the cube has been transformed into a regular octagon (eight-sided figure).

What happens if we continue grinding further and go so far that the cube's square walls finally shrink down to nothing at their mid-points? Figure 3.7.9 shows the final solid. Figure 3.7.8 shows a transition stage (not so easy to draw!) where the ground surfaces consist of regular hexagons (six-sided figures).

From the cube arose an octahedron, whose 6 corners are the mid-points of the cube's 6 squares, and whose 8 triangular faces came from the 8 corners of the cube. We understand now the relationship between the cube and the octahedron in the table. We are in addition completely clear over the fact that the octahedron is regular, since the grinding was done symmetrically.

Someone wonders: what do we get if we grind down corners of an octahedron? Perhaps a new Platonic solid will emerge?

What happens? Let us sand down the corners in our imaginations, so far that only the mid-points of the eight triangular surfaces are left.

We are going to get a solid with 8 corners. A few pupils talented in geometry seem sure of themselves. (It is too bad when they say at once what they have found; it takes away their classmates' joy of discovery.)

It turns out that the cube emerges as the final form! We might have expected that, with a clue from the table. Halfway through the transformation we get the same solid as along the way from cube to octahedron, the solid in Figure 3.7.7 consisting of 8 triangles and 6 squares. It is called naturally enough a cubo-octahedron.

A new question: what arises if we grind down the corners of the tetrahedron so that only the mid-points of the triangular faces are left?

This question is easily answered: we get a new tetrahedron, turned upside down to the original.

In order to proceed we begin to systematize the Platonic solids in a table:

Cube:	6 squares, 3 squares per corner;	6 faces, 12 edges
Tetrahedron:	4 triangles, 3 triangles per corner;	4 faces, 6 edges
Octahedron:	8 triangles, 4 triangles per corner;	8 faces, 12 edges

What ought to come next? A solid with 5 triangles per corner? Or one with 4 squares per corner? "No, that is impossible. Then we couldn't fold up the cardboard walls." "Six triangles per corner won't work either; they fill out the whole cardboard surface around the corner." "But five triangles — that might just work!"

"What might the network for that solid look like?" I wonder aloud. Some suggestions come up. The starting point is that we draw two rows of five triangles. But how should these five-sided pyramids be put together? We help each other reach the conclusion that a "zig-zag girdle" of 10 triangles is necessary. Now proposals for networks start coming up: they consist of 5 + 5 + 10 = 20 triangles. (See Figure 3.7.10).

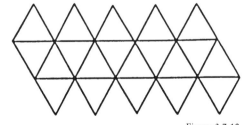

Figure 3.7.10

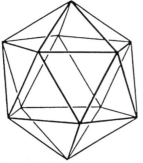

Figure 3.7.11

Scissors go into action; certain edges are scribed so that the faces can more easily be folded up. The pupils help each other in pairs with the tape.

Look, it fits together; it makes a roundish sort of solid!" It is called an icosahedron (icos = 20). It would be going too far to guide the class through an "existence proof" for this solid. For the majority, the successful taping together of the network is "proof" enough, and much more tangible than any theoretical arguments — at the 15 year-old level. (Figure 3.7.11)

We add to the table:

Icosahedron: 20 triangles, 5 per corner; 20 faces, 30 edges, 12 corners

What happens if the icosahedron is ground down, so that only the mid-points of the triangular faces are left? "It *has* to be solid with 20 corners and 12 faces."

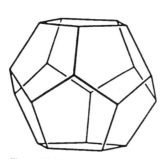

Figure 3.7.12

But what kind of faces? We think this over. How many corners must each face have? Its corners must lie within the five triangles which form each corner of the icosahedron. This means that the new solid's faces have five corners. Why yes, they must be regular five-sided polygons, so arranged that three meet in each corner.

The new solid is a sibling to the icosahedron in the same way that the cube and the octahedron are siblings to each other.

This fifth Platonic solid is called the regular dodecahedron, or more exactly, the regular pentagonal dodecahedron (Figures 3.7.12 and 13.)

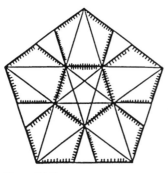

Figure 3.7.13

Can a solid be constructed with 4 pentagons meeting in each corner? The answer is no, since four pentagons drawn on cardboard must overlap each other. Can we succeed with a solid of three hexagons? No, they fill up the surface of the paper around a corner exactly.

We now realize that we have found all of the regular polyhedra which exist; that is, five.

Let us once again look at the pairs of opposites, the cube-octahedron and the icosahedron-dodecahedron:

Cube: 3 *four*-sided faces per corner; 6 faces, 12 edges, 8 corners
Octahedron: 4 *three*-sided faces per corner; 8 faces, 12 edges, 6 corners
Icosahedron: 5 *three*-sided faces/corner; 20 faces, 30 edges, 12 corners
Dodecahedron: 3 *five*-sided faces/corner; 12 faces, 30 edges, 20 corners

We say that these paired solids are duals of each other, two by two. The tetrahedron, which when grounded down, is transferred into itself, is said to be self-dual:

Tetrahedron: 3 three-sided faces/corner; 4 faces, 6 edges, 4 corners

3. Crystals — once more

Now, what do these five Platonic solids have to do with crystal forms? Are there crystals which take on the forms of Platonic solids? Yes. The very simplest example is kitchen salt. It can crystalize in cubic form. It can also have one side longer than the others and become box-shaped. The kitchen salt crystal is completely transparent and as colorless as ice. One might even mistake it for a piece of cut and ground ice. "Who ground it?" someone asks. "Nature herself — but she has not actually *ground* it; it *grew* that way, by itself!"

We take a look at gold-glimmering, cube-shaped crystals. On some of them sits, slightly crooked, a much smaller cube, "like a little crystal child," someone adds. "Is it gold?" — "No, it is pyrite, which is formed when iron and sulfur combine with each other in certain proportions."

I bring out some other gold-shimmering crystals, octahedra this time. Yes, these are also found in magnetite or lodestone (magnetic iron oxide, mined in Kiruna, Sweden, among other places) and in fluorite (calcium fluoride). The latter forms yellow, blue, violet, or yellow-violet octahedra. An almost pure regular tetrahedron form is found in Fahlband (German: Fahlerz); the crystals have a metallic glance and contain copper and sulfur, as well as one or several of the elements iron, antinomy, arsenic, silver, and mercury (Figure 3.1.14).

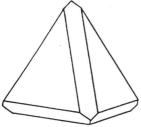

Figure 3.7.14

Figure 3.7.15

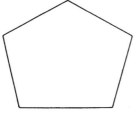

Figure 3.7.16

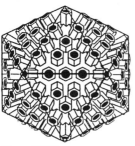

Figure 3.7.17
Schematic picture of a Herpes virus

Pyrite can also form dodecahedra. In Figure 3.7.15 we see a group of pyrite crystals in which nature has formed partial dodecahedra. If one looks very closely at such crystals, one should find that they are not completely regular. The five-sided figures which make up the faces of the crystal are a little bit too wide in one direction (Fig. 3.7.16).

4. Other Areas of Nature

Strangely enough nature does not produce crystallic *regular* icosahedra or dodecahedra. (We will not go further into this here). But in another area of nature, an area which could be said to lie between the mineral and plant kingdoms, there do exist regular icosahedra and in some case dodecahedra: among the viruses. Figure 3.7.17 shows a schematic picture of a Herpes virus, drawn according to pictures taken with an electron microscope. In Figure 3.7.18, also taken from *Scientific American* (Number 1, 1963), may be seen "shadows" of an insect virus, obtained by irradiating it in an electron microscope. The shape of the shadows reveals that the virus body itself has the form of a regular icosahedron!

Even radiolarians (an order of single-celled sea-animals with long slender pseudopodia "radiating" outward) show the regular icosahedron shape — namely in their skeletons, built of silica. They were studied by Ernst Haeckel, who made many interesting and detailed drawings of them (Fig. 3.7.19).

Numbers 2, 3, and 5 in the figure have the forms octahedron, icosahedron

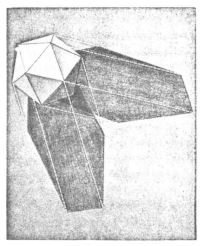

Figure 3.7.18 (From Scientific American 1963/1.)

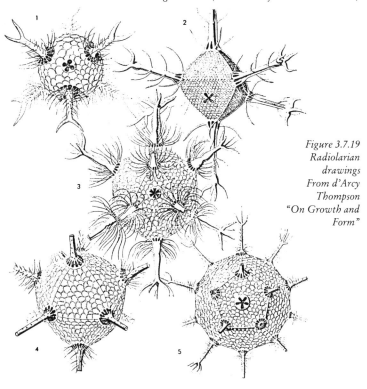

*Figure 3.7.19
Radiolarian
drawings
From d'Arcy
Thompson
"On Growth and
Form"*

and dodecahedron respectively, and take their Latin names from these shapes: Circoporus octahedrus, Circogonia icosahedra and Circorrhegma dodecahedra. So here we have even been able to find examples of dodecahedra in nature.

When we say, for example, that fluorite crystals are octahedra, we must keep in mind the fact that the crystal-faces are not perfectly flat surfaces. The same applies to he radiolarians. But it is nonetheless obvious that the architectural form is regular. With the radiolarians the surfaces are somewhat curved, but the vertices form the corner-points of the Platonic solids. That pyrite crystals are not completely regular dodecahedra, however, has other explanations.

In *Scientific American*, Jan. 1983, there is an article "Platonic Chemistry" (p.59), where the synthesis of a molecular dodecahedron is reported. It was achieved by researchers at Ohio State University. In 1982, according to a Swedish newspaper, an Israeli physicist, Dan Schechtman, then working at the American National Bureau of Standards had succeeded in developing a technique, by which he manufactured dodecahedric crystals containing aluminum and iron atoms. The atoms took the positions of the twelve corners of a regular octahedron.

3.7.2 Curves and Curve Families

In spite of the richness in variety we would have to say that crystal forms are characterized by very strict "laws." For example, two surfaces always meet in a straight line edge. In plants and animals we find completely different principles of form. There the curve, the curling form, plays a major role. We shall soon pick out a few important examples, but let us first begin with the straight line and the swerving curve as basic elements for two-dimensional forms. Among curves we can consider the circle as the primary opposite of the straight line. This reflects itself in the tools we use: straight edge and compass, tools which trace their roots to Euclidean geometry's childhood 2500 years ago. If we wish to see curves as the result of one or more *movements,* we can begin with two basic motions: movement along a straight line and rotation.

Example 1: We can ask ourselves: what sort of curves arise when we combine the straight-line movement with a rotation? We let a point simultaneously surrender itself to both a straight-line and a turning movement. What kind of motion does the point describe?

We can imagine that a little ball moves straight out from the center of a phonograph turntable, at the same time as the turntable rotates. How does the ball's motion look relative to the stationary table? It is a spiraling curve.

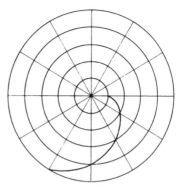

Figure 3.7.20

But we need to be more precise about how the movements come about. To begin with we choose the simplest motion: the point moves with constant velocity of 1 cm/second on the disc and the disc rotates at a constant speed of, let us say, ½ revolution per second around its center.

We can now draw a number of concentric circles a cm apart, and divide these into sections with 12 rays from the center. If the moving point starts at the center, it will describe the type of spiral motion shown in Figure 3.7.20. Irrespective of what values we give to the straight-line speed and the speed of rotation, just as long as they are constant, this kind of spiral is called an Archimedean or arithmetic spiral.

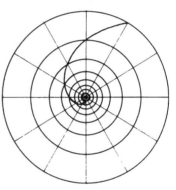

Figure 3.7.21

We now vary the theme and let the straight-line movement be accelerated, such that the moving point's distance from the center grows second by second according to the geometric sequence 4, 6, 9, ... (each radius = 1.5 times the next smaller radius). The

Figure 3.7.22

Figure 3.7.23

Figure 3.7.24
(From Adams/Whicher, The Plant between Sun and Earth, *1952).*

Figure 3.7.25
(From d'Arcy Thompson, On Growth and Form*)*

rotation is the same as before. We get now a considerably more dynamic spiral (Figure 3.7.21), the so-called logarithmic spiral.

Figures 3.7.22-23 show families of the Archimedean and the logarithmic spirals respectively. Diagonally situated "parallelograms" are colored in. What happens with these parallelograms, in their respective figures, as we move outward from the center? The class has no difficulty seeing how the Archimedean parallelograms change shape, they become wider, while the logarithmic spiral's parallelograms seem to keep their shape. They only get bigger. This observation is correct: one can show that the logarithmic spiral continually curves at the same rate, while the Archimedean spiral, as we have seen in the pictures, adapts itself more and more to the circle's form. Do these spirals exist in nature? Let us look at Figures 3.7.24 and 3.7.25 which show respectively the elegantly formed sea conch Nautilus (in cross section) and the mollusk Solarium perspectivus.

Figure 3.7.26 shows Italian cauliflower — an architectural masterpiece, where the spirals in the head of the cauliflower are found again in successively smaller scale in the miniature "heads" which make up the spirals. Here we have spirals of spirals of spirals, as far as the eye can reach. In this example and the next the spirals are logarithmic.

In Figure 3.7.27a we see a photograph of a sunflower (from Adams/Whicher, *The Plant between Sun and Earth*, 1952). Adams points to the sunflower's center, the glomerule or seed cluster, as an example of spiral formation in the plant kingdom. Let us study the photograph closely and notice that the cluster contains *two* systems of spirals: on the one the spirals turn clockwise in toward the center, in the other counter-clockwise. The latter spirals are longer.

Cauliflower   *Figure 3.7.26*

Take a magnifying glass and try to count how many spirals there are, going clockwise and counter-clockwise respectively! We have counted spirals in sunflowers growing on the school grounds and come to the same result: 55 clockwise spirals and 34 counter-clockwise! Let us now reflect upon the following quotation from *Scientific American*, 3/ 1969:

> The most striking appearance of Fibonacci numbers in plants is in the spiral arrangement of seeds on the face of cer-

Sunflower   *Figure 3.7.27a*

tain varieties of sunflower. There are two logarithmic spirals, one set turning clockwise, the other counterclockwise, as indicated in the illustration on the next page. (Figure 3.7.27b)

The number of spirals in the two sets are different and tend to be consecutive Fibonacci numbers. Sunflowers of average size usually have 34 and 55 spirals, but giant sunflowers have been developed that go as high as 89 and 144. In the letters department of *The Scientific Monthly* (November, 1951) Daniel T. O'Connell, a geologist at City College of the City of New York, and his wife reported having found on their Vermont farm one mammoth sunflower with 144 and 233 spirals!

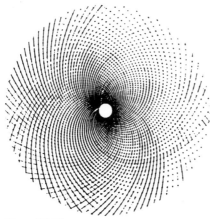

Figure 3.7.27b

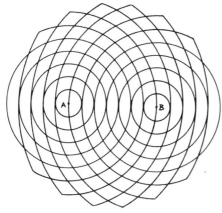

Figure 3.7.28

Example 2: In Figure 3.7.28, points A and B are the points of origin for two motions (each represented by a family of circles) expanding outwards at a constant rate. What is the motion of a point which:

a) goes outward from A and simultaneously inward toward B?

b) removes itself simultaneously from A and B?

We first have to extend the families of circles so that they intersect one another. Then it is only a matter of drawing beautiful curves through appropriate intersections, and we get the family of curves in Figure 3.7.29:

the one family (a) shows us ellipses,

the other (b) portrays arcs off hyperbolas.

Such a pattern occurs when circles spread out in water from two centers; one speaks of interference patterns. Such patterns play a very important role in all technology which utilizes wave motion.

Example 3: Two straight-line motions can interact to create a curve in a completely different manner than they did in Example 2. We let two points A and B move along two respective lines a and b with, let us say, the same constant speed. We mark off the position of the points at evenly spaced intervals of time (Figure 3.7.30).

How do the connecting lines move? We have only to connect successive A-points with their respective B-points and see what emerges: Figure 3.7.31. The line rotates and creates a beautiful curve, a close relative to the ellipse and the hyperbola, namely a parabola. (The figure shows only an arc of the parabola.) The interesting thing is that the figure obtained gives the impression of three-dimensionality: one can see a saddle-like surface. In actual fact the figure may be seen as the plane projection of a so-called parabolic hyperboloid, a saddle-shaped surface with many beautiful characteristics.

The main importance of Example 3 is, however, that the pupils experience a completely different way of generating a curve than

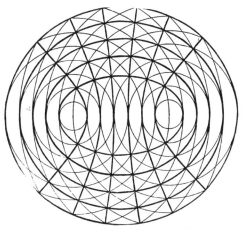

Figure 3.7.29

Figure 3.7.30

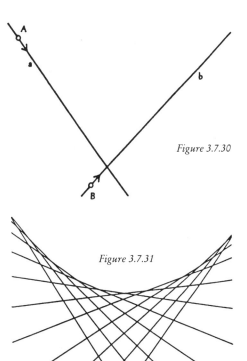

Figure 3.7.31

Figure 3.7.32

the otherwise common method — that which we used in Examples 1 and 2: constructing points on the curve. In Example 3 the curve is *shaped* by straight lines which will become tangents to the curve. There are many beautiful examples of constructing curves as the boundary of a group of lines or planes, so-called envelopes. Such figures are especially beautiful if they are drawn with white pencil or ink on black or blue paper. Figure 3.7.32 shows an example.

3.7.3 Exercises

1. Figure 3.7.4 shows networks for the cube and the regular tetrahedron. How many *different* networks exist for these respective solids? Two networks are considered identical if the one can cover the other. Rotation and mirror imaging thus do not give new networks. In addition, adjacent faces in the network must have a common edge (it is not sufficient for them to have only a common corner).

Finding *all* possible different networks for the cube is a combinatoric-geometric problem which usually really gets the students engaged (ninth grade). One of the questions which comes up is, "How do we know when we have actually found *all possible* networks?" We have to figure out some sort of system for numbering the networks. Yes, we have to order them in a logical sequence. This is a task which fits well together with the problem we discussed in Section 3.2.

2. Make a regular dodecahedron and study it to see if you can find 8 points which would form the corners of an "inscribed" cube.

3. Draw a sketch showing how four of the corners of a cube determine a regular tetrahedron.

4. Start with the same drawing as in Exercise 3 and continue by drawing in the tetrahedron which is determined by the other four corners of the cube. Finally, try to make clear the solid which these two tetrahedra together form (Kepler's twin tetrahedra).

5. Draw networks for the solids shown in Figures 3.7.6 and 3.7.8, and do this in such a way that these solids may be inscribed in a cube of chosen size.

3.8 Curve Transformations

Transformation is one of the most important principles of nature's forms. Perhaps the most striking is the butterfly's development from the larva by way of the pupa stage. Other great transformations can be found, for example, in the tadpole's growth into frog and, in human beings, in the embryo's transformation into fetus. Even plants give us beautiful examples. Take, for example, the different stages of the dandelion. Simply observing the changing form of the flower central base, we can discover a surprising sequence of phases. The central base transformation makes possible the changes in the various stages: bud, flower, seed bud, seed sphere (Figure 3.8.1).

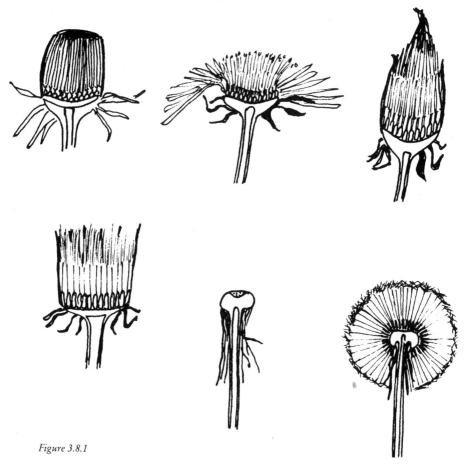

Figure 3.8.1

We will, by and large, here have to limit ourselves to the simpler type of transformation, which can be called curve transformation. So-called projective geometry, which in this book can only be hinted at, lays the foundation for more sweeping metamorphoses.

Next we shall look at four examples, which do not require special prerequisite knowledge.

3.8.1 Four Examples

Example 1: From classical Greek geometry we take the curve which is called cochoid. (It was introduced by Nichomedes as a tool for solving the classical problems of doubling the cube and trisecting the angle.) A line a and a point P (not on the line) are the starting elements. Through P a line p is rotated. We call the intersection with line a X. From this intersection X we now mark off line p a given distance XY, in the direction of point P. To begin with we take XY as a small distance. What curve is described by the point Y, as line p rotates about point P? And in particular: how does the curve change as the length XY grows?

(Nichomedes took interest primarily in the case where the distance XY is marked off so that Y lies on the opposite side of P from line a.)

Point P can be suitably placed at 50 mm away from line a. A few symmetrically chosen lines through point P can then represent the family of lines p. For the dis-

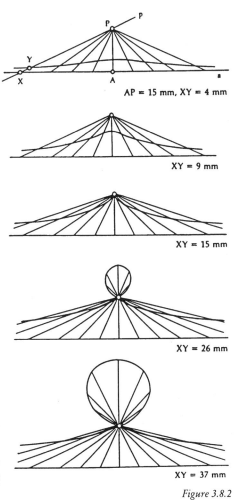

Figure 3.8.2

tance XY we might choose the following: 15 mm, 35 mm, 50 mm, 60 mm and 80 mm. Each of these values corresponds to a conchoid curve (see Figure 3.8.2). (Other values have been chosen for the lengths XY and AP in the figure, with consideration to the size of the book.)

We may now in our imaginations follow the conchoid's continuous transformation as we increase the length XY. In the vicinity of point P arises a little "hill" which changes into a pointed peak, when XY = AP. When XY becomes larger than AP, the conchoid makes a loop above P; P becomes a so-called double point. After that the loop grows without limit as XY is allowed to grow beyond all bounds. To the sides away from point P the conchoid comes nearer and nearer to line a and arbitrarily close to the line on both the far left and the far right. Line a is the asymptote for all the curves in the family.

If instead of drawing conchoids with respect to a single line, we draw conchoid arcs in relation to three line segments forming an equilateral triangle with point P at its center, we obtain the closed curves as shown in Figure 3.8.3. The transformation then makes an even stronger impression.

Figure 3.8.3

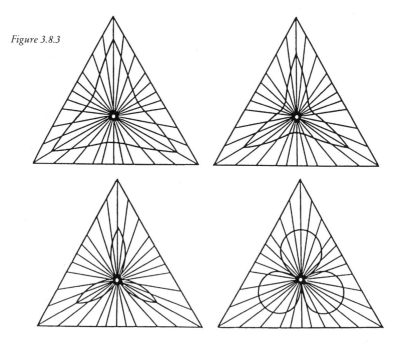

Example 2: We are given a circle C and a line passing through the circle center M. On a line a we are also given a point F, which coincides with M to begin with. What does a curve K look like, whose points lie equally distant from circle C and point F? (F and M are the same, to start with.)

A moment's reflection shows it to be a circle with center at F, concentric to C and with half the radius of C.

We now let C grow by moving M to the right on line a, keeping the left-hand intersection of the circle with line a fixed at point V. We get a family of larger and larger C-circles, all passing through V. How does the curve K change as C grows? (We note that V is fixed.)

The pupils make free-hand sketches, before we begin to analyze the problem more closely. While they are drawing their figures, they come upon the fact that K must be a closed curve. Is it oval, egg-shaped, or what? In Figure 3.8.4 X is a variable point on curve K. We extend the line segment MX out to intersect the circle at N, obtaining the length MN. In order for X to lie as far from F as from the circle's edge we must have FX = XN. From this it follows that

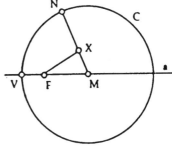

Figure 3.8.4

FX + XM = NX + XM = NM = C's radius FX + XM is thus constant for all points X on curve K. The curve is determined by the equation

FX + XM = constant (= radius)   (1)

From (1) it follows further that the points F and M play identical roles for the curve; it must be symmetrical not only about the line a but also around the line bisecting FM: the perpendicular bisector of FM. The curve is thus doubly symmetrical. Equation (1) specifies, quite simply, the curve which we call an *ellipse* (Fig. 3.8.5).

How does the ellipse change as the circle continues to grow? A few constructions show that the ellipse not only becomes larger but also more drawn out. Finally, if we imagine a circle C with an

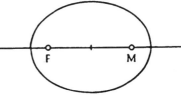

Figure 3.8.5

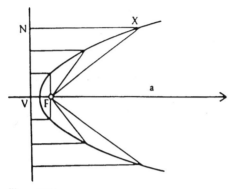

Figure 3.8.6
Parabola. FX = XN

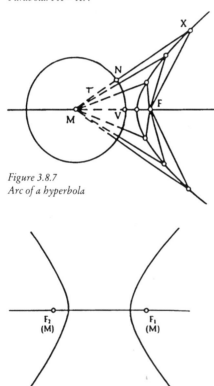

Figure 3.8.7
Arc of a hyperbola

Figure 3.8.8
Hyperbola

"infinitely large radius", then M must be "infinitely far away," and of the circle we see only a straight line passing through V at right angles to a. Our curve is then stretched out infinitely and is called a parabola (Figure 3.8.6).

What happens with our curve if the straight-line circle "flips over" and becomes a circle with center to the left of V?

Once again we have that XF = XN, but this equality can now be written XF = XM - r (Figure 3.8.7) or XM - XF = r, where

r = the circle radius   (2)

X now describes half of a *hyperbola.*

If we change equation (2) to   XM - XF = ± r
that is |XM - XF| = constant

(3)

then the points M and F once again play equal roles, and we have the complete hyberbola (Figure 3.8.8).

As the value of the constant in (3) goes toward zero, the hyperbola straightens itself out; its two branches come nearer each other and close in from both sides on the line bisecting and normal to the segment MF.

If the circle C we start with is very small, we can

summarize the curve trnasformation as going from a little circle through growing ellipses to an infinite parabola to flatter and flatter hyperbolas and finally to a double line. (Figure 3.8.9.)

The equations

(1) XM + XF = constant   and

(2) |XM - XF| = constant

show us that ellipses and hyperbolas can be seen as addition and subtraction curves, respectively. This characteristic of the ellipse and the hyperbola as pictorial representations of addition and subtraction, respectively, is brought out clearly in the curve families (ellipses and hyperbolas) which we constructed earlier in Example 2 of the previous section 3.7.2.

Example 3: What curves might be said to correspond to multiplication and division, respectively?

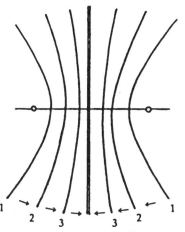

Figure 3.8.9

If A and B are two given fixed points and u and v are the respective distances to them from a point X, then equation

u + v = const.   gives an ellipse and the equation
u - v = const.   gives a hyperbola.

We now form the equation $uv = k$ (4)

and $\dfrac{u}{v} = F$ (5)

where k and F are positive constants and u and v are positive variables. Let d stand for half the distance between points A and B and to begin with we choose

$k = d^2$ in equation (4).

Our task is now to draw the curve corresponding to the equation $uv = d^2$,

or in another form $\dfrac{u}{d} = \dfrac{d}{v}$ (6)

With u=v=d we get a point on the curve which lies halfway between A and B on the line through them. How to obtain other pairs of values for u and v is shown in Figure 3.8.10, which builds upon equation (6).

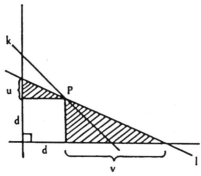

Figure 3.8.10
The shaded triangles are similar. From this follows that $\frac{u}{d} = \frac{d}{v}$ or $uv = d^2$. All lines 1 through point P satisfy $uv = d^2$. The 45° line gives $u = v = d$.

With the aid of a compass we can as a rule obtain 4 points on the curve for every pair of values u, v, symmetrically located with respect to A and B (we could, of course, have let u and v exchange places in equation (4)). The curve we get is the so-called lemniscate (Figure 3.8.11).

For k-values less than d^2 the curve is divided up into two ovals. For k-values greater than d^2, we obtain simple closed curves which are bean-shaped to start with but which straighten out and thereafter turn into curves which remind us of ellipses (Figure 3.8.12).

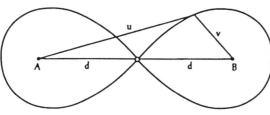

Figure 3.8.11
Lemniscale $uv = d^2$

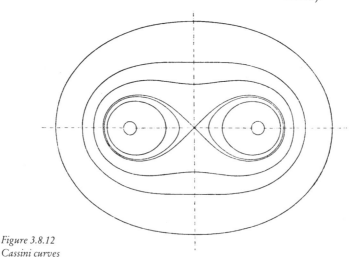

Figure 3.8.12
Cassini curves

The curves obtained are called Cassini curves after the French astronomer J. D. Cassini (1625-1712).

A simple construction, at least when the k-value is not fixed, can be obtained in a manner like that which gave us the ellipse and hyperbola families in Section 3.7.2. Instead of letting the circle radii grow by constant amounts as we did there, we let them increase as a geometric sequence. For example, we might choose the values

$$r = d, kd, k^2d, k^3d, \ldots \; (k = \text{a positive constant, e.g. } k = 1.2)$$

$$\text{and} \qquad r = d. \; d/k, d/k^2, d/k^3, \ldots$$

for both u and v values. We then get two sets of geometrically expanding circles about A and B. The A and B circles intersect one another. By connecting points which simultaneously move outward from A and inward toward B we get Cassini curves: the product uv remains constant, since v is divided by k at the same time as u is multiplied by k.

On the other hand, if we connect points which simultaneously move outward from both A and B, we obtain the curves corresponding to equation (5),

$$\frac{u}{v} = \text{a constant F.}$$

How do these curves look? With our construction method here it is not easy to make them perfect but the suspicion definitely arises that it could be a question of circles. A closer analysis shows that equation (5) actually does give circles for $F \neq 1$. They are called division circles, harmonic circles or Apollonios circles (after the Greek who studied them closely). (See further Figure 3.8.13 and Exercise 5.)

When the constant F inequation (5) grows from small values toward infinity, the division circles grow from tiny circles containing A to larger and larger circles, flip over to the other side of the perpendicular bisector of segment AB (when F = 1) and become circles "containing" B[1]. The constant value F = 5 gives the same circle "containing" B as the circle for F = 1 / 5, containing A. Finally, as $F \longrightarrow \infty$, the circles enclose B all the more tightly . Figure 3.8.14 shows this family of division circles.

But where is the quality of transformation in this family of circles? Is it not simply a matter of a circle which grows in size and simultane-

---

[1] It is actually more natural to continue seeing B as an external point to the circle. Refer to the discussion of the "projective" viewpoint which follows later in this section.

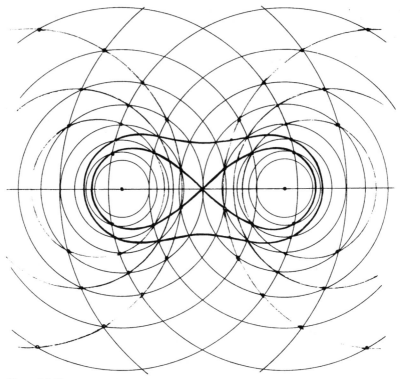

Figure 3.8.13
*Two circle families with radii in geometric proportion (k ≈ 1.27). The figure shows two
Cassini curves and three pairs of harmonic circles.*

ously changes position? The transformation quality first comes to light
when we see the family of circles from the perspective of projective
geometry. In the next section we will take up some of the basic ideas of
projective geometry. Here we begin simply by introducing the *projective*
line.

In classical Euclidean geometry a line extends infinitely in both
directions. On any given axis numbers can be as large as we like, both in
the positive direction and in the negative. The line is endless in both of its
directions. A projective line arises from the Euclidean when we append
an infinitely distant point to the line (and on the line).

Figure 3.8.15 shows a given line a, a given point p above the line
and a line p through point P, intersecting line a at point X.

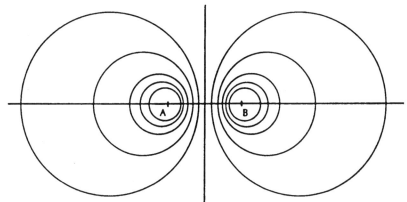

Figure 3.8.14
Harmonic circles corresponding to
$F = \frac{1}{5}, \frac{1}{4}, \frac{1}{3}, \frac{1}{2}$ *and* $\frac{2}{3}$ *(the largest circle) and to*
$F = 5, 4, 3, 2$ *and* $\frac{1}{2}$ *respectively.*
$F = 1$ *gives the perpendicular besector to AB.*

    As line p turns counter-clockwise about P, X moves to the right on line a. When p becomes parallel to a, then according to classic geometry, there exists no point of intersection between them. According to the concepts of projective geometry, we ascribe a common point to lines a and p, namely a "point at infinity," A_∞, on line a. Line p too has a point of infinity, P_∞. Since p runs parallel to a, P_∞ and A_∞ coincide and make up the common point of p and a.

    As P contin-
ues to rotate, the
intersection   X
passes through A_∞
and then returns
from the left at a
finite distance. X

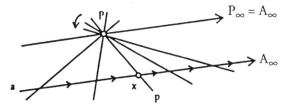

Figure 3.8.15

moves continuously in the same direction, always to the right. That X goes off to the right, through line a's point at infinity and thereby comes back from the left is an idea we cannot clothe in physical form. Considered physically the idea is grotesque. In projective geometry, however, it has a function to serve.

Reconsidering now the growth of harmonic circles from a projective viewpoint, we can describe the circle center's movement with increasing factor F as follows: the center moves to the left on line AB, becomes the point at infinity on AB when F = 1, and returns in from the right as F grows larger than 1.

For F = 1 the circle's interior is the half-plane to the left of the perpendicular bisector to AB. When the circle "flips over," its "interior" — if we wish to maintain continuity — is that which we would normally call the region exterior to the circle. Compared to our habitual way of looking at things, this implies an essential transformation. Let us go a little further with the following example.

Example 4: We let A and B be two fixed points. To begin with, they are the end-points of the diameter of two coincident circles c_1 and c_2. Let n denote the normal bisector to segment AB, and let M_1 and M_2 be the centers of the respective circles (Figure 3.8.16).

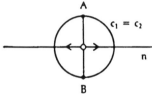

Figure 3.8.16
$M_1 = M_2 = Center$

We now let the centers move at the same rate in opposite directions along line n, M_1 to the left and M_2 to the right. We also require that the circles c_1 and c_2 always pass through the fixed points A and B. This leads to the growth of the circles and to their beginning separation from one another. The plane is hereby divided into various regions.

Let us classify these regions as follows:

No shading: area not covered by either circle
Horizontal shading: area covered by c_1 only
Vertical shading: area covered by c_2 only
Checkered shading: area covered by both c_1 and c_2
(See Figure 3.8.17).

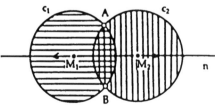

Figure 3.8.17

How does this picture of the plane change, as M_1 and M_2 move away from each other along the projective line n?

Figure 3.8.18 shows the result. In this transformation there arises a quality reminiscent of the negative-positive quality we are familiar with in photography. But it entails considerably more than that, namely, a questioning of the whole concept of inner and outer, shaking up our time-worn habitual thinking that "inside is that which lies within."

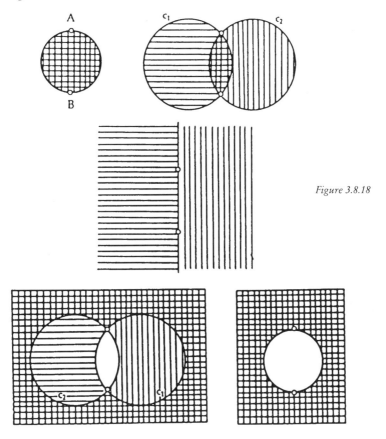

Figure 3.8.18

3.8.2 Forms in Nature, Revisited

Familiarity with geometrical metamorphoses can sometimes shed light on differing natural forms where we would least expect similarity or polarity. This is exemplified by Figure 3.8.19 showing craniums of a

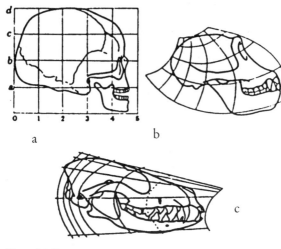

human, a chimpanzee, and a dog. (The figures are taken from d'Arcy Thompson's classic work "On Growth and Form".)

An interesting polarity may be found comparing man's head and his thigh bones: in the head we find the hard bone on the outside and the soft parts inside the cranium. With the leg it is just the opposite: the soft parts are found outside around the bone. The cross sections of the head and thighbone are very nearly circular and hyperbolic, respectively. A comparison with Example 2 above lets us guess that skull and legbone are each other's opposites in the same way that the interior of a circle is on the inside while interior of a shallow hyperbola, in contrast, is found on what we would normally tend to call the hyperbola's "outside."

Figure 3.8.19

3.8.3 Exercises

1. We look back at Example 1 in Section 3.8 and choose the distance XY so that Y is on the other side of P, opposite line a. Draw the curve which point Y describes, for different values of segment length XY.

2. What limiting *form* does the conchoid curve of Example 1 approach, as segment XY grows unboundedly?

3. Is there also a limiting form in Figure 3.8.2, as length XY grows unboundedly?

4. Generalize the transformation of Figure 3.8.18 to include three circles. At the start all three circles coincide. If A, B and C are three fixed points on the common circle, located such that they form the corners of an equilateral triangle, then the three circle centers are to move outward along lines passing through the mid-points of the triangle's sides.

The sketch below shows how the "overlapping figure in the middle" (the intersection of the three circles) changes in form.

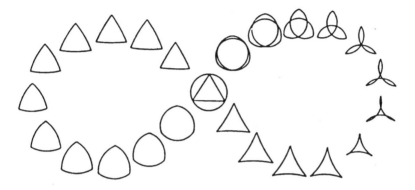

5. Prove that the division curves of Example 3 in this section truly are circles.

3.9 Spherical Geometry

3.9.1 A Few Basics

In their tenth year in the Waldorf schools pupils learn to command the triangle theorems of trigonometry which make it possible to calculate the sides and angles of any triangle we like, when sufficient data is known. Work in geometry is crowned with a period of triangulation out in the field. We have, in a certain sense, complete "command" of plane geometry: we are no longer bound to right triangles and other special cases.

In the eleventh year, during a shorter period, we tackle the sphere — preferably after an introductory study of some important surfaces of revolution, such as the cylinder and the cone. We go here directly to the question: how can one determine distance on a spherical surface (e.g. a ball)? Two points on the sphere are given to us. How do we find the distance between them? Here no pupil has to pay the price of asking, "What use is all this?" — a question which usually is a sign that the teaching has not been concrete enough. The application lies clearly before us: how does one determine the distance between two places on the earth?

Another important problem concerns navigation. On board a ship or airplane, how can one with simple tools determine one's current position?

Other important applications of spherical geometry have to do with the position and movements of the sun, the planets, and the fixed stars, or with the areas of regions on earth, e.g. rounded polar caps or four-sided zones on the earth. Some of the American states have borders which fall along meridians (longitude circles) and latitude lines (circles of constant latitude); the earth's climatic belts also form zones and caps whose areas are of interest.

In short, here are problems and methods worth studying. The most important question during this partial study period, however, is this: how should one build up a geometry for the sphere, starting from basics?

On a plane we measure the distance between two points along the straight line joining them. Such lines serve as "distance lines," so-called geodesic lines. What kind of curves would geodesic lines form on a sphere?

We take a large rubber ball to our aid and mark off two points on its surface with chalk. We turn the ball so that one of the points comes to the top. Now, how does the shortest path to the other point go? "Along a circular arc which you draw straight down — south, if the first point is the North Pole." We draw this arc in by free-hand and discuss whether the answer is correct. The discussion doesn't give a strict proof, but it casts so much light on the problem that the solution with the southerly circle arc seems unique and clear. What radius does the circular arc in question have? The same radius as the sphere. If we lay a circular arc with *smaller* radius through two points, will the distance along this arc be greater? Yes, we see that. (Figure 3.9.1).

The shortest distance between two points on the sphere follows the so-called great circle arc between the points (the great circle has maximum radius = the radius of the sphere). If the points lie diametrically opposite one another, as the Poles, for example, then infinitely many half great circles go between them. They are called meridians when they go from pole to pole. Through Greenwich near London goes the zero-meridian, which everywhere marks off 0° longitude. Together with the equator, which

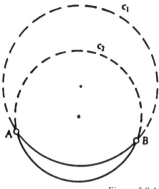

Figure 3.9.1
c_1 = great circle

marks 0° latitude, the zero meridian forms the basis for a network which allows us to specify the location of places and the position of ships and airplanes. As shown in Figure 3.9.2a, longitude λ increases eastward to 180°, while westward we let it decrease to -180°. The meridians for 180° and -180° lie precisely opposite the zero meridian.

Latitude φ increases northward to 90° at the North Pole and decreases southward to -90° at the South Pole (Figure 3.9.2b).

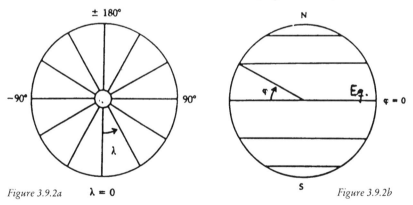

Figure 3.9.2a $\lambda = 0$ Figure 3.9.2b

With relatively good accuracy we can represent the earth as a sphere whose great circles are 40,000 km long. On the meridians and at the equator, 1 degree corresponds to

$$\frac{40000}{360}$$ km, or rounded off, 111.1 km.

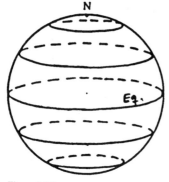

Figure 3.9.3

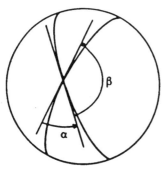

Figure 3.9.4
The angle between two great circles.
$\alpha + \beta = 180°$

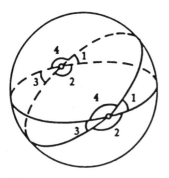

Figure 3.9.5
Two great circles form four
"2-cornered" polygons.

The nautical mile, a measure of distance at sea, is 1/60 of a degree of length, i.e.

$\frac{111.1}{60}$ km, which gives 1.852 km.

Let us now return to the building up of spherical geometry and ask: how do parallel lines run on the sphere? First of all we must observe that all great circles intersect one another. No pair of parallel great circles exists. What curves could then run parallel with a great circle? Clearly a family of circles with smaller radius than the great circle.

Curves parallel to the equator are called naturally enough parallel circles and give constant latitude. They are also called lines of latitude (latitude = width) in this context. (Figure 3.9.3.)

When two great circles intersect each other, they do so with a specific angle, which is determined by the angle between the tangents to the two circles at the point of intersection. This is a definition which quite naturally corresponds to the definition of angle between two curves in a plane (Figure 3.9.4.).

Where two great circles intersect, two adjacent angles are formed which together make up 180°.

When 3 great circles intersect one another there arise a number of *spherical triangles*. How many? There are actually 8 formed. Two great circles form four "2-cornered" polygons (Figure 3.9.5-6).

For angles in a spherical triangle, we allow all values under 180°, just as in plane triangles.

Figure 3.9.6
Three great circles c, c and c form 8 triangles:
On the front:
1. ABC
2. ĀBC
3. AB̄C
4. ĀB̄C
On the "back" side:
1. ĀB̄C̄
2. AB̄C̄
3. ĀBC̄
4. ABC̄

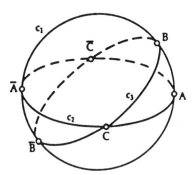

3.9.2 What is the Sum of Angles in a Spherical Triangle?

Isn't it 180°? This is asserted by some pupils, referring to the old familiar statement that "the sum of angles in a triangle is 180°."

But does this theorem apply even on a sphere? Soon an opposition gathers strength and points out the spherical triangle comprising a one-eighth section of the whole sphere and which has right angles in all corners (Figure 3.9.7). Its sum of angles is in fact $3 \cdot 90° = 270°$! "Of course, one of the angles can go up to almost 180°. The triangle is then close to a quarter sphere (Figure 3.9.8) and the sum of angles is near to $180° + 2 \cdot 90° = 360°$."

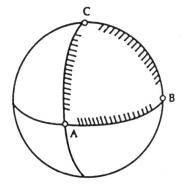

Figure 3.9.7
Sum of angles 270°

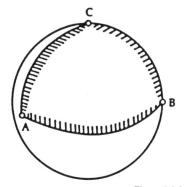

Figure 3.9.8
Sum of angles near 360°

Can the sum be even larger? (The teacher is never satisfied!) Fantasy now begins to wane, but someone suggests that our "almost-1/4-sphere" ought to be able to be expanded downward and include even more area. What happens if we extend the two short sides of our near-360°-triangle downward and let the long bottom edge instead shrink smaller (Figure 3.9.9)?

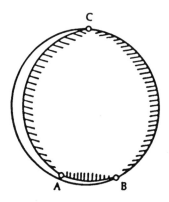

Figure 3.9.9
Spherical triangle with sum of angles near 3.180° = 540°. It covers almost half the globe.

Will not the two 90° angles then become larger?

We make a paper pattern as in Figure 3.9.10. It is simply a circular disc where we have cut away an insignificant "pie" wedge, whose center angle is small, say 3°.

We divide this big circular disc into three equal parts (each with 89° angle), draw radii and fold up the paper into a shallow cone (with point downward). The three arcs then form the sides of a spherical triangle, which in our imagination is a hemisphere arching high over the cone corner (Figure 3.9.11). As can be seen, the angles in the spherical triangle we get will all be nearly 180°.

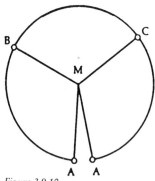

Figure 3.9.10

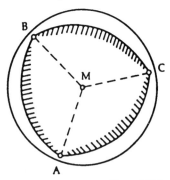

Figure 3.9.11

The conclusion is that the sum of angles in a spherical triangle can vary from slightly more than 180° (a small triangle, flat and almost in the plane) to slightly less than 540° (a triangle covering almost half the sphere).

Closer investigation of the sum of angles (V degrees), which we do not go into here, reveals a linear relationship with the area of the triangle (A):

$$A = \frac{\pi R^2}{180} (V - 180) \qquad \text{where R = the radius of the sphere} \qquad (1)$$

and turned around

$$V = \frac{180\ A}{pR^2} + 180 \qquad\qquad (2)$$

Formula (2) can be written

$$V - 180 = \text{constant} \cdot A$$
$$\text{or} \qquad E = \text{constant} \cdot A,$$

where E = excess above 180° of the sum of angles, and the constant is

$$\frac{180}{\pi R^2}$$

The excess angle E is thus proportional to the triangle area. (See diagram 3.9.12).

On a sphere, therefore, a triangle's sum of angles is uniquely determined by the area or vice-versa, the area is determined by the sum of the angles. In the plane, the sum of angles is fixed at 180°, while the area can vary freely.

Do similar triangles exist on the sphere (other than congruent triangles)? In other words, for a given triangle, does there exist a smaller or a larger with the same angles? The answer which we readily see is "no." The greater the area is made, the greater is the sum of angles.

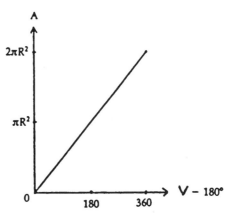

Figure 3.9.12

$$A = \frac{\pi R^2}{180} (V - 180)$$

Area is proportional to the "excess" angle sum, V - 180°

That the sum of angles in a plane is always 180° is closely related to the axiom on parallel lines, i.e. with the existence of a unique parallel (b) to a given line (a) and passing through a specific point off of the line a (Figure 3.9.13).

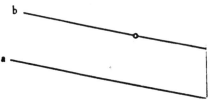

We do not have this axiom for the sphere, and thus the conditions are completely different in regard to the sum of angles in a triangle.

Figure 3.9.13
$\alpha + \beta + \gamma = 180°$ *in the plane.*

3.9.3 Distance

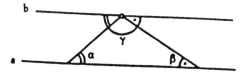

We shall now take up a practical problem: determining the distance between two points on the earth.

With the aid of spherical trigonometry the calculation is quite easy; we would simply use the longitudes and latitudes of the points and plug their values into a formula. But it would take time to develop trigonometry up to the point of the formula which is needed .... We will here use a method with compass, protractor and straight-edge which provides a beautiful example of how one with the aid of methods taken from plane geometry can master the surface of the sphere. As we will see, the method depends upon the simple fact that the circle is a plane curve and that every plane cutting the sphere does so along a circle.

Let us determine the distance between Stockholm-Arlanda Airport (long. 18°; lat. 60°) and Buenos Aires (long. -58°; lat. -34°).

To begin with we note that the longitudinal difference between Stockholm (S) and Buenos Aires (B) is:

$$18° - (-58°) = 76°;$$

S lies thus 76° east of B (not, of course, straight east).

We draw a circle C according to Figure 3.9.14 and let it represent the great circle through B, the western most of the two cities.

The horizontal diameter of the circle is a projection of the equator, and at the top we have the North Pole, N.

We now imagine replacing half of the globe (the half with Stockholm on it) back over the circle so that the North Pole coincides with N and Buenos Aires with B. Stockholm finds itself somewhere in the air above the paper.

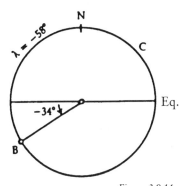

Figure 3.9.14
B = Buenos Aires, N = The North Pole

We shall project Stockholm straight down onto the paper and mark the point with an S. How do we do this? First we draw the horizontal chord corresponding to Stockholm's circle of latitude, i.e. we project the 60° circle of latitude down to the paper — line k in Figure 3.9.15.

With k as an axis we rotate half of the latitude circle into the plane of the paper (above the North Pole) and obtain the half circle c. On c we move 76° east from B's meridian (the left half of the great circle) by placing the protractor at the mid-point of chord k and marking off 76° to the east.

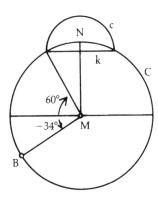

Figure 3.9.15

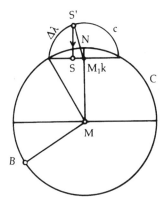

Figure 3.9.16
$$\Delta\lambda = \lambda_S - \lambda_B = 76°$$

We then come to the point S´ on c, which is Stockholm's location on the paper after we have flipped the latitude circle down onto the plane of the paper. If we now rotate the latitude circle back up to its position on the globe, how will S′ move projected onto the paper? At right angles to k.

We therefore draw a line at right angles to k and obtain Stockholm's position (projected onto paper). (Figure 3.9.16)

We next imagine the shortest route from Stockholm to Buenos Aires. In reality it is a great circle arc. What we want to know is: how many degrees of arc, how large an angle, does this great circle take up at the center of the globe, at M? On paper its projection would be an ellipse arc, but could we possibly rotate the globe so that the route comes to lie in the plane of the paper?

Yes, it can be done. We rotate the globe about the diameter BM, i.e. we use BM as an axis! Stockholm then moves down toward the paper, and in projection, along that line which goes through S at right

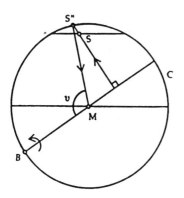

angles to BM, coming to the new position S″ on the great circle C (Figure 3.9.17).

Finally we measure the angle

$$v = BMS''$$

and obtain the *angular distance* for the shortest route BS. The protractor shows 113°.

Since each degree on a great circle corresponds to 111.1 km, the distance is 113 · 111.1 km, or approximately 12,600 km.

Figure 3.9.17

If both cities have the same longitude, the construction is unnecessary. The one city then lies straight south of the other, i.e. the cities lie on the same meridian and we can directly calculate the distance. For example, the distance between Stockholm and Cape Town (18°; -34°) is quite simply the latitude difference times 111.1 km, i.e.,

$$\{ 60 - (-34)\} \cdot 111.1 \text{ km} \approx 10,400 \text{ km}.$$

In class, exercises on this theme are somewhat embellished with flight times, local times, landing times, etc. The class has available a world atlas showing local times, and he pupils may easily be left to themselves finding the longitudes and latitudes of the cities in question.

3.9.4 Exercises

1. How far is it from Stockholm Arlanda (18°; 60°) to the poles? (A great circle is 40,000 km.)

2. How far is it *straight east* from Cape Town (18°; -34°) to Sydney (151°; -34°)?

3. Determine with compass and protractor the distance between

a) Stockholm (18°; 59°) and Mexico City (-99°; 19°)
b) Philadelphia (-75°; 40°) and Dar es Salaam (39°; 7°)

4. Calculate the distance from New York (-74°; 41°) to Hanoi (106°; 21°).

5. A person who is *not* at the North Pole travels first 1000 km south, then 1000 km straight east and finally 1000 km north. He is then back at his starting point. Where on earth is he?

6. How large must the sum of angles in a spherical triangle be for the triangle's area to cover 25% of the sphere?

7. The formula for calculation of the area of a spherical cap or a ring zone on a sphere is

$$A = 2\pi Rh,$$

where R is the radius of the sphere and h is the width of the ring zone or the cap.

What per cent of the earth's area is taken up by the temperate zones, whose latitudes vary between 30° and 60° (both above and below the equator)?

3.10 A Little About Projective Geometry

3.10.1 A Projection Problem

In Euclidean geometry, the lengths of line segments and the size of angles play a decisive role. To become aware of this we have only to remember a few basic problems and theorems from Euclid's work, the *Elements:*

- Constructing a given angle with compass and straight edge, i.e. reproducing an angle
- Constructing a chord of given length through a specified point within a circle
- Sides opposite equal angles in a triangle are equal; opposite a larger angle is a larger side
- The Pythagorean Theorem: in a right triangle, the square of the hypotenuse is equal to the sum of the squares of the sides which form the right angle (Figure 3.10.1).

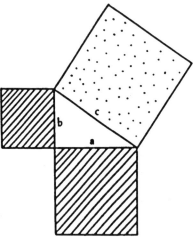

There are, however, many problems in geometry which do not concern quantities but rather put the stress on relations. We know that if an object casts a shadow, then the shadow, the object and the light source stand in certain relation to one another. The shadow picture is dependent on the contours of the object, on the location of the light source, and on the surface upon which the shadow is cast.

Since shadow plays a role in drawing and painting, there eventually developed a "theory" of shadows. Earlier still the need for drawing objects in perspective had arisen among artists. The first attempts at this in the 1400's sometimes appeared rather comical: one can see a table with plates drawn such that they seem to be sliding down the table toward the viewer. People sitting in chairs appear to be half-sitting, half-standing; the seats of the chairs slope steeply forward in the picture.

Figure 3.10.1
The Pythagorean theoriem: dotted area =
sum of the shaded areas. $c^2 = a^2 + b^2$

Many other examples could be presented. Stronger and stronger was the desire to be able to draw pictures in perspective so that the objects appeared natural. And from the theory of perspective developed projective geometry. In fact, it was projective geometry which laid the foundation for the theory of shadows. The name itself, projective, comes from the Latin word *projicere,* to throw or cast, verbs which we in fact use concerning shadows: the sun throws (casts) a shadow.

The simplest projections prevail between two plane figures, e.g. two triangles. Let us pose the following problems:

1. Can a point-source of light and a small equilateral triangle be positioned so that the triangle's shadow precisely covers a given larger triangle with a specified position in the room?

2. Can one find a plane and a vantage point for the eye in the room, where the perspective picture of a given triangle on the plane will be a specified small equilateral triangle?

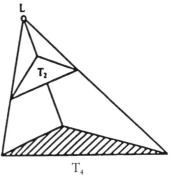

Figure 3.10.2 illustrates these problems (with the desired conditions fulfilled). What we see directly in this figure is that the two problems are the same from a geometrical point of view. They can be formulated in the following way: A triangle T_1 is given in 3-dimensional space. Further, a small equilateral triangle is given. Can this latter triangle and a point be so positioned that both triangles become perspective in relation to the point, regardless of T_1's shape?

Figure 3.10.2
T_1 = *given triangle*
T_2 = *equilateral triangle*

To begin with, we put a large paper triangle (of arbitrary shape) on the classroom floor and cut out small equilateral triangles of cardboard. Some pupils instead cut equilateral triangular holes in the cardboard. We now experiment with our eyes to see if we can place the little triangle (cut-out or hole) in front of the eye in such a position that the cut-out "exactly" covers the triangle on the floor or that the large triangle could "just barely" be seen through the hole. We have to move around a bit but finally seem able to find positions in which we succeed. No one doubts that the problem has a solution. On the other hand, our method could certainly not be called a proof.

How could we achieve a proof for the positive answer?

We begin to simplify the problem by choosing a special case. Why not first let the given large triangle on the floor be equilateral, i.e. have the same form as the little triangle?

Now it is easy to devise a construction which proves the answer: the small triangle can be placed "horizontally" directly above the triangle on the floor so that corresponding corners of the two triangles can be connected to form a three-sided pyramid. We draw a picture of this reasoning and conclude not only that the idea holds water but also that the little triangle can be as small as we like. The point source of light (the center of perspective) comes to lie at the pyramid's top (Figure 3.10.3).

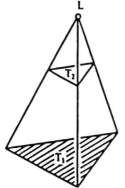

Figure 3.10.3
A simple case: T_1 and T_2 are equilateral.

We return now to the original problem and try to find other simplifications. Since the size of the small triangle does not matter, could we exchange it for a larger equilateral triangle whose sides are just as long as one of the sides of the triangle on the floor?

"Then it'll be easy! We can put the equilateral triangle right down on the floor with one of its edges alongside an edge of the floor triangle."

We implement this thought-process with paper triangles and draw a sketch. But where shall the lamp be put? It takes a good while in fact before the class figures this out. They are apparently locked in by the thought that there is only *one* place where the light can be put and are surprised that it can be placed anywhere at all along the "upper" part of line a in Figure 3.10.4.

And once again, we return for a fresh attack on the original problem. Now there is use for the knowledge gained earlier that the small triangle is unimportant in the first simplified case.

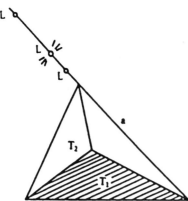

Figure 3.10.4
Sketch of the solution for a simplified case.

Might not the equilateral triangle of Figure 3.10.4 be traded for a smaller? "Naturally, it can be replaced by a smaller one nearer the light."

"How, in that case?"

"We can shift the triangle, parallel to itself, toward the light. It will then get smaller but keep its equilateral shape."

The problem is solved and we draw the final figure (Figure 3.10.5).

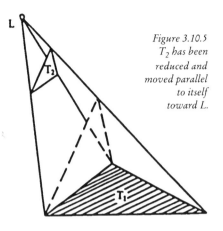

Figure 3.10.5
T₂ has been reduced and moved parallel to itself toward L.

3.10.2 Désargues' Triangles

Gérard Désargues (1593-1662), a French architect who was active in Paris and Lyon, published a theorem on perspective triangles in the year 1648. This theorem has come to be one of the foundations of geometry and plays a decisive role particularly in the theory of perspective. Désargues wrote a dissertation on perspective. His work was appreciated by none less than Pascal, who investigated closely related problems. Désargues' and Pascal's names have come to be associated with respectively the theorem on perspective triangles and an important proposition on hexagrams (6-sided polygons), inscribed in conic sections. Both of these mathematicians laid important foundation stones in projective geometry but their contributions were not given any particular attention by their contemporaries. No trace was left of Désargues' old publications, but in the nineteenth century an English mathematician, Arthur Cayley, succeeded in finding an old transcript of Désargues' manuscript. Not until then were Désargues' and Pascal's methods properly appreciated.

We shall not go into the structure of projective geometry further than to acquaint ourselves with Désargues' triangles and with a few simple so-called dualities. We can begin directly from our introductory study with the paper triangles.

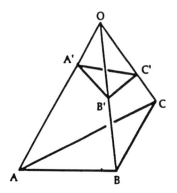

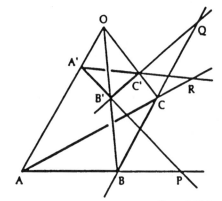

Figure 3.10.6
Two triangles perspective with
respect to the point O.

Figure 3.10.7

We draw two triangles in perspective positions (Figure 3.10.6). ABC is perspective with A´B´C´ (and vice versa) when lines AA´, BB´ and CC´ go through a common point, the center of perspective (O).

Problem: Have these triangles any other lines in common than the three lines through O? Obviously not, if we keep to the figure we have drawn. But let us consider the triangles as each being formed by three *lines* and ask if these lines meet somewhere in space (Figure 3.10.7).

Some pupils answer "yes" immediately, for one pupil perhaps based on a feeling, but others can motivate their answers: lines AB and A´B´ must (in general) intersect each other, since they lie in a common plane, the "wall" OAB of a pyramid. For the same reason must BC and B´C´ intersect, as well as lines AC and A´C´ (in general). We have thus obtained three points of intersection in space, which we will call:

$$P = AB \times A´B´$$
$$Q = BC \times B´C´$$
$$R = CA \times C´A´$$

The class extends the lines and marks the intersections; for some pupils one or more of the intersections turn out to be off the paper. With time one becomes better at choosing the lines so that everything stays on the paper.

How do P, Q and R lie? "In a straight line," someone answers. "Almost a straight line," say others. "Not mine." "Well, maybe they do." Et cetera.

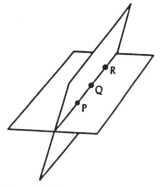

Figure 3.10.8

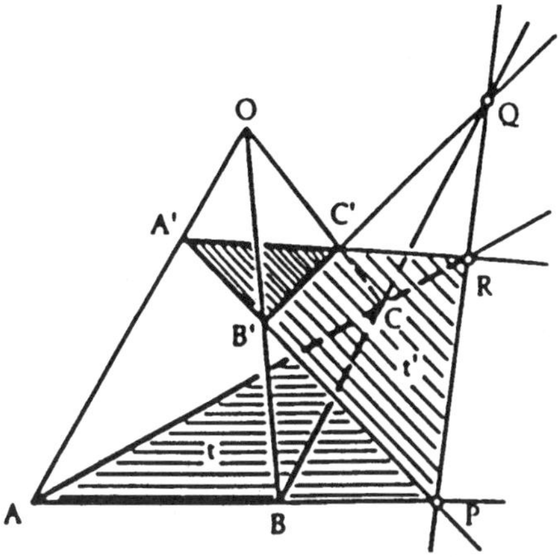

Figure 3.10.9
PQR = the intersection of the
planes of the triangles.

We compare each other's figures and find that in those cases where P, Q and R do not lie in a line, they at least form a very long thin triangle.

Can it be that P, Q and R always lie on a line?

We direct out attention to the *planes* t and t´ in which the two triangles lie. The planes t and t´ meet at the three points P, Q and R. How do two planes meet, if they have a point in common? Always along a straight line! (Figure 3.10.8)

We have herewith reached the conclusion that P, Q and R lie on a line and one that particular line which makes up the intersection of the planes t and t´. All that is left to do is to draw a clear figure illustrating this, for example, as in Figure 3.10.9. "But what about if AB is parallel to A´B´?"

Let us draw this case! (Figure 3.10.10a). What happens now with P, Q and R? We can still, in general, construct Q and R. The line QR, or the so-called Désargues' line, will still be determined by these two points. But how does it lie?

Look very carefully! — "It goes parallel to AB and A´B´ " (Figure 3.10.10b).

It is not difficult for us to focus on the triangle planes t and t´ and *see* that QR must be parallel to AB. And P? "It doesn't exist," some say. "We could say that P is infinitely far away," say others.

It is precisely the latter way of looking at things which Désargues, and later all work in projective geometry, builds upon.

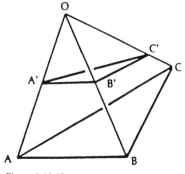

Figure 3.10.10a
AB ‖ A'B'

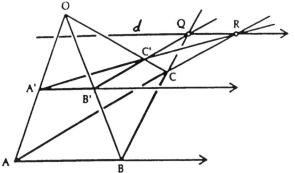

Figure 3.10.10b
d is Désargues' line

The point is that it is our free option to give every line one point at infinity, but not two. Were that the case we would violate the axiom that two lines have *one* point of intersection. We would of course be removing the unsatisfying exception that two lines *sometimes* lack a common point (intersection) but land instead in a new exception: parallel lines would then have *two* common points at infinity.

The classical Euclidean line has infinite extension in two directions. The new concept which projective geometry introduced was that every line has *one* point at infinity in addition to the finitely distant Euclidean points. One then speaks of the projective line. With the addition of the point at infinity, a line closes on itself: any point X which moves in the one direction of movement along the projective line "returns" from the other side with the same direction of movement at a finite distance, if it passes through the point at infinity, P_∞. Consider, for example, the point of intersection between a line and a rotating line as shown in Figure 3.10.11. This continuity aspect was introduced by the Frenchman Victor Poncelet (1788-1867).

But back to Désargues' triangles. If *two* sides of the one triangle are parallel with two sides (respectively) of the other, what happens to Désargues' line? As one easily sees, all three sides in T are then parallel with the corresponding sides in T´. The triangles then lie in parallel planes and all of the points P, Q and R will be points of infinity. (Figure 3.10.12)

Can one here speak of Désargues' line? Yes! Analogous to two parallel lines having a common point at infinity, we give parallel planes a common line at infinity.

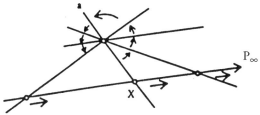

Figure 3.10.11

The question now is to see whether points at infinity and lines at infinity can be fitted in with the basic axioms concerning points and lines which form the foundation of classical geometry and which we do not wish to throw overboard. Here we must skip over most of such a study for reasons of space. However, it turns out that

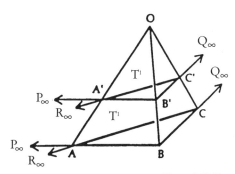

Figure 3.10.12
AB ‖ A'B', BC ‖ B'C'.

- two points have *one* line connecting them;
- two planes have *one* line of intersection.

We observe that two lines (in space) in general lack a common point. But *if* the lines have a common point, then they also have a common plane, and vice versa.

These basic phenomena let us guess that points and planes form a duality in *space*. In the plane, the dual elements are points and lines:

- two points have one line connecting them (common line)
- two lines have one point of intersection (common point)

A particularly important result of this is that plane geometry may be built up with complete symmetry: for every configuration of points and lines, there is a corresponding opposite dual configuration where the role of the points is taken over by the lines and the role of the lines by the points.

Here follow a few simple examples which exhibit such duality:

Three points A, B and C form corners of a triangle, if they do not lie in a straight line (if they lack a common line). Figure 3.10.13a.

Three lines a, b and c form a three-sided figure, if they do not pass through a single point (if they lack a common point). Figure 3.10.13b.

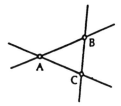

Figure 3.10.13a

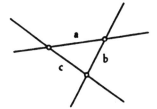

Figure 3.10.13b

A complete quadrangle (four-angled polygon) is formed by 4 points (of which no 3 form a straight line) and the 6 connecting lines determined by the points (Figure 3.10.14a).

A complete quadrilateral (four-sided polygon) is formed by 4 lines (of which no 3 pass through a point) and the 6 points of intersection determined by the lines (Figure 3.10.14b).

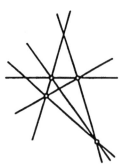

Figure 3.10.14a
Complete quadrangle

Figure 3.10.14b
Complete quadrilateral

A circle may be considered as made up of infinitely many points (Figure 3.10.15a).

A circle may be considered to be formed by infinitely many lines (Figure 3.10.15b).

Figure 3.10.15a
A circle can be formed of points.

Figure 3.10.15b
A circle can be formed by lines.

The fundamental duality in the plane comes out even with the two basic elements themselves, the point and the line. Not only the point, but also the line may be considered as the basic building block, an indivisible fundamental element. But *if* we consider the line as made up of infinitely many points (a row of points), then we may by the same token consider the point as made up of infinitely many lines, see Figure 3.10.16.

Figure 3.10.16
The line as a row of points and the point as a bundle of lines.

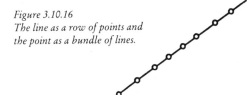

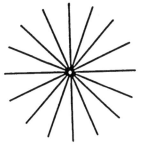

The functional duality between the row of points and the line bundle shows clearly that

— a point-line is created when a line bundle is intersected by a line not in the bundle.

—a line bundle is created when the points of a point-line are connected with a point not on the line.

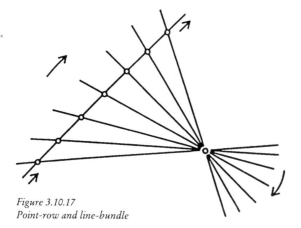

Figure 3.10.17
Point-row and line-bundle

In *space*, as mentioned earlier, the plane and point are dual elements of each other.

Projective geometry's dual structure was not discovered until the 1820's, but it then became the most important factor in further develop- ments which gave projective geometry a dominant position over earlier geometries.

Besides perspective and the theory of shadows, applied projective geometry plays a role in photogrammetry (with applications in the tech- nology for producing maps via photographs, for example). Its great value in the school is that it gives pupils entirely new aspects of what geometry is all about, it allows them to experience considerably more of geometric qualities than the usual quantitative problems, and it for work and dis- covery to youngsters who do not have, or believe themselves not to have, ability for the equation-solving of analytical geometry.

Particularly the duality exercises are aimed at developing agility in thinking and to mould will power into thinking. One learns to see prob- lems from opposing directions and gains a much broader concept of space than the usual idea of space as a great empty container of points.

3.10.3 Exercises

1. We start with the Désargues figure in Figure 3.10.9 and investi- gate whether point O, the center of perspective, may be placed at A: does the figure have two triangles which are perspective in space with respect to A?

The answer is actually yes! Find the triangle and determine their Désargues-line.

2. The same problem as Exercise 1 above but with C as the new center of perspective.

3. (a) We draw Figure 3.10.9 once again and investigate whether the figure contains two perspective triangles with the line OB'B as the Désargues-line. Where do the triangles lie and where is their center of perspective?

(b) Show that any one of the 10 points in Figure 3.10.9 can be the center of perspective and that any one of the 10 lines cane be the Désargues-line.

4. In the complete quadrilateral of Figure 3.10.14b we denote the quadrangle's corners by A, B, C and D and the other points by E and F (E = AB x CD, F = BC x DA).

If the pairs of points (A,C), (B,D) and (E,F) are connected we get three lines which form a triangle. What triangle would correspond to this in Figure 3.10.14a?

5. We label four of the points of the dot-circle in Figure 3.10.15a as A, B, C and D (for example, clockwise). Corresponding to this in Figure 3.10.15b will then be four tangents a, b, c and d in cyclical sequence. These four elements determine a complete quadrangle and a complete quadrilateral respectively. Study how these correspond to each other in the two figures. Among other things, compare the positions of the points and the lines in the triangles which, according to Exercise 4, may be associated with the figures.

6. Draw a Désargues-figure with the center of perspective infinitely far off.

7. Draw a Désargues-figure where one of the corners of the triangles is a point at infinity.

3.11 George Boole and Set Theory

3.11.1 A Pioneering Contribution

It is said that Gottfried Wilhelm von Leibniz (1646-1716), one of mankind's universal geniuses, read Latin fluently and began to study

Greek before reaching the age of 12. As a 20-year-old he published the work "De Arte Combinatoria" (On the Art of Combination) which he later considered "a schoolboy's essay" but which came to be the starting point for a new level of abstraction in pure mathematics.

If we hold ourselves to arithmetic and algebra, we can separate out three levels of abstraction, as far as numbers are concerned:

1. The numbers are drawn or written as pictures, for example the hieroglyphs of ancient Egypt. The number 9 is indicated by nine pictures of the symbol which stands for 1, and so on. The degree of concreteness is very high here.

2. Numbers are indicated by quite abstract characters, for example, as in the base-ten system. The symbol 1 for one is concrete, but already the symbol 2 for two is quite different from the ancient Egyptian or Roman II.

The number ten is indicated by a combination of two digits, 1 and 0, and so forth. In position systems, among them our 10-system, each added digit gives different sized contributions depending on its position (place) in the number. For example, the first five in 5157 contributes 5000 to the number, the second five contributes 50.

3. Numbers are represented by letters. For example, 2n means any even number if n is a whole number (an integer). 2n + 1 means any odd number, a variable x can take on any real value — and so on.

Thanks to representation of numbers by letters it is possible to carry out proofs in arithmetic as in Section 3.5.

Leibniz lifted arithmetic up to a considerably higher level of abstraction by letting letters represent *intervals* on the number line (the x-axis of real numbers). He developed, as we shall soon see, the basis for a kind of interval arithmetic in his work "The Art of Combination." His purpose in this was not small: he wished to create "a general method in which all truths of the reason would be reduced to a kind of calculation. At the same time this would be a sort of universal language or script, but infinitely different from all those projected hitherto, for the symbols and even the words in it would direct the reason..."[1]

The purpose and the wording remind one of Descartes' declaration of program in a work on geometry and methods from 1637, where

---

[1] From E. T. Bell, *Men of Mathematics*

Descartes saw himself as laying out "a completely new science, which will come to admit a general solution of all such problems as can be put as questions of quantities, continuous or discontinuous, each in accordance with its own nature... so that almost nothing will be left to discover in geometry." (Descartes, "Dissertation on the Method of Geometry.")

Following Section XX in one of the fragments of Leibniz' essay, which is still in existence, we let A, B C etc. denote intervals on a line and the symbol ⊕ denote an operation which we can call "addition of intervals." A ⊕ B will then denote the interval which contains points from the interval A or from B or both. We will say, henceforth, "points from A or B" (Figure 3.11.1).

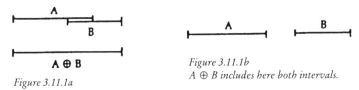

Figure 3.11.1a

Figure 3.11.1b
A ⊕ B includes here both intervals.

Leibniz notes such basic rules for interval arithmetic as

"Axiom 1"   B ⊕ N = N B   (as we see, the commutative law for the newly introduced addition)

"Axiom 2"   A ⊕ A = A   (quite "revolutionary" compared with a + a = 2a!)

"Proposition 5" If A is contained in B and A = C, then C is contained in B.
"Proposition 7" A is contained in A.
"Proposition 9" If A = B, then A ⊕ C = B ⊕ C
"Proposition 10" If A = L and B = M, then A ⊕ B = L ⊕ M.

Apart from "Axiom 2" we recognize these rules from our own arithmetic and algebra. But let us ask, how can we or should we solve an interval equation of the type

$$A \oplus X = A \oplus B \quad \text{where X is unknown?}$$

Can we simply apply our usual equation-solving procedures and write X = B?

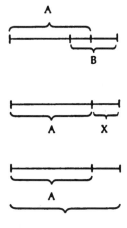

Figure 3.11.2
Here we have
$A \quad X = A \oplus B, but \ X \neq B$

The pupils are allowed to think awhile and are then welcomed to give examples for or against. The results are unanimous: we cannot draw the conclusion that X = B, as Figure 3.11.2 gives witness to.

"Did Leibniz also think of that?" Yes, in a "scholium" (explanatory note) he asserts that "Proposition 10" cannot be reversed. He also gives other examples of non-reversible theorems in interval algebra.

Even if "all the truths of reason" do not quite fit into Leibniz' interval arithmetic, we must still honor him for having developed a new system of calculation, a system in which the symbols no longer represent numbers but rather intervals or sets of points.

3.11.2 Some Applications

"May we see an example of such interval arithmetic in practice?" Of course! Suppose that 100 people are asked about their knowledge of French and German: "How many consider that they have tolerable command of these languages, in the event of a trip to France or Germany or for reading a book?"

Suppose that 35 people considered that they knew French tolerably and 75 German. We now ask ourselves:

a) How many of the 100, as a minimum, are certain to command both French and German?

b) How many at most could command both French and German?

c) How many at most could speak neither language?

d) How many must, as a minimum, have a knowledge of at least one of the languages?

We let F be the set of people who know French, G the group of people who know German. The letter n will denote number, e.g. we will write

$$n\,(G) = 75, \qquad n\,(F) = 35$$

In order to answer question (a) we must separate G and F as much as possible; the overlapping part is to be minimized. We work forward to Figure 3.11.3 and find that

Min n (G and F) = 10

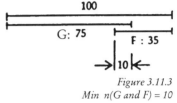

Figure 3.11.3
Min n(G and F) = 10

The number 10 can be obtained in different ways, e.g. as 35 - 25 or as 75 + 35 - 100 (the overlapping portion is left over in this subtraction).

In a similar manner we get the answers to the other questions:

b) Max n (G and F) = n (F) = 35, when F is included in G.
c) Max n (neither G nor F) = 25, when F is contained in G.
d) Min n (at least one lang.) = 75, when F is contained in G (Figure 3.11.4).

If we were to formulate similar problems concerning 3 or 4 languages, the difficulties would increase considerably. And yet abstract combinatorics came to be more and more concrete and manageable for 19th century mathematicians in England. We shall next turn to the most prominent of these mathematicians, George Boole.

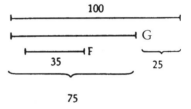

Figure 3.11.4
b) Max n(G and F) = 35
c) Max n (neither G nor F) = 25
d) Min n (at least 1 language) = 75

3.11.3 Boole (1815-1864)

George Boole can be seen as the most prominent founder of abstract algebra. He was, like Leibniz, a genius at language (translating poems by Horatio at 12 years of age). Besides Latin and Greek, he interested himself in modern languages such as French, German, and Italian. Boole, who was more or less self-educated, was employed as a teacher at age 16 and started his own school in Lincoln, England, at age 20.

On the recommendation of a mathematician in London, A. de Morgan, Boole was appointed professor of mathematics at a newly started college in the town of Cork, Ireland, where he worked until his death in 1864. Boole achieved great success both as a lecturer and as a mathematician. His best works were an analysis of logic from a mathematical point of view (1847) and in particular "An Investigation of the Laws of Thought..." with applications in mathematical logic and probability theory (1854). The importance of his contributions was first recognized by Bertrand Russell (about 1910) who considered Boole's Investigation of the Laws of Thought to be the first truly pure mathematics.

When Boole was alive, Leibniz' essay on combinatorics had likely been completely forgotten. It seems improbable that Boole could have known of that work, partly because Leibniz' basic ideas were not even given attention in Germany, partly because Boole's concepts in many ways differed from Leibniz'. Boole, for example, used "addition" in another sense, letting a + a remain 2a and interested himself primarily in his new arithmetic's applications in logic. He investigated, for example, how such conjunctions as "both-and," "either-or," "if, then" and other expressions could be given arithmetic analogy. He formulated a number of axioms for his symbolic logic. We will here look at the set theory which Boole's followers, primarily John Venn (1834-1923), came to develop.

After serving as a clergyman and lecturer, Venn switched over to mathematics. From him we have the rather well-known Venn or set-diagrams which made set theory considerably more understandable and easily manageable than interval arithmetic. In 1864 W. S. Jevons (1835-1882) published the work "Pure Logic, or Qualitative Logic as Distinguished from Quantitative" and made a clear distinction between the concepts "a or b is true" (meaning a or b or both are true) and "either a or b is true" (one or the other is true). Venn accepted and used Jevons' system. We can put this in the terminology of set algebra in the following way: we let a Roman one, I, denote a set of something, for example 1000 people. We let A denote a subset of I (perhaps including the whole set), for example, those people who read American fiction, and let B be another subset, those who read British fiction.

We introduce further

A′ = the complementary set of A
    = the set of people who do *not* read American literature = all
      those among the 1000 who are not members of A

and analogously B′,
as well as two operations ∪ and ∩

A ∪ B = "the union of A and B"
           = the set of people who belong to A or B or both:
           = the set of elements which belong to A *or* B
A ∩ B = "the intersection of A and B"
           = the set of elements which belong to A *and* B
           (i.e. to both A and B)

     We can consider the forming of a complement as an operation
which acts upon one set while union and intersection are operations
which affect two sets. Figure 3.11.5 shows the graphical equivalents as
Venn diagrams.

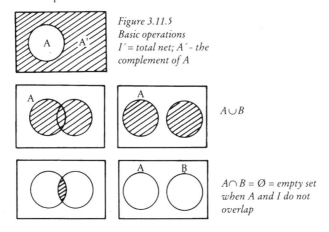

Figure 3.11.5
Basic operations
I′ = total net; A′ - the
complement of A

$A \cup B$

$A \cap B = \varnothing = empty\ set$
when A and I do not
overlap

     If A and B have no common elements, they are said to be separate
or non-overlapping or disjoint. Their intersection then has no elements,
which is usually expressed: the intersection = "the empty set" and one
writes

$$A \cap B = \varnothing$$

Here it is interesting to compare this with a · b = 0 in arithmetic.
Further examples of ground rules are:

$$\varnothing \cup A = A \qquad \varnothing \cap A = \varnothing \qquad A \cup A' = I$$
$$I \cup A = I \qquad I \cap A = A \qquad A \cap A' = \varnothing$$
and $\qquad (A')' = A$

Before going one with further development of the system of axioms we should, as in the classroom, first have a look at a concrete example of calculation with sets in a Venn diagram.

Suppose that the number of people in sets A and B above are 560 and 780 respectively and that 200 people belong to neither A nor B. Can we, from these figures, obtain other information? Can we, for example, determine that a certain number of people must have read both American and British literature, i.e. must belong to the intersection?

We draw a Venn diagram as in Figure 3.11.6. Obviously there are people in the intersection, in the lens-shaped overlapping area. But how many? There are different ways to proceed.

We can let x be the number of people in the intersection and then add up three separate sets which together make up the union:

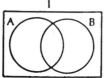

Figure 3.11.6
$n(I) = 1000$
$n(A) = 560$
$n(B) = 780$
$n(\text{neither A nor B}) = 200$

$$(560 - x) + x + (780 - x) \text{ that is } 1340 - x.$$

This expression must be equal to $1000 - 200 = 800$, so we get the equation $1340 - x = 800$ and $x = 540$.

A direct method:

The number of people in the intersection is $(560 + 780) - 800 = 540$.

The line of thought is that adding 560 and 780 gives a double booking of the intersection amount. There are not so very many problem variations with 2 sets, but 3 sets give more. Of much greater interest, however, is set theory's structural foundation.

3.11.4 Rules of Arithmetic

With the aid of Venn diagrams, the class can investigate which rules should be included in the foundations of set theory. Which of the following ought to be considered "laws" in set theory?

$A \cup B = B \cup A$
$A \cap B = B \cap A$ commutative laws?

$\begin{cases} (A \cup B) \cup C = A \cup (B \cup A) \\ (A \cap B) \cap C = A \cap (B \cap A) \end{cases}$ associative laws?

$A \cap (B \cup C) = (A \cap B) \cup (A \cap C)$ distributive law for intersection?

Corresponding laws apply in normal arithmetic if we let addition correspond to union and multiplication to intersection.

We find that all of these laws must be accepted. The last of them is illustrated in Figure 3.11.7a-b:

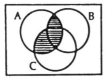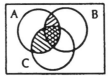

Figure 3.11.7a
$A \cap (B \cup C)$

Figure 3.11.7b
//// $A \cap B$
\\\\ $A \cap C$

Might we expect that there is also a distributive law for union, i.e. that the following should apply?

$$A \cup (B \cap C) = (A \cup B) \cap (A \cup C)?$$

We compare with normal algebra: is $5 + 2 \cdot 8 = (5 + 2) \cdot (5 + 8)$?
No!

We draw Venn diagrams for both the left and the right side of (·). See Figure 3.11.8.

What do we see in the figures? We see that if A, B and C overlap one another, then (·) is true!

It is not difficult to check that this result applies completely generally.

It turns out that the operations union and inter-

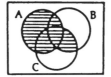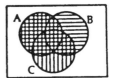

Figure 3.11.8
Three overlapping sets
$A \cup (B \cap C) = (A \cap B) \cup (A \cap C)$

section ($\cup$ and $\cap$) participate entirely symmetrically in the system of axioms. They take on dual roles. For every general relation which is true there corresponds a dual relation which is obtained from the first by interchanging $\cup$ and $\cap$ everywhere, and interchanging I with $\emptyset$, while these are dual elements. For example, the relation

$$A \cap A' = \varnothing$$

transforms into the true relation

$$A \cup A' = I$$

More examples are given in the exercises. The basic dualities are, in summary, the following: For all A and B,

$$\begin{cases} A \cup B = B \cup A \\ A \cap B = B \cap A \end{cases} \qquad \begin{matrix} (1a) \\ (1b) \end{matrix}$$

$$\begin{cases} (A \cup B) \cup C = A \cup (B \cup C) \\ (A \cap B) \cap C = A \cap (B \cap C) \end{cases} \qquad \begin{matrix} (2a) \\ (2b) \end{matrix}$$

$$\begin{cases} A \cup (B \cap C) = (A \cup B) \cap (A \cup C) \\ A \cap (B \cup C) = (A \cap B) \cup (A \cap C) \end{cases} \qquad \begin{matrix} (3a) \\ (3b) \end{matrix}$$

$$\begin{cases} A \cup \varnothing = A \quad (4a) \\ A \cap I = A \quad (4b) \end{cases} \qquad \begin{cases} A \cup I = I \quad (5a) \\ A \cap \varnothing = \varnothing \quad (5b) \end{cases}$$

$$\begin{cases} A \cup A' = I \quad (6a) \\ A \cap A' = \varnothing \quad (6b) \end{cases}$$

We have thus come upon an example of duality even in an algebra, one of a more abstract nature than the duality we found in plane projective geometry (Section 3.10).

3.11.5 Some Applications of Set Theory

From set theory the step is not far to problems in probability and combinatorics.

For example, the basic set I could be the outcome of throwing 4 casts of a dice,

A might be the subset of outcomes containing at least one six
B might be the subset of outcomes containing at least one one, etc.
Then A′ would be the subset of outcomes with no sixes
      B′ would be the subset of outcomes with no ones

and   A ∪ B = the subset of outcomes which contains at least one six or at least one one

A ∩ B = the subset of outcomes which have both one or more sixes and one or more ones.

The number of possible outcomes, elements, in this basic set I is

$$6^4 = 1296.$$

We leave combinatorics and instead ask ourselves how Boole and other mathematicians went about formulating the outlines of a mathematical logic.

A, for example, could represent the statement "It is raining"
A´ would then mean the opposite of A (not-A): "It is not raining"
B could be another statement: "It is warm"
B´ would be its opposite: "It is not warm"
A ∪ B corresponds then to: "It is raining or warm"
A ∩ B corresponds to: "It is raining and warm"

From daily life we are familiar with logical statements of the type, "If the animal is a cod, then it is a fish." One can illustrate this statement using a Venn diagram (Figure 3.11.9): we let a C-circle represent the set of cods and an F-circle the set of all fish. The rectangle can represent the set of all animals species. Since the C-circle lies within the F-circle, every point in C is also a point in F. That the C-set is contained in the F-set is usually written

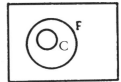

Figure 3.11.9
C ⊂ F: "If C, then F".

$$C \subset F.$$

Reversing "If C, then F" to "If F, then C" is, of course, wrong: if we have a fish, it need not be a cod. A point in the F-circle need not belong to the C-circle.

The illustration of the propositions of the type "If A, then B" with an A-set inside of a B-set is thus completely natural. Besides the expressions OR, AND and IF-THEN, the expressions EITHER-OR and NEITHER-NOR also have their graphical representations (Figure 3.11.10):

Either - or   Neither - nor

Figure 3.11.10 (overlapping sets)

3.11.6 Logic and Truth

It is interesting to follow the development of the "IF-THEN" relation in presentations of symbolic logic. At one point of time the implication, "If A, then B" or "A implies B" took on the following form: "A is false, or A is true, and in this case B is true." For example, we can reformulate the statement, "If the voltage in the line exceeds 100 volts, the transistor will be destroyed" to read: "The voltage in the wire is no more than 100 volts, or it is more than 100 volts and then the transistor will be destroyed." One can say more laconically: "The voltage in the line does not exceed 100 volts, or the transistor will be destroyed." The word "or" has here a broader meaning than the exclusive "either-or": nothing has actually been said about the transistor if the voltage is less than 100 volts, e.g. 99 volts. We have a tendency in everyday language to interpret 100 volts as the maximum safety limit for the transistor, but this is a generalization.

The laconic formulation of the implication can be represented by the expression

$$A' \vee B \quad \text{(not-A or B)}$$

where A stands for "the voltage exceeds 100 volts," and B stands for "the transistor is destroyed" (Figure 3.11.11).

The implication of "If A, then B" which in set theory's symbolic language is written $A \subset B$ (A is contained in B), has therefore been replaced by the expression $A' \vee B$.

In this version logicians went so far trying to simplify that they even allowed false if-statements to "imply" true then-statements. In *The World of Mathematics* (ed. James R. Newman), chapter XIII, section 4, Alfred Tarski gives the "implication"

"If 2 times 2 is 5, then New York is a large city"
as an example of a true statement.

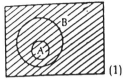 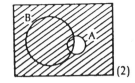 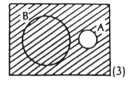

A´ ∨ B in three cases as Venn diagram (shaded area)   Figure 3.11.11
A represents "The voltage exceeds 100 V"
B represents "The transistor is destroyed"
A´ represents "The voltage does not exceed 100 V"
B´ represents "The transistor is not destroyed"
Only the first case can illustrate the statement
"If the voltage exceeds 100 V then the transistor will be destroyed"

The motivation is, by analogy with the transistor example, that the statement reformulates to: "Not-P or Q is true," where P stands for "2 · 2 = 5" and Q is "New York is a large city."

Even the following statements are considered to be meaningful:

(1) "If 2 · 2 = 5,   then New York is a small city"
(2) "If 2 · 2 = 4,   then New York is a large city"

Because one of the clauses in the reformulations

(1) "2 · 2 is not 5   or New York is a small city"
(2) 2 · 2 is not 4   or New York is a large city"

is true (the first in (1), the second in (2)), the statements are accepted as meaningful. One need not be surprised to hear the following words by Alfred Tarski (from *The World of Mathematics*, Ch. XIII, 4, edited by James R. Newman): "The divergency in the usage of the phrase "if ..., then ..." in ordinary language and mathematical logic has been at the root of lengthy and even passionate discussions..."

> The two facts, that a false clause can imply any arbitrary clause and that a true clause is implied by any arbitrary clause, are sometimes referred to as the implication paradoxes. Yet they are not real "paradoxes." They express a discrepancy between the concept of material implication on the one hand and the concepts of conditional relation and consequence relation on the other. The stated discrepancy is remarkable. One can say that it has not been paid sufficient attention by the classics of modern logic.
>
> (G. H. von Wright in "Logik, filosofi och Språk")

In the simplified so-called material implication, exemplified above, the clauses can, as we have seen, completely lack any meaningful relation. But the material implication has been an effective tool when its purpose was to solve problems in logic with rational methods patterned along the lines of arithmetic. Here there are many opportunities in a higher class to take up discussion of logic, reality, and language as different modes of expression, and thereby bridge over into other school subjects.

A particularly meaningful theme is to train the pupils in understanding the difference between "If ..., then ..." and "If and only if ..., then ....." We ought to be aware of the meaning of necessary conditions, sufficient conditions, and conditions which are both necessary and sufficient. We will return to this subject in Section 6.2. See also Section 3.5.3.

The laws which were originally formulated for set theory can strangely enough be used in such divergent areas as logic, probability, and the theory of electric circuits. The algebra which presides over these areas was given the name Boolean Algebra.

3.11.7 Switching Networks

We shall conclude the chapter by looking at a few examples which give a hint as to how Boolean algebra can be applied in switch circuits.

In the theory of switch networks there exist only two values, 0 and 1. We define two operations, "addition" and "multiplication" with the tables

$$
\text{(A)} \quad
\begin{aligned}
0 + 0 &= 0 \\
0 + 1 &= 1 \\
1 + 0 &= 1 \\
1 + 1 &= 1
\end{aligned}
\qquad
\text{(M)} \quad
\begin{aligned}
0 \cdot 0 &= 0 \\
0 \cdot 1 &= 0 \\
1 \cdot 0 &= 0 \\
1 \cdot 1 &= 1
\end{aligned}
$$

The only unusual thing, as we see, is that $1 + 1 = 1$.

The value 0 is applied to an open switch or a non-glowing lamp. (Figure 3.11.12).

Summarizing, we can say that 0 corresponds to an open circuit and 1 corresponds to a closed circuit.

In the function table below we establish that (A) above corresponds to all possible situations with two switches in parallel, while (M) corresponds to two switches in series.

Figure 3.11.12

	Switch x	Switch y	Functional value f for the lamp
(A)	0	0	0
	0	1	1
	1	0	1
	1	1	1
(M)	0	0	0
	0	1	0
	1	0	0
	1	1	1

These tables are illustrated in Figure 3.11.13.

We can now step by step test the structure of switching network theory in relation to set algebra (we take only a few steps here):

For a switch value x, there corresponds the opposite switch value x´ such that

x = 1 is true whenever x´ = 0 and

x = 0 is true whenever x´ = 1

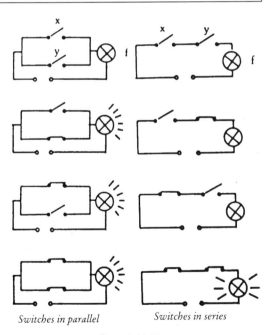

Switches in parallel   Switches in series

Figure 3.11.13

We note that

$$x + x' = 1 \qquad \text{(always)}$$
$$x \cdot x' = 0 \qquad \text{(always)}$$
$$(x')' = x \qquad \text{(always)}$$

See Figure 3.11.14.

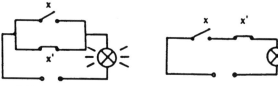

Figure 3.11.14
$x + x' = 1$ *and* $x \cdot x' = 0$ *respectively. Illustrated here for the case $x = 0$.*

Let us now ask: do the following laws hold:

(1) $\qquad x + y = y + x \qquad$?
$\qquad\qquad x \cdot y = y \cdot x \qquad$?

(2) $\qquad (x + y) + z = x + (y + z) \qquad$?
$\qquad\qquad (x \cdot y) \cdot z = x \cdot (y \cdot z) \qquad$?

(3) $\qquad x + (y \cdot z) = (x + y) \cdot (x + z) \qquad\qquad$?
$\qquad\qquad x \cdot (y + z) = (x \cdot y) + (x \cdot z) \qquad\qquad$?

It is easy to verify (1) and (2). That laws (3) also hold is taken up in one of the exercises below.

In this way one can discover that the axioms of Boolean algebra have their analogous counterparts in the theory of electrical circuits.

3.11.8 Exercises

1. In a class of 30 there were 25 who were positive toward Germany and 21 toward France as the country to visit on a school trip.

a) Determine the maximum number of students who could be in favor of both countries.

b) How many, at a maximum, could be against both countries?

c) How many, at least, must be in favor of both countries?

2. In an opinion survey the people polled were asked among other things to answer yes or no to questions of the type: "Do you have confidence in politician A?" 80% showed confidence in A, 70% in B, and 60% in C, when the questions were asked separately one at a time. What per cent of the people asked, as a minimum, must then have confidence in all three of the politicians?

3. Draw two Venn diagrams with two overlapping subsets A and B. Show A' ∪ B' in the diagram and (A ∩ B)' in the other. Are these two sets identical?

Investigate if this result holds even when A and B are non-overlapping sets and when A is wholly contained in B.

4. The same as Exercise 3 but with the sets A'∩ B' and (A ∪ B)'.

5. Exercise 3 and 4 illustrate de Morgan's laws:

$$\text{and} \quad \begin{matrix} (A \cup B)' = A' \cap B' \\ (A \cap B)' = A' \cup B' \end{matrix}$$

Show that these two identities are each other's duals (refer to Section 3.11.4).

6. Verify with the aid of an x - y - z table the identities

$$x + (y \cdot z) = (x + y) \cdot (x + z)$$
$$\text{and} \quad x \cdot (y + z) = (x \cdot y) + (x \cdot z)$$

(the distributive law for the operations + and · respectively, in switch circuits).

7. Show with the help of a function table that the switch circuit in figure (a) below is equivalent to the the simpler circuit in figure (b), that is, show that f ≡ g.

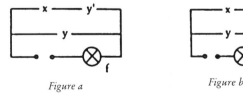

Figure a   *Figure b*

3.12 Exercises in Concept Formation

3.12.1 From Galileo to Riemann

From Galileo's study of little balls rolling down the inclined plane, a red thread leads forward to our time and technology with moon landings, satellites and missiles. This thread goes via the method of calculation which Newton and Leibniz separately developed concerning derivatives.

Galileo wished to determine how far a ball rolls down a plane during a specified given time. In order to arrange the experiment methodically he began with an assumption about the ball's velocity in motion down the plane: that the velocity is proportional to the time. From this assumption Galileo succeeded in *calculating* the distance rolled. He could then arrange his experiments such that they gave him information concerning the correctness of his theory. He found complete confirmation that his intuition had been right.

Newton turned Galileo's approach around. At that time, thanks to Galileo, the rolling ball's distance was well known. Newton posed the reverse problem: how can we calculate the instantaneous velocity of the ball, if we know the rolling distance as a function of time?

Out of this question Newton formed the concept which nowadays is called the derivative and which is the mathematical tool for calculation of instantaneous velocity, among other things. This work was one of many famous achievements made by Newton at the age of 22-23 while Cambridge University was closed in 1665-66 due to the plague. It is an exceptionally important exercise in thinking to start from the well-known concept of average speed and think through the thoughts which lead up to the notion of instantaneous velocity, and more generally, to the derivative of a function, i.e. to the rate of change of a function.

From several simple examples (taken out of train timetables, diagrams, the t^2-expression for the ball's rolling distance, etc.) we soon arrive at the formula

$$\frac{f(b) - f(a)}{b - a}$$
for the average velocity of a function in an interval $a \le x \le b$

To get at the instantaneous velocity at x = a we need to make b variable and let b both decrease toward a and increase toward a. The interesting thing is that we cannot simply put b = a in the expression above. We would then get

$$\frac{f(a) - f(a)}{a - a} = \frac{0}{0} \qquad \text{which doesn't tell us anything.}$$

We must carry out the passage to the limit b → a and investigate what happens to the quotient

$$\frac{f(b) - f(a)}{b - a}$$

It turns out that thanks to our knowledge of algebraic rules for factoring expressions, squaring and so on, we can derive a limiting value for the quotient above, as b goes toward a. (I must leave out this development and refer those readers who are not familiar with the concept of derivatives to a high school text.)

It is very interesting from a theory of knowledge point of view, that with the derived limiting value we are able to define what we mean by instantaneous velocity, a notion that we managed to grasp intuitively long before now.

It takes considerable time and requires much careful work before a heterogeneous class of 17-18 year old pupils can feel they have mastered the concept of derivative. It is interesting for them to experience how simple the actual arithmetic is in problem applications, compared with all the effort required to develop the concept.

3.12.2 Galileo

Let us return to Galileo and ask: how did Galileo calculate the rolling distances theoretically, before he went to experiment? He assumed — at least after a certain amount of "playing around" — that the velocity of a ball which starts from rest and begins rolling down a plane is proportional to the time, i.e. that

$$v = k \cdot t$$

where   v = velocity

k = a constant

and   t = time from start

How would he now calculate the accumulated distance rolled? Galileo began by assuming *uniform* motion, i.e. motion where the velocity is constant, say v_0. For this case the distance is of course $s = v_0 t$, that is, simply the product of velocity and time.

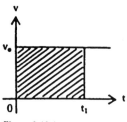

Figure 3.12.1

In a plot of velocity as a function of time (Figure 3.12.1), the identity $v = v_0$ is a straight line parallel with the t-axis. The value of the distance up to a time $t = t_1$ is $v_0 t_1$. In the plot this value corresponds to the area of the shaded rectangle. For uniform motion then, the accumulated distance is proportional to the area of the corresponding rectangle in the v-t-plot.

Galileo assumed now that an analogous relation holds even for the case where $v = kt$, i.e. he assumed that the distance is proportional to a triangular area (Figure 3.12.2) enclosed by the lines $v = kt$, $t = t_1$ and the t-axis.

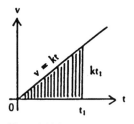

Figure 3.12.2

This area is easily calculated to be:

$$A = \frac{kt_1 \cdot t_1}{2}$$

or

$$A = \frac{kt_1^2}{2}$$

The distance rolled should thus be proportional to the square of the time, which Galileo, as we know, confirmed through his experiments.

Was it audacious of Galileo to assume that the area calculation would give correct results even for accelerated motion? Hardly. We can imagine the linearly increasing speed approximated by a staircase function, where velocity increases in small jumps and is constant during the sub-intervals of time. With such a velocity the distance would be the sum of a number of products (Figure 3.12.3), and it is not hard to imagine passing toward the limiting velocity function

$$v = kt$$

in such a way that one lets the length of the sub-intervals of time go toward zero. The staircase function then approaches the straight line v = kt everywhere, and the sum of the sub-areas nears, as closely as we like, the triangle's area

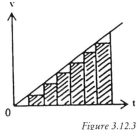

Figure 3.12.3

3.12.3 The Riemann Integral

Ought not this method of approximation with a staircase function work with other velocity functions than the linear? If the answer is yes, we would command a method of calculating distances for any given velocity function.

What must we require of the velocity function, when it is more "difficult" than a linear function? We note the requirement "positive continuous function," which primarily means that the velocity curve must be in one unbroken piece.

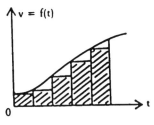

Figure 3.12.4

The graphical picture might then look as shown in Figure 3.12.4, where an approximating staircase function is drawn in under the curve. In order to do the approximation more accurately, we also draw in a staircase function with oversized values (Figure 3.12.5).

We can now begin to develop the integral which Bernhard Riemann (1826-66) gave form to in a dissertation the year 1850 (at the age of 24!).

To begin with we subdivide the interval $a \leq t \leq b$, over which the function $f(t)$ is continuous, into n sub-intervals

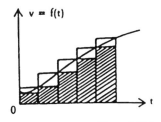

Figure 3.12.5

$$\Delta t_1, \Delta t_2, \Delta t_3, \ldots, \Delta t_n, \qquad \text{where}$$

$$\Delta t_1 = t_1 - a, \Delta t_2 = t_2 - t_1, \ldots, \Delta t_n = b - t_{n-1}.$$

We denote this subdivision as A.

According to a well-known theorem of analysis, and intuitively understandable to the pupils, a continuous function always has a largest value and a smallest value on a closed interval. We can therefore introduce the nomenclature

$M_k = \max f(t)$ on the sub-interval Δt_k

$m_k = \min f(t)$ on the sub-interval Δt_k

We imagine now that $f(t)$ represents a velocity function. The distance during the time interval t_k can then be approximated by

the product $M_k \Delta t_k$ (too large a value)

and by   the product $m_k \Delta t_k$ (too small a value).

The accumulated distance for the whole time interval $a \le t \le b$ can then be approximated by the slightly high *upper sum*

$$S_A = M_1 \Delta t_1 + M_2 \Delta t_2 + \ldots + M_n \Delta t_n \text{ and by the slightly}$$

lower sum $s_A = m_1 \Delta t_1 + m_2 \Delta t_2 + \ldots + m_n \Delta t_n.$

The question is now how dependent these values are on the particular subdivision of the interval, and what happens to them if we let the length of the sub-intervals approach zero. These two questions are interwoven with each other, as we shall see. We shall investigate them step by step and come to the concept of the Riemann integral.

1. What happens to the upper and lower sums if one adds subdividing points and thus gets more but smaller intervals?
     Answer: The upper sum cannot increase, but may decrease;
           the lower sum cannot decrease, but may increase.

2. Can any lower sum be larger than any upper sum?
     Intuitively the answer is easy: no.
There is an elegant proof of this (given by the Frenchman G. Darboux in the 1880's):

We are to prove that if A and B are two arbitrarily chosen subdivisions of the interval, then

$$s_A \leq S_B \qquad\qquad (1)$$

In order to prove this we let C be the subdivision of the interval which the A-points and the B-points together create (Figure 3.12.6). Then, by the conclusion of step 1. above,

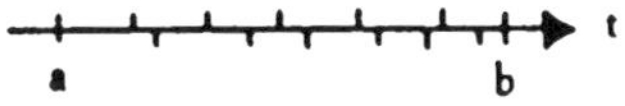

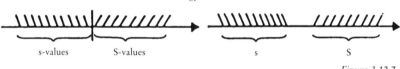

Figure 3.12.6
Marks pointing upward: A
Marks pointing downward: B
All marks: C

$$s_A \leq s_C \leq S_C \leq S_B$$

from which $s_A \leq S_B$.

(The inequality $s_C \leq S_C$ we have understood earlier.)

3. From (1) it follows that the set of all values given by lower sums lies to the left, on the real number axis, of the set of all values given by upper sums. The question now is: have these two sets of numbers a common limit (upper limit for the lower sums, lower limit for the upper sums) or are they separated by an interval? (Figure 3.12.7).

or

![s-values S-values   s   S]

s-values   S-values   s   S

Figure 3.12.7

4. Darboux' version of the Riemann integral now is: If both value sets have a common limit, we define that number as the integral of f (t) on the interval $a \leq t \leq b$ and denote it by

$$\int_a^b f(t)\ dt$$

If the sets are separated by an interval, the function f (t) is said to be non-integrable on the interval $a \leq t \leq b$.

5. We shall now check whether the continuity of f (t) insures the existence of the integral. Do the lower sums and upper sums have a common limit? It is sufficient to show that the difference S-s can be made arbitrarily small, if the interval subdivision is made finer and finer, i.e., as the sub-interval length goes toward 0.

We get

$$S - s = (M_1 - m_1) \cdot \Delta t_1 + (M_2 - m_2) \cdot \Delta t_2 + \ldots + (M_m - m_n) \cdot \Delta t_n \tag{2}$$

We must here recall a theorem (on so-called uniform continuity) that *all* the differences $M_k - m_k$ can be made smaller than an arbitrarily small number Σ, by insuring that all Δt_k are shorter in length than a sufficiently small number δ.

Then this holds true:

$$S - s \leq \Sigma (\Delta t_1 + \Delta t_2 + \ldots + \Delta t_n) = \Sigma (b - a).$$

Since Σ can be chosen arbitrarily small, S-s can be made arbitrarily small, and we have insured the integral's existence.

6. One of the results of step 5 is the extremely important conclusion that the reasoning there holds independent of how the interval subdivision is done; the only thing that matters is that the length of the sub-intervals approaches zero.

The value of the integral, as a common limit for the lower and upper sums as max Δt_k approaches zero, is thus independent of how the sequence of finer and finer subdivisions is made.

7. If f(x) changes sign, one analyzes separately these sub-intervals determined by the zero points. The construction may then be carried through as before.

We now know that the integral of a continuous function *exists*, but how can it be calculated? We have little use for knowing of its existence if we cannot calculate its value!

This situation is rather typical in the area of mathematics dealing with limits: first comes proof of existence ("the limit exists") and after that calculating its value. (In the derivation of the derivative these occurred simultaneously: the existence of the limit value was shown along with its calculation.)

3.12.4 How the Riemann Integral is Calculated

We study the function

$$I(x) = \int_a^x f(t)dt \qquad \text{for } a \le x \le b$$

and ask ourselves: does the derivative of this function exist? To investigate this we form the difference quotient (from our knowledge of the definition of the derivative):

$$\frac{I(x+h) - I(x)}{h} = \frac{1}{h}\left(\int_a^{x+h} f(t)dt - \int_a^x f(t)dt \right).$$

From the construction of the integral, it follows directly that the difference on the right hand side is equal to the integral over the interval $x \le t \le x + h$, and we get

$$\frac{\Delta I}{h} = \frac{I(x+h) - I(x)}{h} = \frac{1}{h}\left(\int_x^{x+h} f(t)dt \right).$$

Let m_h and M_h denote the minimum $f(t)$ and maximum $f(t)$ over the interval $\quad x \le t \le x + h$. Then the following inequalities hold:

$$\frac{1}{h} \cdot m_h \cdot h \le \frac{\Delta I}{h} \le \frac{1}{h} \cdot M_h \cdot h$$

that is

$$m_h \le \frac{\Delta I}{h} \le M_h$$

Now when $h \to 0$, both m_h and M_h approach the value $f(x)$, since $f(x)$ is assumed to be continuous.

We therefore let h go toward zero and find that the limit of $\frac{\Delta I}{h}$ exists and is equal to $f(x)$.

In other words: the function $I(x)$ has a derivative (and thereby is also continuous):

$$I'(x) = f(x) \qquad \text{or} \qquad \frac{d}{dx}\left(\int_a^x f(t)\, dt \right) = f(x) \quad [1] \tag{3}$$

---

[1] Right-derivative or left-derivative at the endpoints x = a and s = b respectively.

A beautiful relation between integral and derivative!

But still we do not know how to *calculate* an integral!

Assume that we have found a function F (x) such that its derivative is f (x):

$$F' (x) = f (x)$$

Example: If f (x) = x²   we can note that F (x) = $\dfrac{x^3}{3}$.

We call F (x) a *primitive* function of f (x) (a source function for f (x) in the sense that f (x) is the derivative of F (x)).

We then are in the situation that our function F (x) and the integral I (x) both are primitive functions of f (x); both have the same derivative, f (x). Here we refer to "The Fundamental Theorem of Integrals," which says that

if I' (x) = F' (x)   then   I (x) = F (x) + a constant.

Popularly interpreted: if two trains, the one following the other, always keep the same speed, then the distance between them remains constant.

According to this theorem we have

$$I (x) = F (x) + C \quad (C = \text{constant})$$

x = b   gives us $I (b) = \int_a^a f (t) \, dt = F (b) + C$ (4)

x = a   gives us $\int_a^a f (t) \, dt = 0 = F (a) + C$ (5)

Subtracting (4) minus (5) now gives us

$$\int_a^b f (f) \, dt = F (b) - F (a)$$

and we have found the formula for calculating an integral.

If we continue with our earlier example, f (x) = x², and let a and b be 1 and 4 respectively, we get

$$\int_1^4 x^2 dx = F (4) - F (1) \quad \text{where} \quad F (x) = \dfrac{x^3}{3}.$$

The value of the integral becomes $\dfrac{4^3}{3} - \dfrac{1^3}{3} = \dfrac{64 - 1}{3} = 21.$

3.12.5 So Much "Theory"?

Are not sections 3.12.3 and 3.12.4 an overambitious load on the students? If one knows the class and makes this judgment, then one might let construction of the Riemann integral be a voluntary chosen extra work project. But isn't it so "beautiful" that it is worth the time required? And doesn't it give valuable *exercise*, both in abstraction and ability to hold several trains of thought together?

The calculation F (b) - F (a) does not require any particular thought when simpler functions are concerned. The calculation becomes a routine procedure. Here, as in the section on derivatives, the formula is very simple to use compared with the thought behind it.

3.12.6 A Word about Karl Weierstrass

We have seen that *all* functions which are continuous on a closed interval are integrable.

What about derivatives: have all continuous functions also a derivative? Are there functions which are continuous on an interval $a \leq x \leq b$ but which somewhere lack a derivative?

We only have to look at Figure 3.12.8 to see the plot of a function which is continuous but lacks a derivative at four places, at the points x_1, x_2, x_3 and x_4.

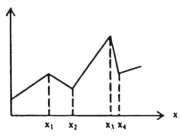

Figure 3.12.8

Can a continuous function lack a derivative everywhere? It seems to be completely impossible. It therefore aroused enormous surprise when Karl Weierstrass (1815-1897) presented a continuous function which does not have a derivative for any value of the variable.

The curve of such a function lacks a tangent everywhere, i.e. at every point it lacks a direction in spite of being continuous (unbroken).

One example of a function with this property is the sum of the infinite series

$$f(x) = \frac{\sin 3x}{2} + \frac{\sin 9x}{4} + \frac{\sin 27x}{8} + \ldots + \frac{\sin 3^n x}{2^n}$$

After Weierstrass other mathematicians have contributed new examples of continuous functions which everywhere lack derivative.

Weierstrass' results are a good example of how mathematics sometimes shows that "the impossible might be possible."

As another example of this, one can study the research results of Georg Cantor (1845-1918), who showed that there exist many different degrees of "infinitely many."

3.12.7 A Little on Georg Cantor's Research

Georg Cantor was a pioneer who developed set theory in the true sense of the term. What we today even in grade school call "set theory" (as part of the 'new math') is by and large completely apart from the area in which the real problems in set theory have their beginnings. In the school's "set theory" one speaks namely about problems involving *finite* sets. Cantor took on the study of *infinite* sets.

One of his very first questions was: are there greater, "bigger" infinities than the infinity we associate with the natural numbers 1, 2, 3 ...?

Let us call this set N and the set of all integers (N plus zero plus the negative integers) H. Is H a larger set than N? Cantor introduced a notion of size which meant that two sets A and B (containing elements a and b respectively) are equally large (equivalent in size) if the a- and b-elements can be put together in pairs, either finitely or infinitely in number. For example, the set of 6 oranges is equally large as the set of 6 apples. (It is such examples, among other things, which some school children are put to work on at a much too early age.)

And now returning to our question: are N and H equally large? It is quite clear that the set N is contained in H. We might perhaps say that it makes up about "half the amount." But maybe the N-elements can be put into pairs with the numbers in H. For example, the set of all numbers counting by "tens", 10, 20, 30, ... is equivalent in size to N, in spite of the fact that the terms make up only a fraction of N: we can namely form the following series of pairs, which runs through both sets equally quickly:

$$1 - 10$$
$$2 - 20$$
$$3 - 30$$

etc.

Two infinite sets can thus be equally large or equivalent even though the one is a part of the other. Here it is possible that the parts are as large (in the Cantor sense) as the whole.

Now , can the integers of H be put into pairs with the natural numbers? Yes, we can construct the following table:

H	N
0	1
1	2
-1	3
2	4
-2	5
3	6
3	7

etc.

H is therefore not greater in size than the set N; H and N are equivalent. We say that H is *countable* — forming pairs with the numbers of N is actually counting.

Is the set of all corner points (lattice points) in an xy-coordinate system countable? (Figure 3.12.9) Here we have an infinite number of horizontal lines and each contains infinitely many points. Is this set countable?

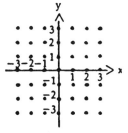

Figure 3.12.9

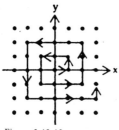

Figure 3.12.10

The answer is easily found: we begin at the origin (point 1) and number the points in an endless spiral outward (Figure 3.12.10). Each lattice point will get counted exactly once. Insight tells us now that the lattice points can be numbered so their set is equivalent to N.

Cantor went on to ask: can all the numbers between 0 and 1 be numbered, i.e. are they countable? They can be illustrated by the interval I of points between 0 and 1 on the x-axis. By way of introduction we can establish that any sub-interval of I, *however small*, must have the same size as I! Take for example, a fifth of the interval and place this one fifth-interval I′ perspectively opposite I, as shown in Figure 3.12.11. Each point x on I has a perspective point x′ on I′. We see that the points on I′ can in this manner be linked in pairs with the points on I. Their equivalence in size is then a consequence of the perspectiveness between I and I′ relative to the center A (Figure 3.12.12).

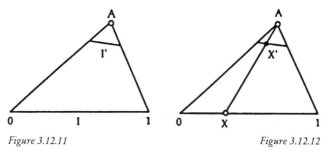

Figure 3.12.11   *Figure 3.12.12*

But now to I: are its points countable?

Let us with Cantor assume, *on a trial basis*, that I's numbers are countable. This means that the numbers could be written down on an infinitely long list, which would contain *all* the numbers between 0 and 1. Then, during the list's "printout," every number between 0 and 1 must sooner or later come out on the list. The numbers between 0 and 1 can be expressed as decimal fractions of the type:

$$0.abcd\ldots,$$

where the letters stand for digits. In order to avoid double accounting of fractional numbers we agree that decimal fractions such as

0.34699999... (where all decimals after a certain place are nines)

will be written as in principle completed fractions with infinitely many zeros, in our example

0.34700000...

This is entirely correct, since 0.9999... must be assigned the value 1.

Let us now assume, for example, that the numbers in the beginning of the list look as follows:

number no. 1	0.35649225...
number no. 2	0.20017180...
number no. 3	0.01587621...
number no. 4	0.77389285...
etc.	

Cantor now shows that we can form a new number x (or even several), which cannot possibly be included in the endless list which has been given us and which claims to contain all the numbers between 0 and 1. He forms, for example,

$$x = 0.4\ 137...$$

according to the following principle:

the 1st decimal in x	shall be different from the 1st decimal in the 1st number of the list (here 3)
the 2nd decimal in x	shall be different from the 2nd decimal in the 2nd number in the list (0 above)
the 3rd decimal in x	must differ from the 3rd decimal in the 3rd number in the list (5 in our list)

and so on.

We can choose the first four decimals to be, for example, 4, 1, 3, and 7. In continuing th nth decimal of x is to be different from the nth decimal in the nth number in the list

Number 1	0.	③	5	6	4	9	2	2	5...
Number 2	0.	2	⓪	0	1	7	1	8	9...
Number 3	0.	0	1	⑤	8	7	6	2	1...
Number 4	0.	7	7	3	⑧	9	2	8	5...

. . . .

Through this Cantor insures that x cannot possibly be the same as any number in the list. To be on the safe side we must avoid giving x an infinite row of nines (if this were the case an x would appear to be different from the number 0.124000000... in the list but actually have the same value as that number, and thus x would be included in the list). Avoiding an infinite row of nines is no problem since for each decimal position we have 8 other digits than 9 from which to choose.

The assumption that all the numbers between 0 and 1 could be listed in an infinitely long list leads therefore to the contradiction that the list would not include all the numbers. This contradiction shows that the set of numbers from 0 to 1 *is not countable*. This set is "larger" than the set N. Cantor called this greater size the continuum.

This result concerning the real numbers was one of Cantor's first discoveries. In an impressively energetic and persevering manner Cantor continued his journey of discovery through the realm of infinity. The work required much mental energy and alertness and took its toll on Cantor's health. On top of this his methods received strong criticism from some mathematicians. The strenuous intellectual work as well as opposition from authorities in the field eventually broke Cantor, and he was obliged to seek help at a mental hospital on several occasions.

When it was later shown, not least through Bertrand Russell's work, that set theory, as it had developed, led to paradoxes of a difficult nature, efforts were aimed at introducing greater stringency into the concept of set. Cantor's contributions maintained their great importance, and it is no exaggeration to see him as one of the boldest and most pioneering of all mathematicians.

3.12.8 Exercises

1. What average velocity does the Intercity train "Tiziano" have between Hamburg (departure 7.45) and Hannover (arrival 9.08)? The distance is 178 km.

2. With what average rate, expressed in volts/milliampère (V/mA), does the voltage in the diagram below increase, as the current increases from 1.1 mA to 5.5 mA?

3. A ball starts from rest and rolls down a plane. The slope of the plane is such that the function for the distance rolled is

$$s(t) = 1.6t^2$$

where t is measured in seconds and s in meters. Calculate the ball's average speed during the time from t = 2 to t = 4 · 5.

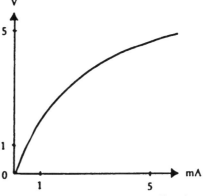

Exercise 2

4. Toward what limit does the expression

$$\frac{(t + h)^2 - t^2}{h}$$

go, as h goes toward zero?

First rewrite the numerator, and then show that the h in the denominator can be factored out and the fraction reduced.

5. The result in Exercise 4 is the limit of the average velocity during the time interval from t to t + h, for motion whose distance as a function of t is t^2.

More generally: the result of Exercise 4 is called the derivative of the function t^2. Try to derive in an analogous manner the derivative of the functions t^3 and t^4 respectively, and generally of the function t^n, where n is a natural number.

6. Show that if f (x) is continuous on an interval a ≤ x ≤ b and

$$\int_a^b f(x)^2 dx = 0,$$

then it must be true that f (x) ≡ 0, i.e. , f (x) = 0 for all values of x.

7. Calculate
   a)
   $$\int_0^1 (x - x^2)\, dx$$

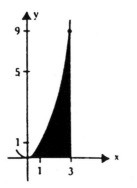

Figure 3.12.13

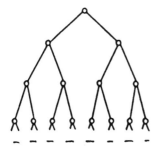

Figure 3.12.14

b) the area in the xy-coordinate system (with 1 cm scale units) bounded by the x-axis, the parabola $y = x^2$ and the line $x = 3$ (Figure 3.12.13)

8. Show that the set of square numbers,

$$1, 4, 9, 16, \ldots$$

is equivalent in size with the set of natural numbers.

9. Show that the set of rational numbers a / b between 0 and 1 is countable, i.e. has the same size as the set of natural numbers.

10. Figure 3.12.14 shows the start of an endlessly dividing "street network." Each road divides into two roads. Show that the infinite set of separate roads contained in this network is not countable. (None of the roads cross one another.)

4

MATHEMATICS AS A FIELD OF PRACTICE FOR THINKING

4.1 To Achieve Sureness in Thinking

Participants in mathematics courses for adults have sometimes said such things as "Mathematics can actually be interesting," "I was afraid I wouldn't be able to keep up in mathematics," "I can't remember that we worked with this kind of arithmetic in school," or "Mathematics was boring in school."

Perhaps teaching is considerably better nowadays than when we sat behind school desks. But the pupils' relation to mathematics is certain to be just as strongly individual now as then. There are students who like the subject, and there are others who find it painful. Quite apart from their attitude toward mathematics many students make an interesting comment when they have just conquered a difficulty in the subject, small or large: "It was really difficult until I understood it, but then it was nothing!"

Later we will go into the psychological aspects of the fact that problems are difficult until one understands them, but for the moment we direct our attention to another phenomenon which teachers may find in virtually all of their pupils: their joy when they have understood something, or even more so, when they have discovered something and best of all, when they have discovered something all on their own.

It is important to give the pupils free space in which they can think on their own. Some pupils gain a very strong motivation to work, when they just once solve a problem without help. "Don't tell us," "wait a minute," "no, I don't want any help" and other such expressions bear witness to the fact that many students wish to test their own abilities in mathematics (just as in other subjects)

This is not a question of achievement-oriented pupils, super-ambitious types, but rather of completely ordinary children and youth who want to keep watch over their own inner workings.

An experience which does not so easily come to expression in the classroom but which I would place highest of all, is the feeling of sureness, when one has attentively, consciously solved a problem. Sureness comes the moment we see that the method we have chosen will lead to the goal. It is not dependent on those calculations or manipulations we might need to do in order to get the answer. It is based on the experience of having found a way to reach the goal and of knowing that one can follow a thought process just as surely as one can walk down a road.

There are some interesting observations on how the experience of sureness can come to be present just as suddenly as when lightning strikes. In *The World of Mathematics* (edited by James R. Newman, New York 1956) we may read personal descriptions of this experience by the French mathematician Henri Poincaré (1854-1912). Poincaré had wrestled with the problems in the theory of functions for a time while in Caen, but left town and mathematical work to participate in a geological expedition:

> The changes of travel made me forget my mathematical work. Having reached Coutances, we entered an omnibus to go some place or other. At the moment when I put my foot on the step, the idea came to me, without anything in my former thoughts seeming to have paved the way for it, that the transformations I had used to define the Fuchsian functions, were identical with those of non-Euclidean geometry. I did not verify the idea; I should not have had time as, upon taking my seat in the omnibus, I went on with a conversation already commenced, but I felt a perfect certainty.

On a later occasion, after unsuccessful efforts with a problem in algebra:

> One morning, walking on the bluff, the idea came to me, with just the same characteristics of brevity, suddenness and immediate certainty, that...

And still another time, when the solution popped up while Poincaré strolled along a street:

I did not try to go deep into it immediately, and only after my
service did I again take up the question. I had all the elements and
had only to arrange them and put them together.

(*The World of Mathematics*, Chapter XVIII, section 2.)

I have quoted Poincaré in considerable detail not because we can
expect pupils to have such marked experiences in school. But the same
quality of knowing, even in a considerably humbler form, *can* be experienced in elementary mathematics. That the solution often comes suddenly, unexpectedly, is an experience shared by many inventors. But even
in those cases where the sureness comes gradually, it is of the very greatest value as an inner experience. It appears most valuable when we succeed in finding the solution to a problem all by ourselves. But even when
the sureness slowly grows while we follow another person's reasoning,
for example in a proof of the Pythagorean theorem, it is an experience of
inner clarity and control.

A person who actively experiences the soundness of the proof of
the Pythagorean theorem *knows* that the theorem is true. This inner conquest cannot be taken away from him. It is not the case that we believe
the Pythagorean theorem because it has been presented to us by mathematicians or because a teacher has explained the proof to us: we know it
is true the moment we can grasp it with our own thought.

Mathematics is therefore a practical field which, despite the dependence of pupils upon their teacher, serves to free them from bonds to
authority.

The teacher can of course encourage or restrain such a process of
independence. The more the student is allowed to orient himself within a
subject, the better. For this reason the students ought, at least in the
beginning of each new topic, to be given opportunity to gain first hand
experience with the arithmetical or geometrical subject materials which
belong to the particular section of study.

The opportunity to try new materials is, as we know, exciting for
most people in subjects such as chemistry and physics. What is more fascinating than trying out a new apparatus, simple though it may be? To
test and to seek are ways of getting answers to questions which one
poses. It is part of our instinct for knowledge to behave in this manner,
to go on scouting expeditions hoping if possible to make discoveries. Big
or little — it does not matter.

In arithmetic I have given examples of how students can go "scouting" and testing with numbers which they themselves have chosen and arranged (for example, Section 3.5.2). In geometry everyone can draw figures and look for what possible relationships can be uncovered in the figures, both in their own and in their classmates'. This element of looking, seeking, appeals to the individual and awakens his or her interest to see what others have come up with and to let others share in one's own results. In short experimenting and searching are a phase in which individualized activity goes hand in hand with social activity.

A truly stimulating tension occurs when one group of pupils have found one result, and another group reports the opposite results. No one need feel beaten when it is seen later which group has made the right judgment, because all the experiences were used in the search for knowledge.

It also gets exciting when we have the choice of proving a supposed lawful relation or of looking for counter-examples. Which horse do we bet on? A proof is often demanding, while a single counter-example is sufficient to throw out a theory we have. The art sometimes lies in being able to find a counter-example. If the new examples instead confirm the theory, shall we then change horses and try to prove the theory?

Once at the end of a lesson where we had treated the theorem that two triangles are similar when their corresponding angles are equal, I gave the following homework assignment:

Investigate whether or not the theorem applies to other figures than triangles, i.e. to polygons with four or more sides. You might draw four- or five-sided polygons and feel your way forward. The angles A, B, C, etc., in the one polygon must be the same size as the angles A´, B´, C´, etc. in the other polygon. The question is: are the polygons then similar? I assured myself that the students had understood the task and was then curious to see what they would come up with the following morning.

A large group in the class had found that the theorem for triangles was true even for polygons. One boy who usually was not particularly outspoken in class said that the theorem did not apply to other polygons than triangles. A smaller group of pupils had not come to any conclusion. The pupils in the large group felt very sure of themselves. Underneath that obviously lay a sense of belonging to the majority. One could note their attitude of superiority toward the boy who claimed the opposite. It was an experience to see their reactions, those in the majority, when the boy went up to the blackboard and drew differently shaped polygons approximately as in Figure 4.1.1 below. (The simpler counter-example of the square and rectangle even this lad had missed.)

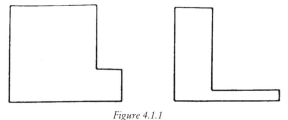

Figure 4.1.1

4.2 The Formal

The more time we use for heuristic, seeking out kind of work, the better. But to training in mathematics also belongs the learning of standard methods and becoming capable of using them and knowing when to use them. It would be an unhappy situation if students should begin searching for a new solution to a problem which they ought to recognize from previous examples. The inexperienced person often takes a long path to get to a solution where the person with experience only needs to make use of a simple idea to re-make the problem in terms of known technique. The risk here is that the mathematics disappears, as routine makes its entry, especially if strong emphasis is placed on memorized formulas or formula reference tables. At higher levels in school many students become conscious of the fact that calculation and mathematics are not identical occupations. They know that calculation can be done by machines. Even further from mathematics lies formalism, the stenographic dress of mathematics. How this and that are written is in reality a matter of convention. That the set of real numbers between 0 and 1 can be written $0 < x < 1$ or E $(x: 0 < x < 1)$ or $(0;1)$ corresponds approximately to the fact that the symbol for four can be written IV or 4. It was a bad pedagogical mistake of "the new mathematics" to put so much emphasis on the rote learning of such things of convention. Through this, teaching took on the character of authority; the pupils became dependent on recipes just as the inexperienced person in cooking does.

Particularly fateful does this kind of teaching become if a number of students want quick results and get support for a question such as: "Tell us how we're supposed to do it" or, as sometimes occurs in physics experiments, "Why do all this experimenting? Tell us now how it's supposed to be." These students have momentarily only results before their eyes and do not want to take the trouble of seeking it out. That comfort can tempt pupils into making the book or their teacher into an authority was once demonstrated by the argument of a high school class to their teacher: "Why do you go through all these proofs? We believe you anyway."

And yet there may lie something admirable in such a student expression, a sound protest against a monotonous going through of proof after proof, often with the good intention of achieving stringency. Euclid's long list of definitions, theorems, and proofs (a pattern which repeats in many mathematical presentations) is not any example of *teaching* excellence. In school the theorems can easily become ends in themselves, just as much as formality. The fight against formalism must naturally not be allowed to lead to a fight against the requirement for logic in reasoning, points out the Russian mathematician A.J. Chintschin in an essay on formalism in teaching mathematics in school. In the same spirit are the words of the American Morris Kline:

> To teach thinking we must let the students think, let the students build up the results and proofs even if incorrect. Let them learn also to judge correctness for themselves. Let's not push facts down students' throats. We are not packing articles in a trunk. This type of teaching dulls minds rather than sharpens them.
>
> (From "Why Johnny Can't Add: The Failure of the New Math," 1973.)

Many who have studied the literature in mathematics or in other subjects, or who have listened to lectures, have noticed that passive, receptive study is tiring. Of course there are captivating books and lecturers, but sooner or later one longs to grapple with a problem on his own. It feels just as necessary as breathing out after a long breath in.

We are not especially creative when we with our thoughts follow after a train of thought laid down by another. School is unfortunately built entirely too much upon such "after thought." Mathematics in particular, where the study of proofs of propositions and theorems places upon us a thinking-after which can negatively effect our fantasy and ability to come up with ideas.

It is important for the mathematics teacher to have the personal experience that the deductive side of mathematics belongs to the synthesizing, concluding phase of problem-solving. It is preceded by an analytical, inductive phase where we seek after ideas, simplifications, associations to previous experience, etc., in order to get at the problem. Yet very first of all we must acquaint ourselves with the problem, listen to its different sides, try to put it into some context or familiar perspective. These phases of thinking we might call preparatory thinking, a "pre-thought" as opposed to "after-thought." We maintain an attitude in this stage much like an architect trying out various sketches of a building, which must meet certain specifications. It is interesting to experience again and again that such "pre-thinking" does not seem too tiring. Often it feels stimulating, it pushes away any possible tiredness, naturally enough because we are creative while we are doing it. Early on in the lives of children, we can develop a sense of this kind of thinking without forcing upon them any precocious intellectuality. Here, too, we include the art of making up and guessing riddles, of finding the right word, and other forms of play which emphasize thought. The students' appetite for "cracking" problems, such a valuable resource in teaching mathematics, must be awakened (but carefully!) while they still have the natural desire to seek out answers themselves.

4.3   Gaining Confidence in Thinking

Even those who do not solve many problems on their own in mathematics will come, sooner or later, to have *one* important experience: learning to be careful with what we happen to have in the way of subjective ideas, our own usually half-conscious opinions. It happens often that a pupil writes 90° by an angle which appears to be right-angled in the given figure and obtains a very simple solution which, unfortunately, is incorrect, because the angle in actual fact is perhaps 87° or thereabouts. And how many are there not who calculate $7 + 3 \cdot 5$ to 50 because they want to do things "in proper order"? Mathematics requires attention, not just with numbers and geometrical figures, but above all with one's own thinking.

It is necessary to have command of certain knowledge, e.g., that the multiplication 3 · 5 precedes the addition in the example above — only a convention, of course, but still important — or that the sum of the angles in a triangle is 180°. Some pupils who want to be on the safe side memorize a great deal of such things, but I usually emphasize that one can often help oneself with a simple example.

Those who memorize a bunch of things run the risk of mixing up the memorized facts and can be psychologically limited to the need to find the right thing in memory on different occasions.

A few examples: "What do I do here as the last step in the equation?" (The pupils point to the equation 16.3x = 0.5.) "Should I divide 0.5 by 16.3 or the other way around?" — "What is the solution of the equation 5x = 20?" — "x equals 4." — "Then you see what you should do in your equation; do the same thing there." — "Thanks, I know."

The simple example with the same structure as the more difficult problem has the pedagogical advantage that it is self-instructive — it contributes to the emancipation from authority.

Another example: "I forget, sir, is the area of a circle πr^2 or $2\pi r$?" — "What does r stand for?" — "Radius." — "What would 5r be, for example, a length or an area?" — "A length." — "And what would 6.28r be?" — "Thank you, I've got it now."

"Is the sum of angles in a triangle 180° or 360°?" — "Draw a right triangle." (The pupil makes a sketch.) "How big is the biggest angle?" — "90°" — "Is it 180° then?" (Hesitation.) "Double your triangle to make a rectangle; then I'm sure you can judge for yourself." (Figure 4.3.1).

I hold as very essential such practice in finding simple examples or simple ways back to basics. When pupils succeed with this, they gain a confidence which is invaluable. They may even go so far as to have confidence that they could reproduce, if necessary, the proof of an important theorem, without needing to use any special tricks of memory.

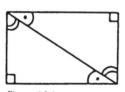

Figure 4.3.1

We then perhaps begin to achieve the most important goal of teaching mathematics: to have confidence in thinking itself.

5

MATHEMATICS AND SCIENCE

5.1   The Natural Sciences

The philosopher Immanuel Kant, and many others with him, claimed that "there is only as much true science in the natural sciences as there is mathematics in them."

Anyone who has had the experience of insightful knowing in mathematics, which was exemplified in the previous chapter, can easily understand Kant's statement. Long before Kant, Descartes emphasized that during his school studies he found the contents in only one subject to be certain beyond doubt: arithmetic and algebra.

Preparing for the study of physics, many have likely heard the advice to lay a good foundation in mathematics: "Then you won't have any problems in physics." It is both natural and a fact that mathematics has become a cornerstone of science. Yet one may ask himself whether it perhaps has not obtained too strong a position in school — at the cost of the study of nature itself — not least through the kind of tests which are given. In recent years even chemistry has become more and more "mathematical." Of course it is a question of balance. The question is if the balance is good today. An exaggerated mathematizing in physics and chemistry places obstacles to the natural studies: observation, the devising of experiments for systematic exploration, looking for answers and, if possible, finding a theory are cornerstones in natural science.

Let us compare mathematics and the natural sciences, beginning with the latter. As the name implies, natural science is a summary of knowledge of nature. Any understanding we might obtain must come first from observation of nature, observation which we do with our senses, possibly intensified with a microscope, telescope, oscilloscope, or other instruments. The observations collected must then be compared

234 |MATHEMATICS AND SCIENCE

with each other and ordered. Certain observations seem to support each other; others seem to go against each other. In both cases thinking about the observations leads to new questions which guide the researcher to new experiments, new perceptions, new measurement; when research has come so far that an agreement, a common factor, a regularity has been found, then it is time for a theory which summarizes.

As a criterion for the value of a theory, we require that the theory be able to predict the outcome and results of new experiments; that the theory can be "confirmed" experimentally. When this has been done in a *large* number of cases, when it has been shown that the results are reproducible and not just the result of chance, the theory can expect to be recognized. Statistical methods must often be used to demonstrate that the obtained results are not the product of the researcher himself and his experimental arrangement but correspond to objective reality. In cases where results have not been able to be verified by other researchers, the theory gains only limited interest.. Natural science cannot be dependent on the practitioner; it is by its nature general. From Michael Faraday, considered by many to be the greatest experimental physicist throughout the ages, we have the following words worth thinking about, made in response to letters asking him to comment on the value of various new discoveries described by researchers:

> I was never able to make a fact my own without seeing it, and the descriptions of the best works altogether failed to convey to my mind such a knowledge of things as to allow myself to form a judgment upon them. It was so with new things. If Grove, or Wheatstone, or Gassiot, or any other told me a new fact and wanted my opinion, either of its value, or the cause, or the evidence it could give in any subject, I never could say anything until I had seen the fact.
>
> (Faraday to Dr. Becker, Oct. 25, 1860,
> see L. Pearce Williams, *Michael Faraday*, London, 1965, p.27.)

Galileo himself, the father of modern physics, had, by and large, written a declaration of policy for the natural sciences when he wrote: Let us rely on demonstration, observation, and experiment. As early as in Galileo's works we find the role of mathematics pointed out: those who wish to solve scientific questions without the help of mathematics are taking an impossible task. One must measure that which is measurable and make measurable that which is not.

Apart from whether natural science leads to knowledge in mathematical or other form, it is firmly anchored in a large number of observations. That the distance a ball rolls grows with the square of the time, as Galileo found, has been confirmed many times over by others than Galileo, and yet there is still an enormous difference between the formula $s = at^2/2$ for the rolling ball and a formula in mathematics such as the closely related $1 + 3 + 5 + \ldots = (2n - 1) = n^2$.

5.2 Mathematics

The formula $s = at^2/2$ has been arrived at empirically and means: under present laws of nature every ball rolling down a plane will cover a distance given by $s = at^2/2$. Almost all scientists consider scientific statements as statements of probability: the probability is currently very very low that a ball will show some other function for the distances covered.

The formula that the sum of odd numbers gives a perfect square is an entirely different matter. We have available a *proof* that the formula is true. The formula will always remain true. We are obliged with the physical "law" to time and again check its applicability — the mathematical formula is demonstrated once and for all. Anyone who has thought through the proof has it as an inner, conquered certainty.

It is of great pedagogical value to let pupils in a sixth grade class clip out angles in a paper triangle and see how big the sum of the angles comes to be. Thousands of people can do this and, measuring with protractors, obtain values around 180°. One could calculate averages and other statistical measures and come to the result 180.00° for the sum of angles. If mathematics were natural science, we would accept such a way of going about things. But it is clear to us that mathematical statement cannot be formulated as the result of experimental works. Intuition leads us to a way of proof, perhaps first of all to a supposed conclusion, but after that logical reasoning must confirm with certainty the correctness of the proposition. The contents of a statement are, of course, a function of the axioms which we have taken as a base.

Mathematics is a science which differs from natural science but which plays a large role in it and which has been and is a model to follow in many scientific contexts. It is relatively easy in a mathematical system to have the axioms which lie at the base of the system, but it was shown during this century that certainty is not absolute, even in so clearly delimited an area as the whole numbers. To obtain a sufficiently complete system for the whole numbers, such an encompassing system of axioms would be needed that (according to a proof 1931 by the German mathematician Kurt Gödel) one could make propositions, formulated in terms of the system's language, which can neither be proven nor disproven with the system's axioms themselves. Paul Finsler gave examples of these kinds of statements, where, nonetheless, our thinking can determine the correctness of the statement. In a certain sense, then, logical systems are relative. This might now be taken as a proof that mathematics cannot be based upon that certainty which logical thinking, according to our experience, leads to and therefore does not differ in character from other science. This, however, would be to underrate thinking. That mathematics cannot be "mechanized" into a logical system need not cast suspicion on thinking. On the contrary, it shows that thinking is the (only) archimedean point which mathematics has available. Gödel's conclusion demonstrates that mathematics is a conquest by our thinking. This does not stand in opposition to the entirely true statement that our sensory perceptions play a large role in our ability to pursue mathematics. They awaken concepts to awareness within us. We never really *see* a line, or a circle, not even a point, but any things stimulate us to arrive at these concepts and become conscious of them.

Here lies the fundamental difference between mathematics and natural science: mathematics springs forth from thinking itself; natural science must be based on observation. I do not, however, hereby consider myself to support Kant's statement that the laws of arithmetic are "a priori" truths. They are conquered by thinking in an alternating play with the senses.

5.3 The Two Parallelograms

During a lecture on natural science, Rudolf Steiner gave a good example of where the border line goes between mathematics and physics. I would like to relate it briefly here:

The principle of the velocity parallelogram and the principle of the force parallelogram seemingly have the same form: a particle which is given two simultaneous velocities as in Figure 5.3.1 obtains a velocity which in direction and size is determined by the vector sum of the given velocities (one of the diagonals in the parallelogram formed). Analogously: if a particle is acted upon by two forces as in Figure 5.3.2, then the resulting force is determined by the vector sum of the given forces.

Figure 5.3.1   Figure 5.3.2

These principles come about in completely different ways: the resulting velocity can be derived from the concept of velocity and the given velocities, while the resultant force principle has been arrived at empirically. Where the force is concerned, it is *nature* which shows us that the parallelogram principle holds (at least so far...).

5.4   Descartes, Newton, and Gauss

Just as there are and have been thinkers who describe mathematics as a natural science, so are there researchers who have held physics, for example, to be a form of mathematics. The most prominent representative of this position was probably Descartes. The human spirit, according to him, brings forth the natural sciences the same way that it creates mathematics. Descartes presented his ideas on "universal mathematics" in the posthumously published work *Regulae ad directionem ingenii* (Rules for the guidance of the genius).

His method is basically to pursue physics according to the deductive methods of mathematics. Phenomena ought, according to Descartes, to be able to be derived from axioms, just as theorems in mathematics build upon axioms. The results of physics should be capable of being derived.

Here we must point out, in all fairness to Descartes, that he wanted to reduce physics to a theory on the motion of bodies. Through this he came to be one of the physicists who pioneered the way for the 19th century mechanistic world view: that natural science must be based in the end on mechanics (space, time and movement).

Descartes tried to practice what he preached. He claimed with determination, thereby directing himself primarily against his contemporary Pascal, that vacuum cannot possibly exist — except in Pascal's head. Pascal, who was certainly just as philosophically inclined as Descartes, let a number of experiments be carried out which confirmed the existence of vacuum and which led physics further along the right track concerning the question of vacuum. Descartes' deductive way of doing research fouled up here in a very noticeable way.

As further examples showing that the roots of mathematics lie in our own thinking, I would like to briefly mention

— that Pascal all on his own by the age of 12 had achieved a knowledge of geometry equivalent to a number of the theorems in Euclid's Elements;

— that Newton, at the age of 24 during the 1½ years when Cambridge University was closed due to the plague, conceived the major part of the pioneering mathematics which he later published;

— that Poncelet with no help and while a prisoner in Russia for over a year (during the Napolean war) laid an important foundation for the new projective geometry.

It is worth mentioning that Newton, in his famous work on the mathematical principles of natural philosophy, lays a system of definitions and axioms as a groundwork for his presentation. This axiom system has a different character, however, than a mathematical system of axioms. Newton describes among other things a number of research principles, for example:

One must as far as possible account for similar effects with similar causes. For example, breathing in humans and in animals, the fall of a stone in Europe and in America, the light of the kitchen fire and from the sun, the reflection of light from the earth and from the planets.

And:

in experimental physics one must hold true those propositions which have been won through induction from phenomena, until other phenomena come to be known...

We see that Newton clearly understood that the conclusions of physics are inductive, in contrast to the deductive results of mathematics.

I would also like to describe Gauss' measurement of the sum of angles in an enormous triangle whose points he placed on three mountain tops, namely the peaks of Hohenhagen, Brocken, and Inselberg, where the shortest side was 69 km. Gauss made the angle measurements with the aid of optical signals and found that the sum of angles was so close to 180° that the measurements could not possibly have shown any other value.

Did Gauss do this measurement in order to convince himself that the sum of angles in a triangle is 180°? Of course not. His question — in all certainty — must have been: when we measure an angle using optical instruments, for example, is it correct to use plane Euclidean geometry on the measurements? As we discovered in Section 3.10, in a spherical triangle the sum of angles is greater than 180°. Gauss' question concerned: which of the known geometries should we apply in a specific practical context? Only through experiment, Gauss rightly believed, could he get an answer to such a question.

Let us lastly return to the starting point for this chapter. It would be to the good of education in both mathematics and the natural sciences if the natural subjects did not become overgrown with mathematics. Because of the essential difference between research in the natural sciences and in mathematics, experiment ought to be given much room in physics and chemistry. The applications which currently are given so much attention in the natural sciences are often a numerical processing of previously given formulas. This does not invite creative mathematical training. I say this fully aware that problems in the natural sciences have been a superb source of inspiration, one of the very best, to mathematicians in pioneering new roads.

6

MATHEMATICS
AS A SCHOOL SUBJECT

6.1 Alternating Between Practice and Orientation

Mathematics is a practice-subject, above all. If we limit ourselves to calculation, it is entirely a field of practice. On first going through an arithmetic formula or a theorem in geometry, pupils generally obtain only a first acquaintance, so to speak. They know the theorem no better than we know a person after a short first meeting. Even the first solved practical example in a new area gives for many only a hint of how the solution is actually done. Here is verified the old expression: "repetition is the mother of learning." Repetition in the form of practice, with as much independent exercise as possible.

We as teachers, on the other hand, may contribute to a first meeting with a mathematical topic by helping it to penetrate more deeply into the pupils than if we spoke like a book. Verbal teaching with its dialogue between class and teacher and between students themselves provides excellent opportunity for creating an atmosphere of excitement around a new element, so that an air of receptive readiness pervades the classroom. Such attentiveness means that impressions are stronger than otherwise. We forget more easily those things which we have placed somewhere without thought. If we want to help ourselves remember where we put a key, we should pay careful attention to and describe for ourselves what the surroundings look like where we place it, and imprint in our memory the picture of the key and the nearby surroundings.

By orienting a class on the "environment" surrounding a mathematical area, either by way of introduction or on a suitable later occasion, we give the pupils the opportunity to have living memories of knowledge. In Chapter 3 I have given a number of examples of such orientation: on a cultural epoch such as ancient Egypt for introducing arithmetic in different number bases, on the positions of leaves in plants for the discussion on Fibonacci numbers, and so on. Many problems can be made to come alive through glimpses from the biographies of the great mathematicians. Such orientation serves not only to strengthen the pupils' motivation; it leads also to the problem formulations which seem natural to the students.

It often seems natural to let historical presentations dominate in such orientation, but experience shows that historical descriptions seldom are particularly accessible in mathematics when they come right in the beginning. The pupils often have not yet made acquaintance with the mathematical material. An historical exposition generally has much greater effect when the pupils have become familiar with the new chapter's content and methods. In the case of number systems it is more effective to let the pupils themselves invent another system, different from the 10-system (with new symbols, see Section 3.1.1) than to directly present the ancient Egyptian system or the modern binary system. When introducing logarithms we made for ourselves a "table of logarithms," however primitive, with the aid of a curve and reading off the graph, and discovered some of the basic rules before the class was told about Jost Bürgi, a Swiss clockmaker who as early as 1588 constructed a shrewd logarithmic system, long before the Scotchman Napier published his work on logarithms in 1614. The classes are usually very interested in hearing about Bürgi's life and ideas and *can* listen actively, because by then they know roughly what a logarithm is and what difficulties come up during construction of a logarithmic system.

If one has something beautiful or exciting to show, which can awaken the children's astonishment, one should be careful not to bring it forth too early. "Astonishment should come last in the art of teaching," recommended Rudolf Steiner to teachers at a conference. Elements which are intended to make strong impressions on the pupils, which can surprise or fascinate them, must be prepared during classwork so that the pupils have the prerequisites for experiencing the intended item as a climax.

The ideal is probably an alternation between orientation in a subject and practical exercises in problem solving (including constructions in the case of geometry). From the orientation, problems spring up naturally and, if the problems are well chosen, give the class reason to ask questions around the problem-solving itself.

In geometry, orientation can quite nicely include elements from non-Euclidean geometries, so that pupils do not live with the same view as scientists did in Kant's time, that there is only one geometry, the Euclidean.

At university and schools of engineering one can meet students who in high school have gotten the impression that there are two kinds of series — arithmetic and geometric. They limited their knowledge of infinite series, one of the most fascinating chapters in the history of classical mathematics, to a dutiful calculation of the sums of a few series of arithmetic or geometrical type. They did not know that the concept of convergence of number sequences is the doorway to broad areas in mathematics.

An orientation on the creativity of some of the great mathematicians, Archimedes, Euler, Gauss, Pascal, Hilbert, and many others, ought certainly to be included in the mathematics course.

Concerning the importance of axioms, an introduction to Hilbert's contributions would have its rightful place. Hilbert held a series of lectures in the winter of 1898-99 on elementary geometry. He built up geometry from the ground floor, and it is interesting in an end-of-term class to compare Euclid's way of defining point, line, and plane with Hilbert's introduction. While Euclid, for example, describes a point as that which "lacks length and width," Hilbert introduces the basic elements point, line and plane directly in relation to each other through a number of so-called incidence-axioms (incidence = relationship). In Hilbert's work one can also find simple, instructive examples of how an axiom system can be tested for the three qualities which characterize an ideal system of axioms: the axioms should be complete when taken together, they should be independent of each other (so that one axiom cannot be derived from the others), and they should be non-contradictory.

Often in the literature, for example on Boolean algebra, one comes upon axiom systems which do not meet the requirement for independence, because the author preferred a more pedagogic presentation.

Proof that students may be interested in the foundations of mathematics, or in any case, that they can become interested, showed up clearly in the following episode which a teacher once witnessed in a sixth grade class. During a geometric lesson a couple of boys stood sharpening their pencils over the wastebasket to give them razor-sharp points. They stood unusually long but finally one said to the other, "You know, this is really hopeless. No matter what, we can never get a point!"

6.2 Quantity and Quality — Teaching on Different Levels

We live in a society in which quantities are given primary importance in many areas. We need only to think of the controversial subject "grade-point averages in the competitive school" to become aware of how society is organized such that quantities quite simply play a major role. In recent years quality has begun to be placed first; especially in the debate on our environment the expression "quality of life" has gained acceptance, even though often weak and though it surely invites the broadest interpretation.

Galileo's challenge to "measure that which is measurable and make objectively measurable that which is not" gained a following which he himself could hardly have guessed at. In various areas of society today there may be found a number of tests which make the claim to measure people's qualities and present the results in the form of easily interpreted point totals. Not long ago we could see examples in the newspaper of employment tests which showed how arbitrarily far the process of making the unmeasurable measurable has gone. In the 1800's there prevailed a widespread optimism within science, which by that time had already celebrated great triumphs — for example, the finding of the planet Neptune through a number of mathematical calculations. Qualities such as heat, color, and taste were thought to be amenable to "explanation" in terms of quantities in space and time. Even medicinal effects were thought to be capable of prediction by calculation along the lines of mechanics.

The quantity-minded stream of thought actually goes back to Descartes more than to Galileo. I cite from the introduction to a book on Descartes by Paul Valéry:

Descartes is certainly one of those who bear responsibility for the life style of our times, in which everything is judged quantitatively. When the diagram was replaced by numbers, when all knowledge was put in the form of comparative measurements from which followed a de-valuing of anything which could not be expressed in arithmetical relationships, then something occurred which has had the greatest import in every area. To the one side is put everything measurable, to the other everything which cannot be measured.

(From P. Valéry, *Les pages immortelles de Descartes*, Éd. Corréa, Paris 1941).

It would be carrying things too far to make Descartes scapegoat for the onesidedness in our culture. But it is apparent that he overestimated the value of arithmetic and that there is much truth in Valéry's words.

A child who has learned arithmetic and done an addition according to the rules of arithmetic has in all certainty found in that doing all of what human thought is capable of finding,

claimed Descartes.

We can appreciate Descartes for his influence on natural science, which in its turn contributed to an impressive technology and through that to a strongly developed intellectual acuteness. Analytic geometry, introduced primarily by Descartes and used later in developments in the theory of functions, gave technology the prerequisites which were needed to apply the laws of causality to the construction of instruments, machines, and ships of all kinds.

But there has also existed and still exists another *scientific* direction apart from the quantitative, although not so patently successful in outward appearance as the mechanical-mathematical stream of thought. Atomic physicist Walter Heitler expressed in a lecture:

One directs attention to qualitative phenomena... to qualities which have something to do with the observed object's wholeness. One of the most important of this philosophy's founders is Goethe, with his writings on natural philosophy.

Heitler gives "The Theory of Color" and "The Metamorphosis of Plants" as examples of Goethe's efforts in these directions.

> He directed his attention to the figure's unity and to the qualitative context. Goethe is the founder of modern comparative morphology within botany.

Why, you may ask, this long introduction before we get into the theme of teaching mathematics? Because I want with these historical examples to give a background to problems of which the mathematics teacher ought to be aware. What do we want from our teaching? Is it the main thing that students train up a given measure of tools and knowledge, i.e. do we place the quantitative aspect first? Schools of engineering want certain prerequisites, business school others, schools of medicine theirs, and so on. But should the high school or corresponding levels in our schools be preparatory for technical schools? If we want once again to place qualitative aspects and the individual's development during his and her time in school in the forefront, should teaching then be pressed into forms which are determined by point totals on exams?

Of course, all higher education demands a measure of preparatory knowledge, but it ought to be just as clear that every pupil should not need to go through the same preparation as candidates for higher education. The solution to the problem of how one can meet the needs of all the students must then be sought in some form of differentiation.

Should this differentiation mean that every pupil studies at his own pace and perhaps follows his own individual program, or can differentiation be best done within the framework of the school class? As far as I can see, the solution ought to have its emphasis in the latter alternative but with some elements of individual programs.

It ought to be possible to keep the class together and active around a basic framework of teaching and exercise materials.[1] But this requires that students be allowed to work at different levels of understanding and

---

[1] In the Waldorf School and other Swedish high schools the class is a fixed group which takes virtually all subjects together and which stays as a unit year after year until graduation.

achievement. In some of the teaching examples in Chapter 3 I have shown that there are problems which allow different levels of abstraction. They can be solved in an illustrative way, they can be treated more abstractly, or they can be approached in an elegant manner which directly leads to general results (see triangle numbers in 3.1, the rabbit problem in 3.3, the bottom sum problem in 3.5.2, the successive number sum in 3.5.5 and exercises in projective geometry in 3.10). A number of results can be achieved with very simple graphical methods *or* with possibly very demanding algebraic approaches (e.g. the problem of the meeting trains, the solution of simultaneous equations, etc.).

For spatial geometry an analogy to graphical solution can be that some pupils construct paper models and measure them, while others apply Pythagoreas' Theorem, solve an equation and so on. In such cases the pupils do not solve the problem on different levels as in the Fibonacci rabbit problem but rather in different ways. It is a matter of different choices of method. The advantage with problems of this type is also that groups which are working in different ways can become interested in each other's results. In this way a social element comes, without special effort, into the lessons.

But does not such choice of methods mean that groups work at different speeds? Of course it may happen, but then the groups can unite around the work which remains to be done. The teacher is also available the whole time and can help those who are having difficulty with methods which are more demanding than the illustrative or graphical.

This question of differentiation is often limited to mathematical problems of a quantitative nature, which concern calculating a length, an angle, an area, an intersection, etc. Would there not be reason to give considerably more time to *problems of a qualitative nature*?

What do such problems look like?

In principle they are problems in which investigation, not results in numeric form, plays the primary role — in which one needs to use a little fantasy, where the urge to build, construct, and shape can be put into action.

It is not difficult to find problems of this type in geometry, where quite naturally a whole category of problems involve constructions of different types. I need not here to further into this kind of problem but choose instead a few other types:

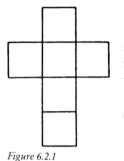

Figure 6.2.1

1. Exercise 1 in Section 3.7.3.

As we know, the network pattern for a cube may have the following shape (Figure 6.2.1): the four squares in a row can form the walls, the other two the top and bottom.

Problem: How many *different* networks can a cube have?

By two different networks we mean that the one network cannot be covered by the other even if the networks are cut out and turned around or turned over. For example, the following networks are the same (Figure 6.2.2):

Further, two adjacent squares in the network must have a common side, not just a common corner. A network such as in Figure 6.2.3 is not allowed.

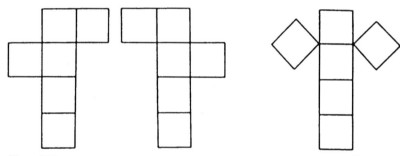

Figure 6.2.2   *Figure 6.2.3*

We have here a geometric-combinatoric problem of a constructive nature. It usually occupies the whole class, and quite intensively. Groups often form, and I emphasize that the problem concerns, above all, the question: how do we know when we have found *all* the networks which are to be found? How shall we determine that there are no other networks than those we have made? We come to the conclusion that some form of systematic ordering must be used in order for us to be able to determine how many networks the cube has. With this formulation it is not decisive how many networks any particular individual or group comes up with. The work is not meted out on a performance basis.

The person who finds "only" two networks may intellectually be the one who succeeds in finding a useful systematic ordering. (Concerning the number of networks, see Section 9.7).

This problem type can have many variations, even problems where the task is to find a single correct network for a solid. For example, we can ask: do both of the patterns in Figure 6.2.4 work for making a tetrahedron (three-sided pyramid)?

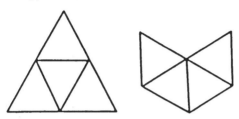

Figure 6.2.4

2. We wish to investigate how the diagonals of a quadrilateral can be determining for the quadrilateral shape. We take one of the following alternative conditions as a starting point:

(1) The diagonals are equal in length, at right angles, and bisect each other.
(2) The diagonals are equal in length and bisect each other.
(3) The diagonals are at right angles and bisect each other.
(4) The diagonals bisect each other.
(5) The diagonals are at right angles. One is bisected by the other.
(6) One diagonal is bisected by the other.
(The list could include still other alternatives.)

What kind of four-sided figures do we get? Can we summarize the results in a table which gives us an overview?

This problem might lead into a discussion on what are necessary and/or sufficient conditions concerning the diagonals such that we get a rectangle, for example. How is a romb characterized by its diagonals? And so on.

3. (From a period in spherical geometry.)
A person travels 1000 km south, then 1000 km east and finally 1000 km north. He is then back at his point of departure. Is there any starting point where this is possible, other than the North Pole?

This problem may seem to have the character of a trick-problem, but it gets the imagination into action (Solution in Section 9.9, exercise 5).

And finally I want to mention that projective geometry provides an excellent arena for problems of a qualitative nature. I limit myself to one example here and refer additionally to Section 3.10.

4. We begin with three corners A, B, and C, which form a triangle with sides AB, BC, and CA. We choose a fourth point, P, anywhere in the plane (but not on the sides of the triangle) and draw connecting lines to the three corners. We then get three points of intersection, X, Y, and Z with the lines AB, BC, and CA. What does the dual to this figure in the plane look like? (The dualization is based upon points and lines exchanging roles in the plane.) And what occurs when P happens to lie on one of the triangle sides?

It is much more difficult to find qualitative problems in arithmetic and algebra, since the material there is in fact numbers. The following examples may serve to illustrate:

1 a) Is it necessary for a whole number to end in 5 in order for it to be divisible by 5?
b) Is it sufficient for divisibility by 5 that a whole number ends in 0?

The students may motivate their answers with the aid of examples.

c) Is it necessary for a whole number to end in 0 in order for it to be divisible by 10?
d) Is it sufficient that the number ends in zero for it to be divisible by 10?

Please note: the purpose of these exercises is for the pupils to learn the concepts of necessary and sufficient conditions and to give them experience of what it means when conditions are both necessary and sufficient. We see often, in the most widespread contexts, necessary and sufficient conditions being mixed up with each other in everyday argumentation. What one person emphasizes as necessary conditions are understood by the other as a statement concerning sufficient conditions. For example, a person responsible for taking on a new history teacher says: "X is very knowledgeable in his field. He has even published books which have received good acclaim." The statement may be interpreted that the speaker considers X's knowledge of history as a sufficient merit for appointment. But what of X's teaching abilities? Isn't knowledge in the subject area only a necessary prerequisite, just as the ability to teach? And what would be sufficient conditions in this case?

2. The Pythagorean Theorem is one of geometry's most important theorems.

As we know it says that

$$c^2 = a^2 + b^2$$

where a and b are the sides of a right triangle and c is the hypotenuse. Is the condition "right triangle" here a necessary or sufficient condition for $c^2 = a^2 + b^2$ to hold? It is obviously a sufficient condition, as always in statements of the form "If..., then ...." Is "right triangle" also necessary? As a rule we overlook this question.

3. There exist, as some may know, so-called prime twins, i.e. pairs of prime numbers which are successive odd numbers: apart from the lowest we have

(11;13), (17;19), (29;31)   etc.

(As far as I know, it is still not proven that the numbers of prime twins is infinite.)

Are there prime *triplets,* apart from (3; 5;7)? That is, can three successive odd numbers, larger than 3, all be prime?

A problem of this kind requires hardly any prerequisite knowledge at all and can immediately engage everyone in the class.

4. The well known problem of how a ferryman would get a wolf, a sheep and a head of cabbage across a river in his rowboat. (Or a fox, a chicken and a sack of grain.) He must assume that the wolf will eat up the sheep if left alone and that the sheep will eat up the cabbage. Further, he has room in his boat for only one thing at a time besides himself.

How does he get them across? This is a kind of combination problem, which can give the teacher good insight into the pupils' abilities, especially if they hand in their solutions in written form.

A simpler variation on the theme: A truck can carry 2000 kg. It is to move 5 heavy machines weighing 300 kg, 400, 500, 1200, and 1600 respectively. How should the transport be organized in the simplest way?

5. As qualitative problems I would also include combinatorics problems in which one needs to draw "decision trees" or similar diagrams, or where groups of letters are to be ordered alphabetically:

Which letter sequence comes just after and which sequence just before the sequence

BCDAE

when all possible sequences made up of the five letters A-E with each letter appearing only once are arranged alphabetically?

The majority of students have an easy time finding the next following letter sequence in such examples but rather more difficulty in finding the preceding sequence. Exercises in going *backwards* in letter groupings, number sequences etc. require more effort of will in one's thinking.

6.3   Should All Pupils Take Mathematics

Based on experience in the Waldorf schools, my recommendation would be that all pupils be given the opportunity to take mathematics during the whole of their school education, which in the Waldorf schools is twelve years. But the answer to the question in the heading is naturally dependent on the kind of school, the organization of the courses, possible integration of school subjects, and above all on the goals of the school form. A school which seeks to give pupils a basis for specific kinds of vocational training differs essentially from a school which seeks to give general preparation, an education apart from the pupil's later education or choice of work. The Waldorf schools belong to the latter type of school. Their primary goal is to develop insofar as possible thinking, feeling, and will into an inner harmony so that youth can go into life with ability to observe and listen to the surrounding world, weigh pros and cons, form mature judgements and make well-grounded decisions. The Swedish elementary schools have the same goal.

In my opinion mathematics has so many opportunities for giving pupils valuable inner abilities that place for it ought to be found in the schedule during all the years in school, even if possible in schools which prepare for specialized work. For safety's sake, I would like to emphasize that I do not consider mathematics to be more important than other school subjects. Every activity has its special value. Any one of the subjects in school may be the field of activity which at a given moment is the very most important for a particular pupil.

Mathematics contributes with qualities which cannot be replaced by exercises in other subjects. The particularly valuable aspect of mathematics is its opportunity, properly pursued, to develop confidence in one's own thinking, a confidence which is built up through experiences of inner certainty of knowing.

Mathematics is, of course, not by any means the only area where one can develop powers of thought, but it would be going too far in the other direction to consider courses in foreign language as an equivalent to training in mathematics. In translating from one language to another, even to or from Latin, considerably more elements of "convention" come into play than in solving problems in mathematics. Those who would draw a parallel between the rules of mathematics and grammar should consider that grammatical rules often have exceptions, which clearly shows that a language does not build upon logic in the same manner as does mathematics. The exceptions, in fact, are what bring language alive, and the fascinating aspect of language studies is perhaps trying to speak and write so that it sounds truly genuine.

That there are pupils who are gifted in language but have great difficulty in both arithmetic and geometry — even drawing the figures — confirms that language and mathematics direct themselves to youngsters in different ways. In a class which I was fortunate to know especially well, the best student in French could not, even after help, see depth in a two-dimensional drawing of a three-dimensional solid (Figure 6.3.1).

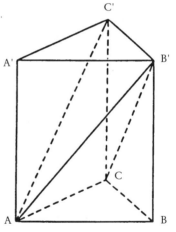

Figure 6.3.1
Can you see the three-sided prism with triangle ABC as the bottom and A'B'C' as the top? Can you also see three tetrahedrons which this prism is divided into by the three diagonals AB', AC', CB'?

In other individuals the allocation of talents is just the opposite.

What we shall now take up is the question of how mathematics instruction can be organized so that it becomes, as far as possible, rewarding for *all* pupils.

For this, the beginning, the first year of school, should be given great importance. It is in the first year of school that children are asked to use their number concepts systematically and where they meet the four basic operations of arithmetic. From colleagues in the lower grades I have understood that even the natural numbers in sequence work to bring order to the inner life of the child. The child meets something objective and great in the world of numbers. Here as in the drawing of shapes and forms, the basis of geometry, an inner certainty begins slowly to sprout, completely unconsciously, of what later will become a conscious asset.

As we know, proportionality is a very important area in everyday mathematics. Its applications are so common that familiarity with proportions should be included in every child's education. Weighing something over, choosing between alternatives is, in fact, seeing the proportions on a qualitative field between action and effect.

When we now take up the question of how mathematics instruction can reach and work on *all* pupils, we usually think primarily of those youngsters who lack aptitude or interest or both, and who experience the daily lessons as a repetitive failure, regardless of how competent the teacher may appear to be. What can we do for these so-called underachieving pupils? Shall they suffer through a subject, perhaps during their whole school lives, because the subject has much to offer other pupils who manage to "keep up" or even to be "clever"?

This question is enormously important. A school subject must naturally not be allowed to be a cause of suffering! Nor to be something dreary or indifferent.

It is a fact that usually more than one pupil in an undifferentiated class, perhaps as many as 5 or 6, have substantial difficulty with mathematics and with logical operations (it usually shows up even in grammar exercises). We know of two measures for solving this:

(1) to give the pupils in question very simple examples in addition to the basic course, a course which hopefully they will get something out of together with their classmates — nota bene: assuming the basic course can be pursued on different levels of abstraction (see e.g. Sec. 3.3.2);

(2) that these pupils get help from a special teacher as early on as possible.

In most cases both measures are needed. It is difficult to predict how successful the results will be for the individual pupil.

I have had pupils who had great difficulties when I received them in the beginning of the eighth grade but who went slowly and steadily forward. During the twelfth grade, the last year, a pupil awakened and developed his talents (still within the framework of easy problems) unexpectedly well. Such pupils were not as untalented in the subject as one might have thought; they were dreamers and first awakened to consciousness in their thinking at the end of their schooling. One of these pupils, as I heard from his elementary class teacher, had as a newcomer been looked for by his parents a few hours after the end of the first day of school. He came home four hours late. What had he been doing? Standing and watching an excavating machine digging, without noticing the time.

Other pupils have not awakened nor come to the kind of involvement in the subject which one had hoped for. A failure? Perhaps. Without it being taken as a kind of general excuse I would like to mention here that it has happened in *rare* cases that the pupil, several years after leaving the school, has succeeded in working up an ability in mathematics or in some other area where logic plays an important role. One pupil who had difficulty getting better than just satisfactory results on his exams, studied several years later at university and got his degree in mathematics with good results.

The awakening of thinking *can* come suddenly. A Norwegian teacher told me once about a boy in the fourth grade who was still very weak in arithmetic and drawing geometrical forms. He had, on the other hand, a rich fantasy, so flowing that its effects often spread out over the whole class and caused the teacher problems. One morning during the spring of fourth grade the boy came to the teacher and said: "Teacher, I can do arithmetic now!" The teacher could not help being doubtful, but it turned out that after that day the boy worked practically every problem correctly and became the best in class. He later took the highest academic honors and is now active as a professor of mathematics!

Perhaps the most famous example of success after leaving school for a pupil with low grades is Einstein, who is said to have "failed" in arithmetic. These sunshine examples are not intended to cover up the difficult problem we face, but they have their place in showing that we need not resign ourselves if the efforts to help weak pupils seem to give little result. We need not have the expectation that they will become professors of mathematics, but we can sharpen our powers of observation in order to better notice the progress.

A number of experiences seem to show that the greater variation we can achieve in our mathematics problems, and the more we can pose problems which do not require very much previous knowledge, the greater are the chances that weak achievers will be stimulated to learn important basic examples.

The need for concreteness in education seems to increase over the years. Some pupils in the sixth through eighth grades color geometric figures in order to get a clearer grasp of the concept of area. Areas can also be compared by cutting a shape into pieces and putting it together again in a new form.

And how many pupils mix up πr^2 (the area of a circle) and $2\pi r$ (its circumference — despite the mathematics teacher's efforts with the dimensional comparison of r^2 and r respectively! Not until in metalwork, where the task is to make a bracelet with copper sheet as the material, do the work and the calculation problem become concrete enough for some students so that the formula for the circle's circumference takes on reality.

There is no doubt that desk calculators have a part to play in high school, but in elementary school I think that the advantages of the calculator in certain parts of the curriculum do not outweigh the disadvantages which follow in the calculator's wake: above all, a growing dependence on tools whenever calculations have to be made. I am convinced that a sound ability to carry out numerical calculations with pen and paper is a necessary prerequisite for achieving the capability and assuredness which is desired in mathematics.

The majority of pupils in the eighth and ninth grades at Kristoffer School have not shown unwillingness to do their own numerical calculations. Many pupils in these grades have the need to recapitulate and practice elementary procedures in arithmetic. I fear that the pocket calculator would have covered up this weakness, and likely made it worse, if it had come into general use.

7

BEING A MATH TEACHER AMONG TEENAGERS

7.1 Phases of Development before Puberty

There are three prerequisites for successful teaching. The teacher must:

1. be familiar with developmental psychology and knowledge of man for the stage the pupils are at;

2. know the subject;

3. get to know the pupils in the class as soon as possible, not only in the classroom but outside as well.

Here are some reflections on the first point, primarily with respect to teaching mathematics.

The child goes through a number of thresholds during his development. Not the first is maturity for school which in general comes along with the loss of milk teeth about the age of 7. Approximately in the third grade, about the ages of 9-10, the child wakes up considerably to his surroundings after earlier having been woven into them. But first at the age of 12-13 does the ability to form independent judgements begin to awaken; the child then also has significantly expanded intellectual resources. Prior to sixth grade most pupils have no interest in mathematical proofs: the inner resources for this are quite simply in a latent state .

Children in the lower (1-3) and middle (4-6) grades want to come into concrete contact with the problems which are to be worked. Experience from practical participation plays a decisive role for them. Piaget's research confirms in large measure the basis for the curriculum which Rudolf Steiner outlined for the first Waldorf school in Stuttgart in 1919. Piaget demonstrated that the child lives "in the stage of concrete operations" up to at least the age 12 years, possibly a few more years. First from about 15 years of age is there a maturity for tasks within the realm of "the stage of formal operations."

Piaget has described a number of examples which show how much in vain it would be to begin with abstract things before children have the maturity which is required.

In the Waldorf schools physics is taught beginning in the sixth grade, chemistry from the seventh. The pupils are trained methodically to follow attentively along in what is happening in simple experiments and afterwards to describe them. Eventually some of the experiments can be contrasted with each other and give rise to thoughtful reflection, so that the class learns to compare and, in time, to draw simple conclusions. The important thing is for the work to be based upon experience of a process. The class is asked to draw the experimental apparatus and to try to describe "what happened" as simply as possible.

This drawing and describing (first verbally, then written) comprises an important preparatory stage for a more concept-oriented penetration of natural phenomena.

In a corresponding way teaching in mathematics should go from the concrete experience to concept and context. The rule "from hand to heart to head" makes possible meeting the children on their own level. That which the hand does — draw, cut out, form, shape or build — gives the child inner experience. In mathematics these can include many impressions of beauty in symmetries or other relations. The activity has then entered into the child's emotional life, gone from hand to heart. Experience shows the best, most fruitful questions for concepts and explanations come from pupils who do not ask out of a quick intellectuality but out of a need to go on from their emotional involvement to a clarity in thought.

In an article in *Scientific American* (No. 11/1953) "How Children Form Mathematical Concepts," Jean Piaget writes:

> A child's order of development in geometry seems to reverse
> the order of historical discovery. Scientific geometry began with the
> Euclidean system (concerned with figures, angles and so on), devel-
> oped in the 17th century the so-called projective geometry (dealing

with problems of perspective), and finally came in the 19th century
to topology (describing spatial relationships in a general qualitative
way — for instance, the distinction between open and closed structures, interiority and exteriority, proximity and separation). A child
begins with the last: his first geometrical discoveries are topological.
At the age of three he readily distinguishes between open and closed
figures: if you ask him to copy a square or a triangle, he draws a
closed circle; he draws a cross with two separate lines. If you show
him a drawing of a large circle with a small circle inside, he is quite
capable of reproducing this relationship, and he can also draw a
small circle outside or attached to the edge of the large one. All this
he can do before he can draw a rectangle . . . Not until a considerable
time after he has mastered topological relationships does he begin to
develop his notions of Euclidean and projective geometry. Then he
build those simultaneously.

I permit myself to doubt strongly the last statement. Need and
ability in Euclidean geometry, according to my experience, come before
abilities in projective geometry. But as a whole the quotation points
clearly to the importance of a didactic road from the child's personal field
of experience to concept formation via the inner experience.

7.2 Puberty

When the children near puberty and enter into it, their repertory of
feelings is considerably broadened. Interest knows no bounds, and it is
important to direct it out into the world, away from the ego. Even the
intellectual powers increase strongly in many pupils. It is striking how
willingly 15-year-olds like to get into discussions with teachers. There
are pupils who always get the last word in. The will to discuss sometimes
gives an impression of an instinct to sharpen the intellectual tools.

In this phase of life, of liberation from the adult world, of confrontation with the problems of the ego, with questions of life's meaning,
and while filled with sensitivity, uncertainty, aggressiveness but also ide-

alism, it is distasteful for many pupils to "plod" through technical concepts of arithmetic, to recapitulate and firm up so-called basic skills. Some become so out-of-tune by such routine exercise that they can't even bring themselves to get started during a lesson period. On the whole, youth in this age want to test their powers on new problems and use their cleverness in a *conscious* way.

Recapitulation and practice must be done in such a way that one simultaneously brings in something new. The young child's unconscious demand for concreteness in work has been transformed to a demand that the content of instruction be motivated, anchored in reality, not necessarily material reality. Youth have a right to get motivation from adults. They want in fact precisely to train up their ability to formulate motives for their own actions. The more the theme of classwork can come alive within the youngsters' own thinking without the teacher having to make introductions and expositions during the work, the better.

For some pupils, ability in mathematics seems to take a step backward during puberty. Self-confidence leaves them during classwork, just as it probably fails them during their free time. It is enough for a problem to "look hard" for some form of resignation to set in. During some lessons it is more important to find the right psychology than to speak to the subject. Mathematical activity requires, as we know, both time and patience, and it must be truly difficult for a pupil to bring himself to solve a problem if his self-confidence fails him in the classroom. A humorous word from somewhere can break the spell, or perhaps a pause with a few folksongs? Such a pause shows in any case that the teacher is not so incredibly intent on "making use" of every minute. Classes usually notice the teacher's degree of seriousness in different situations. Many can also separate teacher and subject, but not all. Some like a subject but do not have much sympathy for the teacher, for others it can be the other way around. A 16-year-old who had difficulty in mathematics said of his teacher, "NN is all right, but he has the subject against him."

Perhaps mathematics teaching has an important task just in this phase of puberty, when the pupils often have to fight so hard to reach objectivity. If one can relate stories from the pupils' own early years in school, or from other childrens' first years, it usually gets a good hearing: teen-agers can recognize very much of themselves in a problem situation which *they* have on another level than the young child, without feeling pointed out.

How wishful thinking can lead to mistakes comes forth quite nicely in the story of an episode which occurred during the registration of a girl who was to begin first grade the coming spring. She and her mother sat together with the teacher-to-be. After a while the teacher wished to feel out the girl's abilities in arithmetic a little bit. He said: "Five soldiers stand guard by a road. How many guards are left if four go away?"

The girl thought a minute, then answered "two." The mother paled and looked worried. The teacher could not have considered the answer as any great failure, but the mother later asked her daughter, how could she say "two"? "Why, mama, you know I felt so sorry that the soldier would have to stand all alone that I said two."

The emotions weigh heavily in the 14-16 year-old, making it difficult for many students to *want* to think about a problem. They would like to recognize the problem and be able to solve it without effort using some method they already know. Yet it is just precisely the unusual problems which would jolt the pupils out of their routine and infuse the power of will into their thinking. A few motivational words from the "grassroots' level" on why we take the time to solve a particular type of problem usually fall on good earth, because youth have a sense for any training which concerns them existentially, which might mean something for them even after leaving school.

7.3 Genetic Teaching

There are schools where so-called genetic teaching has been widely practiced. I am thinking of Martin Wagenschein, who in a number of publications emphasizes the importance, even the necessity, of a "genetic" teaching where pupils — sometimes with a little help from the teacher if needed — out of their own activity solve problems from the start by building up experience of simple, but in the given context, appropriate examples (see earlier sections on the importance of examples: 3.5.1-3.5.4 and Chapter 4). Wagenschein describes a project on the question, "Are there infinitely many prime numbers?" Thirteen boys and girls from different countries, between the ages of 14 and 17, who studied at a Swiss "free" school were asked, without special prerequisites or preparation, to take on the prime number problem which Euclid so elegantly solved. Is the number of prime numbers infinite? The group used five 60-minute periods to accomplish the task. Their logbook looked briefly like this:

The first lesson was needed for the group to become completely familiar with the question.

The second lesson was used for a discussion and investigation as to whether
a) $2n + 1$ is a prime number generator
b) $6n + 1$ or $6n - 1$ always gives a prime number
The second question was broadened by turning around: Does a prime number always have the form $6n + 1$ or $6n - 1$?

The third hour: Discussion of the concepts necessary and sufficient conditions was absolutely necessary! Starting point in simple examples such as: Are all inhabitants of Switzerland Bernese (i.e. from Bern)? Are all inhabitants of Bern Swiss?

Continued discussion of the problems connected with $6n \pm 1$, it was a matter of making sure that all in the group fully understood the progress that had been made.

Fourth meeting: The pupils were asked to write down all the results they had come up with so far. One girl was practically at the goal but did not succeed in formulating the analysis of two cases which form the key point in Euclid's proof; at least she did not succeed in making her own insight understandable to the others.

The fifth hour was wholly directed toward the final formulation. The girl in question later wrote a letter in which she said, among other things: "When we after several days had solved the problem, we were so proud, as if the prime number problem had tormented us our whole lives and we were the first to find the proof."

Pride ought certainly not be the goal in itself, but if we look at the essence of this citation we see that it expresses satisfaction over that which one has achieved through his own work and effort.

To use an analogy: there is a difference between reaching the top of a mountain by car and getting there on foot.

The feeling of the joy of working is not easily found for a pupil with difficulties in the subject. If the evaluation of work in mathematics is primarily based upon written tests, then some pupils can feel pre-destined to getting less encouraging evaluations. The method used throughout the Waldorf schools where each pupil keeps his own workbook can be of great help: for some in the class it is considerably easier to sit in peace and quiet at home and think through course material, and they can present it in an individual way, even in mathematics, where the opportunity for personal ideas in the subject must be limited in comparison with orientation subjects. As pointed out in Section 6.1, the opportunities for individual contribution are greater to the same extent that the mathematics lesson includes elements of orientation.

The keeping of workbooks as a rule gives a kind of satisfaction over seeing a finished job, which stimulates continued efforts. But what school can afford to give five whole hours to the solution of a problem like Euclid's theorem on prime numbers, when in addition the result leads to no practical application? In what kind of school can genetic teaching find its rightful place, regardless of how appealing it may seem?

Perhaps it can be included, despite the above, to some extent; or more than is found in our schools today. The time is well spent if it is used as suggested by Wagenschein's examples. The student group familiarized themselves with the problem, put down some simple "theories" in the beginning and were motivated to test them out. After a number of tries down paths which were not successful, they eventually found the right track. It does not matter that "tips" or "hints" from the teacher may be necessary. During the actual seeking the group felt the need to study the concepts of "necessary" and "sufficient" conditions, concepts which are of great importance in mathematics.

Finally, the group did the work of formulating the proof. Doesn't such work often give significantly more for the time than time spent on routine tasks? It ought to be possible to select a few suitable course elements for such a genetic study.

Another counter argument worth considering is: will not genetic education activate primarily those pupils who are already bright? Perhaps some of the pupils will end up being more or less spectators? This risk exists, and it is up to the teacher to contribute to stimulating everyone. Genetic lessons often require more preparation than other forms of work, for the teacher as well as for the class (not least socially). The brightest pupils must develop a sensitivity for the social, more than in normal forms of work, so that they don't speak out too early, unasked by the class, and take away their classmates' joy of discovery on their own. When social relations begin running smoothly, the group work can be organized so that some pupils become "assistants" to others. Those who are the helpers will certainly come to experience that they themselves understand the results better when they are in the position of answering questions from their friends and giving them helpful hints. Even the act of understanding what a question means can be worthwhile exercise.

7.4   On Kinds of Abilities

I have already mentioned (Section 6.3) the occurrence of different kinds of talents for languages and mathematics. Included in point 3 above — knowing the individual pupil — is developing as quickly as possible a picture of his or her prerequisites for the subject. Just as there are pupils with an aptitude for language *or* mathematics, it sometimes occurs that a pupil has a talent for arithmetic *or* geometry. Most mathematically talented pupils find both of these branches easy, but in individual cases a pupil with good aptitude for geometry can have difficulties in work with numbers. This indicates that arithmetic and geometry direct themselves to different fields of ability, just as, for example, language and geometry.

In an eighth grade class one of the pupils was called "professor." This honorable title had been given to him after a number of prominent contributions in arithmetic calculation, and I was naturally interested to follow his further development. It turned out the next year in ninth grade, where the problems more so than earlier appealed to individual initiative, that the "professor" quite often asked for "hints." His ability had mainly been of a reproducing, receptive kind and no great source of heuristic, inductive thinking was to spring up during his remaining school years. In the higher grades this pupil, for better or worse, went in for memorizing methods of solution.

Other pupils who had not shown themselves active in the lower grades, succeeded in coming more into their own in their later school years. Through suitable choice of problems the teacher can entice ideas out of pupils who usually do not express themselves so much. Faced with unfamiliar, perhaps surprising problems, "clever" pupils often look for a rule or other knowledge in memory, while less advanced youth tend more to use their "common sense" to find an opening for the problem solution.

8

ON THE CURRICULUM
AND THE GOAL

The school seeks to develop capabilities of spirit and soul in the pupils so that they can feel an individual responsibility when they are faced with tasks in society. This feeling of responsibility ought to be the inner strength with which we take in a new task, see the prerequisites for its completion, and come up with creative ideas on how the work can best be carried out. Understanding as quickly as possible the problems involved in a task requires thought and insight. All school subjects should contribute to the development of the pupils' power of judgment. In mathematics particularly there are opportunities for practicing on clearly posed problems and on problems of the most varied degree of difficulty. One can emphasize the analytical aspect by putting a problem within a long text, or one may emphasize the problem-solving aspect which puts demands on constructive thinking. Opportunities for differentiating abound and we ought to be able to stimulate each pupil to work.

Should, then, knowledge in a subject such as mathematics be considered as a by-product to the development of inner abilities? No. To be sure we wish to see the child's inner development as the primary goal, but we would not get very far if we did not give the pupils opportunity to achieve knowledge. Acquiring knowledge is in itself an important exercise and a prerequisite for solving complex problems. Learning things is one step in the pupil's development as an individual. And we can be assured that every child comes to school with a strong, though largely unconscious, desire to learn things which have meaning and importance for later life.

The curriculum should be an aid to the realization of the goals. In mathematics the emphasis must be put on problem-solving because mathematics means independent thinking and this requires training. As

266

emphasized in Section 6.1, however, orientation on cultural epochs, on the lives of mathematicians, and on methods has its important place as a source of inspiration for the pupils and the teacher. Pupils get important historical and human perspectives. They can see how mathematicians solved problems and how they developed new methods.

Orientation itself is thereby a kind of practice and can become a stimulating introduction to more routine problem-solving. It is important that pupils get the chance to think through such orientational concepts and methods and to summarize what they have learned, preferably with their own reflections. Experience alternating orientation with problem-solving has been good,

It has often been said that mathematics curriculum should give greater space to modern areas and not take its material exclusively from times before the nineteenth century. The choice of orientational and practical training elements must be made with great care, and Pólya is here the best example. It is a question of using time wisely and simultaneously hitting on that which is pedagogically effective. A problem from ancient Greece can be equally fruitful as one from our times. Naturally there should be room for orientation on new advances in mathematics, but the overriding criteria should be that the material encourages development in the pupil.

If one requires that the last years of school be dictated by the demands of university and college, then examinations and scores will come to restrict the bounds for general developmental elements in the curriculum. Perhaps it would be possible to arrange on a wider scale preparatory courses in the respective subjects at universities for those school children who wish to study further. Through this the pressure on the schools would diminish to the advantage of the majority of students who do not intend to follow the academic path. In his book *Matematik för vår tid* (*Mathematics for Our Times*) Professor Lennart Carleson points out that even within modern areas of mathematics one ought to be able to "give problems which ... measure mathematical abilities in a more genuine sense than tests of traditional type."

9

ANSWERS AND EXPLANATIONS
TO THE EXERCISES

9.1 Exercises in Section 3.1.8

1. a) 734 and 12059 written in Egyptian hieroglyphic:

   and

b) written in cuneiform:

$$734 = 1 \cdot 600 + 2 \cdot 60 + 1 \cdot 10 + 4 \cdot 1$$

which is represented

$$12059 = 3 \cdot 3600 + \text{a remainder of } 1259$$
$$1259 = 2 \cdot 600 + \text{a remainder of } 59$$
$$59 = 5 \cdot 10 + 9$$

which is represented

2. a) $39 = 124_5 \ (1 \cdot 25 + 2 \cdot 5 + 4 \cdot 1)$
   b) $150 = 1100_5$

268

c) $795 = 1 \cdot 625 + \text{remainder } 170$
$170 = 1 \cdot 125 + \text{remainder } 45$
$45 = 1 \cdot 25 + \text{remainder } 20$
$20 = 4 \cdot 5 + 0$

from which we get $\qquad 795 = 11140_5$

3. Yes, every natural number can be uniquely represented in the base-5-system. For example, to count an unknown number of buttons with the help of the base-5-system we arrange the buttons in piles of five. If doing this gives 3 buttons left over, then the last digit of the number must be 3. If no buttons are left, this digit must be zero. If the number of buttons is less than 25, the 5-position is determined by the number of piles. If there are 25 buttons or more, we arrange the fives-piles into groups, each containing 5 five-piles. The number of five-piles left over from the groups then gives the 5-position digit. And so forth. This representation is unique.

4. a) $34_5 = 19$ b) $230_5 = 65$ c) $304_5 = 79$
   d) $10110_2 = 22 \, (16 + 4 + 2)$ e) $11011_2 = 27$

5. The 5-system multiplication table:

	0	1	2	3	4
0	0	0	0	0	0
1	0	1	2	3	4
2	0	2	4	11	13
3	0	3	11	14	22
4	0	4	13	22	31

6. a) $114_5 \, (16 + 18 = 34)$ b) 111_2 c) $1111_2 \, (21 - 6 = 15)$
   d) $1111_2 \, (5 \cdot 3 = 15)$

7. The sum is $\qquad 10;0102_2 = 2 + \dfrac{1}{4} = 2.25$

8. Each new chord is to be drawn so that it cuts all previous chords in new intersections (otherwise the number of areas will not be maximal).

Imagine that we begin to draw the new chord: each time it comes to an earlier chord we have obtained a new area, since an existing area is divided into two parts. When we arrive finally at the circle's edge we get the last new area.

The number of new areas is thus = the number of old chords + 1
= the new number of chords.

No chords gives   1 area
1 chord   gives   1 + 1 areas
2 chords   give   1 + 1 + 2 areas
3 chords   give   1 + 1 + 2 + 3 areas, etc.
n chords   give $1 + 1 + 2 + 3 + \ldots + n = 1 + \dfrac{n\,(n+1)}{2}$ areas.

9. Beginning with the second, each figure can be divided up into a square number plus a triangle number. If $k_1, k_2, k_3, \ldots$ and $t_1, t_2, t_3, \ldots$ are the successive square and triangle numbers respectively, then we get the following simple expressions for the pentagonal numbers $f_1, f_2, f_3, \ldots$

$$f_1 = 1^2$$

$$f_2 = k_2 + t_1 = 2^2 + \frac{1 \cdot 2}{2}$$

$$f_3 = k_3 + t_2 = 3^2 + \frac{2 \cdot 3}{2} \text{ , etc.}$$

The general formula, according to Section 3.1.5 then becomes

$$f_n = k_n + t_{n-1} \qquad \text{for} \quad n = 1, 2, 3, \ldots \qquad (\text{with } t_0 = 0)$$

$$f_n = n^2 + \frac{(n-1)\,n}{2} \quad \text{or} \quad f_n = \frac{3n^2 - n}{2}, \quad n = 1, 2, 3, \ldots$$

from which

10. All the numbers generated by the polynomial $x^2 + x + 41$ ($x = 0$, 1, 2, …) land in the corners of the spiral square. This is because the number of steps from one corner to the next following corner forms the sequence 2, 4, 6, 8, …, which is identical to the increase in the polynomial when x increases by 1:

$(x + 1)^2 + (x + 1) + 41 - (x^2 + x + 41) = 2x + 2 = 2 (x + 1)$
from which it follows that the increase is 2, 4, 6, 8, ... for $x = 0, 1, 2, \ldots.$

9.2   Answers and Explanations to the Exercises in Section 3.2.8

1. The ordering principle for the 10 paths in Figure 3.2.2 is:
"Take an avenue as early as possible and proceed along it as far as possible." (The avenues run downward to the right, see Figure 3.2.1 and corresponding text).

If one wishes, the different paths can be characterized by letter codes. To that end we let an "a" stand for one block length along an avenue (anywhere) and a "b" for the same anywhere along a boulevard.

Path no. 1 can then be denoted by the code   aaabb. (One goes first 3 blocks along an avenue and then 2 blocks along a boulevard.)

Path no. 2 is   aabab.

One can in fact, *without looking at Figure 3.2.2.*, note down all possible paths by letting the letter combination

aaabb

take on all nine of its successors in alphabetical order. (There are to be 3 a's and 2 b's in each combination.)

The sequence then becomes:

1. aaabb   6. abbaa
2. aabab   7. baaab
3. aabba   8. baaba
4. abaab   9. babaa
5. ababa   10. bbaaa

2. According to Section 3.2.4 we have

$$(1 + x)^n = \binom{n}{0}1^n + \binom{n}{1}1^{n-1}x + \binom{n}{2}1^{n-2}x^2 + \ldots + \binom{n}{n}x^n$$

Since all powers of 1 are 1 we can simply write

$$(1 + x)^n = \binom{n}{0} + \binom{n}{1}x + \binom{n}{2}x^2 + \ldots + \binom{n}{n}x^n$$

$x = 1$ now gives $\quad 2^n = \binom{n}{0} + \binom{n}{1} + \binom{n}{2} + \ldots + \binom{n}{n}$

The sums of the binomial coefficients in the rows of Pascal's triangle thus form the sequence $\qquad$ 1, 2, 4, 8, 16, etc.

3. Seven people can be seated in $7! = 7 \cdot 6 \cdot 5 \cdot 4 \cdot 3 \cdot 2 \cdot 1 = 5040$ different ways along the side of a table.

4. $\binom{7}{4} = \dfrac{7 \cdot 6 \cdot 5 \cdot 4}{4 \cdot 3 \cdot 2 \cdot 1} = 35$ committee combinations.

5. The formula $\quad \binom{n}{k} = \dfrac{n!}{k! \, (n-k)!}$

gives $\quad \binom{n}{n-k} = \dfrac{n!}{(n-k)! \, [n - (n-k)]!} = \dfrac{n!}{(n-k)! \, k!} = \binom{n}{k}$

6. Number of chords = number of connecting lines between pairs of points
= number of ways to choose pairs
= number of ways to choose 2 points out of 10 points

$$= \frac{10 \cdot 9}{2 \cdot 1} = 45$$

7. Number of triangles $\binom{12}{3} = \dfrac{12 \cdot 11 \cdot 10}{3 \cdot 2 \cdot 1} = 220$

8. a) The number of letter groupings (permutations) is

$$5! = 5 \cdot 4 \cdot 3 \cdot 2 \cdot 1 = 120$$

b) The number of permutations with repetition allowed is

$$5^5 = 5 \cdot 5 \cdot 5 \cdot 5 \cdot 5 = 3125$$

9. We call the 200 tries the thief makes a trial series. We know from Section 3.2.7 that each try can be done in 1024 ways, of which 1023 will not succeed in opening the lock. 200 tries made randomly each time (so that any given try combination may be repeated one or more times) leads to

$$1024^{200} \text{ possible tries}$$

among which 1023^{200} are unsuccessful trial series.

The possibility that the thief is unsuccessful in 200 attempts is thus

$$s = \left(\frac{1023}{1024}\right)^{200}$$

The desired risk is the probability that the thief succeeds, which is the complementary possibility of s, i.e. 1-s. One obtains $s \approx 0.822$ which gives $1-s \approx 0.175$. The risk sought after is therefore about 18%. Compare this with the risk of succeeding in 200 different tries,

$$\frac{200}{1024} \approx 0.195 = 19.5\%$$

9.3   Answers and Explanations to the Exercises in Section 3.3.7

1. Yes, the drone's family tree produces Fibonacci numbers generation after generation. This follows from the fact that the family tree pattern

corresponds exactly to the rabbit pair chronology (with time running vertically upward):

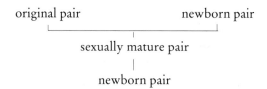

original pair   newborn pair

sexually mature pair

newborn pair

2. If we draw Figure 3.3.18b with a large scale and draw carefully, we find a narrow but obvious gap (space) between the top pieces (A + D) and the bottom pieces (B + C). The gap's shape is a parallelogram and the area is exactly r^2. The area A + B + C + D thus remains the same as in Figure 3.3.18a, $441r^2$. The gap is the "catch." How could this be discovered without drawing?

3. One first obtains

$$f_4 = 8 + 15 = 23, f_5 = 15 + 23 = 38, \ldots$$

$$f_9 = 259 \text{ and } f_{10} = 419.$$

Next,
$$\frac{f_9}{f_{10}} = \frac{259}{419} \approx 0.6181 \approx 0.618$$

4. We have
$$v_{n+1} = \frac{f_{n+1}}{f_{n+2}} = \frac{f_{n+1}}{f_{n+1} + f_n} = \frac{1}{1 + \dfrac{f_n}{f_{n+1}}} = \frac{1}{1 + v_n}$$

If the sequence $\{v_n\}$ is assumed to have a limiting value x, then the recursive formula
$$v_{n+1} = \frac{1}{1 + v_n}$$

leads to the fact that the limit x must satisfy the equation
$$x = \frac{1}{1 + x}$$

The limit must also be positive. The positive root of the equation above is precisely
$$G = \frac{1}{2} (\sqrt{5} - 1)$$

5. The proof is based on the following three points:

1) A monotonic decreasing sequence with a lower bound has a limit. This is an important theorem which we do not prove here.

2) $v_1 = \dfrac{1}{1} > G$ (whose value is 0.618...)

3. $v_{n+2} = \dfrac{1 + v_n}{2 + v_n}$

Relation 3) comes from the recursion formula we used above in Exercise 4:

$$v_{n+2} = \frac{1}{1 + v_{n+1}}$$ in which we put $v_{n+1} = \dfrac{1}{1 + v_n}$

We now show that the numbers v_n with odd index form a monotonically decreasing sequence. Since they are bounded by zero, these numbers must have a limiting value, say A.

From 3) follows

$$v_n - v_{n+2} = \frac{v_n^2 + v_n - 1}{2 + v_n} = k_n \left(v_n^2 + v_n - 1 \right) \qquad (4)$$

where k_n is positive.

If $v_n > G$ we have $v_n - v_{n+2} > k_n (G^2 + G - 1) = 0$ (see Section 3.3.5)

that is $v_n > v_{n+2}$.

In an analogous fashion we can show that $v_{n+2} > G$ (if $v_n > G$).

In other words, 2) implies that $v_1 > v_3 > v_5 > \ldots > G$

If $v_n < G$ we get instead, using the corresponding argument, that

$$v_n < v_{n+2} < G.$$

Since $v_2 = \frac{1}{2}$ we must have $v_2 < v_4 < v_6 < \ldots < G$

Applying 1) means that this sequence $\{v_{2n}\}$ too must have a limiting value, say B. Both limits A and B must according to relation 3) satisfy the following equation

$$x = \frac{1 + x}{2 + x}$$

This equation gives $x^2 + x = 1$, i.e. has G as its positive root. We have thereby shown not only that the sequence $\{v_n\}$ converges to G but also that the numbers with odd index *decrease* toward G while numbers with even index *increase* toward G.

If we represent the recursive relation $v_{n+1} = \dfrac{1}{1 + v_n}$

graphically in x-y-coordinates with the aid of the curve $y = \dfrac{1}{1 + x}$

and the line $y = x$,

we get a picture of the convergence process (see the figure).

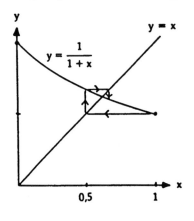

6. Starting with $F_1 = a$ and $F_2 = b$ gives

$$F_3 = a + b$$
$$F_4 = a + 2b$$
$$F_5 = 2a + 3b$$
$$F_6 = 3a + 5b$$
etc.

We see that the F-numbers take the form

$$F_n = f_{n-2} \cdot a + f_{n-1} \cdot b \text{ (for } n \geq 3)$$

where $f_1 = 1$, $f_2 = 1$, $f_3 = 2$, $f_4 = 3$, $f_5 = 5$ etc., are the "original" Fibonacci numbers we became acquainted with in Section 3.3.2.

a) We get $F_{10} = 21a + 34b$

b) If we let $v_n = \dfrac{F_n}{F_{n+1}}$, we get the same recursion formula as in Exercise 5, namely

$$v_{n+1} = \dfrac{1}{1 + v_n},$$

and the proof that the sequence $\{v_n\}$ converges to G works analogously to the proof in Exercise 5. From this it follows that the limit is independent of a and b.

7. We consider the second of the two given sequences,

$$G, 1, \frac{1}{G}, \frac{1}{G^2}, \ldots$$

From Section 3.3.5, equation (2), we know that G satisfies

$$G^2 + G = 1 \tag{2}$$

If we divide this equation by G we get

$$G + 1 = \frac{1}{G} \tag{3}$$

If we again divide by G we get

$$1 + \frac{1}{G} = \frac{1}{G^2} \tag{4}$$

and thereafter in similar fashion $\quad \dfrac{1}{G} + \dfrac{1}{G^2} = \dfrac{1}{G^3} \quad$ and so on. $\tag{5}$

The equations (3), (4), (5) etc., show that the sequence with G, 1, $\frac{1}{G}$, $\frac{1}{G^2}$, ... is a Fibonacci sequence with starting values G and 1. It must therefore be identical to the sequence $\quad$ G, 1, G + 1, ...

8. If the regularity continues infinitely, then it can be expressed

$$f_{2n-1} = f_n^2 + f_{n-1}^2 \quad \text{for } n \geq 2 \tag{1a}$$

$$f_{2n} = f_{n+1}^2 - f_{n-1}^2 \quad \text{for } n \geq 2 \tag{1b}$$

The proof is by induction: we know that the above applies for f_3 and f_4 (n = 2). Assume that the formula applies for n = p.
We then get for n = p + 1

$$f_{2p+1} = f_{2p} + f_{2p-1} \quad \text{by definition and according to} \tag{1a}$$
$$\text{and (1b)}$$

$$f_{2p+1} = f_{p+1}^2 - f_{p-1}^2 + f_p^2 + f_{p-1}^2 = f_{p+1}^2 + f_p^2$$

In other words, the regularity also holds for f_{2p+1}.
It remains to be shown for f_{2p+2}.

$$f_{2p+2} = f_{2p+1} + f_{2p} = \left(f_{p+1}^2 + f_p^2\right) + \left(f_{p+1}^2 - f_{p-1}^2\right)$$
$$= 2f_{p+1}^2 + f_p^2 - f_{p-1}^2$$
$$= 2f_{p+1}^2 + f_p^2 - \left(f_{p+1} - f_p\right)^2$$
$$= f_{p+1}^2 + 2f_{p+1}f_p.$$

From $f_{p+2} = f_{p+1} + f_p$ it follows (after squaring and rearranging items) that

$$2f_{p+1}f_p = f_{p+2}^2 - f_{p+1}^2 - f_p^2$$

Putting this expression into equality (2) now gives

$$f_{2p+2} = f_{p+2}^2 - f_p^2$$

and thereby the applicability of (1b) is shown even for n = p + 1.

Since the equalities (1) and (1b) apply for n = 2 they must according to the above also apply for n = 3, for the same reason also for n = 4, etc. In other words, they must hold for all values of n.

9.4   Answers and Explanations to the Exercises in Section 3.4.7

1. The fourth term is 8.2, the fifth 10.6.

2. The fourth number is $12.8 \cdot \dfrac{8}{5} = 20.48$

   The fifth term is $\dfrac{20.48 \cdot 8}{5} = 32.768$

3. After 3 follows $3 \cdot \dfrac{6}{12} = 1.5$

After that comes 0.75.

4. The angle of the sixth swing is $\quad 15° \cdot \left(\dfrac{12}{15}\right)^5 = 15° \cdot 0.8^5 \approx 4.9°.$

6. The light *remaining* is $0.95^4 I$, where I is the intensity of the incoming light. The percentage of light *lost* is

$$100 \cdot (1 - 0.95^4) \approx 18.5\%.$$

7. The plot shows a discontinuity between 1910 and 1920. This comes from the Russian Revolution in 1917. The population growth is shown plotted below in a semi-log diagram.

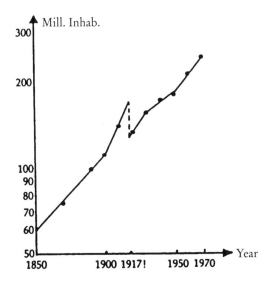

9.5   Answers and Explanations to the Exercises in Section 3.5.8

1. $D - d = 2t,$ $\qquad D = d + 2t,$ $\qquad d = D - 2t,$ $\qquad t = \dfrac{D - d}{2}$

2. Yes, the numbers which turn up there are identical with the numbers in Pascal's triangle (if we put a 1 at the top). Let us look at the case with 4 numbers to begin with: a, b, c, and d.

Starting point	1a	1b	1c	1d
We get first	1a + 1b	1b + 1c	1c + 1d	
Next comes		1a + 2b + 1c	1b + 2c + 1d	

from the addition

```
1       1
        1       1
1       2       1
```

And finally we get   1a + 3b + 3c + 1d

from adding

```
1       2       1
        1       2       1
1       3       3       1
```

The numbers in Pascal's triangle turn up in a corresponding manner:

```
            1
        1       1
    1       2       1
1       3       3       1
```

3. No, because if the square were

$$\begin{array}{cc} a & b \\ c & d \end{array}$$

then the following would hold:

$$a + b = a + c = a + d$$

from which it follows that $b = c$ and $c = d$.

In other words, the numbers would not be different. We would find instead that

$$a = b = c = d.$$

4. We found in Section 3.5.4 that three consecutive integers give a sum which is divisible by 3. If the "middle" number is n, then the sum will be

$$s_3 = (n - 1) + n + (n + 1) = 3n.$$

If we take four consecutive numbers, there is no "middle" number, but if we let the second number be n, then the sum will be

$$s_4 = (n - 1) + n + (n + 1) + (n + 2) = 4n + 2$$

These sums are *not* evenly divisible by 4 (there is a remainder of 2). Taking 5 numbers in a row we get

$$s_5 = (n - 2) + (n - 1) + n + (n + 1) + (n + 2) = 5n$$

which tells us these sums are all divisible by 5.

It is now not difficult to see the following general result:

1) when n is odd, the sum is divisible by n.
2) when n is even, say n = 2p, the sum gives a remainder of p, when divided by 2p

5. Let E represent an even number, T any number divisible by 3 and S any number divisible by both 2 and 3, i.e. divisible by 6. Finally, let O stand for an odd number which is not divisible by 3. Prime numbers greater than 3 can be found only among the O-numbers. Beginning with T, the sequence of natural numbers can be written in the following structured way:

(3) (4) (5) (6) ...   (12)...
  T   E   O   S   O   E   T   E   O   S   O   E   T...

The period SOETEO repeats endlessly because

1) Every other number is even
2) Every third number is divisible by 3
3) Every sixth number is divisible by 6

The O-numbers therefore always lie adjacent to the S-numbers.

We have thereby shown that every prime number greater than 3 can be written in the form 6n + 1 or 6n - 1, where n is a natural number.

6. According to Exercise 5 the prime numbers can all be written in the form $6n + 1$ or $6n - 1$. Assume that $p = 6n + 1$, for some integer n.
Then we get

$$p^2 + 2 = (6n + 1)^2 + 2 = 36n^2 + 12n + 3 = 3(12n^2 + 4n + 1)$$

from which it is seen that the number is divisible by 3.
If instead $p = 6n - 1$, we get analogously

$$p^2 + 2 = 3(12n^2 - 4n + 1)$$

and conclude as before that the number is divisible by 3.

7. We suppose that the formula

$$1^3 + 2^3 + 3^3 + \ldots + n^3 = (1 + 2 + 3 + \ldots + n)^2$$

holds for n = some natural number p.
We know, to begin with, that the formula holds for $n = 1$ (we have, of course, also checked it for a few other values of n). Does it hold now for $n = p + 1$? We need to utilize the knowledge that the right hand side in the formula above can also be written

$$\left\{ \frac{n\,(n+1)}{2} \right\}^2 \quad \text{or} \quad \left\{ \frac{n^2\,(n+1)^2}{4} \right\}. \tag{+}$$

We can now write, adding $(p+1)^3$ to both sides,

$$1^3 + 2^3 + \ldots + p^3 + (p+1)^3 = \frac{p^2(p+1)^2}{4} + (p+1)^3$$

We must now show that the right hand side can be written as

$$\{1 + 2 + \ldots + p + (p+1)\}^2.$$

The right hand side can be written

$$\frac{(p+1)^2}{4} \left\{ p^2 + 4(p+1) \right\}$$

$$= \frac{(p+1)^2}{4} \left(p^2 + 4p + 4 \right)$$

$$= \frac{(p+1)^2(p+2)^2}{4}$$

Comparison with formula ($^+$) above shows that this expression is precisely

$$\{1 + 2 + \ldots + p + (p+1)\}^2.$$

We have thus shown that if the formula for the sum of cubes is true for n = p, then it is also true for n = p + 1, i.e., for the next higher value of n. Since the formula holds for n = 1, it must hold for n = 2, thus even for n = 3, and we have hereby proven the formula for all natural numbers n.

8. We let x be the amount of copper to be added. The equation for x will then be

$$0.6 + x = 0.7 (1 + x)$$

which gives $\quad x = \dfrac{1}{3}$ kg.

9.6   Answers and Explanations to the Exercises in Section 3.6.4

1. The Danes own 2/7 of 50% = approximately 14.3% of Linjeflyg.

2. The increase was 24,000 crowns, which is 4 · 6000 or 400%.

3. Since 0.5% is 1/200, one must mine 200 tons of ore to get 1 ton of copper.

4. Letting p be the original price, the price after the 25% increase is 1.25 p crowns. The discount down to the original price will be 0.25 p crowns which as percentage becomes

$$\frac{0.25p}{1.25p} \cdot 100 = 20\%.$$

5. The 50% decrease means a halving of the contribution. The 100% increase means a doubling, so the school contribution ends up at the original level.

6. If we assume for simplicity's sake that the bones are circular in cross section, then the radius must be enough larger to multiply the area by a factor of 8. The radius must, therefore, be multiplied up by a factor $\sqrt{8}$ or approximately 2.8 times larger.

7. Average speed = total distance / total time.

Total distance = $\dfrac{10}{60} \cdot 20 + \dfrac{30}{60} \cdot 15 = \dfrac{65}{6}$ nautical miles.

Total time = $\quad$ 10 + 30 min = 2/3 hour.

Average speed is thus $\dfrac{65/6}{2/3} = \dfrac{65 \cdot 3}{6 \cdot 2} \approx 16.25 \approx 16$ knots.

8. Let a be the distance travelled outbound in kilometers. We call the airplane in the problem airplane A. A's average speed is the same as the constant speed which another airplane B would need to hold in order to fly the same distance (out and back) in the same time as it took for A.

Let x be A's average velocity = B's constant velocity.

A's travel time is $\quad \dfrac{a}{800} + \dfrac{a}{1000} \quad$ hours.

B's travel time is $\quad \dfrac{2a}{x} \quad$ hours.

$$\dfrac{2a}{x} = \dfrac{a}{800} + \dfrac{a}{1000}.$$

The travel times are to be the same, which gives the equation

The fact that a can be divided out of the equation shows that the answer is not dependent on the distance.

We get $\quad \dfrac{2}{x} = \dfrac{5 + 4}{4000}$

from which we get $\quad x = \dfrac{8000}{9} \approx 889$ km/hour.

9.7 Answers and Explanations to Section 3.7.3

1. Figure 9.7.1 shows that the cube has 11 different networks. These are ordered systematically: in the first row we have the 6 networks which have 4 squares in a row. In the second row are the four networks with 3 squares in a row. The eleventh network is the only one where no more than 2 squares lie in a row. The tetrahedron has only 2 possible networks.

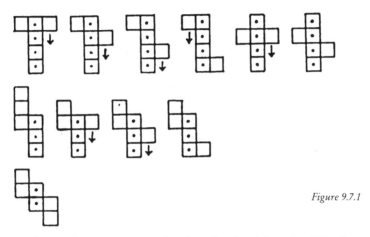

Figure 9.7.1

2. Finding eight corner points for the cube should not be difficult.

3. See Figure 9.7.2.

4. See Figure 9.7.3.

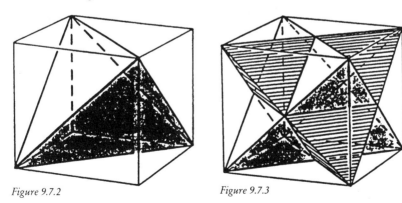

Figure 9.7.2

Figure 9.7.3

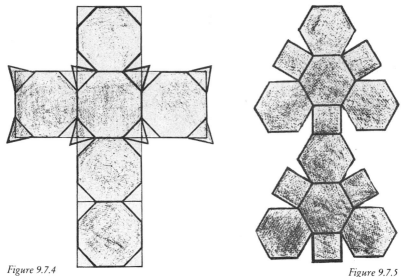

Figure 9.7.4

Figure 9.7.5

5. The network for the solid with octagons and triangles is easy to construct when one knows how to inscribe an octagon in a square: one constructs four quarter circles with centers at the corners of the square and with radius equal to half the diagonal of the square. The square has a side equal to the edge length chosen for the cube. See Figure 9.7.4.

The network for the solid with hexagons and squares is more difficult to find, since one must first figure out the length of the sides which form these figures. It is not too difficult to see in Figure 3.7.8 that the square's diagonal is equal to half of the side of the cube (see Figure 9.7.5). The network for this solid is noticeably smaller than that for the octagon-triangle solid. This is because we are grinding down the solid the whole time.

9.8 Answers and Explanations to Section 3.8.3

1. See Figure 9.8.1. The three curves correspond to XY = 5, 15, and 25 mm respectively.

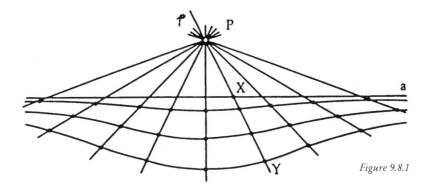

Figure 9.8.1

2. In order to get an idea of the shape of the curve for large XY values, we reduce the figure in size using so-called proportional form reduction with P as the fixed point: each p-line has a point X belonging to line a and a point Y belonging to the curve. The distance PX and PY are reduced to such a scale that the distance between the new points, X´Y´, is for example 25 mm. In this type of reduction (proportional form) all shapes and forms are unchanged, i.e., line a becomes a new line a´ which is closer to P and parallel to a. The curve too (the locus of points Y) will maintain its shape at the same time as it moves nearer to P.

For large values of the distance XY, a´ will lie very close to P. For example, if XY = 1 meter the reduction scale must be 1:40, which means that P lies 40 times closer to a´ than a.

We may now with the same effect imagine line a to be fixed and let P be drawn in toward a while we take XY to be 25 mm. Figure 9.8.2 shows the shape of the curve for the cases where P is at distances 5 mm and 1 mm from a, respectively.

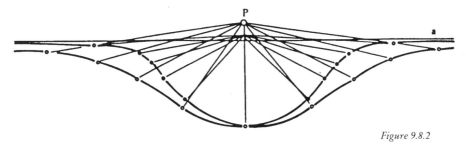

Figure 9.8.2

It is now not difficult to see that Figure 9.8.3 with its half-circle (or radius 25 mm) shows the limiting form of the family of curves as P is moved in toward a. To give a strict proof requires both care and "technique," the latter at university level.

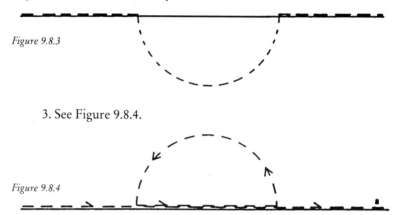

Figure 9.8.3

3. See Figure 9.8.4.

Figure 9.8.4

4. The 9 sub-figures in Figure 9.8.5 show the typical stages of the transformation. In (1) the three circles coincide but begin to move outward away from each other. Each circle's center follows its triangle "height" line (from the base at right angles up to the vertex) in the direction away from the vertex where the height line begins.

> White region: not covered by any circle
> Horizontally shaded regions: covered by 1 circle
> Cross-hatched regions: covered by 2 circles
> Dotted regions: covered by all 3 circles

In drawing (7) the circle radii have become infinitely large. In (8) the circles' inner regions lie "outside" the edge of the circle, in the sense that we used in Chapter 3.8. Drawing (9) shows a stage where we come back to (1) as far as form is concerned, but (9) and (1) are each other's opposites with regard to shading. It would take 9 more stages before we would truly come back to our starting point.

5. With reference to Figure 3.8.14 we let A and B be two given points and F a fixed proportional ratio between the distance from a point X on the curve to A and B respectively:

$$\frac{XB}{XA} = F$$

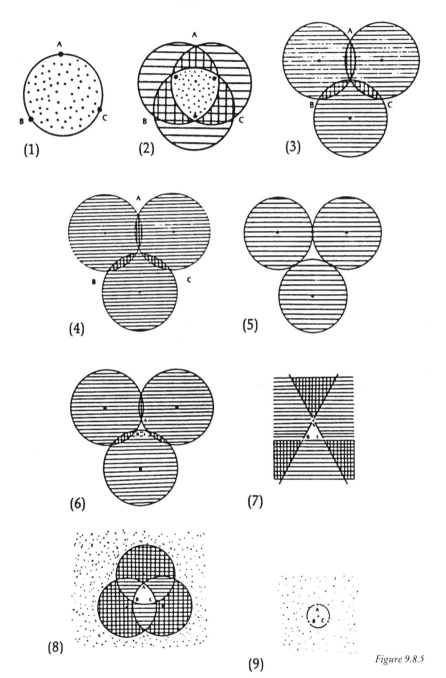

(1)   (2)   (3)

(4)   (5)

(6)   (7)

(8)

(9)

Figure 9.8.5

In Figure 9.8.6 below XP is drawn as a bisector in the triangle ABX. XQ is the bisector of the angle BXr, where r is the extension of AX. XQ is usually called the "external" bisector. According to the so-called Theorem of Bisectors and its correspondence for external bisectors, we have

$$\frac{PB}{PA} = F \quad \text{and} \quad \frac{QB}{QA} = F \tag{1}$$

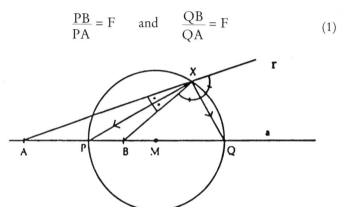

Figure 9.8.6

If we now let X move, then P and Q will be fixed points because of (1). In addition, the angle PXQ is, and always remains, a right angle (= half of 180°).

The conclusion must now be that X describes a circle with PQ as a diameter.

9.9   Answers and Explanations to Section 3.9.5

1. The distance to the North Pole is $\frac{(90-60) \cdot 40000}{360}$ km, or more directly $(90-60) \cdot 111.1$ km   3300 kilometers.

The distance to the South Pole is approximately $2000-3300 = 16,700$ km.

2. The distance is s (151-18) · 111.1 · cos 34° ≈ 12,250 km.

3. The best answers are

a) approximately 9700 km (the angular difference is 87°)
b) approximately 12,500 km (angular difference 113°)

The solution to b) is illustrated in Figure 9.9.1 below.

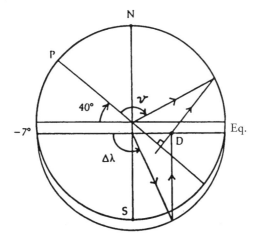

Figure 9.9.1

4. Because 106° + 76° = 180° it happens that New York and Hanoi lie on a great circle which goes through the Poles. The shortest distance passes through the North Pole (first toward the North, then southward) and the distance is

$$[(90 - 41) + (90 - 21)] \cdot 111.1 \approx 13000 \text{ km}.$$

5. Somewhere in the Southern Hemisphere there is a circle of latitude whose length is exactly 1000 km. The starting point can be anywhere 1000 km north of that circle. The person may even start 1000 km north of any circle of latitude whose length is

$$\frac{1000}{n} \text{ km,}$$

where n = 1, 2, 3, 4, etc.

6. According to the formula for the area of a spherical triangle (see Section 3.9.2) we get the equation

$$\frac{\pi R^2}{180} (V - 180) = 0.25 \cdot 4\pi R^2$$

where R is the radius of the sphere. Solving, one gets V = 360°.

7. Letting X be the percentage we are looking for, we have

$$\frac{X}{100} \cdot 4\pi R^2 = 2.2\pi R \ (R \sin 60° - R \sin 30°)$$

from which x ≈ 36.6%.

9.10 Answers and Explanations to Section 3.10.3

The symbol $\overset{=}{\wedge}$ means "perspective with" and d denotes Désargues' line.

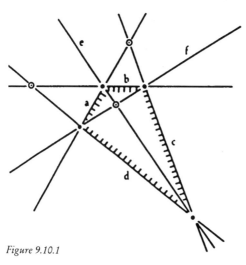

Figure 9.10.1

1. We obtain $A'PR \overset{=}{\wedge}$ OBC   and d = B'C'Q

2. OAB $\overset{=}{\wedge}$ C'RQ;d goes through A', B'and P.

3. a) AA'P $\overset{=}{\wedge}$ CC'Q; R is the center of perspective.

b) The figure is self-dual: on each line lie 3 points and through each point go 3 lines. The statements to be shown are consequences of this fundamental symmetry.

4. In Figure 9.10.1, the sides of the quadrangle are denoted by a, b, c, and d, while the two diagonals are labelled e and f.

The 3-sided polygon we are looking for is determined by the points of intersection

a x c,   b x d   and   e x f.

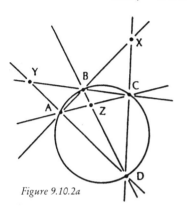

Figure 9.10.2a

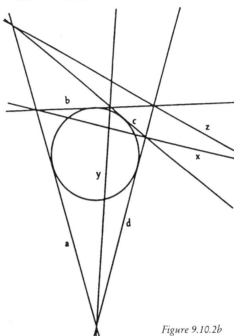

Figure 9.10.2b

5. Corresponding to three points X, Y, and Z in Figure 9.10.2a are the three lines x, y, and z in Figure 9.10.2b.

Z is an inner point of the circular area in Figure 9.10.2a. Analogously, z is an inner line in the circle envelope of lines which surround the circle in the b-figure. X and Y are outer elements, as are x and y.

6. See Figure 9.10.3

7. See Figure 9.10.4

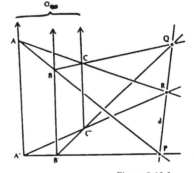

Figure 9.10.3

O_∞ is the center of perspective

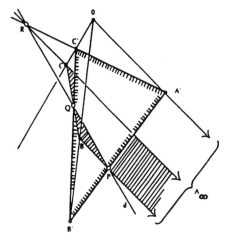

Figure 9.10.4
Triangle ABC has vertex A as a point at infinity

9.11 Answers and Explanation to Section 3.11.8

1. a) 21 pupils (if all those who were positive toward France were also positive toward Germany).

b) 5 pupils

c) 16 pupils (= 25 + 21 - 30)

2. Assume for the sake of simplicity that the numbers of respondents was 100. Let a, b, and c represent the sets of respondents who have confidence in A, B and C respectively. (These sets have common elements, as is clear from the problem statement.)

We are to minimize the intersection of a, b, and c. The minimum intersection of a and b is made up of 50 people (80 + 70 - 100). We now place as much as possible of c outside this intersection, which is 50 people. The remaining 10 c-respondents must be part of the intersection of a and b. In other words, the intersection of a, b, and c must include at least 10 people. The answer is 10%.

3. and 4.
The diagrams for A´ ∪ B´ and (A ∩ B)´ (and respectively for A´ ∩ B´ and (A ∪ B)´) agree regardless of whether the sets overlap one another or not. The figures show A´ ∪ B´ and A´ ∩ B´ when A overlaps B. It can be seen that these sets are identical with (A ∩ B)´ and (A ∪ B)´ respectively.

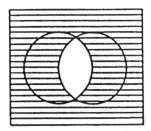

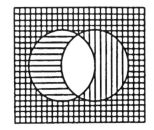

A´ ∪ B´ = the entire shaded area. This corresponds to the complement of the remaining unshaded lens-shaped area, i.e., to (A ∩ B)´.

A´ ∩ B´ = the cross-hatched area. This corresponds to the area outside the circles, i.e., to (A ∪ B)´.

5. Simply follow the rules of dualization.

6. Let f be the left hand side = $x \cdot (y + z)$
and g be the right hand side = $x \cdot y + x \cdot z$
We must calculate f and g for all possible combinations of the values 0 and 1 for x, y and z and then show that f = g.
The table can be condensed down to

x	y	z	f	g
0	arbitrary		0	0
1	0	0	0	0
1	other comb.		1	1

Thus it is true that f = g.
The other identity can be verified with the table

x	y	z	f	g
0	1	1	1	1
0	other comb.		0	0
1	arbitrary		1	1

7. The table is

x	y	f	g
0	0	0	0
0	1	1	1
1	0	1	1
1	1	1	1

from which it is clear that f = g for all combinations of x and y.

9.12 Answers and Explanations to Section 3.12.8

1. The average speed is 129 km/hour.

2. The voltage increases from 2.0 mV to 4.8 V, an increase of 2.8 V.

The average rate of increase is $\dfrac{2.8V}{4.4mA} = 0.64$ V/mA.

3. The average speed is $\dfrac{1.6\left(4.5^2 - 2^2\right)}{4.5 - 2} = 10.4$ m/s

4. $\dfrac{(t+h)^2 - t^2}{h} = \dfrac{(2t+h) \cdot h}{h} = 2t + h$

and the limit is 2t, when h → 0.

5. If one doesn't know the Binomial Theorem for development of the coefficients of $(t + h)^n$,

then the easiest way is to set $\qquad u = t + h$

and consider the ratio $\qquad \dfrac{u^n - t^n}{u - t}$

The numerator can be rewritten

$$(u-t)(u^{n-1} + u^{n-2}t + u^{n-3}t^2 + \ldots + t^{n-1}).$$

The limit we seek is equal to the long expression in parentheses on the right, since the factor u - t can be divided out.

The limit $h \to 0$, that is, is $u \to t$, is

$$t^{n-1} + t^{n-1} + \ldots + t^{n-1} \quad \text{(n equal terms)} = n\, t^{n-1}.$$

6. Let us assume that $f(x) \neq 0$ at some point t within the interval. Since f (x) is continuous there must then be a subinterval where f (x) and thus f (x)^2 is non-zero. Let δ be the length of this closed subinterval and $m > 0$ be the minimum value of f(x)^2 on the subinterval.

Then, since f (x)^2 is non-negative, we have

$$\int_a^b f(x)^2 dx \geq m \cdot \delta > 0$$

which contradicts the given condition that $\displaystyle \int_a^b f(x)^2 dx = 0.$

7. a) One finds the primitive function

$$F(x) = \frac{x^2}{2} - \frac{x^3}{3}$$

and obtains $\quad F(1) - F(0) = \dfrac{1}{6}.$

b) The area in question is

$$\int_2^3 x^2 dx = \frac{3^3}{3} - \frac{0^3}{3} = 9 \text{ cm}^2$$

(The primitive function is x^3/3.)

8. The statement's validity follows from the correspondence of pairs below:

$$1 = 1^2 \leftrightarrow 1$$
$$4 = 2^2 \leftrightarrow 2$$
$$9 = 3^2 \leftrightarrow 3$$
$$16 = 4^2 \leftrightarrow 4$$
etc.

9. The set of rational numbers a/b between 0 and 1 can be numbered in the following way (the list here shows the first 17 numbers):

$$\frac{1}{2}; \frac{1}{3}, \frac{2}{3}; \frac{1}{4}, \frac{3}{4}; \frac{1}{5}, \frac{2}{5}, \frac{3}{5}, \frac{4}{5}; \frac{1}{6}, \frac{5}{6}; \frac{1}{7}, \frac{2}{7}, \frac{3}{7}, \frac{4}{7}, \frac{5}{7}, \frac{6}{7}; \ldots$$

The numbers are arranged in order of increasing denominator and within each group in order of increasing numerator. Fractions such as ¾ and ⅔, ¾, ⅚ are not included since they have the same value as a previously counted rational number.

10. Letting 0 represent the choice of a left branch (seen from below) and 1 be a choice to the right, then every path can be described by an infinite sequence of zeroes and ones. Each path has a unique representation.

We now copy Cantor's diagonal proof and assume that the infinite set of paths, represented as zero-one sequences, is countable. There would then exist an infinitely long list containing all the sequences, for example

1)   0001010...
2)   0110100...
3)   1010100...
4)   1001011...

....

We now form a sequence where

the 1st digit ≠ 1st digit of 1)
the 2nd digit ≠ 2nd digit of 2)
etc.

The infinite sequence we obtain cannot possibly be on the list. We have contradicted our assumption that the list includes all sequences.

The paths are, therefore, uncountable.

BIBLIOGRAPHY

This bibliography does not claim to be as extensive as it could and perhaps should be. It contains a selection of works which I have met. Among the books there are surely numerous ones that are available in new editions with a later year of publication.

A. General References

1. Alexandrov, A. D., Kolmogorov, A. N., Lavrent'ev M. A.: *Mathematics, its Content, Methods and Meaning*, M.I.T. Press, Cambridge (Mass.) 1969.

2. Becker, O.: *Grundlagen der Mathematik in geschichtlicher Entwicklung*, Karl Alber, Freiburg/Br. 1954.

3. Bergamini, D.: *Mathematics*. LIFE - Wonders of Science, Time-Life International, Time Inc., New York 1969.

4. Courant, R., Robbins, H.: *What is Mathematics?* Oxford Univ. Press, Oxford 1981.

5. Davis, J. D., Hersh, R.: *The Mathematical Experience*, Birkhauser, Boston 1981, Penguin, Harmondsworth 1983.

6. Kline, M., *Mathematics in Western Culture*, Oxford Univ. Press, Oxford 1968.

7. Kline, M.: *Mathematical Thought from Ancient to Modern Times*, Oxford Univ. Press, Oxford 1972.

8. Locher-Ernst, L.: *Mathematik als Vorschule zur Geist-Erkenntnis*, Philosophisch-Anthroposophischer Verlag, Dornach 1973.

9. Meschkowski, H.: *Didaktik der Mathematik*, Klett, Stuttgart 1972.

10. Newman, J. R.: *The World of Mathematics*, Simon & Schuster, New York 1956.

11. Otte, M. (Ed.): *Mathematiker uber die Mathematik*, Springer, Heidelberg 1974.

12. Polya, G.: *How to Solve It. A New Aspect of Mathematical Method*, Princeton Univ. Pr., Princeton, N.J. 1945.

13. Polya, G.: *Mathematical Discovery; on Understanding, Learning, and Teaching Problem-Solving*, 2 vols., Wiley, New York 1962-65.

14. Polya, G.: *Mathematics and Plausible Reasoning, 1. Induction and Analogy in Mathematics, 2. Patterns of Plausible Inference*. Princeton Univ. Press

15. Schuberth, E.: *Die Modernisierung des Mathematikunterrichts*, Verlag Freies Geistesleben, Stuttgart,1971.

16. Speiser, A.: *Klassische Stücke der Mathematik*, Orell Füssli, Zürich 1925.

17. Wagenschein, M.: *Ursprüngliches Verstehen und exaktes Denken*, Klett, Stuttgart 1963.

18. Wickelgren, W. A.: *How to Solve Problems*, Freeman, New York 1974.

19. Wittenberg, A. J.: *Bildung und Mathematik*, Klett, Stuttgart 1963.

B. References to Chapters 2 - 8

CHAPTER 2

Appenzeller, K.: *Die Quadratur des Zirkels*, Zbinden, Basel 1979.

Aaboe, A.: *Episodes from the Early History of Mathematics*, Yale Univ. Press, New Haven 1964.

Archimedes, *The Works of Archimedes with the Method of Archimedes*, (ed. T.L. Heath) Dover Publ., New York.

Dantzig, T.: *Number, The Language of Science*, Macmillan, New York/London 1954.

Heath, T.L. (Ed.): *The Works of Archimedes with the Method of Archimedes*, Dover Publ., New York.

Midonick, H.: *The Treasury of Mathematics*, Penguin, Harmondsworth 1965.

Newman, J. R.: "The Rhind Papyrus" (see A 10).

Polya, G.: See A 13 and 14.

Smith, D. E.: *History of Mathematics*, Dover Publ., New York 1958

Waerden, B. L. van der: *Science Awakening*, P. Noordhoff 1954.

Section 3.1

American Math. Monthly, vol. 83, 1976 (see 3.1.7).

Canadian Math. Bull. vol. 18/3, 1975 (see 3.1.7).

Coe, M. D.: *The Maya*, Penguin, Harmondsworth 1977.

Dantzig, T.: *Number, The Language of Science*, Macmillan, New York/London 1954.

Davis, M., Hersh, R.: "Hilbert's 10th Problem", *Scient. Amer.* 1973/11.

Davis, P. J.: *The Lore of Large Numbers*, Yale Univ. Press, New Haven 1961.

Freudenthal, H.: *Mathematik in Wissenschaft Und Alltag*, Kindler, Munchen 1968.

Gardner, M.: "The Remarkable Lore of Prime Numbers" *Scient. Amer.* 1964/3 and see also 1980/12.

Gardner, M.: "Simplicity as a Scientific Concept" *Scient. Amer.* 1963/8.

Locher-Ernst, L.: "Die Reihe der natürlichen Zahlen als Geist-Kunstwerk", *Sternkalender* 1959/60, Philosophisch-Anthro-posophischer Verlag, Dornach.

Menninger, K.: *Zahlwort Und Ziffer*, 2. Aufl. Vandenhoeck & Ruprecht, Gottingen 1957.

Smith and Ginsburg: Article on numbers and digits in A 10.

Ulin, B.: "Beweis oder Gegenbeispiel?" *Mathematisch-Physikalischer Korrespondenz* Nr. 113, Dornach 1984.

Vogel, K.: *Vorgriechische Mathematik*, I und II, Schoningh, Paderborn 1958-1959.

McWhorter, E. W.: "The Small Electronic Calculator" *Scient. Amer.* 1976/3.

Section 3.2

Feller, W.: *An Introduction to Probability Theory and Its Applications,* Vol. 1,2, Wiley & Sons, New York 1965.

Gardner, M.: "The Multiple Charm of Pascal's Triangle:, *Scient. Amer.* 1966/12.

Gardner, M.: "Taxicab Geometry Offers a Free Ride to Non-Euclidean Locale", *Scient. Amer.* 1980/11.

Gnedenko, B. W.: *Lehrbuch der Wahrscheinlichkeitsrechnung,* Akademie-Verlag, Berlin 1965.

Huff, D.: *How to Take a Chance,* Norton & Company, New York 1964.

Locher-Ernst, L.: *Arithmetik und Algebra,* 2. Auflage, Philosophisch-Anthroposophischer Verlag, Dornach 1984.

Polya, G.: *Patterns of Plausible Inference,* (Vol. 2, A14). Princeton Univ. Press 1954.

Weaver, W.: *Lady Luck: The Theory of Probability,* Educational Services Inc. 1963.

Section 3.3

Baravalle, H. von: *Die Geometrie des Pentagramms und der Goldene Schnitt,* Mellinger, Stuttgart 1950.

Coxeter, H. S. M.: *Introduction to Geometry,* Chapter11, 2nd edition, Wiley & Sons, New York 1969.

Gardner, M.: *The 2nd Scientific Amer. Book of Math. Puzzles and Diversions,* Simon & Schuster, New York 1961.

Gardner, M.: "The Multiple Fascinations of the Fibonacci Numbers" *Scient. Amer.* 1969/3.

Locher-Ernst, L.: "Von der Geometrie der menschlichen Gestalt" (in A 8).

Morike-*Mergenthaler, Biologie des Menschen,* Quelle & Meyer, Heidelberg 1963.

Schultz, J.: "Die Blattstellungen im Pflanzenreich als Ausdruck Kosmischer GesetzmaBigkeiten" *Goethe in Unsrer Zeit,* Hybernia-Verlag, Dornack-Basel 1949.

Vorobyov, N.N.: *The Fibonacci Numbers,* Heath & Co., Lexington 1963.

Zeising, A.: *Neue Lehre von den Proportionen des Menschlichen Körpers,* Leipzig 1854.

Section 3.4

Bindel, E.: *Logarithmen fur Jedermann*, Verlag Freies Geistesleben, Stuttgart 1983.

Bindel, E.: *Die Zahlengrundlagen der Musik*, Verlag Freies Geistesleben, Stuttgart 1985.

Boring, E. G.: "Gustav Theodor Fechner" (Article in A 10, Chapter VI).

Locher-Ernst, L.: "Mathematik und Musik" (Article in A 8).

Malthus, T. R.: Article in A 10, Chapter VI.

Pfrogner, H.: *Lebendige Tonwelt*, Langen/Muller, Munchen 1981.

Stumpf, K. (Hrsg.): *Astronomie*, Das Fischer-Lexikon Bd. 4, Chapter "Helligkeiten", Fischer, Frankfurt/M. 1957.

Section 3.5

Becker, O.: *Grundlagen der Mathematik in Geschichtlicher Entwicklung*, Karl Alber, Freiburg/Br. 1954 (speziell Kapitel 2A).

Bindel, E.: *Die Arithmetik*, Mellinger, Stuttgart 1967.

Bindel, E.: *Das Rechnen*, Mellinger, Stuttgart 1966.

Gardner, M.: *aha! Insight*, Freeman & Co., New York 1978.

Schuh, F.: *The Master Book of Mathematical Recreations*, Dover Publ., New York 1968.

Section 3.6

Huff, D.: *How to Lie with Statistics*, Penguin, Harmondsworth 1973.

Physical Science Study Committee: *Physics*, Chapter "Functions and Scaling", Heath & Co., Lexington 1960.

Section 3.7

Baravalle, H. von: *Geometrie als Sprache der Formen*, Verlag Freies Geistesleben, Stuttgart 1980.

Bentley, W. A./Hunphreys, W. J.: *Snow Crystals*, Dover Publ., New York 1962.

Bindel, E.: *Harmonien im Reiche der Geometrie*, Verlag Freies Geistesleben, Stuttgart 1964.

Brun, V.: "On Some Problems in Solid Geometry Inspired by Virology" *Nordisk Matematisk Tidskrift* 1978/3 and 4.

Gardner, M.: "The Multiple Fascinations . . ." (see references for Section 3.3).

Haldane, J.B.S.: "On Being the Right Size" (Article in A 10).

Hilbert, D./Cohn-Vossen, S.: *Anschauliche Geometrie*, Springer, Berlin 1932, Dover Publ., New York 1944.

Schüpback, W.: *Pflanzengeometrie*, Troxler, Bern 1944.

Thompson, D'Arcy Wentworth: *On Growth and Form*, Cambridge Univ. Press, Cambridge 1963.

Unger, G.: *Das offenbare Geheimnis des Raume*, Verlag Freies Geistesleben, Stuttgart 1963.

Weyl, H.: *Symmetry*, Princeton Univ. Press, 1952.

Section 3.8

Adam/Buhler u. a.: *Beitrage Zum Geometrischen Zeichnen*, Verlag Freies Geistesleben, Stuttgart 1967.

Adams, G./Whicher, O.: *Die Pflanze in Raum und Gegenraum*, Verlag Freies Geistesleben, Stuttgart 1960.

Bindel, E.: *Die Kegelschnitte*, Verlag Freies Geistesleben, Stuttgart 1963.Bockemuhl, J. (Ed): Erscheinungsformen des Ätherischen, Verlag Freies Geistesleben, Stuttgart 1977.

Edwards, L.: *The Field of Form*, Floris Books, Edinburgh 1982.

Gordon/Jacobson: "The Shaping of Tissues in Embryos", *Scient. Amer.* 1978/6.

Grohmann, G.: *Metamorphosen im Pflanzenreich*, Verlag Freies Geistesleben, Stuttgart 1958.

Kranich, E. M.: *Die Formensprache der Pflanze*, Verlag Freies Geistesleben, Stuttgart 1976.

Locher-Ernst, L.: *Geometrisieren im Bereiche Wichtigster Kurvenformen*, Orell Fussli, Zurich 1938.

Strakosch, A.: *Geometrie durch übende Anschauung*, Mellinger, Stuttgart 1962.

Thompson, D'Arcy W.: see references for Section 3.7.

Wyss/Adam/Ruchti: *GTZ, Geometrisch-Technisches Zeichnen*, Lehrerhandbuch, Staatlicher Lehrmittelverlag, Bern 1978.

Section 3.9

Baldus, R./Löbell, F.: *Nichteuklidische Geometrie*, Sammlung Göschen, Bd. 970, Berlin 1953.

Balser, L.: *Einführung in Die Kartenlehre*, Teubner, Stuttgart 1951.

Locher-Ernst, L.: *Geometrische Metamorphosen*, Philosophisch-Anthroposophischer Verlag, Dornach 1970.

Zacharias, M.: *Das Parallelenproblem und Seine Lösung*, Teubner, Stuttgart 1951.

Section 3.10

Bernhard, A.: *Projektive Geometrie Aus der Raumanschauung zeichnend Entwickelt*, Verlag Freies Geistesleben, Stuttgart 1984.

Cremona, L.: *Elements of Projective Geometry*, Dover Publ., New York 1960.

Gardner, M.: "Euclid's Parallel Postulate and Its Modern Offspring", *Scient. Amer.* 1981/10.

Kline, M.: *Mathematics in Western Culture*, Kapitel 10, Oxford Univ. Press, Oxford 1968.

Locher-Ernst, L.: *Urphänomene der Geometrie und Projektive Geometrie*, Orell Füssli, Zurich 1937 and 1940.

Locher-Ernst, L.: *Raum und Gegenraum*, Philosophisch-Anthroposophischer Verlag, Dornach 1957, 2. Aufl. 1970.

Rovida, Andes A., *Übungen Zur Synthetischen Projektiven Geometrie*, Verlag am Goetheanum (Schweiz) 1988.

Bell, E. T.: *Men of Mathematics*, Simon & Schuster, New York 1937.

Boole, G.: *An Investigation of the Laws of Thought*, Dover Publ., New York.

Descartes, R.: *Philosophical Works of Descartes*, Dover Publ., New York 1955.

Gardner, M.: *Logic Machines, Diagrams and Boolean Algebra*, Dover Publ., New York 1968.

Hilbert/Ackermann: *Grundzüge der theoretischen Logik*, Dover Publ., New York 1946, 6. Aufl. Springer, Heidelberg 1972.

Midonick, H.: *The Treasury of Mathematics, II*, Penguin, Harmondsworth 1965.

Newman, R. J.: see reference A 10, Chapters XI-XIV.

Whitesitt, J. E.: *Boolean Algebra and Its Applications*, Addison-Wesley, Reading (Mass.) 1961.

Wright, G. H. von: *Logik, filosofi och språk*, Aldus/Bonnier, Stockholm 1957.

Section 3.12

Apostol, T. M.: *Mathematical Analysis*, Addison-Wesley, Reading (Mass.) 1974.

Bell, E. T.: *Men of Mathematics*, see references in Section 3.11.

Goursat, E.: *Cours D'analyse Mathématique*, Paris 1943.

Klein, F.: *Vorlesungen über die Entwicklung der Mathematik im 19. Jahrhundert*. Vol. 1 and 2., Springer, Heidelberg 1979.

Meschkowski, H.: *Probleme des Unendlichen Werk und Leben Georg Cantors*, Vieweg, Braunschweig 1967, 2 Edition as Georg Cantor. *Probleme des Unendlichen*. Bibl.=Inst., Mannheim 1983.

Meschkowski, H.: *Wandlungen des mathematischen denkens*, Vieweg, Braunschweig 1956, edition #4, 1967.

Niven, I.: Numbers: *Rational and Irrational*, Yale Univs., Pres, New Haven 1961.

Reid, C.: *A Long Way from Euclid*, Crowell & Co., New York 1963.

Waismann, F.: *Einführung in das mathematische denken*, Deutscher Taschenbuch-Verlag, München 1970.

CHAPTER 4

Chintschin, A. J.: "Über den Formalismus im Mathematikunterricht der Schule", *Probleme des Mathematikunterrichts*, Volk und Wissen, Berlin 1965.

Gardner, M.: *Aha! Insight*, Freeman & Co., New York 1978.

Kline, M.: *Why Johnny Can't Add: The Failure of the New Math*, St. Martin's Press, New York 1973.

Locher-Ernst, L.: *Mathematik als Vorschule Zur Geist-Erkenntnis*, Philosophisch-Anthroposophischer Verlag, Dornach 1973.

Newman, J. R.: Chapters XI-XIV in A 10.

Poincaré, H.: Article on mathematical creativity. in A 10.

Polya, G.: See A 12-14.

Waerden, B. L. van der: *Einfall und Überlegung*, Birkhäuser, Basel 1968.

Werrheimer, M.: *Produktives Denken*, Frankfurt/M. 1964.

CHAPTER 5

Andrade, E. N. da C.: *Sir Isaac Newton*, London 1954.

Burtt, E. A.: *The Metaphysical Foundations of Modern Physical Science. A Historical and Critical Essay*, Routledge & Kegan Paul, London 1924, 1972.

Caspar, M.: *Kepler*, Kohlhammer, Stuttgart 1958.

Descartes, R.: See reference in Section 3.11.

Dijksterhuis, E. J.: *Die Mechanisierung des Weltbildes*, Springer, Heidelberg 1956.

Dunnington, G. W.: *Carl Friedrich Gauss, Titan of Science*, Exposition Press, New York 1955.

Finsler, P.: "Formale Beweise un die Entscheidbarkeit" *Math. Zeitschr.* 25 (1926) 676-682.

Finsler, P.: *Aufsätze Zur Mengenlehre*, Wiss. Buchgesellschaft, Darmstadt 1975.

Godel, K.: "Über Formal Unentscheidbare Sätze der Principia Mathematica und verwandter Systeme 1" *Monatshefte F. Math und Phys.* 38 (1931) 173-198.

Hemleben, J.: *Galilei,* Rowohlt, Reinbek 1969.

Newman, J. R.: Chapter I in A 10.

Newton, I.: *Mathematical Principles of Nature,* Berlin, 1872.

Pascal, B.: *Aufsätze,* ausgewählt und eingeleitet von R. Schneider, Fischer 1954.

Stegmüller, W.: *Unvollständigkeit und Unentscheidbarkeit.* Die Metamathematischen Resultate von Godel, Church, Kleene, Rosser und ihre erkenntnistheoretische Bedeutung, Springer, Heidelberg 1973.

Steiner, R.: *Geisteswissenschaftliche Impulse Zur Entwicklung der Physik.* Erster naturwissenschaftlicher Kurs. Zehn Vorträge gehalten in Stuttgart vom 23. 12. 1919 bis 3.1 1920. Gesamtausgabe Bibl.-Nr. 320, Rudolf Steiner Verlag, Dornach 1964.

Unger, G.: "Carl Friedrich Gauss" *Sternkalender* 1977/78, Philosophisch-Anthroposophischer Verlag, Dornach.

Valery, P.: *Les pages immortelles de Descartes,* Ed. Correa, Paris 1941.

Williams, L. P.: *Michael Faraday,* Chapman & Hall, London 1965.

CHAPTER 6

Albers/Alexanderson (edit): *Mathematical People,* Birkhäuser, Basel 1985.

Dijksterhuis, E. J.: See references for Chapter 5.

Glöckler, G.: "Zur Frage des Gebrauchs von Elektronischen Taschenrechnern in der Schule" *Erziehungskunst* 1977/6 u. 7, Verlag Freies Geistesleben, Stuttgart.

Heath, T. L. (Ed.): *The Works of Archimedes with the Method of Archimedes,* Dover Publ., New York.

Heitler, W.: *Der Mensch und die naturwissenschaftliche Erkenntnis,* Vieweg, Braunschweig 1961. 1962.

Hilbert, D.: *Grundlagen der Geometrie,* 12. Aufl., Teubner, Stuttgart 1977.

Reid, C.: *Hilbert,* Springer, Heidelberg 1970.

Piaget, J.: "How Children Form Mathematical Concepts" Scient Amer. 1953/11.

Wagenschein, M.: See reference.

Valéry, P.: *Les pages immortelles de Descartes,* Ed. Correa, Paris 1941.

CHAPTER 7

Meschkowski, H.: *Didaktik der Mathematik*, II, III, Klett, Stuttgart 1972.

Piaget, J.: *Die Entwicklung des Erkennens, I: Das mathematische Denken,* Klett, Stuttgart 1975.

Piaget, J.: "How Children Form Mathematical Concepts" *Scient. Amer.* 1953/11.

Steiner, R.: *Die geistig-seelischen Grundkräfte der Erziehungskunst* (Vortrag vom 25. 8. 1922), Gesamtausgabe Bibl. -Nr. 305, Rudolf Steiner Verlag, Dornach 1956.

Wagenschein, M. See reference A 17.

CHAPTER 8

Carleson, L.: "Matematik för vår tid" 1968, Stockholm (Swedish).

NAME INDEX

Name, Section, *Page*

SUBJECT INDEX

Subject, Section, *Page*